D1155064

DOMESTIC ANIMAL BEHAVIOR FOR VETERINARIANS AND ANIMAL SCIENTISTS

THIRD EDITION

Domestic Animal Behavior for Veterinarians and Animal Scientists

THIRD EDITION

Katherine Albro Houpt, VMD, PhD

Iowa State Press

A Blackwell Publishing Company

Katherine A. Houpt, VMD, PhD, is professor of physiology and director of the Animal Behavior Clinic, Cornell University College of Veterinary Medicine. She is a diplomate and president-elect of the American College of Veterinary Behaviorists.

© 1998, 1991, 1982 Iowa State Press
A Blackwell Publishing Company
All rights reserved

Iowa State Press
2121 State Avenue, Ames, Iowa 50014

Orders: 1-800-862-6657   Fax: 1-515-292-3348
Office: 1-515-292-0140   Web site: www.iowastatepress.com

Authorization to photocopy items for internal or personal use, or the internal or personal use of specific clients, is granted by Iowa State Press, provided that the base fee of $.10 per copy is paid directly to the Copyright Clearance Center, 222 Rosewood Drive, Danvers, MA 01923. For those organizations that have been granted a photocopy license by CCC, a separate system of payments has been arranged. The fee code for users of the Transactional Reporting Service is 0-8138-1061-2/98 $.10.

♾ Printed on acid-free paper in the United States of America

First edition, 1982 (through three printings)
Second edition, 1991 (through five printings)
Third edition, 1998

Library of Congress Cataloging-in-Publication Data

Houpt, Katherine A.
    Domestic animal behavior for veterinarians and animal scientists /
    Katherine Albro Houpt.—3rd ed.
        p.   cm.
    Includes bibliographical references (p.   ) and index.
    ISBN 0-8138-1061-2 (alk. paper)
    1. Domestic animals—Behavior.   2. Behavior, Animal.   I. Title.
SF756.7.H68   1998
599.15—dc21   97-44669

Last digit is the print number: 9 8 7 6 5

CONTENTS

Preface   xv

1. COMMUNICATION   3
Introduction   3
   Perception   4
Horses   7
   Audition   7
   Vision   9
   Olfaction   12
Dogs   13
   Audition   13
   Vision   14
   Olfaction   16
   Behavior Problems   19
Cats   20
   Audition   20
   Vision   21
   Olfaction   24
   Behavior Problems   25
Pigs   29
   Audition   29
   Vision   30
   Olfaction   31
Cattle, Sheep, and Goats   31
   Audition and Vision   31
   Olfaction   32

2. AGGRESSION AND SOCIAL STRUCTURE   33
Introduction   33
   Categories of Aggression   34
   The Biological Basis of Aggression   36
Cattle   38
   Free-Ranging Cattle   38
   Confined Cattle   38
   Clinical Cases of Aggression   43

Sheep   44
    Free-Ranging Sheep   44
    Confined Sheep   45
Goats   46
    Free-Ranging Goats   46
    Confined Goats   46
Horses   46
    Free-Ranging Horses   46
    Domestic Horses   48
    Clinical Problems   52
Pigs   55
    Free-Ranging Pigs   55
    Confined Pigs   56
Dogs   61
    Social Behavior   61
    Determinants of Dominance   62
    Territorial Aggression   62
    Fear-Induced Aggression   62
    Aggression and Hormonal Influences   63
    Predatory Aggression   63
    Guard Dogs for Predator Control   63
    Clinical Problems   64
Cats   73
    Free-Ranging Cats   73
    Confined Cats   74
    Clinical Problems   79

3. BIOLOGICAL RHYTHMS AND SLEEP   83
Introduction   83
    High-Frequency Rhythms   83
    Ultradian Rhythms   84
    Circadian Rhythms   84
    Other Rhythms   88
    Annual Rhythms   89
    Sleep   89
Patterns of Sleep and Activity in Domestic Animals   91
    Dogs   91
    Cats   93
    Pigs   94
    Horses   95
    Cattle   98
    Sheep   103
    Goats   105

Clinical Problems   105
    Hyperactivity   108
    Narcolepsay   108
    Nocturnal Wakefulness   109

4. SEXUAL BEHAVIOR   111
Introduction   111
    Physiological Bases of Sexual Behavior   111
    Social and Sexual Experience   114
    The Central Nervous System and the Control of
        Sexual Behavior   117
Cattle   120
    The Cow   120
    Clinical Problems of Cows   123
    The Bull   124
    Clinical Problems of Bulls   128
Sheep   131
    The Ewe   131
    The Ram   133
    Clinical Problems of Rams   139
Goats   139
    Free-Ranging Goats   139
    The Doe   140
    The Buck   140
    Clinical Problems   140
Horses   141
    Free-Ranging Horses   141
    The Mare   141
    Clinical Problems of Mares   143
    The Stallion   146
    Clinical Problems of Stallions   147
Pigs   152
    The Sow   152
    Clinical Problems of Sows   154
    The Boar   154
    Clinical Problems of Boars   156
Dogs   156
    The Bitch   156
    Clinical Problems of Bitches   157
    The Dog   158
    Clinical Problems of Dogs   160

Cats   163
  Free-Ranging Cats   163
  The Queen   164
  Clinical Problems of Queens   165
  The Tom   166
  Clinical Problems of Toms   167

5. MATERNAL BEHAVIOR   169
  Introduction: General Principles of Maternal Behavior   169
    Internal Factors That Elicit Maternal Behavior   169
    External Factors That Elicit Maternal Behavior   171
  Pigs   173
    The Free-Ranging Sow   173
    The Group-Housed Sow   173
    The Confined Sow   173
    Clinical Problems   179
  Sheep   180
    Parturition   180
    Behavior of the Ewe toward the Newborn Lamb   180
    Acceptance of the Lamb   183
    Clinical Problems   186
  Goats   188
    Clinical Problems   189
  Cattle   189
    Parturition   189
    Bonding   190
    Suckling   190
    Weaning   191
    Clinical Problems   192
  Horses   193
    Parturition   193
    Postparturient Behavior   194
    Suckling   194
    Weaning   197
    Mutual Recognition   197
    Clinical Problems   198
  Cats   200
    Parturition   200
    Suckling   201
    Teat Order   202
    Homing and Retrieval   203
    Communal Nests   203
    Grooming   204
    Acceptance of Kittens   204
    Clinical Problems   204

Dogs   205
   Parturition   205
   Suckling   206
   Licking   207
   Weaning   207
   Clinical Problems   208

6. DEVELOPMENT OF BEHAVIOR   211
Introduction   211
Dogs   211
   Critical or Sensitive Periods   211
   Neurological Development   218
   Sleep   219
   Play   220
   Temperament Tests   222
Cats   222
   Critical or Sensitive Periods   222
   Neurological Development   226
   Sleep   227
   Play   227
   Relationships With Humans   230
   Feline Temperament Test   230
   Clinical Problems   231
Horses   231
   The Foal's First Day   231
   The First Year   232
   The Juvenile Period   238
Pigs   239
   Sleep   240
   Teat Order   240
   Feeding   240
   Play   240
   Elimination   241
   Relationships with Humans   242
   Temperament Tests   242
Ruminants   242
   Lambs   242
   Kids   245
   Calves   246
Conclusion   249

7. LEARNING   251
Introduction   251
   Types of Learning   251
   Formation and Strengthening of a Learned Task   257

Rewards   259
Comparative Intelligence   260
      Methods of Measurement   260
      Problems of Cross-Species Comparison   262
      Summary   265
Pigs   265
      Sex, Breed, and Age Differences   265
      Operant Conditioning   266
      Visual Discrimination   268
      Effect of Drugs   268
      Effect of Malnutrition on Learning   268
Dogs   268
      Housebreaking   268
      Canine Geriatric Cognitive Dysfunction   271
      Breed Differences   271
Cattle   275
      Operant Conditioning   276
      Taste Aversion   277
      Visual and Auditory Discrimination   277
      Effects of Age   277
Sheep and Goats   279
Horses   280
      Operant Conditioning   280
      Visual Discrimination   281
      Maze Learning   281
      Observational Learning   282
      Influences on Learning   282
Cats   283
      Discrimination   284
      Rewards   284
      Brain Stimulation   284
      Conceptual Learning   285
      Imitation   285

8. INGESTIVE BEHAVIOR: FOOD AND
   WATER INTAKE   287
Introduction   287
      General   287
Control of Food Intake in Pigs   289
      Meal Patterns   289
      Social Facilitation   290
      Palatability   291
      Environmental Temperature   291
      Gastrointestinal Factors   292
      Estrogen Levels   293

Feeding at Parturition   295
Imbalance of Dietary Amino Acids   295
Glucose Utilization   295
Defense of Body Weight   296
Integration of Factors That Stimulate and
        Inhibit Intake in the Central Nervous System   297
Clinical Problems   299
Control of Food Intake in Dogs   299
Meal Patterns   299
Social Facilitation   299
Palatability   299
Environmental Temperature   301
Gastrointestinal Factors   301
Estrogen Level   302
Glucose Utilization   303
Defense of Body Weight   303
Integration of Factors That Stimulate and Inhibit Intake
        in the Central Nervous System   303
Clinical Problems   304
Control of Food Intake in Cats   307
Meal Patterns   307
Social Facilitation   307
Palatability   307
Environmental Temperature   308
Gastrointestinal Factors   308
Hormonal Effects   308
Glucose Utilization   309
Defense of Body Weight   309
Integration of Factors That Stimulate and Inhibit
        Intake in the Central Nervous System   309
Clinical Problems   310
Control of Food Intake in Horses   312
Meal patterns   312
Social facilitation   312
Palatability   312
Grazing   313
Gastrointestinal Factors   313
Central Nervous System Depressants   314
Defense of Body Weight   314
Clinical Problems   314
Control of Food Intake in Cattle   319
Meal Patterns   319
Social Facilitation   320
Palatability   320
Grazing and Selectivity   320

Environmental Temperature   322
Estrogen Level   322
Rumen Fill and the Products of Rumen Fermentation   322
Humoral and Central Neural Factors   324
Defense of Body Weight   324
Clinical Problems   325
Control of Food Intake in Sheep   325
Meal Patterns   325
Social Facilitation   325
Palatability   326
Defense of Body Weight   326
Rumen Factors   326
Intestinal Receptors   327
Grazing and Selectivity   327
Environmental Temperature   328
Hormonal Factors   328
Glucose Utilization   328
Central Neural Mechanisms   329
Control of Food Intake in Goats   330
Meal Patterns   330
Water Intake   330
Increase in Osmotic Pressure   331
Decrease in Blood Volume   331
Angiotensin   331
Dry Mouth   332
Multiple Causes of Thirst   332
Specific Hungers and Salt Appetite   333

9. MISCELLANEOUS BEHAVIORAL DISORDERS   335
The Interview   335
Behavioral History   336
Environment   336
Early History   336
Training   337
Other Behavioral Problems   337
Instructions   338
Dogs   338
Destructiveness   338
Self-Mutilation   348
Tail Chasing   349
Phobias   349
Running and Car Chasing   350
Backyard Problems   350
Jumping Up on People   352

Fly Catching   353
Vomiting   353
Psychosomatic Problems   353
Cats   353
Destructiveness   353
Withdrawal   353
Horses   354
Behavior Problems in the Stable   354
Trailering Problems   356
Behavior Problems under Saddle   361
Phobias   362
Behavior-Related Problems   362
Physiologically Based Problems   362
Cattle and Other Farm Animals   363
Cattle   363
Pigs   364
Sheep and Goats   364

Appendix 1: Separation Anxiety   365
Appendix 2: Animal Ownership, History, and Behavior Records   371
2A   Canine Behavioral History   373
2B   Cat Owner Questionnaire   384
2C   Behavioral History Form for Horses   388
2D   History form for Foal Rejecting Mares   392
2E   Behavioral History Form for Horses
     with Trailer Problems   394
Appendix 3: Canine and Ovine Brains   397
3A   Ventral View of Canine Brain   397
3B   Median Section of Canine Brain at the Third Ventricle
     and Mesencephalic Aqueduct   398
3C   Transverse Section of Ovine Brain at the Level
     of the Hypothalamus   399

Glossary   401
References   409
Index   479

PREFACE

My principal goal in writing this third edition is to provide an updated version of the text. My goal also is to provide one that is easier to read and, toward this end, has more defined terms.

The areas in which most advances have been made seem to be communication, perception, and cognition. In this edition, there is recognition and decoding of more of the signals exchanged by animals, including visual signals of stallions and cats and vocalization of pigs. As well, animal cognition, which may be overestimated by pet owners to the pets' detriment and underestimated by other portions of the population, is yielding to objective experimentation. Concept formation and individual decision making based on recognition by ungulates is one example that is presented here.

The other area in which progress is being made is that of clinical behavior. There has been an exponential increase in the number of scientific publications dealing with behavior problems and their solutions. Particularly important are advances in the use of psychotropic drugs and in the role of underlying disease in some behavior problems, both of which are addressed here.

DOMESTIC ANIMAL BEHAVIOR FOR VETERINARIANS AND ANIMAL SCIENTISTS

THIRD EDITION

CHAPTER 1

COMMUNICATION

The position of the ears and tail, as well as the overall posture are most indicative of the animal's immediate intentions. Flehmen (tonguing and gaping in carnivores) and the vomeronasal organ function in sexual behavior are discussed. Visual communication is an aid to understanding animals, but both vocal and olfactory communication can be problems. Hypervocalization and marking behavior or failure to eliminate in the proper place are common behavior problems of cats and dogs. Treatments of these problems and those involving inappropriate scratching by cats are addressed in this chapter.

INTRODUCTION

Communicating with animals, in particular, learning to understand the messages the animal is sending, is a most important part of diagnosis. Communication is a vital part of animal husbandry and the art of veterinary medicine and a very useful adjunct to the science of veterinary medicine. Before ordering a complete blood count and liver function tests, the astute clinician already will know that a dog is suffering from abdominal pain because she assumes an abnormal posture with rump high and head low or that a horse that paces in his stall and kicks at his belly is suffering from colic.

Another important aspect of communication between veterinarian and patient or between handler and stock is assessment of an animal's emotional state or temperament. Adequate restraint or, preferably, a quiet tractable patient is necessary for thorough examination and diagnosis. Most practitioners learn eventually to recognize animals that will be aggressive or fearful and, therefore, require tranquilization, muzzling, or more stringent methods. It would be helpful for agriculture and veterinary students to learn in advance how to recognize animals' moods. Learning by experience to recognize behavior problems may be at the expense of a badly bitten hand or kicked leg. For their own safety, as well as for acuity of diagnosis, clinicians should learn to listen to and watch for the messages their patients are transmitting both to them and to each other. Farmers can prevent injury to themselves and to their stock if they can interpret the animals' messages.

Animals communicate not only by auditory signals, as humans do, but also by visual and olfactory signals. Many olfactory messages cannot be detected by humans, although male pheromones, such as those contained in the urine of tomcats and the very flesh of boars and billy goats, are quite discernible to humans. We are all aware of vocal communication by animals, but many of these calls remain to be decoded. It is the visual signals made by ear, tail, mouth, and general posture that are of most benefit in gauging the temperament and the health of the patient.

Perception

Vision

Acuity. Communication in animals depends on their ability to perceive messages. The sensory abilities of domestic animals, with the exception of dogs and cats, have not been studied systematically. The perception of animals is almost always compared with that of humans. Dogs and cats have a higher critical flicker fusion (point at which a flickering light appears to be fused, or a steady light) than humans.[307] This indicates that both species have a well-developed cone system in their retina. Cats can discriminate illumination at one-fifth the threshold of humans, but their resolving power is only one-tenth that of humans.[451] Siamese cats do not have stereoscopic vision; other cats do.[1103] Environmental conditions affect visual acuity. It has been shown that free-ranging cats are hypermetropic, whereas caged cats are myopic.[180] When the relative visual acuity of some other species was compared, the pig ranked highest, followed by sheep, cattle, dogs, and horses.[634] The visual acuity of cattle, measured by using a closed or partially opened circle at various distances from the cow, is inferior to that of humans.[434] Cattle can discriminate objects at 2 lux of illumination. Cattle are also poorer in brightness discrimination than are humans. They have a brightness discrimination threshold of 66 lux in bright light and 4.8 lux in dim light, whereas humans have discrimination thresholds of 105 and 4.2 lux.[1136]

Color Vision. A question often put to a behaviorist is whether animals have color vision. All species of domestic animals have been shown to possess color vision in that they will make discriminations based on color, but color probably is not as relevant to these animals as it is to birds, fish, and primates. For example, it is very difficult to teach cats to discriminate between colors, although they learn other visual discriminations with ease and have two types of cones that absorb green and blue. Nevertheless cats,[1289] dogs,[1070] horses,[579] cattle,[345,550] pigs,[818] goats,[261] and sheep[1055] can all make discriminations based on color alone. Color vision in domestic animals is not identical to that in humans. In the most carefully conducted studies, dogs appear to see the world not in shades of gray, but rather in shades of violet, blue, and yellow. Their vision is similar to that of color-blind or dichromat humans, who see the green light as pale yellow, the yellow as yellow, and the red as dark yellow.

It is interesting that ruminants can discriminate medium and long wavelengths (yellow, orange, and red) better than short wavelengths (violet, blue, and green). Bulls, indeed, can perceive the matador's red cape.[1201] Horses can discriminate red and blue but not green from gray.[1138]

Monocular and Binocular Vision. Eye placement in the skull also affects vision. Horses have eyes set quite laterally and can, therefore, see to the side and well to the rear. They cannot see well right in front of their heads. Lateral vision is necessarily monocular, and horses see binocularly only in the 70° directly in front of the head. The blind area of sheep is 90°. Contrary to many popular and scientific sources, horses do not have a ramped or slanted retina;[1328] the retina is similar to that of other animals. Figures 1.1 and 1.2 illustrate the fields of vision of the cat and horse, respectively.

Audition

Acuity. In hearing, as in vision, cats and dogs appear to perceive more than humans. Humans can hear 8.5 octaves; cats can hear 10. Cats have 40,000 cochlear fibers, and humans have 30,000, although the range of hearing in cats is only 1.5 octaves greater than that of humans because higher frequency detection requires a disproportionate increase in cochlear nerve fibers; the number of fibers needed per octave is not constant but, rather, rises with the rise in frequency.[1217] Cats, despite their mobile pinnae, can discriminate between sounds 5° apart, whereas humans can discriminate sounds 0.5° apart. The pinnae significantly lower the auditory threshold in the cat. The absolute upper limit of hearing in cats and dogs is 60 to 65 kHz (kilohertz = kilocycles per second). Dogs and cats can discriminate one-eighth to one-tenth tones.[451,1069] Sheep also appear to perceive higher frequencies than do humans.[43] The auditory acuity of dogs has been used in silent dog whistles and ultrasonic (but not to the dog) distracting devices.[271] Dogs and adult cats are not known to produce ultrasonic calls, but rodents do make ultrasonic noises[40] that the carnivores use to locate them.

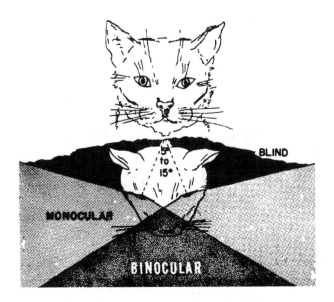

FIG. 1.1. Fields of vision of the cat, showing a large binocular area resulting from the forward position of the eyes (J.H. Prince. Reprinted from M.J. Swenson, ed. *Duke's Physiology of Domestic Animals*. Copyright, 1970, 1977 by Cornell University. Used with permission of Cornell University Press[1163]).

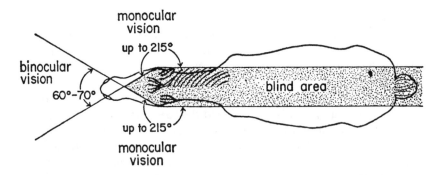

FIG. 1.2. Fields of vision of the horse. Note that there is only a small area of binocular vision but a very wide field of monocular vision[1454] (copyright 1975, with permission of Bailliere Tindall and W.B. Saunders Co.).

Olfaction

Acuity. Olfaction is to animals what writing is to humans—a message that can be transmitted in the absence of the sender. The sender must be present for auditory or visual signals to be sent, but an odor persists for minutes (or days) after the sender has gone. Olfactory acuity is probably the most important sense of domestic animal species because individual odor recognition and pheromonal release are an important part of their communication. Dogs probably have the greatest olfactory acuity, and this macrosmatic species is the one most investigated. Dogs can detect aliphatic acids at one-hundredth the concentration detectable by humans;[1051] they can distinguish between the odors of identical twins[764] and detect the odors of fingerprints 6 weeks after the fingerprints were placed on glass.[809] Dogs frequently are trained to sniff out drugs and natural gas leaks, and bloodhounds have been used for centuries to track people; apparently, no modern invention is as reliable as the canine olfactory mucosa. When administering drugs to odor-detecting dogs or those used in tracking, care must be taken by the veterinarian because a combination of corticosteroids can interfere with olfactory discrimination.[452]

There is considerable controversy about the ability of dogs to match a human scent from one part of the body, that is, the hands, with another part, that is, the elbows. Although dogs can be trained to do so, it appears to be a difficult concept for them to grasp. This is probably because they can easily discriminate odors from different body parts.[247,1301]

HORSES

Audition

Neigh

The neigh (or whinny) is a greeting or separation call that appears to be important in maintaining herd cohesion. It is most often heard when adult horses or a mare and foal are separated. A separated mare and foal will neigh repeatedly. These appear to be nonspecific distress calls, which the mare but not the foal may recognize individually.[1507] Some horses will call to their owners, but usually only when they are in their line of sight.

Nicker

The soft nicker is a care-giving (epimeletic) or care-soliciting (et-epimeletic) call. It is given by a mare to her foal upon reunion and probably is recognized specifically by each.[1418] A horse may also nicker to its feeder and a stallion to a mare in estrus.

Snorts, Squeals, and Roars

Nickers or neighs usually elicit a reply; other equine vocalizations, such as snorts, squeals, and roars, do not. The roar is a high-amplitude vocalization of a stallion usually directed to a mare. A sharp snort is an alarm call. More prolonged snorting or sneezing snorts appear to be a frustration call given when horses are restrained from galloping or forced to work. Snorts and nickers are sounds from the nostrils. The mouth is closed. Other calls are given with the mouth open.

When two strange horses meet, or when horses have been separated for some time, they greet each other by putting their muzzles together nostril to nostril (Fig. 1.3). The nostrils are flared, but if any vocal signals are given, they are inaudible to humans. Usually one, the other, or occasionally both of the horses will squeal and strike or jump back although neither has been bitten or threatened. The squeal is, therefore, a defensive greeting. It is heard frequently when horses are forming a dominance hierarchy and many bites are being exchanged. Mares that are not in estrus squeal and strike when a stallion approaches too closely. A squeal may also be a response to pain.

FIG. 1.3. Greeting. Nostril-to-nostril investigation, in this case by a horse and pony.[679]

Vision

Expression

The horse's ears are probably the best indicator of its emotions. The alert horse looks directly at the object of interest and holds its ears forward. Ears pointed back indicate aggression, and the flatter the ears are to the head, the more aggressive the horse[1413] (Fig. 1.4). Frequently, veterinarians are called upon to examine a horse for soundness for a prospective buyer. If the horse reacts to examination, or even saddling, by swiveling its ears back, it may not be a desirable purchase, even if perfectly sound physically.

Other facial expressions of the horse are more subtle; nevertheless, they can be used profitably to understand a horse's mood. A submissive horse turns its ears outward. Young horses (less than 3 years old) have a more dramatic display, snapping, also called champing or tooth-clapping, in which the lips are retracted, exposing the teeth that are sometimes clicked together (Fig. 1.5). This expression will be shown by a yearling colt to an approaching stallion or toward a mare with a newborn foal.

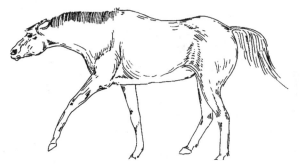

FIG. 1.4. The aggressive posture of a horse. The ears are back, and the horse is striking out with its front leg and lashing its tail.[680]

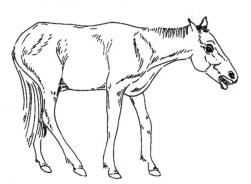

FIG. 1.5. The submissive posture of a horse. The tail is tucked in and the ears are turned outward. The horse is also snapping (opening and closing its mouth while retracting the lips).[679]

The sexually receptive mare shows a unique expression, the mating face, in which her ears are swiveled back and her lips hang loose (Fig. 1.6). The flehmen response, or curled upper lip, of the courting stallion is discussed in the next section. A horse that sees but cannot reach food, or is anticipating food, makes chewing movements and sticks out its tongue (Fig. 1.7). More difficult to identify is the horse in pain. A horse that is exhausted and in pain will show loose lips but clenched masseter or cheek muscles. Before a horse is in such pain with colic that he kicks at his belly, he will repeatedly swivel his ears back as if attending to his abdomen. The various facial expressions of horses have been illustrated by Schafer.[1265]

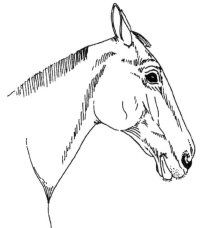

FIG. 1.6.   The mating expression of the mare.[679]

FIG. 1.7.   The food-anticipating expression of the horse[691] (copyright 1978, with permission of Elsevier Scientific Publishing).

Horses tend to position their ears in the same direction that they are looking. Thus, when the horse's ears are pointed straight ahead, he is looking straight ahead. This can be a clue that the horse is about to shy at an object. Usually the rider can identify the frightening object by looking where the horse's ears are pointing. The horse can then be coaxed to investigate and conquer his fear of the object. When the horse turns his ears to the side and back, he is looking to the side. He can see in a wider range to the side and well behind his head. The horse can see best when he is grazing and only his legs impede his vision; he can see in all directions and detect an approaching predator easily. Horses have blind spots, including the area just ahead of them when their heads are high. As has already been discussed, horses appear to have color vision.

Posture

The posture and bodily actions of the horse are also useful in interpreting its moods. The relaxed horse stands quietly while his nervous counterpart prances and chafes at the least restraint. The aggressive horse, which is threatening to kick, lashes his tail and may even lift one of his hind legs. The frightened horse tucks his tail tightly against his rump and stands with his feet close together. Muscle guarding is seen, especially if the animal anticipates pain. A few mares will urinate as they are being chased; because it is rare, it probably does not have a signal value and should not be confused with the frequent urination seen in estrous mares. The stallion moving his mares assumes a unique posture, called herding, driving, or snaking, with head down, nearly touching the ground, and ears flattened (Fig. 1.8). Horses paw the ground not in aggression, but in frustration, if they are eager to gallop or, more commonly, if they want to graze and are restrained by rope or reins. Pawing to eat may be a behavior derived from pawing through snow for grass and might be considered a form of displacement behavior.

FIG. 1.8. Driving posture of the horse. The stallion, *left*, drives a mare. This behavior is also called snaking, herding, or driving (rounding).[679]

Olfaction

Scent Marking

Olfactory communication plays an important part in the sexual behavior of horses. Stallions curl up their upper lip in the flehmen position or "horse laugh" when they smell the urine of a mare (Fig. 1.9). Estrous urine alone does not stimulate more episodes of flehmen by stallions than nonestrous urine,[935,1346] but the frequency of flehmen by a stallion toward a particular mare in his herd increases as she approaches estrus, perhaps because the mare urinates more frequently. After the stallion investigates urine by putting his lip in it, the flehmen position carries the urine into the nasal cavity. When his lips are raised in the flehmen position, the nostril opening is partially blocked and the horse, by breathing deeply, carries the urine into the vomeronasal organ. Information from the vomeronasal organ is transmitted more directly to the hypothalamus via the accessory olfactory tract than by way of the main olfactory system. Although stallions flehmen most frequently, geldings and mares also exhibit the behavior in response to olfactory or gustatory stimuli. Cough medicine or a new bit often causes the horse laugh or flehmen, obviously not a sign of amusement. Stallions usually urinate on the urine or scent mark as they are exhibiting flehmen.

Horses also use olfactory cues, especially from their own or other horses' manure, to find their way home. Wild stallions use manure piles along well-used pathways, possibly to scent mark.[463] These piles may separate bands of horses both spatially and temporally. Even in a pasture, stallions select one place to defecate and then back into the pile to

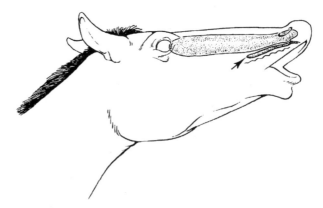

FIG. 1.9. The flehmen response, or lip curl. The location of the vomeronasal organ is indicated by the *arrow.*

eliminate, so the pile does not grow much wider. On the other hand, mares and geldings face outward, gradually increasing the diameter of the pile. Because horses will not eat grass contaminated with feces, a pasture containing mares and geldings rapidly becomes "horsed out" or inedible.[1102] Despite the discrimination of older horses against feces, foals show coprophagia, as discussed in more detail in Chapter 6: Development of Behavior.

DOGS

Audition

The common vocal communications of dogs are the bark, whine, howl, and growl.

Bark

Barking is a territorial call of dogs. It is used to defend a territory and to demarcate its boundaries. Stray dogs, whose resting places may be quite temporary, rarely bark.[168] As a stray dog passes the yards of owned dogs, however, it precipitates territorial barking. The observant owner can recognize various types of barks. The bark to be let in the house differs from that directed at human intruders, which may differ from that directed at canine intruders. Barking occurs in wild canids; a wolf in a seminaturalistic pen will bark at an intruder, but barking has been a trait selected for in domesticated dogs. People obtain dogs because they bark and can serve to warn their owners of the approach of intruders. Unfortunately, dogs are much more likely to bark in response to another dog's bark than to the sound of a human intruder.[10] The barking trait can become a problem in a highly urbanized environment. Two thousand two hundred complaints about barking are made in Los Angeles per year.[1295] For this reason, a dog's barking can be a problem for the owner (see Clinical Problems). A more acute problem is the barking of kenneled or caged dogs in a veterinary clinic. The noise level generated by barking can exceed the 90-decibel limit of the Occupational Safety and Health Act.[1088] Animal hospitals must, therefore, be constructed with very good sound insulation.

Whine and Howl

Whining is an et-epimeletic or care-soliciting call of the dog. It is first used by puppies to communicate with the mother who provides warmth and nourishment. Mature dogs will whine when they want relief from pain or even in a mildly frustrating situation, such as when they want to escape outdoors or to reach a rabbit for which they are digging.

Howling is a canine call that has not been deciphered well. It occurs more frequently in wild canids, coyotes, and wolves and in some breeds of dogs, such as huskies, malamutes, and to a lesser extent hounds. Harrington and Mech[605] found that the incidence of howling in wolves increased 10-fold during the home-site season. As the year's pups mature, the pack becomes more dispersed and the howling apparently takes the place of scent marking in coordinating pack member spacing and activity.

Harrington[604] has also found that wolves can discriminate strange adult from strange pup howls and answer only the former. A different component, lower in frequency, occurs in the answering howls of wolves approaching the source of a strange howl. Whether this is true in dogs as well remains to be determined.

Growl

Growling is an aggressive or distance-increasing call in dogs.

Vision

A dog's emotional state can be determined by observation of its ears, mouth, facial expression, tail, hair on its shoulders and rump, and overall body position and posture (Fig. 1.10). The calm dog stands with ears and tail hanging down. When he becomes alert, his tail and ears are pointed upward. The dog may point with one front foot. As the dog becomes more aggressive, the hair on the shoulders (hackles) and the rump rises and the lips are drawn back. The ears remain forward, and the tail may be slowly wagged. With increasing aggression, the lips are retracted and the teeth exposed in a snarl. The dog stands straight. As the dog becomes frightened, the ears go back until they are flattened against the head and the tail descends until it is between the legs.

Posture

The posture of the fear-biting dog, the one most likely to injure a veterinarian, is that of the frightened dog with tail and ears down and the body leaning away from the source of fear. He will have raised hackles and lips retracted in a snarl, which may expose the molars as well as the canines. Care must be taken when approaching a dog to notice any lifting of the lip because this may be the only prediction of defensive aggression or fear biting. The fear biter will escape if possible; but if he is approached within his critical distance, which may be a yard (approximately a meter or less) from him, he will attack.

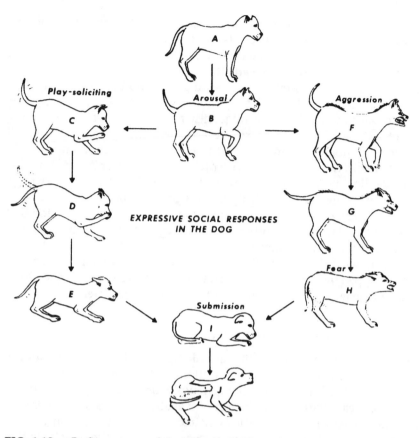

FIG. 1.10. Body postures of the dog. *(A, B)* Neutral to alert attentive positions; *(C)* play bow; *(D, E)* active and passive submissive greeting—note tail wag and shift in ear position and in distribution of weight on fore and hind limbs; *(F, H)* gradual shift from aggressive display to ambivalent fear-defensive aggressive posture; *(I)* passive submission; and *(J)* rolling over and presentation of inguinal-genital region[497] (copyright 1975, with permission of W.B. Saunders Co.).

More common, fortunately, is a dog in which fear is not mixed with aggression. The fearful dog crouches with his tail between his legs and his ears flattened down. If the dog is abjectly submissive, he will lie on his side and lift his hind leg, displaying the inguinal area. He may also make licking intention movements. Finally, he may urinate. This behavior probably represents a reversion to puppy habits in which the puppy lies down on his side and presents the inguinal area to the mother (who is, of course, dominant over the puppy) and allows her to lick and clean it. Submissive urination is discussed under Behavior Problems.

During a submissive approach, dogs will curve their bodies, whereas a dominant dog stands straight with tail and ears erect. This stiffening of posture can be used to predict when an initially friendly greeting is about to become an attack.

Dogs greet their owners as they did their mothers, by licking their faces. As puppies, dogs lick their mothers' faces to beg for regurgitated feed. Although wild canids frequently regurgitate food for their pups, not all domestic dogs do so; nevertheless, the begging behavior is shown by domestic puppies. The behavior persists in the adult dog who either licks the owner, or, if prevented by discipline or its small stature, makes licking intention movements. Licking their own lips, yawning, or even falling asleep sitting up are all signs of ambivalence in dogs.

Dogs have a play signal; it is necessary to signal that the action that follows is play because, otherwise, the recipient of the playful act will consider it genuine aggression or sexual activity and respond in kind. A bowing with the forequarters lowered and the hindquarters elevated and topped by a rapidly wagging tail is the signal for canine play.[178] Often one paw is waved or rubbed at the dog's own muzzle.

Genetic and Surgical Alteration

Ears, tail, and hair position are all important in visual communication between dogs, but communication tends to break down in breeds that have been modified either genetically or surgically. Dogs with dependent ears like hounds can only hint at attentiveness or fear. It is hard to detect piloerection on a long-haired dog. Can an Afghan raise its hackles? Hair also prevents many breeds from seeing the signals of other dogs. The Old English sheepdog is a good example because she has hair across her face, so she cannot see, and because of coccygeal amputation, her tail position must be left to the imagination. Simply cutting the hair that obstructs a dog's vision can improve its temperament. Dogs with docked tails learn to wag their whole hindquarters so that pleasure, if not fear, can still be expressed.

Olfaction

The legendary olfactory acuity of dogs has already been mentioned. Because dogs can smell some olfactants in a concentration one-hundredth to one-hundred thousandth that of the threshold for humans,[170] it is not surprising that dogs use odors as a means of communication.

The importance of olfactory communication to dogs is exemplified by the diligence with which male dogs scent mark vertical objects by urinating. It is believed that the species, sex, and even individuals can be identified by other dogs from the odor of the urine. It is worth noting that dogs scent mark much more frequently in areas where other dogs have marked. The record may be that observed by Sprague and Anisko[1343] of 80 markings by one dog in 4 hours. Even though male dogs rarely empty their bladders completely, such efforts exhausted this dog's supply, and the last urinations were dry.

Elimination Postures

It is appropriate at this point to discuss elimination postures in male and female dogs. Owners are often concerned because their young male does not lift his hind leg but, rather, still squats. Although standing and lifting the hind leg are typical innate male behaviors mediated by testosterone,[188] males urinate in other positions 3% of the time. Bitches assume not only the squatting position (68% of the occasions that they urinate), but also lift their hind legs (2%) and use various combinations of the two postures.

Urine marking is the most common form of scent marking in dogs, but vertical objects may also be marked with feces as any kennel cleaner has observed. Again, males are more likely than females to mark with feces[1343] (Fig. 1.11). Hart[610] found that castration reduces scent marking in

FIG. 1.11. Elimination postures of the dog[1343] (copyright 1973, with permission of E.J. Brill Publishers).

male dogs. Dogs that cannot smell (anosmic) and that, therefore, cannot identify other dogs' urine, mark less frequently and do not urinate on the urine of other dogs as intact dogs do. When dogs scratch after eliminating, they are not making rudimentary burying movements but are spreading the scent and possibly adding the odor of secretions from interdigital sebaceous glands.

Urine

The most powerful means of olfactory communication in the canine species is the urine of an estrous bitch. Doty and Dunbar[393] have shown that male dogs are more strongly attracted to the urine of an estrous bitch than to vaginal or anal sac secretions, although Goodwin et al.[563] present strong evidence that it is the vaginal secretion methyl *p*-hydroxy-benzoate that induces the actual mating behavior sequence in the male (see Chapter 4, Sexual Behavior). Dogs "tongue," that is flick their tongues against the palate just behind their incisor teeth, when they encounter estrous urine. This is the canine equivalent of flehmen.

Dunbar[406] also demonstrated the marked preference by an estrous bitch for male urine as compared with either estrous or nonestrous urine. The urine contains pheromones, substances secreted by one animal that affect the behavior of another animal. In estrous urine, these compounds are probably estrogen metabolites. The urine of a bitch in heat can attract males from great distances. The attractant effect of the bitch's pheromone is usually considered a nuisance, but there are practical applications. For instance, the pheromone could be used to attract stray dogs that could then be easily captured. One might expect that male dogs would inevitably be attracted to the urine of a receptive female, but Beach and Gilmore[155] noted that a dog without mating experience did not investigate estrous urine in preference to anestrous urine as sexually experienced dogs did.

Anal and Aural Secretions

Urine is not the only olfactory means by which dogs communicate. The anal gland secretions normally are eliminated with the feces and, no doubt, give them a unique odor.[393] Dogs, on meeting, usually sniff under each other's tails. This behavior is probably one of identifying the individual by its smell. A very excited dog can express its anal sacs forcefully; the resulting odor is pungent enough to be smelled by humans and may function as a fear pheromone.[388] The secretions of the ears are also believed to function in individual identification,[497] and investigation of one another's ears is a common greeting behavior of dogs.

Behavior Problems

Excessive Barking

Barking by dogs can be a serious enough problem to warrant debarking, but before surgery is undertaken it is important to determine where and when the dog is barking. If the dog barks only when outside by himself, the solution is simple—walk the dog on a leash and keep him inside otherwise. Owners often need to be persuaded that this is not unkind to the dog. Barking is probably not something dogs enjoy; they are defending their territory and not exchanging friendly greetings. In other cases, the dog is barking when left alone. This can be treated as a separation problem (see Chapter 9, Miscellaneous Behavioral Disorders). A more common problem is the dog that barks at guests. The dog is exhibiting territorial behavior just as he would if a burglar were attempting to enter the house. The owner's task is to teach the dog to stop inappropriate barking. Behavioral means can be used. It is usually better to reinforce positively a desirable habit than to punish an undesirable habit. This is particularly true of undesirable behavior that might have originated in fear. When the dog barks, the owner should call him and reward him for coming or command him to sit and reward him for sitting. The owner should never yell at the dog for barking, for loud noises will only encourage him to add more of his own vocalizations.

Dogs should not be encouraged to bark as puppies by praising or by adding to their excitement by saying "What's there?" Teaching a dog to "speak" for food is, of course, the best way to create a problem barker at meal times. Mild punishment, in the form of a water gun or lemon juice sprayed on the dog's mouth, can be used. If mild behavioral methods do not work, conditioning with a collar that sprays citronella on the dog's chin[763] or an electric shock collar could be used to teach the dog that barking should be kept to a minimum. The only type of electronic collar that is humane is that which is activated by the animal's bark rather than by the human. As a final resort there is vocal cordectomy (debarking), which may save the life of a dog that has been a barking problem. There are various procedures described for vocal cordectomy.[65] Dogs are not rendered silent, but the strength and pitch of their voices are lowered.

Submissive Urination

Submissive urination is a frequent behavioral problem. It occurs more often in young dogs and small dogs. It may be difficult to live with a dog that is dominant over his owner, but it is fairly messy to live with a dog that urinates submissively. Punishing the dog for urinating in fear or excitement aggravates the problem. The dog is already afraid of his owner, and punishment will only confirm and reinforce that fear. The wisest course is to avoid overexciting the dog.

Overenthusiastic greetings and overly harsh punishments should be avoided. If the person in the household who most often elicits the submissive behavior generally ignores the animal, the problem may be minimized. Submissive urination often declines as the dog matures.

Urine Marking

Another type of communication by pet animals that is not appreciated by humans is urine marking. Because urine marking is a sign of dominance, methods suggested for treatment of dominant aggressive dogs should also be applied to dogs that urine mark in the house (see Chapter 2, Aggression and Social Structure). The stimulus for urine marking in the house is a vertical object, but the motivation of the dog may be either elimination, marking, or even separation anxiety. Elimination problems are addressed in Chapter 7 under Housebreaking and in Appendix 1, under Separation Anxiety. The marking of every vertical object in a city block is another source of pollution that not only may kill the trees sprayed, but also spread such urine-borne diseases as leptospirosis. Urine marking may increase the level of aggression in male dogs. Smelling the urine of other dogs does, no doubt, excite a dog regardless of whether it might lead to aggression directed toward humans. For a number of reasons, therefore, dogs should be encouraged to urinate on their own territories.

CATS

Audition

Moelk,[1024] in an early study of domestic cat vocalization, listed an extensive vocabulary that may not be recognized by every cat owner. McKinley[985] used sonographic analysis to study feline vocalizations and divide feline vocalizations into pure type calls in which the vocalizations are homogenous and complex, made up of two or more pure types.

Pure Types

Murmur.   A soft, rhythmically pulsed vocalization given on exhalation. Murmurs are the request, or greeting call, which can vary from a coax to a command, and the acknowledgment, or confirmation call, a short, single murmur with a rapidly falling intonation.

Purr. A soft buzzlike vocalization, the purr is easy to recognize. It occurs only in social situations and may indicate submission or a kittenlike state. Remmers and Gautier[1193] have shown that purring is associated with rapid (25 per second), disynchronous contraction of the muscles of the larynx and diaphragm.

Growl. A harsh, low-pitched vocalization,[266] usually of long duration. Given in agonistic encounters.

Squeak. A high-pitched, raspy cry given in play, in anticipation of feeding, and by the female after copulation.

Shriek. A loud, harsh, high-pitched vocalization given in intensely aggressive situations or during painful procedures.

Hiss. An agonistic vocalization produced while the mouth is open and teeth exposed. This vocalization is probably defensive and can be used to gauge whether a cat is defensively or offensively aggressing.

Spit. A short, explosive sound given before or after a hiss in agonistic situations.

Chatter. A teeth chattering sound made by some cats while hunting or more commonly when restrained from hunting by confinement.

Complex Calls

Mew. A high-pitched, medium-amplitude vocalization. Phonetically it sounds like a long "e." It occurs in mother–kitten interactions and in the same situations as the squeak.

Moan. This is a call of low frequency and long duration. The sound is "o" or "u." It is given before regurgitating a hair ball or in epimeletic situations, such as begging to be released to hunt.

Meow. This characteristic feline call "ee-ah-oo" is given in a variety of greeting or epimeletic situations just as the mew and squeak are.

Vision

Posture

The postures and facial expressions of the cat are shown in Figures 1.12A and B. A cat carries its tail high when greeting, investigating, or frustrated. The tail is depressed and the tip is wagged during stalking.

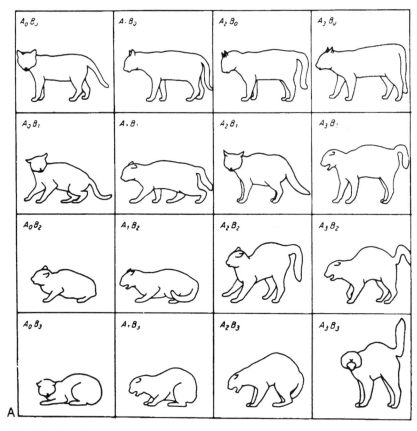

FIG. 1.12A. Body postures of the cat. Aggressiveness is increasing from A_0 to A_3, fearfulness from B_0 to B_3. $A_3 B_0$ is the most aggressive cat, $A_0 B_3$ the most fearful, and $A_3 B_3$ the defensively aggressive cat[873] (English ed., *Katzen—Eine Verhaltensstudien*, 6th ed., copyright 1982, illus 65, p. 146, with permission of Paul Parey, Berlin and Hamburg).

When walking or trotting, the tail is held out at a 40° angle to the back, but as the cat's pace increases, the tail is held lower.[798] A relaxed cat, like a relaxed dog, usually stands with tail hanging, but the cat's ears are usually forward. When the cat's attention is attracted, the tail is raised and both ears are pointed forward and held erect. Raising of the tail might be considered a greeting signal. The aggressive cat walks erect on tiptoe with head down. Because the cat's hind legs are longer than his front legs, he appears to be slanting downward from rump to head. His tail is held down, but arched away from the hocks; it is partially pilo-erected. His ears are held erect and swiveled, so the openings point to the side. His whiskers are rotated forward, and his claws are protruded. Subordinate cats crouch in the presence of a dominant cat.

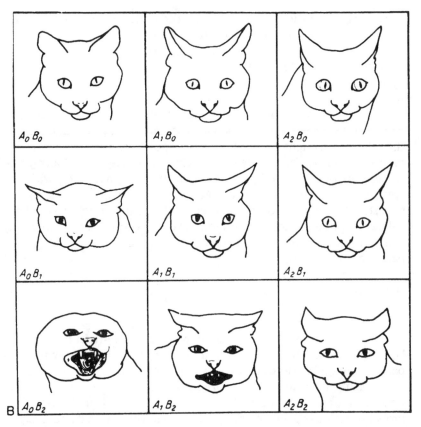

FIG. 1.12B. Facial expressions of the cat[873] (English ed., *Katzen—Eine Verhaltensstudien,* 6th ed., copyright 1982, illus 65, p. 146, with permission of Paul Parey, Berlin and Hamburg). $A_2 B_0$ is offensively aggressive; $A_0 B_2$ is defensively aggressive.

The frightened cat crouches with ears flattened to her head, and she salivates and spits. The pupils of the aggressive cat are constricted; as the animal becomes more defensive, the pupils dilate. The light-colored iris of the cat's eye makes an especially prominent signal of the cat's mood; it is probably an important intraspecific signal and should be used also to advantage by the veterinarian. The eyes of an excited cat appear red because the retinal vessels can be seen through the dilated pupils. Contrary to popular belief, the "Halloween cat" is not the most aggressive one; this cat, with arched back, erect tail, and ears flattened, which is piloerected and hissing, corresponds to the fear-biting dog. The cat is fearful but will become aggressive if her critical distance is invaded. One clue to the cat's emotions is that the hind feet appear to be

advancing while the front feet retreat; the paws are gathered close together under the cat. Cats roll on their backs, but there are sex differences in this behavior. Most female rolling occurs during estrus and is
directed toward males, whereas most male-to-male rolling is exhibited
by young males to adult males and is presumably a sign of submission.[465]

The gape is a response to a strange smell. This expression is most
commonly seen when the cat smells a strange cat's urine, and may be
the feline equivalent of the flehmen response of the ungulates. The
mouth is opened, the tongue is flicked behind the upper incisors where
there is an opening in the hard palate that communicates with the
vomeronasal organ[824] (Fig. 1.13). At the same time, an autonomic response to the odor is occurring whereby the urine brought to the hard
palate is aspirated into the vomeronasal organ during parasympathetic
stimulation. Fluid is flushed from the vomeronasal organ during sympathetic stimulation.[419]

Olfaction

Scent Marking

Male cats scent mark, that is, spray urine, more than females, but
both sexes do it. They spray trees along their most frequently traveled
path. Spraying is also done by cats that are subjects of aggression.[1067]
Free-ranging tomcats spray a dozen times per hour.[1416] Queens spray
once an hour and are more likely to spray when they are in heat. Cats
probably also use scent marking to arrange their activity temporally

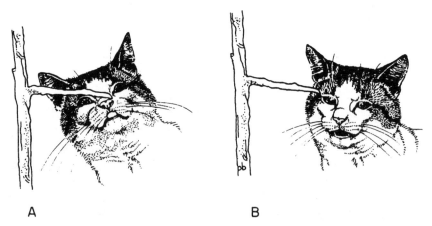

A   B

FIG. 1.13. The gape expression of a cat. *(A)* The cat touches the investigated object with its nose and may lick its nose; *(B)* then opens its mouth
while gazing in a preoccupied fashion (drawing by Priscilla Barrett, Cambridge, UK).

with other cats. De Boer[364] showed that much of the signal value of urine is lost within 24 hours, as evidenced by a comparison of interest in fresh and older urine marks by male cats. Cats can apparently distinguish the urine of familiar cats from that of strange cats.[1065] The smell of male cat urine is quite detectable by humans and usually objectionable to them. The smell of tomcat urine is probably due to the sulfur-containing amino acid felinine, which is present in highest quantity in tomcat urine and may be an important olfactory component in territorial spraying.[654]

Anal Secretions

Cats are well known for their fastidious covering of their feces, but in some situations, such as outside their core living area, cats may leave their feces uncovered. Cats probably use fecal and anal sac odor for communication; two strange cats spend considerable time circling one another attempting to sniff in the perianal area. If the cats are not too antagonistic, they will eventually permit each other to sniff.

Rubbing

Cheek rubbing (bunting) behavior may also be a form of olfactory communication in that glandular secretion from the cat's face is deposited on the object bunted. Cats bunt the objects to which they respond with a gape. Urine up to 3 days old can elicit these responses.[1434] Cats also rub each other. In general, the subordinate cat rubs the dominant one. This behavior serves to exchange odors among all the cats in a group, that is, they all smell the same.

Behavior Problems

Inappropriate Elimination or House Soiling in Cats

House soiling or inappropriate elimination is the most frequent behavior problem of cats.[226,617]

Diagnosis. The first problem in treating elimination problems in cats is to determine exactly what type of elimination is occurring because the various forms of house soiling are treated differently. Cats may be urinating or defecating in the house. If they are urinating, they may be either spraying or squatting. Even if the cat is not observed, the owner can usually differentiate between spraying and squatting by the location of the urine and the amount of urine produced. Cats most often spray on vertical objects, usually areas with some social significance such as windows through which the cat sees other cats; beds and chairs; or microwave ovens and stereo speakers, possibly in response to the sounds they produce. Small volumes are eliminated in spraying. Cats that urinate in the usual posture for elimination, the squatting position, rather than marking, produce larger volumes of urine usually at the edges of a room, on rugs, in sinks or bathtubs, or on beds, especially water beds or mattresses lying directly on the floor. Cats may defecate just outside the litter box, on rugs, in fireplaces, or in the soil of flowerpots; rarely a cat will neither urinate nor defecate in the litter box. Because the causes of the different types of elimination problems are different, the treatments differ, so it is important to determine how the cat is house soiling.

Causes.   Spraying is a normal and frequent behavior of intact male cats and is the reason most male cats, in contrast to most male dogs, are castrated. One of 10 castrated males, and one of 20 spayed females, spray. The primary cause of spraying in castrated males or females is the presence, particularly the odor, of other cats in the environment. For this reason, a cat in a household with more than one other cat is more apt to spray than one in a single-cat household. Male cats with a female housemate are more likely to spray than are males in households with another male.[615] The proportion of cats spraying increases with the number of cats in the household, so homes with a dozen or more cats almost inevitably have problems with spraying.[1091] If the owner is unsure which cat is spraying, it is possible to administer a capsule containing sodium fluorescein (0.3 mL of 10% solution—100 mg/mL), or six fluorescein strips may be placed three to a capsule for easier oral administration, to the most likely offender and then the house examined with an ultraviolet light to detect fluorescence. If no fluorescence is detected, the next most likely cat can be given fluorescein 2 days after the first.[617]

Spraying may occur when a new cat is introduced or when the cat is introduced to a new house and gradually subsides as the novelty of the new cat or new home declines; in other cases, the problem may persist. There also appears to be a seasonal incidence, with more cats spraying in early spring and late summer than at other times of the year.

Nonspraying or squatting urination is apt to be a sequelae of urinary tract pathology. The cat apparently associates the painful urination with the litter box and avoids it. The cat also may postpone urination until the urge is too great for it to reach the litter pan. Cats that use sinks or bathtubs are often suffering from cystitis or urinary calculi. Other cats are reacting to some aspect of their litter—the smell or texture of it, its location or its container, the litter box. Cats that like to perch on the edge of the litter pan while urinating rather than sit in the litter may prefer the floor because it is more stable. There may be a marking aspect to nonspraying urination, too, especially when the cat urinates on a bed or couch or when there are many cats in the household. In multicat households, one cat can also keep another cat from using the litter pan.

Defecation problems are less frequent and more apt to be a response to the litter than to the social environment. Constipation may cause the cat to avoid the litter because of the pain associated with straining in that location. Old cats, who may have difficulty in moving, may be reluctant to enter the box to defecate.

Treatment. Any animal with a behavior problem should be examined for and treated for any concurrent medical problem. This is particularly true of feline house-soiling problems because of the association of urological problems with failure to urinate in the litter box.

Spraying is best treated by castration of intact males. Spraying nearly always ceases if the tomcat is castrated before he is a year old, but it will reoccur in 10% of old cats. The result in mature males is more variable, but 87% of older males also abandon the habit after castration.[614] It may also be treated by reduction of the number of cats in the household. Most owners are unwilling to reduce the number of cats because it is difficult to find homes for adult cats. Therefore, a high proportion will require drug therapy. Spraying synthetic cheek gland secretion in the area in which the cat sprays also is effective. One type of mark can replace the other. In 30% of the cases, synthetic progestins, medroxyprogesterone acetate or megestrol acetate reduce spraying,[612,617] but the side effects of the drugs, particularly diabetes mellitus, preclude their chronic use. Buspirone (2.5–5 mg every 12 hours)[616] and amitriptyline (5–10 mg per cat every 24 hours)[933] are helpful treatments in about 60% of the cases. Diazepam is also successful in reducing spraying, but recently, hepatotoxicity has been identified as a side effect in some cats, so liver function tests should precede administration and a low dose (1 mg per cat twice daily) should be used for the first week. If the cat is not anorexic and is otherwise healthy, the dosage can be increased to 2.5 mg per cat twice daily.

Cats seem to depend most on their sense of smell to recognize other cats. For this reason, olfactory tractotomy is successful in eliminating spraying in most female cats and half the male cats that spray. Appetite and food intake are not adversely affected.[613] Lesions of the medial preoptic area of the hypothalamus (see Appendix 3) are also successful in eliminating spraying, but unlike olfactory tractotomy, which is a reasonably simple procedure requiring no specialized equipment, surgical production of brain lesions requires specialized equipment and training.[622]

Nonspraying elimination should be treated by changing the characteristics of the litter. Different types of litter should be presented: clay, sand, corncob, scented, alfalfa, paper, and so on. Soil can be used and litter gradually added; this will help cats that have been accustomed to going outdoors to eliminate. Some cats like deep litter; others shallow. The majority of cats prefer clumping litter[224] because it is made of fine particles similar to sand. Nonclumping litter should be discarded daily.

Clumping litter should be scooped daily and discarded every 2 weeks. Two boxes should be provided, one for urine and one for feces. There should be as many boxes as there are cats in the household. Several locations of the litter box should be tried. A tray can be substituted for a box, a covered litter pan substituted for an open one, or vice versa. If a social cause is suspected, the same drugs used to treat spraying can also be used.

Defecation problems should also be treated by changing the litter substrate or location. A low-sided box or tray is often successful.

In general, punishment is not successful; the cat learns only to avoid house soiling in the owner's presence. Furthermore, if stress is part of the underlying cause of the misbehavior, punishment will only further stress the cat. Retraining the cat by restraining it in a cage or bathroom may lead the cat to use the litter box while restrained, but such action seldom affects the cat's behavior when it is free to roam the house. One method of gradually reintroducing the cat to freedom is to keep it on a leash whenever the owner is home and confine it when the owner is away or asleep. The owner can prevent the cat from soiling and can take it to the litter if it starts to squat or to spray. If a discrete area such as a bathtub is used, keeping a few inches of water in the tub will prevent the cat from using it.

Repellents may be effective in discouraging a cat from using a particular area to eliminate, but the cat usually will choose another area. Effective repellents are citrus-scented air fresheners and strongly scented soaps, both of which are effective longer than commercially available animal repellents. When dealing with this problem, as when dealing with any behavior problem, the best solution is to change the animal's motivation by making the litter attractive. It is important to remove the odor from a soiled area. There are various commercial preparations available for this; the best combine enzymes and bacteria.[998] Owners should be discouraged from using ammonia because the odor is part of the typical urine odor and might actually encourage the cat to eliminate.

A few zealous owners may be willing to train the cat to eliminate on command, that is, to housebreak the cat as most dogs are. The owner should be aware of the times the cat usually eliminates and call the cat to the litter box at those times, encourage it to eliminate by stirring the (clean) litter and praising it for eliminating. Toilet training the cat has also been suggested.

Clawing and Scratching

Clawing or scratching behavior may be considered grooming behavior because the cat is loosening old layers of the claw, but seems to be primarily a form of marking behavior. Whatever the motivation for scratching may be, it is often an undesirable behavior, especially if the new sofa or draperies are chosen for a scratching site. Many cats are declawed to eliminate the problem, but bad scratching habits can be prevented from developing. If kittens are encouraged to use a scratching post, they usually will not abuse furniture. A good scratching post should have loosely woven material or sisal, not a firm rug, on it because the purpose of the scratching post is to allow the cat to hook its claws in the fabric. Placement is important; cats scratch the most prominent vertical surface. Carpet-covered, floor-to-ceiling cat trees may help, as will real bark-covered logs. Catnip-impregnated corrugated cardboard is also used for scratching. Because cats scratch more often when they awaken, along with stretching, one post should be placed near the cat's usual sleeping place. The other location should be a prominent spot. The best teacher of a kitten is its mother, so kittens should be obtained from queens that use a scratching post.[611]

If all else fails, the cat should be declawed rather than euthanized or sent to a shelter for eventual euthanization. There is no question that the cat will experience some pain at the time of declawing, so every effort should be made to improve analgesia, but there do not appear to be any long-lasting behavior sequelae to onychectomy.[187,1039]

PIGS

Audition

Vocal signals are probably the most important means of communication in pigs. Twenty calls have been identified,[591] and half a dozen are easily recognizable to humans. Kiley[796] has analyzed the vocalizations of ungulates in depth.

Grunt, Bark, and Squeal

The common grunt is 0.25 to 0.4 seconds long and is given in response to familiar sounds or while a pig is rooting. The staccato grunt or short grunt is, as the name implies, shorter (0.1–0.2 seconds) and is given by an excited pig and may precede a squeal. A crescendo of staccato grunts is given, for example, by a threatening sow and may precede an attack on anyone who disturbs her litter. The bark is given by a startled pig. The long grunt (0.4-1.2 seconds) appears to be a response to pleasurable stimuli, especially tactile ones. The squeal is a more intense vocalization, and the pig that is actually hurt screams.

The various grunts and combinations do not appear to have specific meanings, but the intensity of the vocalization varies with the intensity of the situation. A common sequence is to proceed from common grunts to staccato grunts to repeated grunts without interruption to grunt squeals to screams as the animal is approached, chased, picked up, and injected. Staccato greeting grunts are given by pigs that are reunited after a separation, and a series of 20 grunts with no pause may be given by the hungry pig. Nursing calls are described in Chapter 5, Maternal Behavior. Changes in the frequency and length of calls can indicate need. When separated from the sow, hungrier piglets call more frequently and at a higher frequency than satiated ones.[1457]

Isolation in a strange place causes pigs to vocalize. Short grunts are followed by screams. At the same time, the rate of defecation increases.[509] Mature pigs often react to restraint by tantrum behavior accompanied by very loud calls, but with no increase in heart rate.[931] When disciplining a subordinate pig, a dominant pig will give a sharp bark as it feints with its snout. The pig in chronic pain grinds its teeth.

Vision

Posture

Possibly because the vocabulary of swine is so large, visual signals do not appear to be as important. One can learn something about pig thermoregulatory problems, if not about their moods, by observing their posture. Newborn pigs are relatively deficient in fur or fatty insulation and their surface:volume ratio is large; therefore, maintaining body temperature is difficult. Pigs have compensated for their poor physiological abilities with several behavioral strategies to reduce heat loss. A warm piglet lies sprawled out, but a cold one crouches with its legs folded against the body. The surface area is thus reduced, and contact with a cold floor is minimized.

Tail Position

The tail, particularly in piglets, is a good index of general well-being in most breeds. A tightly curled tail indicates a healthy pig in most breeds, and a straight one indicates some sort of distress. The pig's tail is elevated and curled when greeting, when competing for food or chasing other pigs, and during courting, mounting, and intromission. The tail straightens when the pig is asleep or dozing, but curls again when the pig arouses unless the animal is isolated, ill, or frightened. The tail will twitch when the skin is being irritated. Amputation of pigs' tails removes a valuable, if crude, diagnostic aid.

Group Behavior

Group behavior is even more important. Pigs, especially newborn pigs, huddle when they are cold. They thereby convert several small bodies into one large one, both decreasing their surface area and using one another for insulation. Pigs can select an optimal temperature when a gradient is present, both in the laboratory and on the farm. Therefore, heat lamps are provided, and newborn pigs, except those brain damaged by anoxia at birth, stay under the lamp at a comfortable 29°C (85°F). Adult pigs still huddle when they are cold, but their thermoregulatory problem is more apt to be one of hyperthermia. Pigs do not sweat, and although they pant, it is not sufficient for cooling. Again, behavioral thermoregulation takes over and pigs wallow in mud, which is more effective than plain water for evaporative heat loss.[1052]

Olfaction

Boars may use behavioral signs more than pheromones to determine the sexual receptivity of the sow. Boars are the only male ungulate who do not exhibit flehmen. Instead, they gape as a cat does when they encounter sow urine. Females can identify intact males, probably by the strong boar odor produced by the androgen metabolites present in both the saliva and preputial secretions of boars.[1325] There are sex differences in the ability to detect androstenone.[391] Boars may habituate to this odor because it is present in their saliva. Females can detect the pheromone at one-fifth the concentration that intact boars do.[392]

Olfactory stimuli serve to identify pigs individually, for pigs can distinguish conspecifics by means of odor.[992] Pigs investigate any newcomer or any pig that has been temporarily removed by nosing him. The ventral body surface is a preferred site for sniffing. The ability of pigs to form a dominance hierarchy while blindfolded indicates that olfactory and auditory, rather than visual, signals are important to pigs.[449]

CATTLE, SHEEP, AND GOATS

Audition and Vision

Despite the intimate association of humans and ruminants for thousands of years, very little is known about communication in these species. Kiley[796] has analyzed cattle vocalization phonetically and according to the motivation of the animal. The "mm" call is of low amplitude and is usually detectable only within a few cow lengths. It is a common call given by a cow to her calf or while waiting to be fed or milked. A "mm(h)" call is given in a slightly more frustrating situation, for example, when a cow is isolated. A threatening bull gives a roar of high amplitude and "(M)enh" sound. A very hungry calf will give a high-intensity "menh" call. During copulation, grunting sounds are heard. Some humans can recognize cows by voice, so it would not be surprising if cattle were able to recognize one another. Cattle appear to respond to a vocalization with a vocalization of similar intensity. An excited call is answered by excited calls. Calves have a special moo, almost a baa, or play call.[257]

Vocal communication in a prey species like cattle may be most important in transmitting information about general safety or danger. It may have been more important for cattle (and horses) to be alert and ready to flee than to communicate more precise information in their calls. If domestic animal communication is studied in as great a depth and with the same ingenuity as bird communication has been studied, it may be found that vocal communication is more precise in domestic animals. Careful analysis of the situation in which a call is given, recording of the call, and playback of the call to conspecifics in a naturalistic setting may help to break the code of domestic animal languages.

Vocal communication in sheep consists of bleating in distress or to initiate contact. Ewes rumble to their newborn lambs (see Chapter 5, Maternal Behavior), and rams make a similar call while courting. The snort is an aggressive communication in sheep. Submissive postures are the low neck and the head shake given mostly by small sheep in the presence of larger ones. Sheep have a visual signal for defensive aggression—they stamp. Threats in sheep are the foreleg kick, often repeated several times and sometimes actually contacting the opponent. The horn threat is movement of the head sharply downward. The twist and low stretch involves stretching the neck and twisting the head accompanied by tongue flicks. Some rams threaten by standing stiffly with their heads up, which causes their neck to bulge. Rams rub their horns on one another's face probably spreading preorbital secretions. The other visual signals used in courting behavior are discussed in Chapter 4, Sexual Behavior. Rarely, sheep will huddle facing one another; head-to-head orientation is aggressive behavior in this species.

Adult sheep continue to use vocalizations as contact calls. Sheep also are able to distinguish conspecifics by means of olfaction.[122] The typical aggressive and submissive postures of cattle are described in Chapter 2, Aggression and Social Structure.

Olfaction

Olfactory communication is very important for sexual activity in ruminants. Goats and cattle can distinguish conspecifics by means of urine. Male urine is more easily distinguished than is female urine.[111] The flehmen response is shown by all male ruminants in response to female urine.

AGGRESSION AND SOCIAL STRUCTURE

The social structure of free-ranging domestic animals is that of groups of females with one or several resident males (horses and dogs), separate male and female groups (sheep and goats), groups of females and solitary males (swine and cattle), or groups of females that are flexible from solitary to matrilineal (cats). The determinants of dominance are age, weight, and sometimes sex. Methods of reducing aggression among newly mixed animals rely on environmental and management practices. Methods of reducing aggression of dogs and cats to people, whether their owners or visitors, or to other conspecifics in the household depend on training, management, and pharmacological treatments.

INTRODUCTION

Aggression is not a unitary phenomenon but serves a variety of functions in an animal's life. In some cases, aggression is used to obtain food; in others, it may facilitate access to a sexual partner or establish an animal's place in a social hierarchy. In some situations, aggression is highly desirable, as when fighting takes place to establish a dominance hierarchy. The importance of such fighting is that once the dominance hierarchy is formed, it provides the animals in the group a means by which additional serious combat may be minimized. Within an established hierarchy, subtle threats replace physical violence in competitive situations. For the animal practitioner, the problem is not to eliminate all aggression but to determine the type of aggression with which he or she is dealing; only then can the problem of control be dealt with effectively. For example, castration may stop a tomcat from fighting with neighborhood cats, but it may have little effect on his hunting behavior.

Categories of Aggression

Social or Dominance-Related Aggression

Social aggression occurs when animals live in groups. It serves to establish the dominance hierarchy, that is, who will be dominant over whom. When adult animals that previously have never been penned together are brought together for the first time, intense aggressive encounters may occur for several days until each animal has established its position in what generally turns out to be a hierarchy of dominant–submissive relationships. In this type of social grouping, there is an alpha animal, who is seldom challenged by subordinates; a beta, or second-ranked, animal, who is only challenged by the alpha animal; and so on. Within this organization, the type of aggressive encounter changes once the rank of each animal has been determined. No longer is an attack and subsequent fight needed for an alpha animal to establish its rights over a beta animal. Now, a direct stare, or the threat of a charge, will generally serve to deter the beta animal from further confrontation. This assertion of dominance in the absence of a physical combat is called ritualized aggression. Although perception of an extensive hierarchy is questionable in domestic species, the relationship between any two animals certainly is recognized. The scarce resource over which social dominance is expressed can be food, a comfortable place to rest, a mate, or any action by one animal that is perceived as a threat or a challenge by the other. The best examples are two horses or a dog and its owner. In the case of two horses and one bucket of food, one horse displaces the other at the bucket and obtains the scarce resource. In the case of dog and owner, the dog growls when his collar is grabbed because the owner has challenged his dominance.

Territorial Aggression

Territorial aggression keeps others out of a particular geographical area. This is the type of aggression seen when the domestic dog becomes a snarling menace to the delivery person. In essence, the dog is defending a territory that he considers to be his own, and strangers—canine or human—simply are not welcome.

Pain-Induced Aggression

Pain-induced aggression develops directly out of induced pain or fear of pain. The function, of course, is to reduce the pain by eliminating the source. When an animal breaks a leg, he does not discriminate between the pain that comes from the break and the unavoidable pain induced by the veterinarian who tries to set the broken leg. A defense re-

action of many species, including dogs and cats, is to attack the cause of pain.

The veterinarian must expect to encounter pain-induced aggression frequently, for any animal will attempt to retaliate if he or she is suffering acute pain. Therefore, a general discussion of this type of aggression is appropriate.

Some species and some individual animals are more stoical than others, but most will bite or kick if the pain is severe. To judge when aggression may be induced by pain, we must consider the anatomical area involved. A wound on the face will be more painful to the animal than a similar wound on the back because more receptors per unit surface are present on the face. Other areas that appear to be most sensitive are the ears when afflicted with otitis and the rectum, especially when the anal sacs are infected. Any animal will be in pain if a bone is fractured or if the animal has been subjected to surgery.

Injections usually cause pain, and the veterinarian must be prepared for the animal's reaction. Subcutaneous injections of a nonirritating liquid are virtually painless if placed in a loose-skinned area, but intramuscular injections of an irritating fluid such as tetracycline or ketamine are very painful even if no nerve is struck. Horses can become very "needle shy" and quite unmanageable. Equine practitioners would be well advised to infuse local anesthetic with a small-gauge needle before placing a large-gauge needle (>18) in the jugular vein of a horse and to reward the horse with a carrot immediately afterward. A horse that can be repeatedly injected is worth the few extra minutes and few extra cents involved in rendering the procedure painless. If they cannot escape, horses, dogs, and cats will usually direct their aggression toward the veterinarian; cows, however, usually will direct their aggression toward the painful area. A cow that can scratch her ear with a hind foot is quite capable of kicking someone standing at her shoulder, so one would do well to take advantage of the cow's pain-directed aggression. If possible, injections should be given on the side opposite that on which the injector is standing. The teats are probably the most sensitive area of the cow, with the possible exception of the muzzle, and should be handled from the opposite side of the udder if the procedure is to be painful or if the cow is nervous.

An example of pain-induced aggression involves a cat with no previous history of behavioral problems that became markedly aggressive when admitted for treatment of a retro-orbital abscess. The cat was aggressive not only toward the hospital personnel, but toward his owner as well. The aggressive behavior, almost fury, did not subside until the cat had been in his home environment for several days. The fact that pain increases aggression should help to explain why corporal punishment of an aggressive dog may exacerbate, rather than attenuate, its undesirable behavior.

Fear-Induced Aggression

Fear-induced aggression can be related to pain, but in some cases it is motivated by neophobia (fear of the unknown) or fear of a particular person or animal with no apparent cause. It is usually accompanied by more physiological and visceral signs than those seen in pain-induced aggression, for example, crouching, spitting, and dilated pupils in a fearful cat.

Maternal Aggression

Maternal aggression is directly related to the protection of young. Although the male is generally considered the more aggressive of the two sexes, maternal aggression can equal the ferocity of any male attack.

Predatory Aggression

Predatory aggression is usually directed toward another species, and its purpose is to obtain food. Cats that are fully satiated will often hunt and not eat the catch, indicating, however, that predatory aggression is a behavior that is not entirely governed by hunger. During predatory aggression, the predator adopts an inconspicuous posture and usually does not vocalize.

With this classification scheme in mind, we now can analyze aggression species by species. In subsequent sections, the major forms of aggression in domesticated species are discussed. Some species, such as the canids, show a much wider spectrum of aggression than others, or perhaps we observe a wider spectrum because of our close association with the species. We omit pain-induced aggression because it occurs in all species.

The Biological Basis of Aggression

Genetic Factors: Breed Differences

Breed differences in aggression and the dominance of one breed over another are discussed under each appropriate species. For example, dominance is clearly influenced by heredity, for twin cattle are often of equal dominance.[444] Also, there are breed differences in dominance among dairy cattle. Among beef breeds, which do not differ markedly in weight, there are also definite differences in temperament.[1445] Another clear indication of hereditary influence on aggression is the difference in temperament between dog breeds. These differences, resulting from the selective breeding of dogs, have provided us with some very truculent breeds. Dogs are the best examples of genetic influences on behavior because there are so many breeds of dogs. They also have a short generation time that makes it possible to select for behavioral characteristics easily. Mackenzie et al.[916] have reviewed the numerous studies on heritability of temperament in dogs. See Prevention of Aggression for more examples of heritability of aggression in dogs.

Environmental Control of Aggression

Various environmental factors can increase aggression. Hunger and crowding are the primary ones. Almost every study has found that decreasing enclosure size increases the rate of aggression. This is true of dairy cattle,[1010] beef cattle,[1381] and pigs.[748] Most aggressive interactions occur at the time of feeding. Unpredictability of feeding time also increases the rate of aggression.[275]

Neurochemical Control of Aggression

Serotonin may act to inhibit aggression in cats[775] and is the neurochemical influenced by the psychotropic medication fluoxetine (Prozac) used to reduce aggression in dogs. Serotonin metabolite levels are lower in the cerebrospinal fluid of aggressive dogs than in nonaggressive ones.[1192] The rationale for the use of low-protein diets is that they allow more tryptophan, the amino acid precursor of serotonin, to traverse the blood–brain barrier and thereby increase brain serotonin levels. Oral tryptophan may also be administered.

Hormonal Control of Aggression

In many species, the male is more aggressive than the female, both at the interspecies and intraspecies levels. There are, however, some notable exceptions, such as the female with young. At other times, female aggression is generally limited to discouraging male suitors when the female is not sexually receptive and to maintaining the female's place in a female hierarchy if she lives in a group, such as a dairy herd. Female dogs do show territorial defense, but usually not with the gusto that males do. Through training or experience, however, the female can become an adequate watchdog.

Testicular hormones.   In species in which there is a marked sex difference in levels of aggression, testicular hormones appear to play two distinct roles in the control of aggression. During very early development, the presence of testicular hormones establishes a heightened potential for aggression. In the absence of testicular hormones, this aggression fails to develop. Hence, in the male, the presence of androgens during sexual differentiation enhances the potential for aggression, whereas the female escapes this influence.

The effect of androgens on aggression has been thoroughly investigated in rodents but is not limited to them. Cats show sex differences in aggressiveness that are dependent on the neonatal hormonal environment.[735] Female puppies treated with testosterone in utero and after birth were as adult dogs more successful in competing for a bone than were normal females, but they still were less successful than males.[154] Testosterone administration increased dominance rank in cows[235,236] and aggression in sheep.[1110] Castration has been practiced for centuries to improve tractability as well as to prevent breeding. Bulls, for example, are more aggressive than steers; the difference increases with age. Bulls also mount one another, which may be an expression of dominance rather than homosexuality.[813]

In addition to the developmental effects of androgens, there are the well-known activational effects of androgens upon aggression. Exposure to androgens during adulthood increases the probability that the male will show various forms of aggression (territorial, sexual, social, or dominance), but has little to do with predatory aggression.

CATTLE

Free-Ranging Cattle

In contrast to most other domestic species, cattle are not often found in the feral state. Their large size and nutritional requirements may be reasons. One group of cattle has remained relatively unmanaged on an estate in England for over 500 years. These animals, the Chillingham or White Park cattle, form cow and calf herds, but the bulls live separately either alone or in groups of two or three. They join the cow herds during the breeding season.[859] Their home ranges are stable, but different areas may be used in different seasons. More aggression occurs among the cows than among the bulls or between cows and bulls when they are artificially fed.[595] The bulls maintain a hierarchy through displays with little overt aggression, but bulls from different home ranges rarely breed the same group of cows.

Confined Cattle

Social Aggression

The dominance hierarchy in cattle is known as the bunt order in polled cattle and the hook order in horned cattle. Dominance can be determined by observing the stances of the two cows involved. The dominant cow, when threatening the submissive one, will stand with her feet drawn well under and with her head down, but perpendicular to the ground. The ears will be turned back with the inner surface pointing down and back. The submissive cow also stands with lowered head, but her head is parallel to the ground and her ears are turned so that the inner surface points to the side (Fig. 2.1). Aggressive bulls will turn perpendicular to the opponent and display their full height and length. Some may paw and drop to their knees to horn the ground. Aggression is expressed by bunting or striking the opponent with the head.

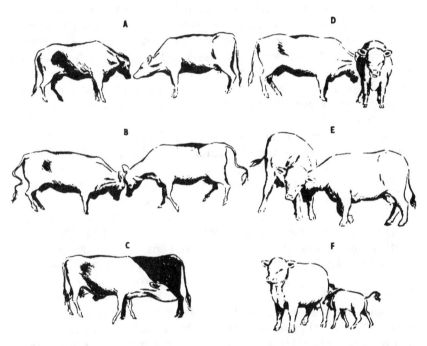

FIG. 2.1.   Patterns of agonistic interactions in cattle. *(A)* Cows meeting after an active approach. The one on the *left* is threatening, whereas the one on the *right* has assumed a submissive posture. Note the head and leg positions of each cow. *(B)* Physical combat: a fight. The cows bunt or push against each other head to head, each striving for a flank position. *(C)* The clinch. One contestant of an evenly matched pair slips alongside the other; the head of the former is pushed between the legs and udder of the latter. In unusually prolonged contests, the cows rest briefly in the clinch between bouts of bunting. *(D)* Flank attack. The animal that gains a flank position is at a decided advantage over the other. The flanked animal either submits and flees or strives to regain the head-to-head position. *(E)* The butt. A dominant animal directs an attack against the neck, shoulders, flank, or rump of a subordinate, which in turn submits and avoids the aggressor. *(F)* Play fighting. The calf butts the mother[591] (with permission of W.B. Saunders Co.).

Determinants of Dominance

The determinants of dominance in cattle appear to be height, weight, age, sex, presence or absence of horns, and territoriality, with horns, age, and weight being most important. The cow with horns will dominate a polled animal. In general, the heavier animal will be dominant over the lighter one,[232,234] but in one study[312] height was found to be negatively correlated with dominance. Success (or failure) in dominance interactions is genetically determined in heifers,[1166] but there is no correlation of dominance with age or weight. In an established herd, the older cows tend to be dominant, probably because initially the older cows are larger than the younger ones; once the hierarchy is formed it remains stable.[1268] If strange cattle are added to the herd, they will tend to be subordinate even if they are older and heavier. The reason is, presumably, that the cattle on their own territory have an advantage over those just introduced.[1268] Bulls were dominant over cows in a study of Holstein cattle.[1337]

The same cow does not "win" every interaction. To be dominant, a cow must win most of the interactions with the other cow.[1482] Dominance hierarchies can be observed simply by noting all agonistic interactions between cattle. However, this can be a slow process. Clutton-Brock et al.[303] noted only 0.1 agonistic encounters per hour in free-ranging Highland cattle, whereas they found that ponies engaged in agonistic behavior 1.9 times per hour. Provision of food to hungry cattle almost always provokes aggressive behavior and can be a technique for determining food-related dominance quickly. Cattle of the same rank feed within 2 m (6.6 ft) of one another at a trough, but the greater the difference in their rank, the farther apart they will be. Presumably, it is the lower-ranked cow who is responsible for the separation.[929] The cow that delivers the most blows and spends the most time controlling the food is dominant. Physical contact is necessary for dominance to be determined, but vision is not. When two cows are in separate pens with the bucket anchored between the pens, both will attempt to eat from it; neither will retreat. If the cows are in the same pen, one will defer to the other, with or without a struggle. The same process occurs even if the cows are blindfolded.[233] A unique way to determine dominance is to put the animals facing one another in a passage that is too narrow for them to turn around. The animal that is forced to back out of the passage is the submissive one.[1365]

In the absence of horns, cattle use their heads as battering rams, pummeling each other's heads and shoulders until one can reach the more vulnerable flanks or simply inflict too much punishment on the other. Equally matched cows may fight for long periods, interrupting active aggression to rest in clinches; one cow will put its muzzle between the hindquarters and the udder of the other, effectively immobilizing it (see Fig. 2.1).

Dominance hierarchies are similar, whether determined by aggressive interactions in many situations or just in feeding situations. Dominance does not seem to be related to access to resting places in a free-stall situation,[524] nor is entry into the milking parlor, although there is a tendency for middle-ranking cattle to come first, dominant cows in the middle, and low-ranking cattle last.[531,804] In most cases, there is no correlation of milk production and dominance,[375] but in at least one study,[241] high-ranking cattle tended to be good producers. Purcell et al.[1167] have found that herds in which the average behavior in the milking parlor is better (that is, less restless and aggressive) have higher milk production than herds with poor behavior.

There are many triangular relationships in large herds because the number of cows is so much larger than that of a natural herd of cattle.[594] Furthermore, hierarchies fluctuate because dairy herds have so many changes in composition due to culling of poor producers, addition of re-placement heifers, and movement of dry and calving cows to separate facilities. Cattle apparently remember one another, so when a cow leaves her herd and then returns within a few weeks, she will assume the same rank.[313] Sick cows and heavily pregnant ones will withdraw from the herd and, therefore, show a change in status—a fact that the stock-person and veterinarian should note.[176]

When cattle are driven, the least dominant animals will be first and the dominant animals will be in the middle of the herd,[175] although when grazing freely, the dominant animals will be the farthest from an observer.[174] During undisturbed grazing the dominant cattle tend to be the leaders, but the individual cow who leads is variable.[1261] When feed is available from a stall where only one animal can eat at a time, animals that are dominant in other situations do not supplant subordinate animals.[1364] Perhaps having protection on three sides allows the subordinates to maintain their position.

Problems can arise when crush gates are used to speed entry into milking parlors because subordinate cows that would enter last are pushed into dominant cows. Aggressive interactions may result. Non-pregnant cows precede pregnant ones in the crush order.[387] Most cattle that refuse to enter a crush do so consistently. More bulls than steers were very agitated and more steers (40%) than bulls (25%) were calm.[567]

The individual distance of grazing cattle is about 20 m (66 ft), and intrusion into this area may be met with threats or bunts.[303] Cattle high in the hierarchy have an interanimal distance smaller than that of cattle low in the hierarchy, that is, they are not reluctant to approach an-other animal. Bulls tend to have greater interanimal distance than steers.[665] Interanimal preferences may also be related to dominance be-cause animals found close to one another in a field are also close in rank.[1374]

The dominant cow will not be the first into the milking parlor, but it will be the first to a feeding area. Once at a feeding area, the dominant cow spends more time eating[988] and less time moving from place to place than a low-ranking cow.[16] This is true both in the feedlot and on pasture. There are several interactions of reproductive status and dominance. Dominance increases with estrus and decreases with pregnancy.[174,1268] Prolactin levels are negatively correlated with dominance.[62]

There are breed differences in dominance. Angus cattle are usually dominant over Herefords and shorthorns.[1445] Brahma-Hereford crosses are dominant over Herefords.[988] Among the dairy breeds, Ayrshires are dominant over Holsteins, and these two breeds are dominant over the smaller breeds, such as Jerseys.[241] These breed differences indicate that dominance may be inherited, and heritability of dominance has been calculated to be 0.4 to 0.5.[172,374]

Stage of lactation and adaptation to a challenging environment can have a destabilizing effect on dominance hierarchies. For example, when pastured in the Alps, Holsteins were subordinate to the native Swiss breeds,[1087] although one would have expected the larger Holsteins to be dominant.

It is at least 4 days after strange cows are mixed that nonphysical (that is, threats) replace physical interactions.[825] Once formed, dominance hierarchies reduce overt aggression; only the lowest-ranking animals may suffer deprivation of food when supplies are scarce or feeding space limited. It is during the initial stages of formation of a hierarchy that most aggression is seen. Stock managers should mix strange animals with care and avoid putting hungry animals together. When a previously unacquainted group of cattle is formed, it is 24 to 28 hours before the hierarchy is formed. In addition to physical injury, cattle may also suffer from lack of rest because of the general turmoil. The normal pattern of standing and lying as a group does not emerge for at least 48 hours after the group is formed. The stress resulting from lack of rest and lack of rumination is added to the stress of transportation that usually precedes the formation of a new group.[1478] These considerations may explain why cattle are more susceptible to such diseases as the shipping fever complex when new groups are formed.

Aggression in Bulls

In cattle, the greatest problem is the notorious and unpredictable aggressive behavior of bulls. Dairy bulls are generally more aggressive, as well as larger, than beef-breed bulls. One reason that artificial insemination of cows has been so enthusiastically accepted, despite the lowered fertility that has resulted, is that it is no longer necessary to keep bulls on the farm. A teaser bull with a deviated penis or a vasectomy can detect cows in estrus, but cannot impregnate them, and will be as dangerous as a fertile bull. The same hormones that motivate him to mount the cow also will induce him to charge his owner. Elimination of those hormones will reduce both mounting and charging.

Bulls are sometimes kept in groups; their behavior in this situation is described by Dalton et al.[348] and Kilgour and Campin.[803] Bulls mount other males, and these animals retaliate by butting.[1025] Mounting in this case is probably motivated by dominance, not sex. Bulls who were hand reared individually were more aggressive toward other bulls than were group-reared bulls.[1159]

Maternal Aggression

Cows will attempt to protect their young, and caution always is warranted when a mother is with her offspring. Angus cows may be especially protective.

Grooming

Mutual grooming occurs, but this only occupies a few minutes per hour. Cattle groom their age mates, their kin, and those of higher rank. When one cow solicits grooming from another, the licking is of the head and neck, but the majority of the grooming is of the back and rump.[508,1262,1263] Older and larger cattle receive and give more grooming than younger cattle. Milk production and milking order (order of entrance into the milking parlor) are also correlated with the amount of grooming received.[1508]

Clinical Cases of Aggression

Aggression toward people, including butting, kicking, and crushing, is most apt to be a problem in dairy cattle that are handled several times a day. Dangerous animals are usually culled, but a high-producing cow may be kept. She will pose most danger to those unfamiliar with her temperament, that is, the veterinarian. Some cows can be handled only from one side so the astute clinician should try to work from the side of the cow where she is milked.

Veterinarians can be the victims of bovine aggression when their treatment is most successful. A cow recumbent with hypocalcemia may, when treated with calcium intravenously, leap to her feet and attack because, whereas severe hypocalcemia results in muscular weakness, mild hypocalcemia can cause irritability.

The disappearance of bulls from dairy farms, as artificial insemination replaced natural service, has reduced the risk of death from bovine aggression. Nevertheless, every precaution is taken to provide escape routes for the handlers at the artificial insemination centers where large numbers of bulls are kept.

An unusual case of bovine aggression demonstrated the importance of visual cues to cattle. A herd of Hereford cattle were bred to a short-horn bull. Most of the calves were red with a little white, but one was mostly white with a few red spots. When the calf and its mother were re-leased into the pasture a few days after the calf's birth, all the other cows attacked the calf. Altering olfactory cues had no effect, but the problem was dealt with immediately by putting the calf and its dam in a corral where they could be seen but not injured; eventually, the problem was solved by adding more cows with white calves to the herd.

SHEEP

Free-Ranging Sheep

Feral sheep in a natural setting form separate ewe and ram flocks. The ewe flocks also include lambs and immature rams. There are seldom more than 20 adult ewes in a flock. Ram flocks are much smaller (about six animals) and less stable. This type of social organization is seen in Soay sheep, which are a primitive form of domestic sheep, and among mountain sheep.[545,576] Domestic, as well as wild, sheep "camp" in one particular area at night. It is theorized that these areas may be useful be-cause information can be exchanged between animals even though en-ergy must be expended traveling from food sources to the night camp.[1083]

Flocking

The formation of large commercial herds of hundreds of sheep is usually accompanied by a cacophony of baaing as the small flocks are lost within the large one and the individual sheep give separation calls. The formation of smaller or larger groups of sheep in farm situations is somewhat unnatural. Three sheep do not readily form a flock, and they tend to disperse; therefore, three sheep are often used in sheepdog trials as a test of the dog's ability to control the sheep. Sheep tend to select sheep of the same breed as flock mates when randomly mixed.[78,1498] Fa-miliarity is very important to sheep. They quickly form associations that are slow to break down.[1499]

There are breed differences in the tendency to aggregate. For exam-ple, Clun Forest sheep gather in large groups, whereas Dalesbred and Ja-cob sheep are more dispersed and form smaller groups.[1313] Similarly, there are breed differences in the tendency to form separate subgroups within large flocks. Merinos rarely form subgroups, whereas Dorsets and Southdowns do. The grouping depends on the sheep's activity.

Some sheep form subgroups only when grazing, whereas others form subgroups both when grazing and when camping.[80] Leadership in sheep is negatively correlated with the tendency to join the flock. In other words, independent sheep lead; the others follow.

Confined Sheep

Dominance

Within an established related flock of sheep, the oldest ewe is dominant over other ewes and is usually the leader in movements. Dominance is not related to body weight in commercial flocks of sheep of similar age. Very little overt aggression is seen among sheep, but dominance can be determined by limiting feeding space. Dominant sheep will push out subordinates.[76,1345] On a large pasture, sheep will divide into flocks with individual, but overlapping, territories. Newly introduced sheep, even offspring separated since weaning, are not allowed to join the original flocks but are relegated to less productive parts of the pasture,[717] which may explain their tendency to wander. Lambs may follow their mothers for up to 2 years after weaning, but this varies from population to population.[856,1230,1278]

One can observe the dominance hierarchy by entering a pen of sheep. The farthest sheep will be the most dominant.[398] Aggressive sheep shoulder push; once the dominant animal has been determined, it displays behaviors toward the subordinate that appear identical to those used in courtship, that is, nudging, head low and nose up, while striking with the front limb.

Age and weight, but not sex, are also important in determining dominance.[1225,1226,1279,1280] Appearance must be important because shearing may reduce the rank of a dominant ewe.

Sexual Aggression

Sexual aggression in sheep can actually interfere with breeding. Among rams, the dominant ram will usually breed more ewes than his subordinates unless he is so aggressive that his battles distract him from the ewes' estrus.[1320] Two rams may breed fewer ewes than one alone if they are often engaged in the butting contests typical of ram aggression. If three rams are present in a flock, two may fight while the third, less aggressive but evolutionarily more competent, impregnates the ewes.

Interspecies Aggression

When goats and sheep are placed in the same pasture, a strange situation can develop. The typical aggressive posture of the goat is to rear and meet its opponent in midair, where their horns clash. Sheep charge with lowered head and butt one another head on. If a goat and a sheep fight, the goat rears, but is butted in the belly by the sheep, which is usually the victor.

GOATS

Free-Ranging Goats

Feral goat herds can range in size from one to 100 goats, but the mean size is four.[1302] Goats form sexually segregated flocks, except in mild climates where breeding takes place throughout the year. As with sheep, goats tend to spend each night in a particular area, a night camp. Home ranges of male goats are larger than those of females and vary with the season. The total range is 10–40 ha (25–99 acres).[1082]

Does live in small, stable groups (heft-groups), each comprising 3 to 4 animals and occupying its own range. Dominance in male feral goats is determined by age until they are 6 years old. Thereafter, the bucks decline in dominance; despite larger horn size, they may not be as strong.

Confined Goats

Dominance

Dominance is much more obvious in a flock of goats than in a flock of sheep. Despite their close relationship, the two species are very different in behavior. Goats are much more aggressive and much more exploratory than sheep. In goats, the presence of horns is an important determinant of dominance, as it is among horned sheep. Because horns confer dominance, most agonistic interactions are brief feints or rushes in which the dominant animal lowers its head and points its horns at the subordinate. When the animals are of equal or undetermined rank, long fights occur in which horns and heads are clashed together repeatedly. The goats that act as leaders have been born in the area and have more kin in the flock than do nonleaders.[436] Syme et al.[1373] found that a novel food will increase the level of aggression within a goat herd.

HORSES

Free-Ranging Horses

A herd of horses is defined as those sharing the same general range. In a free-ranging horse herd, each stallion is associated with a band of mares numbering from 2 to 20 (mean, 6).[190,776,1418,1453] Band size is optimal at 5–7 mares. Stallions with larger bands actually sire fewer foals.[773] An older or larger mare is apt to be the highest ranking female, and it is she who leads the herd in flight and in daily journeys to rest or to a new grazing area. The stallion drives the herd from behind, only going to the front to confront another stallion. Nevertheless, the stallion is usually, but not always, dominant over his harem.[189,690,1465] Mares apparently choose the stallion and the herd that they ultimately join. Fillies usually leave their natal band at puberty, possibly to avoid an incestuous breeding with their sire.[412] They join another band or a bachelor band. The small percentage that do remain in their natal band have very low foaling rates, indicating that inbreeding depression of reproduction does occur.[779] The dominant stallion within the bachelor band will be the one to form a new harem band with the young mare.

The best studies of feral ponies are those on Assateague Island off the coast of Maryland. These are known as Chincoteague ponies. About 20% of mares on Assateague Island change bands. Mares with foals are more likely to change bands than mares without foals. Mares may leave bands for reasons having to do with the stallions. They are less likely to leave larger bands with older stallions who have had a harem for several years.[1250] If a mare enters a new band, the resident mares are aggressive toward her, but the stallion will protect her. The larger the herd the higher the aggression rate per mare. Dominant mares interrupt nursing bouts of subordinates. The highest rate of aggression occurs at water holes.[1251] Dominance is important not only for immediate access to scarce resources, but also for the reproductive success of one's offspring. Stallions born to dominant mares sire more foals.[460] Expressed in order of dominance, the typical hierarchy is adult male, adult female, juvenile male, juvenile female, male foal, female foal.[1352] The dominance hierarchies remain stable in undisturbed feral herds,[781] and age appears to be the most important determinant of dominance.

Bands of horses compete for fresh water; the band drinking will not be ousted by an intruding band.[503] The order of drinking is stallion, mares, and then juveniles. Only juveniles will be displaced by an intruder.

Stallions defend their mares, not a fixed territory. Two herd stallions, upon meeting, usually prance toward each other. Once they are close enough to do so, they investigate each other with their nostrils. They will then defecate and sniff at the manure. Feist and McCullough[463] noted that within bachelor herds, the most dominant animals defecated last. These displays between stallions can lead to a fight, but aggression is much more likely to result in the absence of the displays when a bachelor stallion tries to abduct a mare.

Stallions may evaluate their rivals on the basis of vocalizations. Subordinate stallions have shorter squeals that also begin at a lower frequency than those of dominant stallions.[1234] Rival stallions may be more likely to challenge a subordinate after having heard him squeal.

In large bands, there may be more than one stallion.[190,1018] The dominant stallion in the herd does most of the breeding; the subordinate stallion does most of the fighting with any other stallions that approach the band. These large multimale bands tend to supplant smaller bands at water sources.[1019] Multimale bands have an advantage in that fewer mares leave these bands during the winter when food is scarce.[503]

In interspecies relationships, horses dominate cattle.[1418]

Domestic Horses

What is a band? One or two horses are more restless than a larger group. Judging by the decrease in walking and increase in grazing compared with those of smaller groups, three horses are a band.[846] Although the strict order by sex and age just described may be observed in wild horses, quite a different picture emerges when herds of domestic horses are studied. Dominance hierarchies tend to be linear unless the group is large, in which case triangular and more complex relationships appear (Fig. 2.2). Montgomery[1032] studied one herd of 11 horses and found that dominance, as determined from observation of interactions of the whole herd, was determined by weight, not length of residency, but Clutton-Brock et al.[303] did not find any correlation between size and dominance in Highland ponies. In a larger study of 11 herds of horses and ponies, Houpt et al.[691] found that age and weight were not statistically correlated with rank in dominance hierarchies. All possible pairings of the herd members were made in food dominance tests so that both lower and higher rankings could be determined. Over an 18-month period, the hierarchies in domestic horses were stable. The most aggressive horse was the dominant one. Horses under 3 years of age are never dominant over adult horses and, in fact, display little aggression toward one another even when vying for food, as Grzimek[578] had noted previously.

The position of the stallion in the hierarchy is variable. In three herds of ponies of mixed size and age, geldings and mares were dominant over the stallions, but Arnold and Grassia[75] found that two horse stallions were dominant to the mares of their herds. Both studies used food competition as a measure of dominance.

Przewalski's horses have linear dominance hierarchies even on 30-ha (74-acre) pastures. The stallion is not always the highest ranking horse. Most aggressions were displacements or threats, rather than kicks or bites.[780]

Mothers are not necessarily dominant over their daughters. The daughters of dominant mares tend to be dominant within their own groups.[697] The hierarchy of foals is first related to birth order, but later, when size differences are not so great, the foal's rank is that of its dam.[61] There is little change in rank of mares when they foal.[438]

Within a herd, the mares appear to have preferred associates, "friends," anthropomorphically, with whom they mutually groom, especially when their winter coats are shedding. Preferred associates share resources without competing. When the relationships of the mares are known, the preferred associates often are mother and daughter or siblings[1418] and are usually animals close in social rank.[303]

In summary, the determinants of dominance in horses appear to be more closely related to the animal's temperament and the position of its mother in the herd than to physical characteristics.

Types of Aggression

Fighting can include a variety of responses in horses: running, chasing or fleeing, circling, neck wrestling, biting, and kicking.[816] Horses neck wrestle and nip at one another even in play, but biting or biting attempts with ears flattened to the head and lips retracted are a sign of serious aggression (Fig. 2.3). Kicking is considered to be the horse's most aggressive act[1418] and, although some have hypothesized that kicks are defensive, that is, directed up the dominance hierarchy,[1465] it is more likely that kicking occurs when either the challenge or the danger is from the rear (Fig. 2.4).

Aggression between stallions can take many forms. Prefight behavior includes the arched neck threat, fecal pile display, head bowing, strike, threats to bite, squealing, snorting, prancing in parallel, and pushing. The levade (rearing with deeply flexed hind limbs) is part of

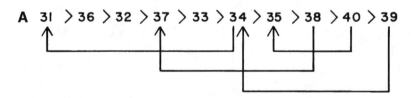

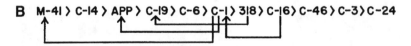

FIG. 2.2.   Examples of dominance hierarchies in mares. *(A)* A herd of 10 thoroughbred mares. *(B)* A herd of 11 mares of various breeds. The *greater than symbol (>)* indicates that the horse on the right is submissive to the horse on the left. *Arrows* indicate direction of dominance in triangular relationships[691] (copyright 1978, with permission of Elsevier Scientific Publishing).

FIG. 2.3.   An aggressive expression in a horse. A threat to bite.

FIG. 2.4.   Threat to kick. The horse on the *right* is threatening to kick the horse on the *left*. Note the lashing tail of the threatening horse and the tucked tail of the threatened horse.

untrained stallions' interactions. This is probably the reason that Lippizaner stallions, rather than mares, are used by the Spanish Riding School. Stallions may rear and box with their forelegs or actually make contact with their forelegs. A horse may avoid the lunge of another horse by swinging his head in a dorsolateral direction away from an apparent threat while the hind legs remain stationary, a posture termed the balk[966] (Fig. 2.5). A less intense version of this is seen when two horses share feed over which they might have been expected to fight. The horses turn away from each other. Avoiding eye contact as well as physical contact may be the goal of this behavior. Elements of sexual behavior such as resting the chin on the opponents rump, rump presentation, and mounting also occur. These behaviors are seen in bachelor herds of stallions where they may be play or determination of the hierarchy. The activities practiced in the bachelor herd are used to defeat a band stallion when a former bachelor acquires mares.

FIG. 2.5. Balking behavior by stallion on the *right*. Photo courtesy of Dr. Sue McDonnell, University of Pennsylvania.

Horses can be severely injured in the process of forming a hierarchy, especially if they are so confined that the loser cannot escape. Horses that have been stalled separately for a few months may show much more aggression than when they have been together on a daily basis. Because neither age, weight, nor sex appear to be important determinants of dominance, one should hesitate to predict a hierarchy in horses. It is, therefore, a good management practice to introduce (or reintroduce) horses to one another across a fence. The horses can investigate and threaten each other, but will be able to escape easily without being kicked.

Grooming

Horses mutually groom one another. Licking of the foal is seen only in the short period after the foal's birth, but horses will stand shoulder to shoulder and nibble at each other's withers and back.

Horses tend to groom animals close to their own rank in the dominance hierarchy, which are also the horses in closest proximity to them.[303] This behavior is more pronounced in the spring when the heavy winter coats are being shed. Horse owners assume the role of grooming partner when they curry their horses. Grooming in the withers area reduces the horse's heart rate.[461] Rolling, which serves to scratch the horse's back occurs more frequently in the spring. Horses rub their rumps and tails against fences. This appears to have erotic properties; males show penile erection. Rubbing can also be a sign of perianal pruritus due to pinworm (*Oxyuris* spp.) infestation.

Summer weather brings horses irritating companions: flies, many of which are biting species. To escape flies, horses spend many of the daylight hours in the shade,[1418] in grassless areas or, if available, in the snow or surf.[778] Some horses have a particularly effective way to deal with flies. They stand side by side, nose to tail, and keep the flies off one another's faces with their tails. Not all horses form pairs, despite the obvious advantages, but most will stand closer to one another during times of high fly density.[411]

Clinical Problems

Treatment of Equine Aggression Toward People

Aggression is an all too common behavior problem of horses. Aggression can be directed to people or to other horses. Aggression toward people is seen most often in the stall, a small, easily defended space. This is probably a form of dominance.

A simple way to acquire dominance over a horse is to gain control over the animal on the ground. Teach the horse to back away from a person on command, to never invade a person's space by touching, and to move to the side when touched. Free lunging, in which the person stops, starts, and turns the horse by moving in front of or behind or toward the horse's balance point, is another method of establishing control. A round pen is necessary to free lunge the horse easily. Another method to obtain dominance is to flex the horse's forelimb and strap it in that position for a few minutes. The horse is, in effect, three legged and should be urged to walk so that it is aware of its helpless situation. Of course, this exercise should be done only on a soft surface so that the horse will not injure itself if it should fall. When the horse has been restrained for 5 minutes, the limb should be freed and the horse walked again. This process is repeated several times until the horse has learned that the person can give him the use of his leg. This technique is most effective for a person with whom the horse has had no prior experience.

For simple cases of aggression (that is, of a horse toward its owner), rewarding nonaggression is usually effective. Alternatively, aggression can be punished, but many owners of pleasure horses are unable or unwilling to inflict appropriate punishment. In addition, the punishment must follow the misbehavior within seconds. If a horse threatens its owner, but the owner must run around the paddock to catch the horse before whipping it, the horse will not learn to stop threatening, it will learn to avoid the owner. The same principle applies to punishing a horse for misbehaving in the show ring after he has left the ring.

Rewards must also be carefully timed, but the timing of rewards is not as crucial as that of disciplinary action. The best reward for horses is food, and grain fed in many small portions can be used to "shape" certain responses. For example, if the horse tends to swing its rump toward anyone who enters the stall, the following course of treatment should be used. On the first day, no grain should be poured into the horse's bucket until it turns 45° or more toward the front of the stall. By feeding grain in measurements of 1 cup or less, the owner will give the horse plenty of opportunities to learn that its movement toward the front of the stall will be rewarded. The next day, the criterion for reward should be raised to a 90° turn toward the front of the stall. This process can be continued until the horse learns that it must turn and face the front of the stall before it will receive its grain. The owner must be rigid about enforcing this, even if it means several days without grain for the horse.

A similar method can be used for treating a horse that lays its ears back in a threatening manner. The horse should be fed only when it puts its ears forward. It may be necessary to elicit the desired response by whistling or throwing pebbles, and then the grain should be given only as long as the horse's ears remain forward. Each day a longer duration of this behavior should be demanded before the horse is fed. Once the horse is responding well to this procedure, the owner can assume dominance by standing over the feed, starting with hay, and waving the horse off. This is the way that dominance is expressed between horses, but care must be taken to ensure that the owner is the "winner."

These simple behavior modification exercises should be performed by the owner because the horse must learn that it cannot threaten the owner. The horse must be rewarded and punished by each person affected by its behavior.

For severely aggressive horses, a more drastic treatment program must be used. The effectiveness of this treatment depends on three factors: a stall or barn that is virtually lightproof, presence of only one source of food, and the absence of other horses.

The horse should be put in the dark alone and handfed, receiving food and light only from people. Food should be withheld as long as the horse approaches aggressively. The animal should be given as many opportunities as possible to earn food and light, but initially the lights should be on for only a few minutes each day. If the horse refuses to eat for several days, hay may be provided, but grain should continue to be used as a reward for good behavior.

The presence of another horse will give the aggressive horse companionship, and the appearance of a person, therefore, will not be as rewarding. Solitude is a form of punishment for the herd-loving horse, and success is near when the horse nickers as people approach.

A final treatment of aggression is pharmacological in nature and much more dangerous to the horse. It should be used only in those horses who otherwise would be euthanized. The horse is immobilized with succinylcholine so that voluntary movement, but not respiration, is prevented. Treatment should be administered by a veterinarian who has resuscitation equipment on the premises. The principle depends on rendering the horse helpless, and once the horse has fallen to the ground, the owner and caretakers should handle the horse thoroughly, including its ears and feet. Its mouth should be opened and closed, and slightly frightening stimuli, such as flapping saddle blankets, should also be used. The horse is to experience all this while fully conscious but unable to move.

In addition to the usual risk of using a muscle blocker, there is also the possibility of intensifying fear in a horse that is already frightened. The only type of horse considered for treatment with succinylcholine should be one that is not afraid of people—a dominant horse.

Treatment of Equine Aggression Toward Horses

Aggression between two or more horses can be treated by changes in management, such as separation of individuals. In other cases, widely spacing feed buckets is the easiest way to prevent aggression between horses by reducing competition over resources; the resource is usually food. Holmes et al.[673] have shown how wire partitions along a feed trough allow a subordinate horse to eat in the presence of a dominant one. If aggression occurs in other circumstances, it is more difficult to treat, particularly if it occurs at pasture. Such behavior may be treated hormonally with progesterone (Depo-Provera) or medroxyprogesterone (Ovaban) 65-85 mg/day orally for a 300-kg horse. The pharmacological basis for the effectiveness of this treatment is not known. It is hypothesized that progestin administration inhibits those areas of the hypothalamus (see Appendix 3) that induce aggression, especially sex-specific types of aggression. Regardless, the benefits achieved by such treatment must be balanced against the side effects. For example, long-term use may affect fertility in stallions. Tryptophan, as a feed additive or paste, should increase brain serotonin and reduce aggression.

A shock collar operated by a remote transmitter may also be used to treat aggression at pasture. Shock, like any punishment, must be delivered at the proper time in order to be effective; specifically, the moment the horse threatens, kicks, or bites another horse. To ensure that the horse does not associate the punishment with the collar, a dummy collar should be worn by the horse for several days before the shock collar is worn. If possible, the transmitter operator should be hidden, and although the signal can be transmitted over great distances, it cannot be transmitted through metal (that is, wire fences). Commercially available collars are made for dogs but are adequate for use on horses if a longer strap is used.

Care must be taken to avoid overusing shock, and only one unwanted behavior should be punished. Shock can also teach place avoidance: if a horse is shocked while crossing a stream, it subsequently may be afraid of water.

Aggression toward other horses that occurs under saddle or in harness is punished more easily, and most of these problems can be solved by a competent horse trainer.

There are a variety of ways to treat aggression in horses. The method chosen should be determined by the type and severity of the aggression, as well as the circumstances under which it occurs. Owners should participate in treatment and should be urged not to breed vicious or unmanageable horses.

PIGS

Free-Ranging Pigs

There are numerous populations of feral swine. They form groups of approximately eight, most commonly three sows and their offspring.

The males are solitary for much of the year, but may form all-male groups in the late winter.[569] The males travel farther than the females. Young pigs do not leave the sow until they weigh 27–32 kg (60–70 lb). The pigs have overlapping home ranges of 121–809 ha (300–2000 acres).[844]

Confined Pigs

Social Aggression

Teat Order.   Pigs have the most intriguing of hierarchies because the ranks are formed soon after birth, not by uncoordinated pushing for a nipple as exhibited by puppies, but by vicious blows with the appropriately named needle teeth possessed by piglets at birth. To reduce the injury and infection from snout lacerations during the neonatal period, most swine producer clip the teeth to the gumline.[511] The resource over which the piglets are fighting is the preferred pair of teats, usually the most anterior pair that produces the most milk and has the lowest incidence of mastitis. In addition, pigs sucking at these teats are much less likely to be kicked by the sow's hind legs. The hierarchy is formed within the first 2 days after birth; the heaviest and first-born pigs are usually dominant.[950] Because the anterior teats produce the most milk, the pigs that suckle these teats grow fastest and remain dominant;[417,951,952] once formed, the teat order of hierarchy remains stable, especially the top and bottom ranks.[448] By the 6th day after birth, the same teat is suckled by the same pig 90% of the time.[652]

Hierarchy Formation.   When strange weanling or older pigs are mixed, a hierarchy must be formed.[951,1182] Even week-old pigs will fight with strange piglets, although the fights are short. In older pigs, the process of hierarchy formation takes several days.[993] Although most aggression is seen in the first 24 hours after mixing strange pigs, the food intake and weight gain of pigs is inhibited for more than 24 hours after mixing. Females fight longer than castrated males.[1362] Boars are presumably dominant over sows, but when barrows (castrated males) and sows are penned together, the males might not be dominant.[993] There are breed differences in aggression. Yorkshires are more aggressive than Berkshires.[951] There are also breed differences in dominance by sex; more Hampshire males are dominant over females than are Durocs.[173] Small pigs and newcomers to an established group are usually subordinate.[508]

Removal of pigs for as long as 25 days does not affect their rank upon return to the group.[448] The dominant pig will lie down and its belly will be nosed by the other pigs. This behavior, the function of which remains unknown, is seen most often when the dominant pig has returned to its group after a separation.[511]

One consequence of social hierarchy can be stress. Subordinate pigs have higher levels of cortisol after fighting for 30 minutes, but both winner and loser pigs have elevated levels of catecholamines.[468] Both dominant and subordinate pigs show immunosuppression, especially if the pigs are in a hot environment.[1050]

Dominant sows give birth to more male piglets than do subordinate ones.[997,1004] The biological significance of this is that the offspring of the dominant animal is more likely to grow large and strong and to be dominant itself. The male offspring of a dominant sow has a better chance to dominate other boars, to gain access to estrous sows and, therefore, to sire many piglets. If the sow is not dominant, there is a greater risk that her sons will not sire any piglets. A safer way to ensure that her genes are passed on is to produce daughters, all of whom will have at least some offspring.

As in other species, once the hierarchy is formed, fighting is replaced by threats that consist of a sharp loud grunt and feint with the snout by the dominant pig. Aggressive behaviors include thrusting the head upward or sideways against the head or body of the opponent. These activities may be accompanied by biting. Levering, in which the snout is put under the body of the opponent, usually from behind, also occurs. The submissive gesture in pigs consists of twisting the head away from the opponent.[747] The subordinate pig quickly gives ground (see Fig. 2.6). Leadership on a novel pasture is not correlated with dominance.[994]

Not surprisingly, the sows that received the most aggressive acts showed the least estrous behavior. They did not mount or nose other sows as much as high-ranking sows.[1114] The level of aggressiveness is not correlated with age, weight, or in the case of sows, parity.[1053] Pigs in their home pen have some advantage over intruders.[1377] The greater the number of unfamiliar, as opposed to familiar, pigs the more fighting there will be.[67] Newly added sows rest apart from the original residents in a less desirable area of the pen, the dunging area. Only after 3 weeks is there integration of the new and resident sows.[1033] Low-ranking sows will lose weight unless feed is provided ad libitum.[253]

Vision is not necessary for dominance hierarchy formation in pigs because pigs temporarily blinded with opaque contact lenses form a hierarchy, although overt aggression is reduced.[449] Advantage can be taken of the effects of lack of vision; there is less aggression if strange pigs are mixed at night.[141] Anosmic pigs are not as aggressive as normal pigs, perhaps because they have difficulty discriminating among pigs.[990]

Most aggressive behavior is seen in relation to food. Pigs may show aggression over entrance to an electronic feeder and queue up in order of dominance. A subordinate sow may produce small litters or low birth weight piglets,[1003] not because she lacks the genetic potential, but because she cannot obtain adequate nourishment.

The importance of confinement as a factor in dominance is that there is less aggression and milder consequences for subordinate pigs when they are housed outside.[938] Crowding increases aggression in most species, and pigs will show more aggression when the stocking rate is increased.[259]

A classic example of porcine sexual aggression is the confrontation of two boars. Pigs tend to use loud vocal communications in general, but two boars threatening one another are eerily quiet. They strut shoulder to shoulder champing their jaws, from which fall clumps of thick, white saliva containing an androstenol pheromone (Fig. 2.6). When they face each other they often paw the ground, a sign of aggression in many artiodactyls. The animals meet in frontal assault. They slash at each other's shoulders with their well-developed tusks, inflicting severe lacerations. The stronger pig will achieve a flank attack and, consequently, victory. The winner of a conflict usually chases the loser. Aggression between sows and barrows is similar to that of boars, except that champing and strutting are restricted to intact males.

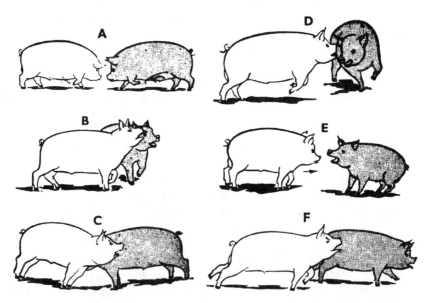

FIG. 2.6. Agonistic behavior in boars. *(A)* Pawing the ground during initial encounter, *(B)* strutting, *(C)* shoulder-to-shoulder contact and slashing, *(D)* perpendicular biting attack, *(E)* submission of pig on the *right*, *(F)* pursuit of the loser[591] (copyright 1969, with permission of W.B. Saunders Co.).

Preventing Aggression Among Newly Mixed Pigs.   There are two circumstances when strange pigs are mixed: when pigs are grouped by sex
or size after weaning or after sale to a feeder pig operation and when
sows are grouped for breeding and gestation.

There are many methods for minimizing aggression among newly
mixed pigs: tranquilization, provision of shelters, and boar pheromone.
The aggression can be reduced by tranquilizers such as azaperone 2.2
mg/kg[1375] or amperozide 1 mg/kg[203], but these are not readily available.
McGlone and Curtis[977] used hides, small recesses in the pen into each of
which a pig could put its head. It is not clear why hiding the head
should reduce aggression. Perhaps the sight of the head stimulates aggression in the more aggressive pig. A simpler explanation is that the
pig's head is protected. Furthermore, his weapons are the teeth and
snout, and if they are in the hides they are not being used on other pigs.
Simple masking odors have no effect on aggression.[140] Spraying or dab-
bing the boar pheromone 5-α-androstenol on the pigs reduces aggression among young pigs,[974-976] but not older ones. Perhaps the explanation is that young pigs are always submissive to adult boars, whereas an
older pig may challenge an adult male or a pig that smells like one.
Lithium has also been used to reduce aggression but probably does so by
causing malaise.[978]

Even simple physical factors such as a draft can cause pigs to be
more aggressive.[1267] Providing toys such as tires or pull toys can reduce
aggression among newly weaned pigs.[1027] Pen size and shape can also affect aggression when strange pigs are mixed. There is less aggression in
a rectangular pen.[139]

When strange pigs are mixed, size disparity reduces the initial fighting; but not surprisingly, the smaller pigs do not gain as well or remain
as healthy as they do in groups of similar size pigs.[1034] When pigs of dis-
parate size must be mixed, however, smaller pigs fare better if the larger
pigs are added to the pen so that the small pigs have the advantage of an
established territory.

Tail Biting

Another type of aggression that is mostly seen in penned pigs
housed on artificial floors is tail biting.[445] Crowding encourages the outbreak of tail biting,[446,1282] but the main cause appears to be lack of opportunity for oral stimulation in a species that normally spends 7 hours
a day rooting on pasture.[591] Quite possibly, bored pigs begin to nibble on
each other's tails for lack of anything else to do. Once a tail has been bitten severely enough to bleed and the bleeding aggravated as the victim
swishes its injured tail, the pigs become much more aggressive and bite

in earnest.[1426] The blood itself appears to be the stimulus for play to become true aggression.[514] Some pigs are killed outright, but more often losses occur as the result of infection of the wounded tails. If only one or two pigs are responsible for most of the biting, they should be removed. Often, simply giving the pigs corn on the cob to chew will stop an outbreak of tail biting that has not progressed to the cannibalistic stage. Tail biting can be reduced by providing a rooting source such as soil, which is also a source of iron[60] because iron deficiency can be a cause of tail-biting.[515] The incidence of tail biting increases when pigs are housed without bedding on slatted floors and are fed automatically. Feeding by hand and providing straw bedding decreases the incidence. Docking of the pigs' tails at birth is performed on many hog farms. This approach eliminates the target, but not the vice, and ear biting may arise instead.[1121]

Grooming

Subordinate pigs groom dominant ones. The dominant pig lies on its side while the subordinates nibble at its belly. There are areas of the pig that it cannot reach with its own snout or hind feet (Fig. 2.7). These areas, the flanks and back, are groomed by other pigs. Singly penned pigs scratch themselves on inanimate objects instead. If scratching seems particularly prolonged or intense, skin parasites may be present.

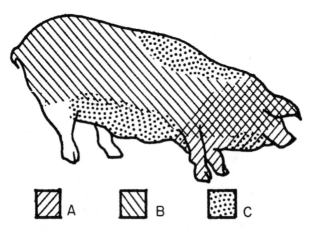

FIG. 2.7.   Grooming behavior of the pig. *(A)* The area scratched by the hind legs, *(B)* the area rubbed on vertical objects, *(C)* the area licked and nosed by other pigs[591] (copyright 1969, with permission of W.B. Saunders Co.).

DOGS

Social Behavior

Urban dogs are either solitary or form small groups, most often of two or three.[168,194,350,497,864] Wolf packs contain two to 12 members.[989] Perhaps, we have selected dogs to be less social, or pack living, than their lupine ancestors. Another explanation is that dogs perceive their owners as part of their pack because the majority of these dogs were free-ranging pets, not strays. Large groups of urban dogs are seen only in association with estrous bitches.[349] Rural dogs form slightly larger packs that often contain two to five dogs.[1286] Dogs in packs are more dangerous than solitary dogs are to humans. This kind of aggression is usually predatory; the pack, usually underfed, chases a person who is on foot or on a two-wheeled vehicle.[225] Regardless of the size of a canine group, a hierarchy is formed. A dominant dog typically assumes a T-position in relation to the submissive dog's shoulder (Fig. 2.8). The submissive dog turns its head away, avoiding the eye contact that might elicit an attack. The submissive animal often remains stationary because running usu-

FIG. 2.8. Dominant and submissive postures in dogs. The dominant dog (*D*) forming an intimidating T-position relative to the position of the subordinate (*S*), who attempts to avoid a confrontation by turning away[495] (copyright 1972, with permission of Coward, McCann, and Geoghegan).

ally elicits an attack or chase.[495] Dogs seem to be able to identify their own breed and will choose their own littermates in a two-choice test.[656] There is not only preference for own breed, but aversion to others. For instance, the attack of fox terriers will be more aggressive toward a dog of another breed.

Determinants of Dominance

Size is important as a determinant of dominance, but territoriality is perhaps more important; a small terrier can attack a Doberman pinscher with impunity on the terrier's own territory.

When breeds of dogs are of similar size, breed temperament is important in determining who will be dominant. For example, fox terriers tend to be dominant over cocker spaniels and beagles. This was true of feeding and mating behavior in a mixed breed group; the fox terriers sired all the puppies.[742] Males are usually dominant over females, although there are breed differences. Female Shetland sheepdogs can be dominant over males, but female fox terriers cannot dominate males of their breed.[1284] Within a litter of puppies, the dominance hierarchy does not seem to be formed until 4 weeks of age, in contrast to the early formation of the hierarchy in pigs. When the dominance relationship of dogs is uncertain, it can be quickly established by placing one bone with two dogs. They will very rarely share it; the dominant one will appropriate it.

Changes in the "pack" membership can result in changes in the hierarchy, especially if a dominant, or alpha, animal or person is removed. An interesting example is a dog over whom dominance was established late in its development by the owner. When the owner left the dog with his wife and children for a prolonged period, the dog quickly reestablished itself as the alpha member of the pack and expressed its newly won position by urine marking inside the home. The urination, rather than the domination, however, was the presenting problem.

Territorial Aggression

This is the most commonly observed type of aggression, but not the one most commonly presented to behaviorists. Dogs barking at one another from their respective territories and dogs threatening or actually attacking either dogs or people that encroach on their territory are examples. The treatment of this as a behavior problem is discussed later in this chapter.

Fear-Induced Aggression

Fear-induced aggression is the type of aggression that most often directly confronts the veterinarian. The fear-biting dog will be most apt to attack when its critical distance has been invaded, which the clinician must do to examine it. Every effort should be made to reduce fear in this type of dog. The astute behaviorist should be able to judge which animal is the fear biter and should not be dominated, and which is the generally aggressive dog that will be more easily handled when made submissive.

Aggression and Hormonal Influences

See Chapter 5 (Maternal Behavior) for a description of pseudopregnancy in bitches that may be accompanied by aggression. There are nonmaternal forms of female aggression. Spayed females tend to be more aggressive than intact ones, possibly because the source of progesterone has been removed.[1086] Lactating bitches may aggressively protect their puppies. Pronounced aggression by a lactating bitch with a large litter may be a sign of lactation tetany (calcium deficiency).

Predatory Aggression

Predatory behavior by dogs is often a clinical problem. Dogs that kill chickens, deer, lambs, or cats are frequently presented for treatment. The easiest approach is proper restraint of the dog. A dog on a leash or in a pen is not only prevented from killing other animals, but is also no longer at risk of automobile-induced trauma. A more drastic treatment may also be used. Many families of mammals, including canids, are able to learn to avoid a food that they associate with illness.[583] If the dog not only kills, but also eats its prey, this phenomenon of taste aversion can be used to eliminate predatory behavior. For example, if the dog kills and eats chickens, it can be allowed to do so and then be injected intraperitoneally shortly afterward with lithium chloride (LiCl) or apomorphine, or a dead chicken can be baited with an equivalent amount of LiCl in a capsule. Two or three exposures to the nausea associated with LiCl should suffice to teach the dog to avoid attacking chickens. Another method is to "socialize" the dog to the prey animal by penning the dog and prey together. At first, however, the prey animal might have to be caged for protection.

Guard Dogs for Predator Control

In the western United States, coyotes are the major predator of sheep; in the eastern United States, dogs are the primary predator. In both areas, guard dogs, not herding dogs, have been moderately successful. The breeds used are the Anatolian Shepherd, Maremma, Shar Planinetz, Spitz, Komodor, and Great Pyrenees. Sixty percent to 70% of sheep producers who use these dogs believed that they were economically beneficial. The major problem is that 25% of the guard dogs injure or kill the sheep themselves, but most dogs can be trained not to chase the sheep.[317,570] Bonding sheep or goats to cattle may be a better means of reducing losses from predation.[46,47,714] The bonding will succeed more often if only one heifer is added to a flock of sheep.[48] If two or more heifers are present, they may not stay with the sheep. Llamas and donkeys have also been used as sheep-guarding animals.

Clinical Problems

Aggression Toward People

It is widely believed that dogs are aggressive because their owners "spoil" them by allowing them on the furniture, feeding them from the table, and so forth, but Voith et al.[1444] showed that spoiling and anthropomorphic behavior toward a dog did not predispose the animal to behavior problems. Obedience training did not protect the dog from behavior problems, although obedience training does appear to improve the human–dog relationship and to decrease separation anxiety.[299] Although some behavior problems may be learned and others exacerbated by the owner, many problems are innate.

There have been many studies of aggression in dogs.[223,228,618,681,930,1518] All these studies concur that males are more likely to be presented for aggression than females and that certain breeds such as spaniels predominate in the dominance aggression category, whereas the more typically aggressive dogs such as German shepherds or rottweilers present as territorially aggressive. Borchelt and Voith,[228] Hunthausen and Landsberg,[719] Schwartz,[1276] Tortora,[1407,1408] Hart and Hart,[617] Overall,[1097a] and Askew[88a] discuss treatment of aggression.

Aggression toward people is probably the greatest canine behavior problem,[466] as can be seen from the following statistics. Two hundred seventy-nine people were killed by dogs in the United States in the years 1979 through 1994. In 1995–1996, 25 people, including 4 infants, died as a result of dog attacks;[718a] it is estimated that half a million people are bitten each year. These figures are similar to those of earlier years.[1500,1254,1037] In 1995–1996, of 20 people killed by pit bulls, 19 were owned by men, seven of whom had convictions for violent crimes.[900] Rottweilers were the most commonly reported breed involved in fatal attacks, but they are the second most popular breed.

Owner-Directed Aggression.   The best or the least arguable way to classify canine aggression is by the victim and circumstances. Although owner-directed aggression may be called dominance, competitive, possessive, social, irritable, or learned, if the case is described as a dog on a couch who growls or bites when the owner approaches, the situation is clear. The most frequent situations when aggression is directed toward owners are the following: (1) when the dog is resting, especially on furniture, in particular beds, and is disturbed or forced to move; (2) when the dog is playing with, eating, or in close proximity to a toy, food, or any stolen object or to a favorite person. The toy most likely to elicit aggression is a rawhide chew toy. The food most likely to elicit aggression is a bone; (3) when the dog is touched, especially its head or paws. There are many circumstances, varying from cutting toenails, which may be mildly painful, to putting on a collar, to stroking; and (4) when the dog is challenged by a direct stare, a visual threat, or punishment.

This behavior is called dominance aggression by most clinical veterinary behaviorists. The hypothesis is that the dog considers himself (and the dog is male in the majority of cases) dominant over the human. Dominance confers access to scarce resources such as food, resting places, and mates. It also means that the dominant animal can touch or invade the personal space of the subordinate, but the subordinate cannot invade the dominant's space with impunity.

The most common signalment is a 2-year-old, neutered male purebred. A typical history is that the dog has always growled when reprimanded while taking food from the garbage or when asked to lie down. He was castrated because of these misbehaviors, but his behavior did not improve. The owners did not seek more help until the dog bit the husband when he approached the bed on which the dog and the wife were lying.

Why do dogs aggress against their owners, literally biting the hand that feeds them? There are probably two main reasons: dominance and fear. Because dogs are pack animals, they have evolved tendencies to vie for rank within their social group. In the case of pet dogs, the social group is the family. Whether dogs really consider people as part of their pack or have separate hierarchies for people and for other dogs is unknown. All dogs do appear to be "running for higher office," especially at the time of social maturity, 2 to 3 years of age. Identification of the member of the family attacked by the dog may give some insight into the human hierarchy. In some cases, the husband is aggressed against, in other cases, it is the wife. The member of the family that is aggressed against is usually the one closest to the dog in rank. If the husband is clearly dominant, the wife may be the victim. If one person never threatens the dog, that person will probably not be aggressed against either.

Prevention of Aggression

The first step in preventing aggression is to obtain dogs of nonaggressive breeds because there are both environmental and hereditary influences on aggressive, as on all, behavior. Mackenzie et al.[917] have shown that traits necessary for good, that is, aggressive, guard dogs are heritable. Fearfulness also is heritable.[559,1284] In fact, it was possible to develop a fearful and a normal strain of pointers within a few generations.[1059] Although dogs of mixed breed and of nearly every pure breed can be diagnosed as aggressive, some breeds are more at risk than others. German shepherds are more likely to show territorial aggression and, because of their popularity as a breed, are at or near the top of lists of breeds presented to behavioral clinics.[166,223,618,681] The same behavior clinics also report that cocker and springer spaniels are frequently presented, usually for aggression toward the owner, either dominance or guarding (possessive) types of aggression. If the public were to stop buying dogs of breeds that tend to be aggressive, breeders would select dogs for suitable pet temperament as well as for conformation and coat.

It is much easier to prevent the development of aggression than it is to cure it. Puppies, rather than adult dogs, should be obtained as pets. The owner can then establish dominance over the dog when it is easy to do so. Puppies should be acquired during the socialization period, 6 to 12 weeks (see Chapter 6, Development of Behavior). The source is important: puppies from pet shops are more apt to have problems.[1299] Both parents should be friendly and approachable. The puppy should be outgoing, but not so assertive that it bites at hands when held or at feet when following people.

Once in its new home, the puppy should be picked up and suspended with its feet off the ground several times a day. This is effective in dominating the dog and should be continued until the dog is mature (or too heavy to lift).

To avoid aggression toward other dogs, try to socialize the puppy to other dogs as soon as it has been properly vaccinated. If the puppy has plenty of pleasant experiences with other dogs while it is at its most playful age, it will be less apt as an adult to be either aggressive or fearful toward other dogs. Veterinarians should offer puppy socialization classes not only to socialize puppies to other dogs and people, but to minimize the dog's fear of the veterinary clinic.

As soon as possible, the dog should be trained to sit, lie, stay, and come on command. This can be taught at home or in special puppy kindergarten classes. It is not necessary to wait until the dog is 6 months old. It can learn much earlier, especially when food rewards, rather than force, are used. Fifteen minutes a day doing obedience work for the dog's lifetime will promote a stronger owner–dog bond, reduce the likelihood of aggression, and provide exercise for the owner and the dog.

The History of Aggression

The complete history (see Appendix 2A) is very useful, not only to the clinician but also to the owner. The examples of the problem behavior indicate the situation that is worrying the owner, whereas the aggression screen gives an indication of the other situations in which aggression is demonstrated.

The 24-hour schedule indicates how much exercise the dog gets, how many times it demands to be let outside, and where it spends the majority of the day.

The reason for acquiring the dog may be revealing. If the dog was purchased for protection, its aggression toward strangers may be both inherent in the breed and encouraged by the owner, but owners do not choose a dog to be aggressive toward the owners themselves. Puppy temperament tests might reveal potential problems of owner-related aggression; furthermore, few clients know whether the breeder performed the test or what score their dog received.

The diet is important both to determine whether food is being restricted and to determine diet composition information. Information on obedience training is important because many owners begin, but do not finish, a course. The owner may not have finished a course for two reasons: (1) the dog's aggression was exacerbated by the owner's attempts to dominate the dog and (2) the trainer may have used harsh methods, such as hanging, scruff hold, or striking under the chin, which either aggravated the dog's aggression or upset the owner. In other situations, the dog may have completed the course but does not obey some or all the family members at the time of presentation. The owner's perception of the percent of compliance to commands may be very much higher than the dog's actual compliance during the consultation.

Treatment for Owner-Directed Dominance Aggression

The first step in the treatment for owner-directed dominance aggression is to explain to the owner what is happening. They should be told that the dog is the leader, or alpha member, of the pack and is the dominant dog. Although most owners are aware of the general concept of dominance, they usually have not realized that every time the dog nudges them to be petted, asks to go out and is let out, and so on, he is winning a dominance interaction. One method of explaining the situation to the owner is to use a military analogy: the dog is the general and the owner is the private. The goal is for the owner to be the general. Techniques such as attention withdrawal or no free petting in which the dog must obey a command before it is petted are helpful.

The owner's safety should be the behaviorist's main concern. To that end, the owner should try to avoid the situations that have led to aggression in the past. Some things are simply a matter of common sense. If the dog is aggressive over bones or rawhides, he should not be given these items. If the dog is aggressive on furniture, he should not be allowed on furniture. The problem may be that the dog will be aggressive if forced off the furniture. In some cases, a short leash can be used to pull the dog off without touching him. In more severe aggressive problems, a halter collar can be used to inhibit aggression or the dog can wear a basket muzzle. Finally, mechanical devices can be used to prevent the dog from getting on the furniture physically or to scare him (see Chapter 9, Miscellaneous Behavioral Disorders).

Punishment should be avoided. Striking the dog could lead to a severe counterattack. The halter collar can be tightened around the dog's nose by pulling on the leash. The principle underlying this treatment derives from the fact that the bitch disciplines her puppy by grasping its muzzle with her jaws and that a dominant wolf does the same to a challenger.

Many dominant dogs are demanding attention from their owners. This trait can be used to advantage by first withdrawing attention and then insisting that the dog do something submissive before it gets any attention. First, the dog is ignored for a few days except for necessities, such as walks and meals, and a training session. Later, the owner may resume petting the dog but only after the dog sits, or preferably lies down, on command. Only a few strokes of petting or one treat should follow the successful response to a command. The dog should also obey a command before it goes out a door, before it is fed, before a toy is provided, and so forth. The dog should not be allowed to run free even in a fenced yard, but it should be walked on a leash for 30 minutes of brisk exercise as well as for elimination. The dog should not go out on demand, and the owner should precede the dog out the door.

Surgical Procedures. Castration should always be advised for aggressive male dogs.[674] The effect of castration is twofold: (1) aggression may be reduced, particularly aggression that is sexual in motivation, and (2) the dog will not be able to pass on its aggressive tendencies. For the same reasons, ovariohysterectomy should also be advised for aggressive females, particularly in cases of maternal or pseudopregnancy-related aggression. Spaying does increase aggression in bitches who were already aggressive,[1086] probably because of the removal of progesterone, but it is more important to prevent reproduction.

Prefrontal lobotomy (see Appendix 3) has been used as a treatment for aggression in dogs.[39,1185] Although it can only be considered to be an experimental procedure, intraspecies aggression was reduced to the point where formerly extremely aggressive malamutes could be used in harness together. Predatory aggression was also reduced. The aggression of household pets directed at humans was not permanently attenuated, so prefrontal lobotomy is not recommended for dogs that bite people.

Drug Treatment. In some cases, the serotonin-enhancing drugs, fluoxetine and amitriptyline, are effective in reducing aggression.

Phenobarbital has been effective, particularly with cases of aggression in springer spaniels.[382] The medication that has been most effective in reducing aggression is the synthetic progestin, megestrol acetate. Progestins can produce serious side effects such as diabetes, especially with long-term use, but should not be overlooked as a means of reducing aggression while behavior modification is being instituted (see Chapter 9, Miscellaneous Behavioral Disorders, for drug dosages).

Predictors of Success and Failure in Treatment. The owner of a dominant dog frequently wants to know the prognosis for curing aggression. Most aggressive dogs can be made better, that is, less aggressive, if the owner institutes the management and behavior modification program. Nevertheless, even if the owners do comply with all suggestions, the dog probably will not be cured; rather, it may, for instance, now growl once a week instead of once a day.

The prognosis depends on a number of factors: severity of aggression, size of dog, and presence of children or infirm elderly in the household. One important prognostic factor is the predictability of aggression. If the dog always growls when approached while eating, the owners will avoid approaching the dog while it eats. If the dog allows itself to be petted most of the time, but occasionally and without warning bites when petted, however, the owner is more likely to euthanize the dog.[1191]

Other factors that have not been proved to worsen the prognosis are the presence of other behaviors, particularly other types of aggression. Dogs that are also aggressive toward strangers are less likely to improve because their threshold for aggression in general is lower. Dogs that resist challenges by the behaviorist, a stranger to them, are worse. If a stranger such as the behaviorist stares at the dog, thereby eliciting a growl or lunge, the prognosis is worse. If neither the behaviorist nor the owner can place a head halter or muzzle on the dog without eliciting snapping, the prognosis is worse.

Aggression Toward Children

Children are a special problem. Dogs can and do kill babies. No dog should ever be left alone in a room with a baby, nor should a baby be on the floor where the dog can reach it, even when other people are present. A dog can kill a baby before the parent can cross the room. In the case of very young babies, the dog may be reacting in a predatory fashion to the baby's high-pitched cries. Aggression that occurs later, so-called jealous behavior, still must be eliminated. Toddlers are most at risk because they make direct eye contact with the dog, a threatening gesture, and are apt to pull hair, ears, and tail or grab for bones or toys. Dominant dogs will react to these challenges with aggression. Older children can learn how to approach dogs. Many owners claim that their children's teasing caused their dog to be aggressive toward children, but the most teasing can do is to aggravate an underlying aggressive tendency.

Fear-Based Aggression

Although high-ranking or dominant dogs are most apt to aggress against their owners, low-ranking dogs may also be aggressive. Their aggression is less frequent and, therefore, surprises the owner. A low-ranking or submissive dog may react to a threatening stimulus, a human hand or head in close proximity, with fear-induced aggression. If the aggression causes the stimulus to be withdrawn, the dog will be encouraged to repeat the aggressive act when next threatened. It has learned that aggression is rewarded. It is important to distinguish fear-induced aggression from dominance aggression because the treatment for each is different. In particular, dominant behavior or punishment would be contraindicated for submissive, fearful dogs. It is usually possible to distinguish a dominant from a submissive dog on the basis of posture and facial expression. See Chapter 1, Figure 1.10. It is perfectly possible for one dog to be dominant over food, but fearful when cornered and approached.

Aggression Toward Strangers

Dogs bite strangers who are coming into the house or yard. This is termed territorial aggression. Most bites occur on or near the dog owner's property, suggesting that the dogs are protecting their territories. Deliverypersons and postal employees are well-known recipients of canine aggression. Stray dogs do not inflict many bites, presumably because they do not have a territory to protect. Dogs chase cats and strangers who are running past. This is termed predatory aggression. Children receive the most bite wounds (60%), some of them fatal.[467] Children tend to run, whether in play or in fear of the dog; running triggers pursuit and attack by the dog. Joggers and bicyclists are also frequent targets of aggressive dogs for similar reasons.

Treatment of Aggression Toward
Strangers—Territorial

Most aggression toward strangers involves territorial boundaries, the front door or the yard. Many owners encourage this type of aggression because they want a watchdog. Unfortunately, they are not teaching the dog to discriminate between people with legitimate business and burglars. Both by their owner's approval and by the flight of the victim, the dogs are being rewarded for barking and lunging at visitors. For example, many dogs have learned that if they bark at mail carriers, the latter will leave. Of course they would leave anyway, but the dog has been operantly conditioned (see Chapter 7, Learning).

Desensitization. Specific behavioral modification techniques must be tailored to each case. In general, the principle is to reward the dog with small bits of very palatable food for good, that is, nonaggressive, behavior in situations that once elicited aggression. If the dog aggressed when petted by strangers, one person should hold the dog on a sit stay while the other, the "victim," approaches the dog. If the dog does not aggress, the holder rewards the dog. Next, the victim approaches closer. Again, the dog is rewarded for good behavior. Next the victim extends a hand over the dog's head. Again the dog is rewarded. Finally the victim touches the dog's head and the dog is rewarded for tolerating that without growling or snapping. Each step in the process should be very small so that the dog is unlikely to aggress. If it does, the holder should be able to prevent the dog from reaching the victim. These simple exercises should be repeated 10 times or more a day, gradually recreating the circumstances in which the dog used to aggress, that is, a rapid approach, many pats on the head, and so forth. If aggression occurred in one room and at one particular time, the exercise should first be done elsewhere, and only when the dog is consistently nonaggressive should it be tested at "the scene of the crime."

Aggression at the Door. The best method of treating aggression toward people other than the owners is to have the owner in proper control of the dog. Because many dogs are aggressive both to their owners and others, this can be difficult. If, as is usually the case, the dog is most aggressive toward people entering the house, then the dog can be taught to lie down and stay when people enter. This should be done gradually by first teaching the dog to lie in a particular place near the door and, then, when it will down stay there, adding cues such as knocking or doorbell ringing, that have signaled a visitor. Once the dog has learned to maintain its stay in the face of those cues, then people (family members at first and then others) can come to the door and enter while the down stay is enforced. This method will work for all dogs, but there are easier methods that will reduce aggression in some dogs. If the dog responds to the sound of knocking or a doorbell, these can be paired with a food or play reward so the dog will associate pleasant things, not challenging or frightening things, with the sound. Playful dogs can be taught that everyone who enters will throw a ball for it.

Dogs that are highly responsive to food rewards can be given food treats by everyone who enters. Keeping a ball or a cup of treats outside the door and a sign advising visitors what to do facilitates the process.

Some dogs will aggress toward visitors after their initial entrance, particularly if they walk across the room. For this reason aggressive dogs should be kept on leashes while company is present. It is best not to isolate the dog or it will not improve; the dog can be muzzled if necessary.

Of course, any dog that is aggressive toward people should not run free and should not be tied outside alone. Tying seems to increase aggression, and if the tether breaks or someone gets within its radius, they will be bitten.

Diet. The diet may be modified to one containing lower protein.[383a] There have been studies in other species, primarily rodents, that indicate that a low-protein diet reduces aggression. There is a common carrier for neutral amino acids across the blood–brain barrier. When the concentration of competing amino acids is reduced when a low protein is fed, tryptophan can enter the brain in higher concentration. Tryptophan is the precursor of serotonin. Serotonin metabolite, as noted above, is lower in the cerebrospinal fluid of aggressive dogs, and higher serotonin is associated with lower aggression.[1192] Presumably combining a low-protein diet with supplementary tryptophan would be even more effective.[288a]

Aggression Toward Veterinarians

Dogs entering a veterinary clinic can be classified as apprehensive and submissive (the majority of the dogs); fear biters; actively defensive (large breed dogs in particular), and friendly and outgoing.[1347]

The veterinarian faces an acute and personal problem in the dog that is aggressive upon examination. Whenever possible, of course, conditioning the dog to fear the animal hospital or the clinician should be avoided (see Chapter 7, Learning). Placing a dog on the table often induces submission. One reason that veterinarians particularly and wisely dislike house calls is that a dog in its own territory is far more apt to be aggressive.

Muzzling. If placing the dog on the table and manual restraint by an assistant or a competent owner is not sufficient, a gauze or fabric muzzle may be placed on the dog. Dogs sometimes are so intent on trying to remove the muzzle that they do not notice the veterinarian's actions. Other dogs, unfortunately, will be driven to frenzy. The other disadvantage of muzzling is that the oral cavity and tonsils cannot be examined. Manual muzzling by holding the mouth shut has been suggested as a method of gaining temporary dominance over feisty small to medium-sized dogs, for instance, Scottish terriers. This method takes advantage of the fact that the abductor muscles of the jaws are not as powerful as the adductors. This method is only successful if the dog makes direct eye contact with the human and resists the restraint. In a few moments, the dog will stop resisting and can usually be examined with minimal restraint.

Sedation. As restraint is escalated, various devices such as choke chain collars or nooses on poles may be used, but the development of fast-acting sedatives, like xylazine, that can be given intramuscularly has revolutionized a practitioner's ability to deal with extremely vicious large dogs. Not only will the dog be calm and manageable in a few minutes following injections, but it will not have had a traumatic and frightening wrestling match with kennel personnel. Therefore, it will probably be easier, or at least no worse, to handle on its next visit.

There are dangers in using such drugs. First, if sedation is too deep, the animal will not respond to pain, and diagnosis, especially of neurological disorders, will be difficult. Second, any powerful sedative is contraindicated in seriously ill or aged animals.

Threat Reduction. What should clinicians do to reduce aggression in their patients? The most effective method seems to be threat reduction. The veterinarian should not threaten or act in a dominant manner toward the dog. This indicates to the fearful dog that it need not fear and to the dominant dog that it need not defend its dominance. By reviewing the actions and postures of a dominant dog, one learns what to avoid. Avoid approaching the dog directly. Squat down while taking a history from the owner, and address some remarks toward the dog. If possible, call the dog by its name so that it will approach you. Avoid direct eye contact, especially at first. Do not place your hand on the dog's head or shoulders. Stroke it under the chin. Because some dogs associate being placed on a table with unpleasant past experiences, they may be much more tractable if examined on the floor. If the initial interaction between the dog and the clinician is not unpleasant, the veterinarian may then perform a routine examination with much more ease.

CATS

Free-Ranging Cats

Feline social organization is very variable. Group size varies from fewer than 10 on most farms to more than 30 in some urban areas where an abundant food source is located in a confined area.[1416]

In a rural setting, cats have territories as large as 200 hectares per female cat and 600 hectares per male cat[914,1416,1506] or one cat per square kilometer, whereas in an urban setting the density varies from 1000 cats per square kilometer (or ~2000 cats/sq mi).[356,1065] Cats were considered to be a nonsocial species because they do not live in groups as adults if they are living on natural prey,[1416] but cats have been able to modify their social organization and live in groups, even multimale groups. Cats can adapt to a concentrated food source such as found in dumps, fishing villages, and farms by living in groups, but these groups are of matrilineal female kin. Females rarely transfer from group to group, although males can. In general, the dominant tom's territory encompasses the females' (Fig. 2.9). Although he will not hunt on the females' territories, he will repel any marauding male and the females will repel any female intruder.

Confined Cats

Social Aggression

When two cats approach each other aggressively, they walk on tiptoe, slowly lashing their tails about the hocks and turning their heads from side to side making direct eye contact. This threat may intimidate a subordinate cat so that it slinks off; evenly matched rivals will continue to approach one another (Fig. 2.10). They will walk slightly past one another before one cat will spring, trying for a grip on the nape of the opponent's neck. The attacked cat throws itself on its back, thus protecting the nape. The two adversaries will both lie on the ground belly to belly while they claw, vocalize, and bite at each other. After a few moments, one cat, usually the original attacker, will jump free. The other cat may adopt a defensive posture, attack, or run away. The victor usually pursues the vanquished.[874] Cats that are aggressive toward other cats are not necessarily aggressive toward people and vice versa.[1417] These differences in personality appear to be inherited because a paternal effect has been noted.[4,458]

When placed together in a home or a laboratory or on a farm, cats will form dominance hierarchies,[142,308,854] but marked aggression may persist in this originally solitary species. Cats may divide up a house: one's territory may be the first floor; the other's the second floor. Roommates may find that the two cats belonging to one person will gang up on the single cat belonging to the other. Urination in the house, especially on beds or rugs, often occurs when strange cats are introduced.

Territorial Behavior

Males maintain nonoverlapping territories in the nonbreeding season, but overlap considerably in the breeding season. Therefore, in both

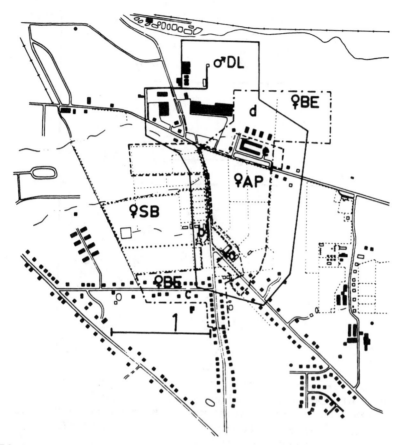

FIG. 2.9.   Ranges used concurrently by three adult females (AP, BE, SB) and one adult male (DL) cats during 1979. AP and BE shared three of four barns at the home farm (*a*) and most of the yards and pastures immediately surrounding the barns. SB was the only female at her farm (*b*). BE used the southern section of her range (*c*) through most of 1978 and 1979 and started using the northern area (*d*) in the late summer of 1979 in a series of foraging excursions with that year's litter. One female kitten remained and eventually reproduced in this section of the range, whereas BE and a male kitten disappeared from this area in late 1979. The two sections of BE's range were connected only by the road between SB's and AP's ranges; BE hunted frequently along the road shoulders, and SB and AP foraged only in the adjoining pastures. AP and BE shared the area around an apartment complex (*e*) but never were noted to contact one another here. DL's range included large areas of each female's range, and on most evenings DL visited each barn complex at least once. The *enclosed line* at the lower end of the map represents 500 m. *Immediately above it,* the north-south axis is indicated. *Dark quadrangles* represent homes, apartments and stores; open quadrangles barns and other out buildings. *Narrow dotted lines* indicate fence lines[1506] (with permission of Veterinary Clinics of North America: Small Animal Practice).

FIG. 2.10. Dominant and submissive postures in cats. The dominant cat is on the *right*. The submissive cat moves slowly away and avoids eye contact[873] (*Katzen—Eine Verhaltensstudien,* 6th ed., copyright 1982, with permission of Paul Parey, Berlin and Hamburg).

free-ranging and pet cats, intraspecies aggression among intact male cats is a very common problem. Many tomcats are presented repeatedly for treatment of bite wounds and abscesses resulting from fighting behavior. Castration is approximately 90% effective in eliminating roaming and fighting in adult male cats,[614] although the disappearance of one may not be associated with a decline in the other.

In highly concentrated populations, where cats compete for food, males are dominant over females. The larger and older males are dominant over younger, smaller ones in competition for food or for females.[1520]

Predatory Behavior

The tall posture of the cat engaged in territorial or sexual aggression is to be contrasted with the stalking posture of predatory aggression.

The predatory cat carries its body as close as possible to the ground. It moves toward its quarry slowly, taking advantage of any natural cover. The closer the cat gets to its prey, the more slowly it advances. Almost inevitably, the cat will pause before leaping to attack. Only the tip of the tail will move as the cat lies in wait. There are usually two or three bounds from hiding to the prey. When attacking a large animal, cats try to make a nape bite to sever the spinal cord.[874]

Predatory aggression is not easily elicited in cats that have not been taught to hunt. Kittens raised with a mother who killed rats in their presence killed at their first opportunity; kittens raised alone seldom did, whereas those raised in a cage with a rat never did.[843] Apparently, kittens learn to direct various innate predatory motor patterns to the prey (see Chapter 6, Development of Behavior) their mother brings to them. She does not simply let them eat the prey, she lets the prey go and catches it again. If the kittens attempt to catch or eat the mouse, the mother will compete with them for it. In this manner, the kittens are stimulated by the hunting game and, apparently, learn by observation. The types of prey brought to the kittens may influence the range of prey hunted by the kittens as adults. Although the mother can influence kittens' predatory skills by bringing prey and interacting with it, adult cats without such learning experience also become competent predators, so a kitten that is not a good hunter can acquire the skills as an adult.[1416]

Most cats will kill rats if fasted for 2 or more days, but they still prefer to eat commercial cat food rather than their prey.[3] One feline characteristic that is distasteful to some people is that cats sometimes play with their prey before and after it is dead. They will catch a mouse, let it go, and catch it again. After it is dead, they will throw it up with their paws and leap upon it. The function of this behavior is obscure, although it may be appetitive or, perhaps, displacement behavior. Truly hungry cats rarely play with prey; they eat it as soon as it is dead and they have recuperated from the predatory effort.

Grooming

Although licking is a very important part of maternal behavior in cats and dogs and self-grooming occupies a great deal of their time as adults, mutual grooming, or allogrooming, is not common. Cats sometimes lick one another; this is most likely to occur when a mother continues to groom her adult offspring, but long-term associates also allogroom. Feline grooming is an important part of daily activities. One of the simplest types of grooming is licking the nose and lips. These are two distinct motions that rarely overlap. Licking the nose occurs after gaping, for example, and the tongue goes dorsally on the midline and then is pulled immediately vertically and into the mouth. A common licking problem is an exaggeration of this behavior in which the nose is chronically irritated by the abrasive tongue. Licking the lips involves movement of the tongue along the edge of the upper lips to the corners of the mouth; this behavior is seen after eating or drinking.

Feline face washing is a stereotyped behavior. The cat is in a sitting position and applies saliva to the medial aspect of the front leg, which is held horizontally. The paw is rubbed from back to front over the nose with a circular upward motion. This motion is repeated a few times; each time the paw reaches out a little farther until it reaches behind the ear (only after three rubs) and then travels downward over the backside of the ear, forehead, and eye. Other areas of the body are cleaned, but not in the stereotyped order that the face is washed. The tongue is drawn over the coat in long strokes, mostly in the direction of the hair. During normal grooming, the saliva applied is licked up again, but in a hot environment the saliva is allowed to remain to aid in thermoregulation. The neck, chest, shoulders, and front paws receive the most grooming. The stomach, rear legs, back, croup, tail, and anal areas receive less attention. All the former regions are licked as the cat sits. The cat lies to lick the sides, stomach, rear legs, tail, and front paws[874] (Fig. 2.11). A cat may sit like a bear on its haunches to lick the penis; this can be a sign of urethral obstruction.

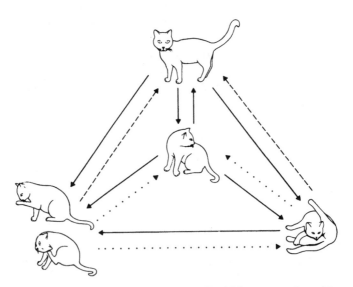

FIG. 2.11.   Grooming postures of the cat. *(Top)* Non-grooming. *(Center)* Flank grooming. *(Clockwise)* Grooming of hindquarters; scratching ears; first step of face grooming, licking the front leg. *Solid arrows* indicate grooming sequences of normal cats. *Dotted* and *dashed arrows* indicate sequences in cats with tectal lesions[1372] (copyright 1977, with permission of *J. Comp. Physiol. Psychol.*).

Clinical Problems

Pathological causes of aggression in cats are more common than in dogs.[1189] Meningiomas, feline ischemic syndrome,[367] and toxoplasmosis have all been associated with aggression. Although some meningiomas can be removed so that the cat's behavior returns to normal, the other conditions cause tissue damage in the limbic system that is diffuse and, therefore, untreatable. Sudden onset of severe aggression is a poor prognostic sign. Euthanasia should be recommended because the cats may climb up the owner and attack the face. The aggression is usually well directed. For example, a cat with toxoplasmosis attacked dogs, but only when she had been with her kittens. Beaver,[165] Borchelt and Voith[228] and Hart and Hart[617] have addressed the problem of aggression in cats.

Aggression Toward People

Predatory Aggression. Feline aggression toward people can be subdivided into predatory or playful, redirected, territorial, and dominance. The predatory type of aggression is preceded by stalking and pouncing and is usually directed toward the feet of a moving person. If the cat is young and has no other kitten with whom to play, the aggression is probably play. In that case, the bite and scratches are usually inhibited; however, if the owners have not reprimanded the cat for biting too hard in play, he may not have learned to inhibit his bite. Playful aggression should be redirected toward swinging toys. The owner may swing a toy from a string and praise the cat verbally for attacking the toy but punish the cat for biting at people's feet. The best way to punish a cat is to startle him. Water guns or aerosol sprays are very effective. Even a loud noise, such as a whistle, can be used. The best punishment is one that the cat does not associate with the owner. A similar strategy is used for true predatory aggression except that neither toys nor rewards are used, but the cat is sprayed for attacking.

Dominance. Dominant or irritable aggression usually occurs when the cat is being stroked. The cat, particularly a male cat, may be giving the nape bite that he gives during copulation. If a verbal reprimand does not suffice, the cat's nose can be flicked with the thumb and forefinger. The owner should also pet him more gently and for less prolonged periods. Dominance over the cat can be attempted by getting up and thus forcing the cat to jump to the floor and by training; for example, by reward, for sitting on command. The cat can be given a reward, a food treat, for allowing two strokes, then for allowing three, and so forth, to increase its tolerance for petting. Holding the cat down so he cannot move for a few minutes a day may also help to eliminate the problem.

Redirected Aggression. See Aggression Toward Other Cats. Cats may become aroused when they see a strange cat through a window, and they may attack their own owner. Owners should be advised not to pick up or pet an aroused, caterwauling cat.

Fear- or Territory-Related Aggression. Attacks may be sudden and explosive, with or without vocalization (caterwauling), and may occur only once or recur. Common stimuli are the odors of other cats. Consequently, owners of cats that have become aggressive toward them after they have handled a strange cat should be careful to wash and change clothes before interacting with their own cat if the owner has had any contact with other cats.

If no stimulus is identified, the cat may be exhibiting idiopathic aggression. If identified, the initiating stimulus should be removed. Severely aggressive cats should be isolated from the owner in a dark room for several days. Food and light are brought by the owner for brief periods several times daily until there is no indication of anxiety or aggression. Drug therapy with serotonin reuptake inhibitors may be indicated in cases of refractory or severe aggression. Benzodiazepines should be avoided because of the potential for disinhibition.

Aggression Toward Other Cats

Aggression among cats in the same household is the most common feline aggressive problem. Male cats are more likely to initiate aggression than females. In contrast to dogs, male cats are aggressive toward females as well as males.

Territorial Aggression. Introducing a new cat stimulates territorial aggression. The aggressor may be the newly introduced cat. Males are more likely to exhibit this behavior than females. The victim may be either sex.

Redirected Aggression. Redirected aggression occurs when a cat sees another cat outside a window and attacks its housemate. This type of aggression can be short-lived, but it may persist if the victim continues to flee whenever it encounters the aggressor. Apparently, the fearful behavior stimulates continued aggression by the other cat. If a new adult cat is introduced, aggression is to be expected, but aggression can also occur between cats that have lived peacefully together for years. In some cases, a physical change and/or a change in odor can precipitate the aggression. For example, if one cat is hospitalized she may be attacked when it returns either because she is weak or because she smells different. This could be termed nonrecognition aggression.

Treatment of Aggression Toward Other Cats

Separate the two cats for the entire day except at mealtimes, but rub each cat's checks and tail and then rub the other cat with the same towel. At mealtimes, feed the two cats in their separate environments for a few meals. Then gradually introduce them back into the same room, but feed them at opposite ends of the room with the aggressor on a leash or in a cage so that he cannot reach the victim. Because the cats are together only at mealtimes, they will associate the reward of food with each other. If there is no hissing or fighting at one meal, you may gradually bring the victim's dish closer to the aggressor at the next meal. Feed the cats three times a day instead of twice because this will allow the cats to have more time in each other's presence. When the victim and aggressor can eat right next to each other with the aggressor confined, begin again to feed the cats at separate ends of the room but with the aggressor free (but with a leash and collar for easy catching). Gradually, day by day, move the bowls closer and closer and increase the length of time they are together after mealtime.

Drug therapy is directed at reducing aggression in the attacker and reducing fear in the victim. Place a bell on the attacker to warn the victim. Close draperies and discourage visits from outside cats to prevent further episodes of redirected aggression. For severe territorial aggression, separate the cats with "invisible fences." Holding down the aggressor in front of the victim can help.

CHAPTER 3

BIOLOGICAL RHYTHMS AND SLEEP

The activity patterns of the domestic animals vary considerably both between species and within a species depending on the diet and environment. Horses spend the majority of their time grazing on pasture, and ruminants spend the majority of their time grazing and ruminating; this changes considerably in confined animals on high-concentrate diets. Cats and dogs rest or sleep for a large percentage of the time. Pigs with free access to food also sleep many hours a day. Circadian and other rhythms are important in determining activity and sexual cycles, as well as physiological responses.

INTRODUCTION

One should be aware of the activity and sleep patterns of animals so that abnormality can be detected. A horse that is lying down at night is probably sleeping; an adult horse that lies down during the day (especially a cold, cloudy day) is abnormal and should be observed carefully because this is an unusual time for a horse to be recumbent.

The patterns of behavior, especially those of activity and sleep, reflect internal rhythms. There are several types of rhythms of differing duration. The circadian rhythms, occurring in approximately 24-hour periods, are the best known and best studied. The activity cycles of most animals are circadian in that the periods of activity and inactivity add up to approximately 24 hours. Other types of rhythms are high-frequency, ultradian, infradian, and annual cycles.

High-Frequency Rhythms

High-frequency rhythms, for example, heart and respiration rates, occur in periods of less than 30 minutes. Heart rate varies inversely with body weight, so the heart rate of a cat (110–130 beats per minute) is considerably higher than that of a horse (28–40 beats per minute). Respiratory rate does not vary linearly with body size, so the cow breathes 10 to 30 times per minute, and the pig, 8 to 18 times per minute. Respiratory cycles have an effect on cardiac rate; it increases during inspiration. This effect, called sinus arrhythmia, is more marked in dogs than in other domestic species. The endogenous nature of biological rhythms can be best appreciated by a consideration of the contraction rate of the embryonic heart, especially that of the chick embryo, which does not have even the maternal heart rate to influence it.

Ultradian Rhythms

Ultradian rhythms are more frequent than 24 hours; for example, the fluctuations of growth hormone output from the pituitary (see Appendix 3), which in cattle occur in cycles of 3.5 hours.[214] Body temperature also varies in ultradian cycles of approximately 1 hour in cats.[630] The physiological bases for, or influences upon, these short cycles are unknown but are believed to be the result of oscillations of cells in central pattern generators. A most interesting ultradian behavior rhythm is that of feeding. When food is available ad libitum, nearly all species eat 9 to 12 meals a day. This pattern is seen in dogs and cats,[765,1054] sheep,[290] horses,[855] pigs,[197] and cattle.[1168]

Circadian Rhythms

A circadian rhythm is self-sustaining, maintained under conditions of constant light or dark, with a cycle of approximately 24 hours.

Zeitgebers

Circadian rhythms are endogenous, that is, they persist under conditions of constant light or constant dark, but usually are influenced by, and entrained to, external factors, which set the biological clock. Some of these factors are temperature, barometric pressure, various drugs, hormones, and light. Of these factors, the most important is light. These factors are called zeitgebers (German, "time givers") because they set the rhythms just as one might set a clock.

Light. Circadian rhythms are entrained to light; that is, although under conditions of constant illumination a rhythm may have a period of approximately 24 hours, under naturally occurring light, the rhythm will be that of the light–dark cycle. The light must be present during a specific portion of the endogenous rhythm. Hamsters, for example, entrain to a 12-hour–light/12-hour–dark day and to a 6-hour–light/12-

hour–dark day, but not to a 6-hour–light/30-hour–dark day.[431]

Considerable practical advantage has been taken of the entraining function of light to bring mares into estrus early or, conversely, to avoid injuries during hierarchy formation by keeping pigs in the dark. Even the simple act of putting a cover over a parrot's cage makes use of the effect of light on avian activity.

Not all types of light are equally effective in entraining circadian rhythms. Green light is most effective, and red light is least effective; therefore, red light may be used when visibility is desirable but interference with an animal's circadian rhythms and dark activities is not.[984] When zeitgebers are removed, the resulting desynchronization of internal rhythms may have deleterious results; for example, thermoregulation may be impaired.[525] When one travels, zeitgebers are removed or are not present at the proper portion of the endogenous rhythm, consequently, jet lag results; rhythms are not synchronized.

Barometric Pressure. The influence of other factors on circadian rhythms has not been as well studied, but barometric pressure has been shown to influence activity patterns. Mice show higher activity levels when barometric pressure is increasing.[1344] Although the phenomenon has never been quantified, farm animals, such as horses and dogs, show high levels of activity before storms, and tail-biting episodes often occur in swine just before storms.

Drugs. Drugs can affect rhythms. Examples are caffeine and theophylline[425] and lithium.[368,759,971] The action of lithium on circadian rhythms of humans as well as of animals may be the basis for its amelioration of depression and aggression. The two compounds that may be useful for treatment of jet lag and sleep disturbances of shift workers are melatonin (see Pineal Gland) and benzodiazepines.[1291,1414]

More important from a clinical standpoint is that a given drug may have a greater effect and/or have lower toxicity at one time of day than at another. Hypoglycemic agents, if administered while liver glycogen and plasma glucose are low, are more apt to precipitate hypoglycemic convulsions than if administered at another point in the cycle.[758]

Pineal Gland. The pineal gland is probably an important intermediary in the synchronization of circadian rhythms because it demonstrates marked rhythms of output of several hormones and neurotransmitters.[534] Melatonin is produced by the pineal and is present in higher quantities in plasma and cerebral spinal fluid at night[635] (Fig. 3.1). Melatonin has an antigonadotropic effect in long-day breeders and a progonadotropic effect in short-day breeders. It may be the means by which the hypothalamus is appraised of day length, the link between circadian rhythms and annual sexual cycles, for melatonin would increase as dark-period length increases, and the increased melatonin levels would depress gonadal activity. The practical application of this role of melatonin is that short-day breeders such as sheep may be brought into estrus earlier by oral administration of melatonin in the afternoon for several months.[66]

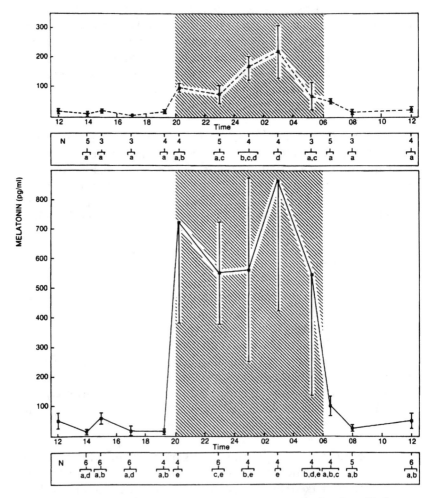

FIG. 3.1. Nyctohemoral cycle of melatonin concentration in calf plasma *(upper)* and cerebrospinal fluid *(lower)* at selected times of the day. The lights were on from 0600 to 2000 hours. Data points are means ± 1 standard error of samples from number (N) of calves. *Letters* indicate that when two means do not have any letter in common, they are significantly different[635] (copyright 1977, American Association for the Advancement of Science).

The peak level of the neurotransmitter serotonin is 180° out of phase with melatonin. Serotonin is a precursor of melatonin. The activities of the enzymes catalyzing the reaction (serotonin to melatonin) are influenced by light.[972] Serotonin has been implicated as a sleep-inducing neurotransmitter. Day length will influence the relative amounts of serotonin and melatonin present in the pineal, thereby influencing the organism's sleep–wakefulness and reproductive condition. Aggressive behavior also may be influenced. Most owners are bitten by their dogs at night, when serotonin levels are low. Drugs that increase serotonin activity decrease aggression (see Chapter 2, Aggression and Social Structure).

In addition to gross activity, a number of cellular and endocrinological parameters vary in a circadian rhythm. Many hormones have been demonstrated to have circadian rhythms. Corticosteroids, including both cortisol and corticosterone, increase during the day in pigs and horses, with peak levels in late morning[231,1471] (Fig. 3.2). Both pigs and horses are diurnal (day-active animals) and show diurnal peaks in adrenocortical activity and adrenal responsiveness to ACTH during the day, whereas cats[1277] show increased adrenal activity at night. In stallions and boars, testosterone levels are highest during the day.[427,810] Some hormones, such as vasopressin in the cat, show circadian rhythms in the cerebrospinal fluid but not in the blood.[1195]

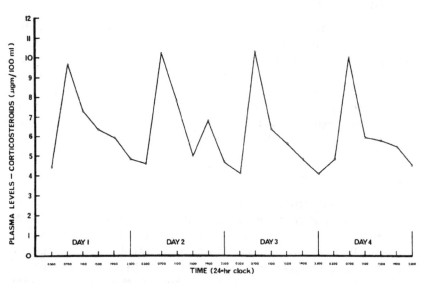

FIG. 3.2. Circadian rhythms of corticosteroid secretion in pigs. The graph is based on the mean results from six pigs. The peak of corticosteroid output occurs in late morning[129] (copyright 1973, with permission).

There are age-related effects on the expression of circadian rhythms. For example, although there are circadian rhythms that occur in cortisol concentration in adult dogs, rhythms are not observed in either puppies or old dogs (>12 years of age).[1105] These age-related changes may be related to some of the problems of restlessness seen in older dogs.

Not all hormones are secreted in greatest quantity during the day in diurnal animals; growth hormone, for example, decreases in output during the day in pigs.[1406] Circadian rhythms of heart rate, body temperature, white blood cell number, metabolic rate, liver glycogen and glucose, and glucose absorption from the gut have also been identified.[87,439] Dogs show circadian rhythms of body temperature with a period of 23.7 hours.[766] There is a circadian rhythm of body temperature in calves peaking in the afternoon and reaching nadir in early morning.[913]

Feeding.   An important rhythm is that of feeding. When, instead of freely feeding, an animal has meals imposed on it (the situation for most domestic animals), the animal anticipates the meal with an increase in activity. In order to entrain this rhythm, the meal must contain calories.[1022] There are daily variations in core body temperature of sheep, but these are entrained by feeding.[1026] Rhythms of intestinal enzymes are also secondary to feeding.[473]

Parasitic Rhythms

Still another facet of biological rhythms of importance to the clinician is the rhythm of parasites. Perhaps the best example is that of *Dirofilaria immitis,* canine heartworm. Microfilaria are most active and most likely to be found in the peripheral circulation in the evening.[629] The activity peak nicely coincides with that of the insect vectors, the mosquitoes, that will carry the microfilaria to a new host. The clinician may have more success in making a positive diagnosis of the presence of the parasite by examining blood taken in the evening.

Other Rhythms

Infradian rhythms have cycle periods less frequent than 24 hours. Circatrigentian rhythms are those of approximately 30 days. The sexual cycles of polyestrous domestic animals show periods of approximately 3 weeks. The sow and cow come into heat every 21 days. Examples of species that are seasonally polyestrous include the mare, which comes into heat every 17 to 24 days in the spring, and the ewe, which comes into heat every 16 to 17 days in the fall.[455] These cycles may represent rhythms of hypothalamic, pituitary, or ovarian activity.

Annual Rhythms

Annual or seasonal cycles are somewhat better understood. Horses and sheep are seasonal breeders. Horses are anestrous in the fall and winter and begin to show estrus as day length increases in late winter. Sheep show an opposite response in that they begin to be sexually active when days shorten in the fall. The evolutionary advantages to both species are obvious: the offspring of horses and sheep are born in the spring when food is abundant. Dogs now show sexual cycles of approximately 6 months' duration, but there is reason to believe that they, too, were once annual breeders. Basenjis, for example, still show an annual fall breeding season.[1284] Domestication, abundant food, and selective breeding also may have caused cattle and swine to become polyestrous throughout the year rather than during one season.

Not all annual cycles are reproductive. Cats show annual cycles of corticosteroids, thyroxine, and epinephrine levels. Peak levels of these three hormones occur during the winter.[1179] More familiar are the cyclic changes in hair coat. Hair-follicle activity in cats is highest in late summer and lowest in late winter, and as a result, fur is 0.5 mm (0.2 in.) longer in winter than in summer.[1253] Adult ewes show a seasonal variation in heart rate, with a minimum in winter. Horses show seasonal rhythms in carbohydrate metabolism, but these may be related to training.[551]

Sleep

Sleep occupies one-quarter (ruminants) to one-half (dogs) of the lifetime of animals, but the function of sleep remains unknown. One possible function is replenishing of neurotransmitters. A device to conserve energy, a means of remaining inconspicuous, a period for consolidation of memory, or simply a way to fill up time not needed for foraging are other hypothetical functions of sleep.[1531]

Types of Sleep

Sleep can be classified into two types: the "sleep of the mind," slow wave sleep (SWS), or quiet sleep; and the "sleep of the body," paradoxical, active, or rapid eye movement (REM) sleep. The two types can best be differentiated from wakefulness and from one another by means of electroencephalography.

The electroencephalogram (EEG) of the alert animal is characterized by low-voltage, fast waves that are not synchronized. Slow wave sleep is characterized by synchronous waves of high-voltage, slow activity. During paradoxical sleep, the EEG shows low-voltage, fast activity similar to that seen in the wakeful state (Fig. 3.3), but there is very little

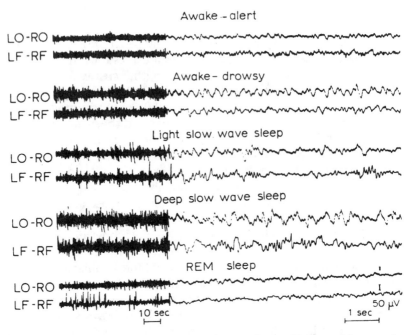

FIG. 3.3. The stages of vigilance and sleep in the cat. Polygraphic record at the speed of 30 mm per second showing the stages identified. LO-RO electroencephalographic record from the left and right occipital area; LF-RF electroencephalographic record from left and right frontal area[1419] (copyright 1968, with permission of Pergamon Press).

muscular activity; therefore, this type of sleep is called the sleep of the body. The animal is more difficult to arouse than when it is in SWS. Although overall muscle tone is very low during paradoxical sleep, the muscles of the eyes frequently contract; hence the term *rapid eye movement*. The low voltage, fast activity of REM sleep does not result in many body movements because there is an inhibitory area in the medulla that, in effect, paralyzes the muscles of the body.[1044] Humans awakened from REM sleep report that they have been dreaming; the twitching of the face and legs (which are not completely inhibited) and whining during canine sleep indicate that dogs may also be dreaming. We can only speculate as to the presence or content of animal dreams. It is REM sleep, however, that appears to be the most critical or necessary component of sleep. Deprivation of REM sleep results in behavioral abnormalities in all species tested, and rebound or extra REM sleep occurs during recovery from deprivation.[1236,1237,1421]

PATTERNS OF SLEEP AND ACTIVITY IN DOMESTIC ANIMALS

Sleep varies considerably among species.[269] The activity patterns of the various species described here are reported under specific environmental conditions. The behavior patterns may be different under different environmental conditions, and, therefore, the numbers given should not be considered applicable to every animal under every condition. Allison and Cicchetti[41] hypothesize that sleep time is inversely related to the danger of predation for a given species. Roughly speaking, predators sleep more than prey animals, and large animals, more than small animals. See Fig. 3.4 for activity patterns of three species in the same pasture.

Dogs

Sleeping dogs often lie in a characteristic posture with their hind legs tucked up and their heads turned caudolaterally. Their eyes may be open or closed. Rapid eye movement sleep may be accompanied by leg movements, vocalizations, and either polypnea or apnea. A dog awakened abruptly from REM sleep may bite; so it is best to let dreaming dogs, or at least sleeping dogs lie, or awaken them gently. Dogs show short periods of activity interspersed with periods of rest when free ranging,[168] when tethered outdoors,[369] and when caged.[630] Pet dogs appear to sleep at night, but their behavior may be entrained to that of their owner. Active (REM) sleep occupies only 6% of their time.[9,10] Dogs sleep in cycles of 16 minutes asleep and 5 minutes awake. House dogs and caged dogs sleep more than free-ranging dogs, but even the latter sleep 60% of the night.[8] Although sleeping dogs are relatively easily roused, owners who wish their dogs to guard property must be aware that dogs will sleep where they are most comfortable, that is, on soft surfaces, and not necessarily at the property line.[11] In the case of caged dogs, 30-minute to 2-hour periods of activity alternate with longer quiescent periods. Dogs in groups in kennels spend 7% to 24% of their time in active behavior (walking, trotting), 5%–10% socializing with another dog, and the majority of the time inactive (sitting, standing, or resting, with 20% of inactive time in light slow wave sleep (LSWS), 25% in deep slow wave sleep (DSWS), and 10%–12% in REM sleep).[662,667,707]

Free-ranging, feral, urban dogs are most active in early morning and in the evening. Foraging for food, usually garbage, socializing with other dogs, and traveling from alley to street to park are their major activities and are usually interspersed with periods of rest (Fig. 3.5).[168] Similar periods of activity occur just after sunrise and an hour or 2 before sunset in Huskies tethered to their doghouses by 240-cm (8-ft) leads. Under these conditions, dogs spend over 80% of their time, night and

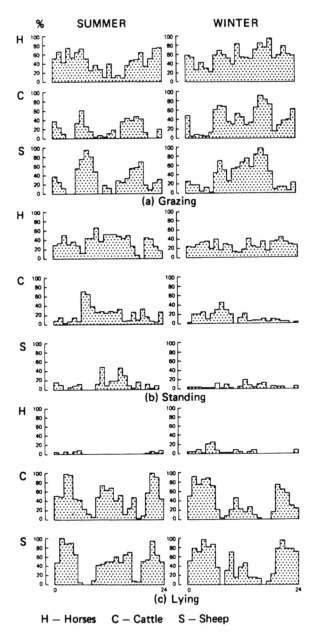

FIG. 3.4. Time budgets of domestic herbivores. Grazing, standing, and lying behavior of cattle, sheep, and horses living on the same pasture during two seasons[72] (copyright 1984, with permission of Elsevier Scientific Publishing and CSIRO).

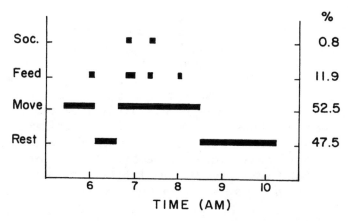

FIG. 3.5.   Activity patterns of feral urban dogs. The activities of two dogs during a morning in the fall. Soc = social behavior with other dogs, including sniffing and chasing. Feed = feeding behavior, including rummaging through garbage and eating. Move = walking and running. Rest = resting or sleeping[168] (copyright 1973, with permission of York Press).

day, lying down and sit only 2% of the time, mostly while observing another dog or a person. A pet's activity is controlled by its owner. Dogs are walked or let out early in the morning when their owners arise and in early evening when their owners return from work.[194] A peak can also occur at noon in communities where people apparently return home for lunch.[864]

Cats

General Activity

In the laboratory, cats, like dogs, show short bursts of activity (1 to 2 hours of activity distributed throughout the day) and are 1.4 times more active during the day than at night.[1359] During the day, group-housed laboratory cats spend 36% of their time in maintenance behaviors such as resting, sitting, ingesting, or eliminating and another 30% in comfort behaviors such as grooming and stretching. A quarter of their time is spent in motion.[1142]

Farm cats spend 40% of their time asleep, most of it at night. Although active at dusk and into the evening, cats are not really nocturnal. The rest of the farm cat's time is divided into 22% resting, 14% hunting (although this will vary from cat to cat), 15% grooming, 3% traveling, and 2% feeding.[1107] Urban cats are most frequently seen and are presumably most active at night between dark and dawn.[194]

Sleep

Caged cats spend 10 h/d sleeping. Slow wave sleep occupies 39% of the day and REM sleep, 8%. During REM sleep, the nictitating membrane covers the eye. Ursin[1420] found that SWS of cats, like that of dogs, can be subdivided into LSWS and DSWS, based on electroencephalographic characteristics and ease of arousal. The usual sequence of sleep stages in the cat is from wakefulness to LSWS to DSWS to REM to either LSWS again or to wakefulness. The two major sleep epochs occur at night. Drowsiness varies with feeding schedule. Cats fed three times a day drowse more than those fed once a day, and fasted cats drowse even less.[1240] Older cats (>10 years of age) show less REM sleep and more SWS, as well as more brief episodes of wakefulness than do young cats.[237]

Pigs

Pigs usually are kept in confinement so that the 7 hours of rooting (for example, food searching) noted on pasture[1512] fall to 2 h/d of eating in a pen. The more hungry pigs are, the more time they spend rooting and the less time, lying. Hunger, too, may predispose to tail biting.[363] Pigs spend more time resting than any other domestic animal.[627] They are recumbent 19 h/d. They drowse 5 h/d. Slow wave sleep occupies 6 h/d, and REM, 1.75 hours in 33 periods. Pigs are characterized by extreme muscle relaxation during sleep (See Fig. 3.7). It is difficult to evaluate muscle tonus in a 400-lb sow, but when a sleeping piglet is picked up, it is as relaxed as a rag doll. Only 1–3 h/d are spent in other activities, such as drinking, walking, playing, or fighting.[512] Domestic pigs are diurnal, and most activity takes place during the day.[1045] Pigs, like most diurnal species, have higher melatonin levels during the scotopic, or dark phase, of the day; however, particularly bright light is necessary for entrainment.[574] Although motor activity and food intake increase during the day in pigs, these rhythms disappear in constant light, indicating that they are not circadian. Rhythms of body temperature depend on feeding and do not occur in pigs fed ad libitum.[731] Pigs can entrain to 9 hours of light and 9 hours of dark, as well as to 12:12 cycles.[732]

Feral pigs and wild boars are more nocturnal in habit during the summer months, probably to avoid predation. Even wild pigs are not very active. Trapping and retrapping indicated that sows were usually found within 0.3 km (0.19 mi) of the point at which they were originally trapped; boars were within 2.0 km (1.24 mi).[939] The sex difference may be in range rather than in activity.

Circadian rhythms are disrupted by changes in physical, social, or reproductive conditions. For example, putting a pig into a group after it had been housed individually, or tethering a pig, will disrupt circadian rhythms of cortisol for 1 to 4 days, as will surgery or estrus.[169]

To avoid stress and its detrimental effects on meat quality, animals are often placed in pens after transportation and before slaughter. There are large species differences in how rapidly they lie down. Swine lie down within an hour and cattle within 2 hours, although this will vary with the amount of space per animal and number of disturbances by people.[746,808]

Horses

Horses may be able to drowse and even to engage in SWS while standing by means of the unique stay apparatus of the equine hind legs, but they lie down[593] when engaging in REM sleep. The horse in REM sleep lies either in lateral recumbency or in sternal recumbency with its muzzle touching the ground so that its head is supported during the atonia accompanying that phase. A healthy horse seldom remains lying when it is approached, probably because a standing horse is better able to flee or to defend itself. It is interesting that a dominant stallion lies down first,[1238] that is, before subordinate horses lie down.

During the day, the horse is awake 88% of the time, and most of this time the animal is alert. Even at night, the horse is awake 71% of the time, but it drowses for 19% of the night.[1236] Stabled horses are recumbent 2 h/d in four to five periods. Ponies are recumbent 5 h/d and donkeys even more.[899] Slow wave sleep occupies 2 h/d, and REM sleep occurs in nine 5-minute periods. Horses, unlike ruminants, show tachycardia, leg movements, and an increase in respiratory rate during REM sleep.[1236] Management practices can affect equine sleep patterns. Previously stabled horses sleep less on pasture; they do not lie down during the first night, and total sleep time remains low for a month. If horses are tied short in a straight stall so that they cannot lie down, they may not have REM sleep. The horses compensate by sleeping while free during the day. Care must be taken not to deprive horses of sleep inadvertently. This is most apt to occur when horses are transported long distances or when they must be tied in straight stalls. Diet also affects length of sleep in horses, as it does in ruminants. An increase in lying is seen when the protein content of grasses increases in the spring;[410] a similar trend, an increase in lying, occurs when oats are substituted for hay; fasting has the same effects.[346,347]

Ponies on pasture lie down 7% of the night, 2% in lateral recumbency. Lying in lateral recumbency occurs only in the hours just after midnight. In stalls, the same ponies lie down 12% of the time, probably a response to the drier stall environment. Horses can be seen lying in lateral recumbency during the day, usually after bad weather has kept them from lying down.

Horses feed, lie, stand, and travel, but the main activity is feeding, either eating hay when that is available free choice or grazing. Grazing

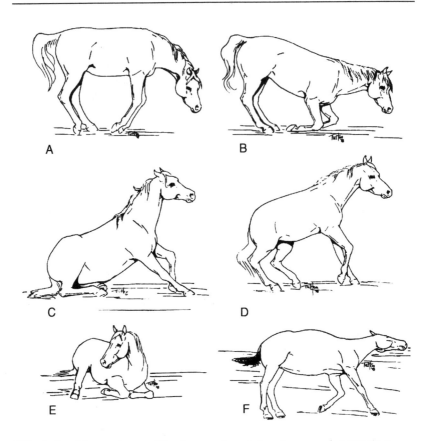

FIG. 3.6.   The postures of the horse when lying down and getting up.
(A, B) Lying down; *(C, D)* getting up; *(E)* sternal recumbency, the horse not
lying symmetrically but with the lateral surface of one foreleg on the
ground; *(F)* lateral recumbency, the upper foreleg anterior to the lower
one. The horse exhibits rapid eye movement (REM) sleep in *(F)* or in *(E)*
with the muzzle touching the ground.

time varies from 50% to 80% of the 24 hours. Grazing takes place both
day and night. The time of day when grazing takes place varies with the
presence of biting insects. More time is spent grazing during the day in
the winter when forage is scarcer (more steps between bites of food) and
when biting insects do not drive the horses to refuge in snow, water, or
barren areas.[411,778] Horses graze 15 minutes less for every extra hour of
sunlight per day in the spring. Lactating mares graze more than barren
or pregnant mares, reflecting the greater energy demands of milk production. Table 3.1 lists the percentage of time spent grazing by different
populations of free-ranging or pastured horses and ponies. The circadian rhythms of most of our domestic animals are influenced by the

TABLE 3.1. Percentage time grazing by various populations of free-ranging horses

				Time of Day								Season	Population	Reference
0600	0800	1000	1200	1400	1600	1800	2000	2200	2400	0200	0400			
98	95	33	85	70	80	80	81	—	—	—	—	Winter	W. Alberta	1254a
100	65	45	90	70	80	92	80	—	—	—	—	Summer	W. Alberta	1254a
80	77	83	83	75	72	75	—	—	—	—	—	Summer	Assateague Is. MD–VA	776a
—	—	—	—	—	—	—	63	53	53	40	70	Summer	Assateague Is. MD–VA	779a
						55-64[a]						Winter	Camargue, France	409
						51-60[a]						Summer	Camargue, France	409
						25-50[a]						Summer	Grand Canyon, CO	189
80	80	75	85	80	85	70	75	70	—	—	—	Winter	Shackleford Is. SC	1234a
70	70	60	55	50	60	65	65	60	—	—	—	Summer	Shackleford Is. SC	1234a
—	40	—	31	—	38	—	41	—	68	—	60	Summer	Front Royal, VA	240

[a] Average percentage of time spent grazing over total time period.

provision of meals. This is especially true of herbivores, such as horses and ruminants, who normally would have access to grass at all times and whose hour-to-hour behavior would not depend on access to food. The large amount of time occupied by oral behavior indicates the reasons for the appearance of oral stereotypies, such as wood chewing and cribbing, in stalled horses on low roughage diets (see Chapter 9, Miscellaneous Behavioral Disorders).

The amount of traveling a horse does depends on two things, the availability of nutrients and the horse's social status. Young bachelor stallions travel more than harem stallions or mares, but otherwise, the distance that must be traveled to procure water or enough forage determines the amount of movement.[190] Isolated horses walk considerably more than those in sight of other horses. This type of walking, like that of the bachelor stallion, is presumably a search for companions. Camargue horses grazing on a large pasture walk 7%–10% of the day, as do horses in a grassless corral.[409,687] Rapid traveling, that is trotting and cantering, occupies a very small fraction, less than 1%, of the time budgets of adult horses studied in a variety of environments. Pasture or paddock design influences activity. Horses in rectangular paddocks make many more abrupt turns than those in square ones. The abrupt turns and stops could lead to leg injury.[847] Horses do not spend as much time traveling at night as they do during the day. Ponies on lush pasture walk 3% of the night, whereas stalled horses and ponies walk less than 1% of the night.

Standing is the behavior that occurs when horses are not engaged in acquiring food, socializing, or sleeping deeply. Standing increases when feeding decreases (see Table 3.2 for time budgets of confined horses and ponies on a variety of diets). Standing is also influenced by weather conditions in pastured horses. Horses stand rather than lie when it rains and stand 20 minutes more per day for every Celsius degree drop in environmental temperature.[410]

Cattle

Cattle are essentially diurnal (day active). Their major activities are grazing, ruminating, and resting. Cattle lie down to sleep, to ruminate, or to drowse. Lying occupies nearly half the cow's day; when deprived of the opportunity to lie down, she will compensate by lying for longer periods when she is free to do so. This compensatory behavior indicates that rest is necessary. In fact, when both rest and food deprived, cattle lie down rather than eat when given the opportunity to do either.[1008] Lying occupies 13 hours of a dairy cow's day in a loose housing environment, but this will be reduced if there are not as many cubicles as cows.[1483] Lying time is affected by the environment. Loose housed cattle spend less time lying than cattle in tie stalls, although their feeding

TABLE 3.2. Time budgets of stabled horses and ponies

Time/Equid	Environment	Diet	Feeding	Standing	Moving	Drinking	Lying	Reference
Day/Horse	Metabolism cage	Limited hay	50	45	0	1	4	1487a
		Concentrates	32	62	0	1	5	1487a
Day/Pony	Box stall	Ad libitum hay	76	19	3	2	1	1370
Day/Horse	Corral	Limited hay and grain	43	27	6	—	0	687
Night/Pony	Box stall	Limited hay and grain	15	17	1	—	13	705
Night/Horse	Box stall	Limited hay and grain	27	67	0.3	—	6	1306
24 hr/Pony	Pen	Ad libitum grain	17	—	—	2	—	855
24 hr/Pony	Pen	Ad libitum pellets	31	18	—	—	—	1172a
Day/Horse	Small pen		68	18	12	1.5	0	239a
Day/Przewalski	Large pen		44	45	8.5	1.5	0	239a

Percentage of Time

times are similar.[837] Tethered cattle also spend more time in the process of lying down instead of quickly lowering themselves to the ground as pastured cows do. Stall design has a large influence on the ease with which the cow lies down.

Grazing

Most grazing takes place during the day.[89] Cattle on pasture spend anywhere from 5 to 8 hours grazing. Grazing time is inversely proportional to the quality of the pasture. Cattle on moderately good pasture spend 5 hours actually gathering food with their prehensile tongues and 2 hours walking.[760] As herbage is sparser, more walking between mouthfuls is necessary.

Grazing usually occurs in bouts and is engaged in by the entire herd; social facilitation is strong in cattle. There are two major grazing bouts, one just after sunrise and the other during late afternoon until sunset.[710] Midmorning and midafternoon are resting and idling times.[1446] By an hour after sunset, most cattle are lying down, although they will usually arise to graze during the night.[543] Night grazing may increase during warm weather.[1288] Cattle drink two to four times a day during the summer on the range, but only once or even every other day during the winter.[321] Grazing is covered in more detail in Chapter 8, Ingestive Behavior: Food and Water Intake.

Distance Traveled

The distance traveled by cattle or sheep can be measured by a rangemeter, a device that is similar to an odometer.[49,326] The distance covered by a grazing cow varies from 0.3 to 9 km/d (0.19–5.6 mi), depending on the size of the pasture or range and the abundance of forage. Beef cattle on range in Montana walk 3 km/d (1.9 mi) while spending 11–12 hours grazing per day at a bite rate of 50–60 per minute.[529] Dairy cows with free access to a pasture, a yard and a barn, traveled 3 h/d in summer but less than 1 h/d in winter.[838]

Housing Conditions

Cattle live not only on the range, but also in varying degrees of confinement. Dairy cattle are milked at least twice a day, and their grazing habits are organized around the milking schedule. The most intense grazing activity follows each milking. There are two bouts after the afternoon milking, and one before the morning milking. A brief bout of rumination follows each grazing bout. A total of 5.5 hours during daylight is spent grazing, and an equal time is spent ruminating. In contrast, cattle in a loose housing situation spend only half as much time eating and ruminating as do cattle on pasture. They spend 6–7 hours loafing, that is, standing neither grazing nor ruminating, and 12 hours resting. Time spent walking decreases with the size of the idling area available to the cattle.

Cattle on feedlots are in a highly unnatural environment, as reflected in their activity patterns. Grazing bouts are replaced by nine to 14 feeding periods, 70%–80% of which occur during daylight hours. If hay and/or silage is fed, a total of 5 h/d are spent eating, but the time decreases as the percentage of concentrates in the diet increases or if the roughage is ground.[1168] Standing increases at the cost of lying during rain or snow.[562]

In hot climates, moving into the shade is another activity that appears to be a response to light rather than to temperature per se. Cattle should, of course, have access to shade, and shading behavior should be considered when management plans are made.

Elimination

Cattle defecate 7 to 15 times a day and urinate 5 to 13 times. The frequency of both excretory activities decreases in hot weather.[318] Rumination time also decreases under these conditions.

See Table 3.3 for activity patterns of cattle in different environments.

Sleep

The presence or absence of true sleep in ruminants has been controversial,[109,182,1005] but the extensive studies of Ruckebusch[1236-1239] indicate that cattle show both REM sleep and SWS. Rapid eye movement sleep occurs in 11 periods, so the total of 45 minutes of REM sleep and 3.5 hours of SWS is divided into many short naps. When cattle are in REM sleep, they usually are lying down with their heads resting on the ground and turned back into the flank. Cattle sniff the ground before lying down, and on arising, lick and scratch themselves. Cows in slings are sleep deprived, as are cows that have not yet adjusted to stanchioning or newly mixed groups of cattle. The stress of sleep deprivation should be considered by clinicians and stock managers. When kept in a corral at night, cattle, or Zebu cattle at least, tend to sleep in areas that remain constant for each individual from night to night. The resting places do not appear to depend on dominance.[1188] Most characteristic of ruminants are the extensive periods of drowsiness usually associated with rumination. Cattle are in a drowsy state 7.5 h/d, divided into 25 periods that precede and follow sleep. Rumination and sleep are inversely

TABLE 3.3. Activity patterns of cattle in different environments

Grazing (hr)	Number of Grazing Bouts	Ruminating (hr)	Lying (hr)	Walking (hr)	Standing (hr)	Idling (hr)	Type of Cattle	Reference
								Cattle on pasture
5.5–7.5	6 (2 at night)	—	13	—	4	—	Dairy cows	89
5.5–10	—	—	—	—	—	8.25	Beef steers (Hereford)	105a
6.5	5–7 (1 at night)	5.5	9.25	—	—	3.50	Dairy cows (shorthorn)	283a
8		5.5	9.25	—	—		Dairy cows (shorthorn)	318
7–9	2	4	—	—	—	2	Beef cattle	341a[a]
7.25–7.5	4–5		8.25	—	—	9	Dairy calves	348
6	3		—	—	—	4	Beef cows (Charolais)	543
10–12	6 (1 at night)	8	—	—	—		Dairy cows	599a
9	4 (1 at night)	8.5	9	—	15	6	Dairy cows (Holstein)	600a
7–8		4.5	5	0.25	3.25		Zebu cattle	600b
9(8–11)	2	—	2	—	2	—	Steers	637a
9–10	4	8	—	2–3	—	—	Beef cows (Hereford, Santa Gertrudis)	657
11.50	5	8.50	—	—	4	9	Dairy cows	671a
8	—	8	—	—	—	—	Beef steers (Hereford)	710
7–8	5	7	12	1.25	—	5	Beef cattle (Hereford)	260
11.50	2	7	—	—	—	—	Beef and dairy heifers	839a
7	4	7	—	—	—	—	Zebu and grade steers	852a
—		—	—	—	—	7–12	Dairy cows (Brown Swiss)	852b
6–8	4–8	—	9–11	—	2–3.5	—	Steers	900a
10–10.5	3	—	10–14	—	1.25–4	—	Beef cattle (Hereford)	1037a
9.5–12	4	—	—	—	—	—	Nonlactating cattle	1085a
8–9.5	2	—	—	1.5	—	—	Dairy cows	1288
10	3	—	—	—	—	—	Beef cattle	1333a[a]
—		—	—	—	—	—	Beef cattle	1446
7(5.5–8)	5	6.25(4.5–9.5)	—	—	—	—	Dairy cows (Ayrshire)	145a
9	6	7	5	1[a]	6	—	Beef steers (Hereford)	1531a
								Cattle in confinement
3–5	—	—	—	11	—	—	Dairy cows[b] (Holstein)	524
4–5	9–12	—	—	8–11	—	—	Dairy cows[c] (Brown Swiss)	871a
3.5–5.25	4	—	—	—	14	1.25	Steers[d]	1168
3.5		7.5	9.5	6.5	—	—	Beef cows[d] (Hereford)	1266
3–4		—	—	12.25	—	6–7	Dairy cows[c] (Holstein)	1270a
5	10	—	10.5	—	8.5	—	Dairy cows[d] (Ayrshire)	1417a
6.25	18	—	—	—	—	—	Dairy cows[e] (Guernsey)	1457a

[a] Daylight observation only.
[b] Free stall.
[c] Loose housing.
[d] Feedlot.
[e] Cowshed.

related, so sleep time decreases with rumen development (see Chapter 6, Development of Behavior) and decreases as the percentage of roughage in the diet increases.[109]

Environmental Influences

Social changes can disrupt activity rhythms in ruminants as well as in pigs. For example, calves in a stable group show definite diurnal activity patterns; those in continually changing groups do not.[826] Calves are affected by the lighting regime. They prefer a lighted area and spend more time lying.[1459]

In summary, the normal bovine day depends on the diet and on the housing conditions and, in general, consists of alternating periods of eating and ruminating interspersed with resting or loafing and short periods of sleep (Fig. 3.4). Activity patterns in cattle have been studied by other investigators in addition to those listed in Table 3.3.[480,599,601,1183,1308]

Sheep

Grazing and Traveling

Until recently, sheep were seldom kept in confinement, so that studies of their activities have dealt with range or pasture conditions (Table 3.4). Sheep on the range spend 50% of the daylight hours grazing,[321] of which 7 hours are spent grazing and 2 hours traveling.[390] On the range, sheep travel 6–14 km/d (4–9 mi), but they travel only 0.8 km/d (0.5 mi) on pasture.[327] Two factors determine how the range is used: familiarity with the area and social integration into the flock. Newly introduced animals may wander 14 km (9 mi), for example, when introduced to a new flock in an unfamiliar environment.[1456] On pasture sheep spend 9 to 10 hours grazing in four periods, and they spend an equal amount of time ruminating in 15 bouts. Sheep allowed to graze only during daylight hours also spent 9 hours grazing,[192] similar to the time spent by sheep allowed 24 hours to graze. As has already been noted in the case of cattle, more time is spent grazing on a poor pasture (up to 12 h/d) and twice as much distance traveled as on a good pasture.

Sheep in particular are synchronized in their behavior in that all or most of the sheep will be doing the same thing at the same time. Sheep may all begin to graze at the same time, but there is much greater variation in the end of a grazing bout. The satiety factors (discussed in Chapter 8, Ingestive Behavior: Food and Water Intake) are more important in ending a meal, whereas the behavior of the other sheep is more important in starting it.[1214] Even on pasture, the type of feed can affect behavior patterns. For example, sheep grazing clover spend less time grazing and ruminating than those grazing grass.[1120]

TABLE 3.4. Mean values of comparative data of sleep-wakefulness states and attitudes in four species of farm animals (three subjects of each species)

Species and Time Period	Duration and Percentage					
	Wakefulness		Sleep		Attitude	
	AW	DR	SWS	PS	Standing	Recumbent
Horse						
24-hr period	19 hr 13 min 80.8%	1 hr 55 min 8.0%	2 hr 05 min 8.7%	47 min 3.3%	22 hr 01 min 91.8%	1 hr 59 min 8.2%
Nighttime (10 hr)	5 hr 14 min 52.4%	1 hr 54 min 19.0%	2 hr 05 min 20.8%	47 min 7.8%	8 hr 01 min 80.1%	1 hr 59 min 19.9%
Cow						
24-hr period	12 hr 33 min 52.3%	7 hr 29 min 31.2%	3 hr 13 min 13.3%	45 min 3.1%	9 hr 50 min 40.9%	14 hr 10 min 59.1%
Nightime (12 hr)	1 hr 55 min 16.0%	6 hr 14 min 51.9%	3 hr 06 min 25.8	45 min 6.3%	1 hr 30 min 12.5%	10 hr 30 min 87.5%
Sheep						
24-hr period	15 hr 57 min 66.5%	4 hr 12 min 17.5%	3 hr 17 min 13.6%	34 min 2.4%	16 hr 50 min 70.1%	7 hr 10 min 29.9%
Nighttime (12 hr)	5 hr 59 min 49.8%	2 hr 45 min 22.9%	2 hr 43 min 22.5%	34 min 4.8%	7 hr 10 min 59.7%	4 hr 50 min 40.3%
Pig						
24-hr period	11 hr 07 min 46.3%	5 hr 04 min 21.1%	6 hr 04 min 25.3%	1 hr 45 min 7.3%	5 hr 10 min 21.5%	18 hr 50 min 78.5%
Nighttime (12 hr)	4 hr 23 min 36.5%	2 hr 30 min 20.8%	3 hr 52 min 32.2%	1 hr 15 min 10.5%	1 hr 20 min 11.1%	10 hr 40 min 88.9%

Source: 1236.

TABLE 3.5.   Activity patterns of sheep

Grazing (hr)	Number of grazing bouts	Ruminating (hr)	Standing (hr)	Lying (hr)	Walking (hr)	Reference
7	—	5.5[a]	—	—	2	390[b]
9–12	5	9–10.5	2.5	3.5	0.5	433
9	2	—	3.5	11.25	0.25	710
4–5.5	2	—	—	—	—	1344a

[a] Includes idling and resting.
[b] Observed for 14.5 hr/day (daylight).

Sleep

Sheep are awake for 16 h/d. They drowse 4.5 h/d, far less than cattle. Slow wave sleep occupies 3.5 h/d, and REM sleep occurs in seven periods for a total of 43 minutes.[1236] Sleep increases in sheep fed a low roughage diet.[1038] While sleeping, sheep expend 10% less energy than while waking,[1409] so sleep deprivation would be expected to increase energy expenditure. Sheep will stand up eight to 11 times during the night, usually to urinate or defecate.[433] Activity patterns in confined sheep have also been studied.[1450]

The ocular muscle relaxation and other physiological parameters that accompany the various states of vigilance in farm animals are shown in Figure 3.7. The percentage of the night spent in sleep, wakefulness, and various sleep states and postures are shown in Figure 3.7. Comparative data for sleep and wakefulness in farm animals are given in Table 3.5.

Goats

There has been very little work on activity patterns of goats. Adults spend 41%–47% of the time foraging and kids spend 59%–65%, depending on the stocking rate.[1164] Goats graze less and travel more when it is raining or when flies are abundant. These changes are more pronounced in shorn goats.[246]

CLINICAL PROBLEMS

Human mental illness is often associated with disorganization of diurnal rhythms[1074] and with sleep disturbance.[1371] This has not been investigated in veterinary medicine, but owners of elderly dogs, in particular, complain that the animals are restless at night. There has been a case of a cat that lacked the normal inhibition of movement during REM sleep and was very active at that time.[653]

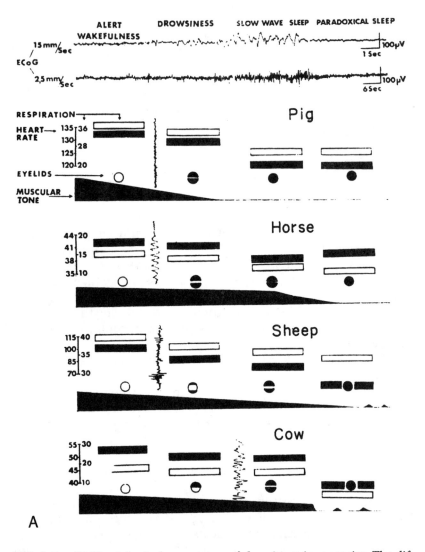

FIG. 3.7. (A) Physiological parameters of sleep in various species. The different electrocorticogram (ECoG) patterns for each species are shown at a speed of 15 mm per second: theta rhythm (horse), delta rhythm (ruminating cow), spindles (sheep), and alpha rhythm (pig)[1236].

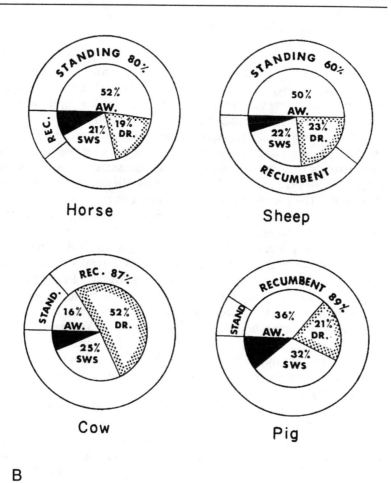

(B) Mean comparative data of sleep and wakefulness states and of attitudes during night time. The *inner circle* shows the relative duration of the ECoG pattern (rapid eye movement, REM, in *black*) and *the outer circle* shows the relative duration of the attitudes) (copyright 1972, with permission of Bailliere Tindall).

Hyperactivity

Hyperactivity is probably the behavior problem most frequently diagnosed by owners and least frequently confirmed by the clinician. Most dogs that owners perceive as hyperactive are simply unruly. The owners usually have unconsciously rewarded the dog for hyperactivity by ignoring it when it is quiet and paying attention to it when it is rambunctious. For these dogs, even negative attention is preferable to no attention. The dogs usually are young dogs of working breeds. Although they actually may require more exercise than smaller breeds, it is more likely that they cause more disruption when they are active than a little dog would. The running, jumping, and mouthing behavior may cause the owner to isolate the dog for long periods, which of course leads it to be more excited when it sees people.

If the dog appears normal in the absence of the owner, then recommending more exercise, a canine companion (two or more owners could leave their dogs together so that extra dogs would not have to be acquired), and obedience training usually suffices to improve the dog's behavior. A specific behavioral exercise, "quiet training," is also recommended. The owner should pet and praise the dog whenever it is lying quietly. More formally, the dog can be given food rewards for lying quietly. If it becomes hyperactive, the owner should either ignore or, if destruction of property is likely, isolate the dog, but only until the dog is heard to be resting quietly.

A few dogs are truly hyperactive. This can be diagnosed by measuring heart and respiratory rate and activity of the dog, giving 0.5 mg/kg amphetamine and measuring the same parameters 30 and 60 minutes later.[906] If the dog is truly hyperactive, dextroamphetamine 0.2–1.3 mg/kg or methylphenidate (Ritalin) 5–20 mg per dog may be prescribed.

Narcolepsy

The number of clinical problems associated with sleep and circadian rhythms is small when compared with those associated with, for example, dominance and aggression. A syndrome involving sleep in dogs,[1023] cats,[820] and horses[490] is narcolepsy. Narcolepsy is characterized by attacks of inappropriate sleep. The affected dog will collapse and fall asleep for several seconds or minutes at a time. Play or food, especially very palatable food, often elicits the attacks. It is interesting that there are species and breed differences in the stimuli that elicit cataleptic attacks. Food is most apt to elicit catalepsy in dogs, but not in narcoleptic horses, who are most apt to collapse when petted or saddled. Young dogs and Labradors are most apt to collapse when playing. Although Doberman pinschers are the breed most affected, the disease appears to be more severe in small breeds of dogs. The best diagnostic test is to space small bits of food a foot or so apart and time how long it takes the dog to consume all the food and the number of times it collapses. The animal can be aroused with auditory stimuli. The drug imipramine reduces the severity and incidence of the attacks. In the Doberman pinscher breed, narcolepsy is inherited as an autosomal recessive gene.[490]

Nocturnal Wakefulness

Much more common than narcolepsy is the problem of dogs and cats that demand attention at night. In most cases, these are animals that are left alone during the day. The usual pattern is that the dog wakes the owner because it has a genuine need to eliminate. The dog learns after only a few nights that it can waken the owner and go for a walk or at least get some attention. Although sedation may be necessary in extreme cases, usually firmly enforcing a down stay command if the dog demands attention, combined with an increase in exercise and attention during the hours the owner is at home and awake, will solve the problem. Nighttime wakefulness usually accompanied by vocalization and pacing is a common problem in old dogs. Sometimes, treatment with a dopamine agonist such as selegiline (1 mg/kg) can be helpful.

The problem in cats occurs either in young cats that are seeking play or in cats subjected to a change in social or physical environment. More play in the evening will help to reduce nighttime, usually dawn, play periods in kittens. Free-choice food often is helpful, especially if the cat is crying for breakfast. Food in the bedroom is more effective because the cat may prefer to eat in the presence of the owner. Sometimes punishing the cat with a water pistol suppresses the behavior, but cats often are able to avoid a stream of water and still disturb the owner. Wrapping the cat in a blanket and restraining it for a moment is effective in some cases.

CHAPTER 4

SEXUAL BEHAVIOR

Sexual behavior includes proceptive and receptive behavior by the female and courting and mate guarding by the male, as well as actual copulation. Tests have been developed to determine which bulls and rams will be successful breeders. There is a relation between management practices, such as separation of the sexes at weaning, and later problems of homosexual or inadequate sexual behavior. Poor libido in breeding males and unwanted sexual behavior in castrated males are the most common behavior problems.

INTRODUCTION

Sexual behavior is important in all species of animals. The importance lies not only in maintaining adequate levels of libido in breeding animals, but also in controlling the various aspects of sexual behavior that persist in neutered animals.

Mating systems have evolved within the framework of the morphological and physiological parameters of the individual species under continual ecological pressures. Although we have drastically reduced the nutritional, climatic, and health stresses to which domestic animals are exposed and have selected heavily for "good breeders," we still find remnants of this long-term evolutionary selection hindering our goal of high reproduction rates. Thus, we may continue to see problems that hinder breeding and production schedules, such as seasonal breeding in housed stock subjected to artificial control of photoperiodicity; poor libido in some of our artificial insemination (AI) programs, despite the absence of organic disease; and mate selection preferences that do not coincide with our ideas of desirable matings.

Physiological Bases of Sexual Behavior

Adult male and female sexual behavior depends on a variety of factors, physiological, environmental, or psychological, for its expression.

These factors are (1) the genetic sex of the animal, (2) perinatal organizational action of hormones, (3) past social and sexual experience, (4) adult activational action of hormone and anatomical status, (5) the attractiveness of the potential mate, and (6) the external environment. Figure 4.1 illustrates the factors that affect sexual behavior, using the stallion as an example.

Genetically Determined Sex

The sex of the animal is determined at the moment of conception, and the chromosomal sex will determine whether the indifferent fetal gonad develops into an ovary or a testis. Nevertheless, the potential for masculine and feminine behavior remains in both sexes. Studies on laboratory animals have revealed that the brain, and therefore behavior, is usually female unless the fetus is exposed to androgen during development. Similarly, without androgenic stimulation, the external genitalia will be female.

Organizational Perinatal Hormonal Influences

The role of sex hormones during ontogeny has been studied extensively in laboratory rodents and also in dogs. Male puppies have been castrated at birth and their behavior compared with that of intact dogs and dogs castrated as adults. Anatomically, these dogs were altered in

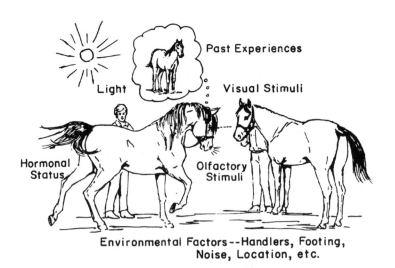

FIG. 4.1. Factors that affect sexual behavior.

that their penes were very small. Despite this anatomical change, they still urinated in the masculine posture and were no different in their response to exogenous testosterone from dogs castrated as adults.[152] They were attracted to estrous females and mounted them; however, their underdeveloped external genitalia prohibited normal intromission. Both sexes have the genetic potential for male or female behavior, but neonatal androgens "defeminize" males, that is, make them less likely to show female sexual behavior. Therefore, when treated with estrogen as adults, the neonatally castrated males show receptive behavior to other males.

Female puppies treated with androgens in utero and neonatally were markedly altered anatomically. They had no external vagina and did have small phalluses. They urinated in the masculine posture half of the time. As adults, they were ovariectomized and treated with either female (estrogen and progesterone) or male (testosterone) hormones. The estrogen treatments revealed that the dogs had been defeminized because they would not stand or show any other signs of sexual receptivity. They were not even attracted to the male as is a normal intact estrous female or ovariectomized, estrogen-treated female. When treated with testosterone, the experimental dogs were obviously masculinized because they were attracted to, and would mount, other female dogs that were in heat. These studies indicate the powerful effects of perinatal androgens on the anatomy and behavior of animals of either genetic sex.[156-158]

Sheep and cattle are also defeminized by their neonatal androgens. The freemartin cow (born twin to a bull) is an example of a naturally occurring manipulation of the perinatal hormonal environment; as is described later in this chapter, these females exhibit masculinized behavior. Pigs appear to be unique in that they are defeminized not during the prenatal or neonatal period, but at puberty.[484]

Activational: Adult Hormonal Status

The most important feature of the hormonal basis of sexual behavior is that hormones have a permissive role; that is, an animal requires a certain level of hormones for normal sexual behavior, but a higher level of hormones will not increase libido or receptivity. Hormonal treatment will not cure a deficiency of sexual behavior unless there is a deficiency of that hormone.

The complex relationships of the central nervous system, gonadal hormones, and behavior are discussed in more detail later in this chapter; in general, however, ovarian hormones result in an attraction to males and receptivity to male mounting. In some species (cats and pigs), the complete pattern of estrous behavior can be elicited by estrogen alone. In others (dogs and sheep),[160,195,1324] progesterone must also be administered. In ungulates, the behavioral action of estrogen is facilitated by a rapid preovulatory fall in progesterone, whereas in dogs a rise in progesterone is important. Progesterone is administered before estrogen in the ewe and after estrogen in the bitch to induce estrous behavior.

Ovariectomy and Castration. Ovariectomy (spaying) usually abolishes estrous behavior in females. Castration (orchiectomy) generally abolishes sexual behavior in males, but there are many exceptions. The more experienced the male, the longer sexual behavior, both arousal and copulation, will persist after castration. Prepubertal castration is, therefore, more effective than postpubertal castration in eliminating sexual behavior. There are species differences in the effectiveness of prepubertal castration. Cats are affected more than dogs.[608,861,1221,1222]

Anatomical Factors. As mentioned previously, anatomical factors are important because a small penis precludes successful intromission. Intact afferent pathways from the penis are also necessary; experimental or traumatic neural damage to the penis results in misorientation in tomcats[84] and failure to ejaculate in bulls.[171] Similarly, desensitization of the vagina inhibits ovulation in the cat, an induced ovulator.[373]

The species differences in the structure of the penis are also important. For anatomical reasons, castration reduces the copulatory ability of male cats and horses more than it does that of male ruminants and dogs. The muscular penis of the horse is more dependent on erection for successful intromission than the fibroelastic penis of the ruminant. Similarly, the penile spines of the tomcat, which atrophy in the castrated male, are important for successful intromission and ejaculation as well as for induction of ovulation.

Social and Sexual Experience

It is much more common to observe animals that have been influenced by lack of adequate sexual experience than animals influenced by abnormal hormone levels. There are social influences on male sexual behavior. Having absolutely no social interactions, that is, being raised in isolation from weaning to adulthood, suppresses sexual behavior. The type of social interaction is important but there are species differences. Whereas rams raised in all-male groups from weaning to 1 year are less sexually active than those raised (or even given brief exposure) with ewes, bulls and boars are independent of female exposure for normal behavior.[1152] Total lack of experience, homosexual experience only, too much sexual experience, or sexual experiences that are too unpleasant can all lead to sexual abnormalities.

Lack of Socialization

The concept of critical or sensitive periods of development is discussed in Chapter 6 (Development of Behavior); it should be emphasized here, however, that the most obvious effect of lack of socialization to conspecifics is on sexual behavior. Dogs raised without physical contact with other dogs from the age of 3 weeks showed normal libido toward estrous bitches but were very poor at orientation; they would mount improperly and seldom achieved intromission. This is probably an effect of lack of mounting experience because mounting forms a large part of male puppy play.[149] Similarly, boars raised in isolation from the time they were 3 weeks old showed very little sexual behavior.[647]

Within all-male groups, individuals may direct their sexual attentions to other males or be subordinate to other males. The dominant males may continue to mount males even in the presence of an estrous female; the subordinate animals often have little or no mounting experience. Some of these inappropriate responses will cease over time, but breeding efficiency is affected (at least temporarily).

Negative Sexual Experience

Unpleasant experiences during mating will have a deleterious effect on future sexual behavior, especially if the animal is young and has not had many (pleasant) experiences. The negative associations can be the result of overt aggression on the part of the sexual partner, rough handling by a stockperson, or an injury sustained during mating.

Effect on Female Sexual Behavior

The effects of experience on sexual behavior have not been as thoroughly studied in females as in males. There may be less effect, in part because the female plays a less active role in mating in most management situations. Sows raised in isolation showed normal sexual behavior and were attracted to males when in heat.[1323] More research should be done on the effects of experience and age at weaning on female sexual behavior. Cats, in particular, may not be adequately socialized under normal rearing techniques and may reject toms, at least initially.

Attractiveness of Potential Mate

The element of attractiveness, or lack of it, of the sexual partner is not often considered, but higher mammals are influenced by this factor as well as by their hormonal levels. The attractivity of an estrous bitch's urine depends on her hormone state. If the donor of the urine is treated with estrogen her urine becomes more attractive to males; if treated with testosterone, her urine is less attractive. Marking behavior of males is apparently an attempt to mask the attractiveness of bitch urine.[405,407]

Females may show individual preferences for one male over another. All females do not prefer the same male, indicating that the differences in attractiveness of males are based on the female's innate preferences and on past experience, rather than on some physiological characteristic of the male, such as pheromone release.

Male preferences can be based on physiological factors. For example, rams prefer unmated ewes, as is discussed later in this chapter. The action of ovarian hormones not only renders the female receptive but also increases her attractiveness, presumably by pheromonal release. Care must be taken when comparing the results of experiments on animals artificially brought into estrus with those involving females in natural estrus because the latter may be more attractive than the former.

Like females, males also show individual mate preferences that are apparently psychological or idiosyncratic. There may be an evolutionary basis for some of these preferences. Males may prefer females that are similar, but not too similar, to themselves so that inbreeding will not occur and yet the same gene pool will be propagated.[145] Evidence for such behavior influencing sexual behavior of sheep,[632,863] cats,[875] and horses[779] has been found. The mechanisms and ramifications of mate selection are topics of considerable interest to researchers; extensive reviews of this work (primarily nondomestic species) may be found in Wilson,[1491] and Krebs and Davies.[834]

External Environment

The external environment is very important for optimal sexual behavior. It is obvious that extremely inclement weather will inhibit sexual behavior, but more subtle environmental factors, such as too many human spectators or a slippery floor, may also inhibit it. Not all additions to the environment are detrimental. In some cases, the presence of another male may stimulate sexual behavior. This has been well documented in cattle and goats.[212,1157] Males generally are more influenced by the environment than are females, so the female is usually brought to the male. Nevertheless, environmental factors in female sexual behavior deserve study. Time of day, for example, is important in cows who show more signs of estrus at night. It is becoming clear that olfactory stimuli from the males, probably androgen-derived pheromones, affect sexual cycles in female ruminants.

The Central Nervous System and the Control of Sexual Behavior

Females

Hypothalamic Factors.   In the female, gonadotropin releasing factor (GnRH) in the hypothalamus (see Appendix 3) stimulates the release of follicle-stimulating hormone (FSH) and luteinizing hormone (LH) from the anterior pituitary. Follicle-stimulating hormone induces follicular development and FSH and LH together stimulate estrogen and progesterone production by the ovary. For most of the estrous cycle, estrogen and progesterone maintain low levels of LH and FSH through a negative feedback action on the hypothalamic–pituitary axis.

Near proestrus, however, the situation is reversed; rising estrogen levels have a positive feedback on LH secretion, resulting in the LH surge that causes ovulation. This preovulatory rise in estrogen is responsible for the hormonally based components of estrous behavior. The brain shows a refractory period during which biochemical changes presumably take place that will result in estrous behavior.[101] For example, radioactive estrogen is taken up by cells in the hypothalamus within hours of administration to an ovariectomized cat, but estrous behavior does not appear for 3 days.

Cyclical Ovulation.   In spontaneous ovulators (bitch, ewe, mare, and sow), the LH surge and, consequently, ovulation take place cyclically; but in induced ovulators, such as cats, external stimuli from the vagina either by natural coitus or artificial manipulation are necessary to trigger the LH surge. The relationship between hormone levels and sexual behavior in the bitch is shown in Figure 4.2. The female's active solicitation of the male is called proceptive behavior.

The seasonal nature of reproductive behavior is also due to central neural variation in responsiveness to gonadal hormones. The same ewes had to be injected with a larger amount of estrogen to induce estrous behavior in the spring than was needed in the fall.

Olfactory Influences and Pheromones.   Odors can have important effects on reproduction. This has been demonstrated in a variety of rodents, mice, in particular.[250] In domestic animals, odor is probably not as important; nevertheless, the age at first puberty is lower in gilts exposed to a strange boar (continuous cohabitation with a boar will not produce the effect). Olfactory bulbectomy (see Appendix 3) eliminates the effect, indicating that the odor is the important stimulus.[81] Ewes show a similar response called the ram effect, but odor may not be essential. Postpubertal cows show another response to odor; they will come into estrus sooner if exposed to the odor of estrous cow urine or

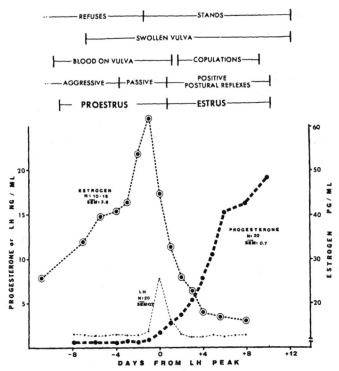

FIG. 4.2. Relationship of hormone levels to behavior in the bitch. Times of onset of behavior and physical signs represent means of a range of times[314] (copyright 1975, with permission of the Society for the Study of Reproduction).

mucus.[737] The best studied farm animal pheromone is the boar odor that stimulates the estrous sow to assume the immobile, rigid posture that permits the boar to mount her and that can be used for estrus detection.

Males

Hypothalamic Factors. The hypothalamic–pituitary axis is also involved in male sexual behavior. Follicle-stimulating hormone is released in response to hypothalamic releasing factor. In the male, FSH stimulates spermatogenesis and LH testosterone release. Testosterone, in turn, acts upon the anterior hypothalamus-preoptic area in conjunction with appropriate stimuli from an estrous female to produce male sexual behavior. Inhibin, a testicular factor produced in the spermatic tubules, acts as a negative feedback on the hypothalamus.[1300]

There is considerable evidence accumulating that it is not testosterone itself, but rather a metabolite of testosterone, estradiol, that acts on the central nervous system to produce male sexual behavior. In general, estrogen and testosterone have similar actions in stimulating male sexual behavior in castrated animals, but an androgen that cannot be metabolized to estradiol, dihydrotestosterone, does not.[286,301] When freemartins are treated with estrogens and androgens, both hormones increase aggressiveness, mounting, and sniffing of the vulva of other cows, but only testosterone stimulates the flehmen response.[571] Testosterone itself, rather than a metabolite, may be responsible for this response (see Fig. 4.3). Dihydrotestosterone does stimulate enlargement of the penis and of the sex glands, so when it is administered in combination with estradiol to castrated pigs, the full sequence of male sexual behavior, including intromission, occurs.[1111]

Stimulation of the brain by testosterone or its metabolites may facilitate appearance of male sexual behavior; it is the appearance of a female, however, that triggers the behavior. The stallion, for example, will

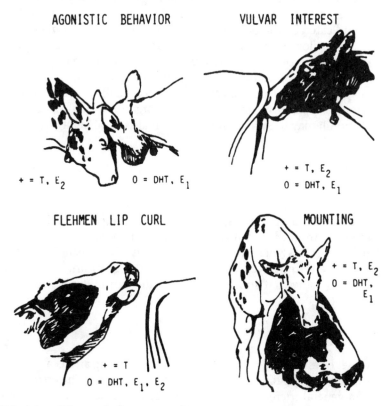

AGONISTIC BEHAVIOR

+ = T, E_2 0 = DHT, E_1

VULVAR INTEREST

+ = T, E_2
0 = DHT, E_1

FLEHMEN LIP CURL

+ = T
0 = DHT, E_1, E_2

MOUNTING

+ = T, E_2
0 = DHT,
      E_1

FIG. 4.3.   Effects of testosterone (T), estradiol (E_2), estrone (E_1), and dihydrotestosterone (DHT) on agonistic behavior, vulvar interest, the flehmen response, and mounting[571] (copyright 1978, with permission of Academic Press).

exhibit the flehmen response and begin the courtship rituals of nibbling the crest and rump of the mare. At the same time, stimulation of the parasympathetic nervous system results in secretion of the accessory sex glands. The combination of penile stimulation after intromission and sympathetic stimulation leads to ejaculation.

Olfaction.   Olfaction is, no doubt, important for identification of the estrous female; but elimination of the sense of smell by olfactory bulbectomy does not impair sexual performance of cats or rams, indicating that olfactory stimulation is not essential and that the male can identify the receptive female by visual or auditory means. The lack of resistance to his mounting attempts may be the most important information to the male.[83,619,890]

Summary

In summary, hormones are important for normal sexual behavior, but the central nervous system can be relatively independent of them. This is indicated by the persistence of copulatory behavior in castrated male cats,[1221] dogs,[150] and rams[302] that had considerable precastration experience. Even prepubertal castration may not eliminate sexual behavior; one third of prepubertally castrated bulls mounted cows.[478] A similar percentage of prepubertally castrated geldings showed sexual behavior.[897] Apparently, once the brain is organized by androgens, external stimuli may be sufficient to trigger male sexual behavior.

CATTLE

The Cow

The cow is a nonseasonal, continuously cycling breeder, but shows peak fertility from May to July and a low from December to February. Puberty occurs anywhere from 4 to 24 months of age, usually at 6 to 18 months. The estrous cycle is 18 to 24 days long (mean, 21 days), although it is somewhat shorter in heifers and in the *Bos taurus* (Zebu) breeds.

Onset of Estrus

Onset of estrus occurs more often in the evening and ceases in the morning. Actual sexual receptivity lasts 13 to 14 hours. The estrous cow shows a general increase in motor activity and a decrease in food intake[720] (Fig. 4.4). Investigative behavior such as flehmen, sniffing, rubbing, and licking increase as does premounting behavior such as standing behind the cow and resting the chin on the back of another, usually

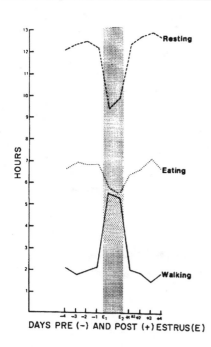

HOURS

DAYS PRE (-) AND POST (+) ESTRUS (E)

FIG. 4.4. Average composition of daily activities of cattle during estrous and nonestrous stages of the reproductive cycle[720] (copyright 1975, with permission of Department of Animal and Poultry Science, University of Guelph, Ontario, and Elsevier Science Publishers).

another estrous cow. She will bellow a great deal, switch her tail and raise or deviate it to one side, and urinate frequently. Estrogen levels peak when the cow stands to be mounted.[437,555] The cow that mounts is usually preovulatory. The mounting cows are also usually dominant over the mounted cows. This represents an interaction between hormonally mediated behavior and social influences.

Because visual cues are received over greater distances than olfactory cues, homosexual mounting may attract bulls who live apart from the cows (see Chapter 2, Aggression and Social Structure). Furthermore, bulls choose cows who are mounting and being mounted in preference to those who are not.[544] Homosexual mounting in cows may have been selected for in dairy cattle because as long ago as medieval times, bulls were not routinely kept with cows; so only those cows that the owner noticed to be mounting were taken to a bull and bred.[107] Beef cows mount much less frequently than dairy cows, confirming this hypothesis.

Aggressive behavior also increases markedly during estrus.[720] Cows mount cows both before and after the period in which bulls would mount[805] (Fig. 4.5). Mounting behavior by cows may not occur if a bull is present.[1269] The intensity of estrus can be measured only subjectively using the degree of restlessness and the conspicuousness of mounting; however, intraindividual and intrabreed differences may be noted. For example, the Brown Swiss breed shows the least marked estrous activity

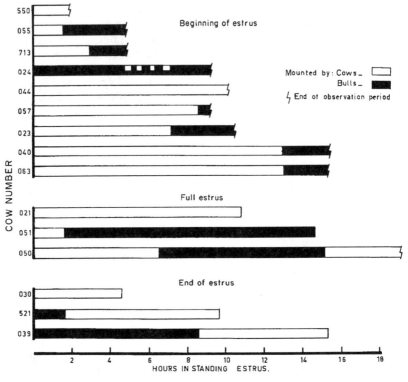

FIG. 4.5. Mounting of cows at different stages of estrus by cows and bulls.[805]

of the dairy breeds, and it has been reported that black cattle may show stronger signs of estrus than red, roan, or white cattle.[27] The physiological bases of these observations are not known but are probably under genetic control.

Detection of Estrus

Detection of estrus becomes more and more critical in the dairy industry as artificial insemination becomes more widespread (over 90% of Holstein calves are conceived via AI). The physical signs of estrus, such as a copious vaginal secretion and vulvar relaxation, may be weak or absent. Estrous behavior may be less noticeable in housed cattle than in yarded or pastured ones.[1135] Traditionally, the bull is the best detector of estrus in the cow, with humans doing a poor to fair second. An estimated 19% of heats may be missed because behavioral signs are absent (silent heats) even though ovulation occurs, and an additional 15% are labeled as estrous periods even though ovulation does not occur (false heats). Some silent heats may have been behaviorally evident but were not observed by the herders. Pregnant cows may also show signs of estrus.[505] Footing can also affect a cow's willingness to mount.[1422]

Mounting behavior is used by the farmer to aid heat detection, but it may not occur around the milking time or when the cows are usually observed. A commercial estrus detection device, which consists of a plastic vial containing a red dye, is glued to the dorsal tail base of a cow suspected to be in estrus. As she is mounted by other cows in the herd the dye is gradually expressed into a viewing chamber; a full chamber supposedly is correlated with a sufficient number of mountings to indicate a full estrus.[1490] Several cows may be in heat at once and one may not be mounted, so even the mechanical heat detector can give misleading false negatives.[1062] Pedometers can be used to detect the increased activity typical of estrus.

A vasectomized bull or teaser bull with a surgically deviated penis may also be used to detect heat. The danger of keeping an unpredictably aggressive bull in close contact with people, however, has discouraged this practice. A freemartin heifer may be used satisfactorily. Fetal exposure to the androgens produced by the male fetus masculinizes the nervous system of freemartins; they are infertile and, thus, of little value as dairy animals. Freemartins are even more likely than the other cows in the herd to mount an estrous cow, and treatment with injectable androgens may heighten this behavior even further. Dogs have been trained to detect estrous cows, an interesting innovation presumably based on canine olfactory acuity.[795] The dogs were 80% to 90% correct in detecting estrous cows.

Clinical Problems of Cows

Silent heats, a phenomenon to which heifers are especially subject, have already been discussed.

Nymphomania is more common in high-production dairy cows than in beef breeds. The cow shows intense estrous behavior either persistently or at frequent, irregular intervals. Milk production drops noticeably as these cows are often the best producers in the herd. Most commonly, the affected cow is 4 to 6 years old and has calved two to three times. Unlike a cow in normal estrus, however, the nymphomaniac does not stand for other cows. She actively seeks out other cows and mounts them. She paws and bellows like a bull and with time also becomes more malelike in voice and body conformation. Nymphomania is usually associated with follicular cysts, and treatment is sometimes successful using a source of LH, such as pituitary extract or chorionic gonadotropin.[1203]

The Bull

In contrast to rams, bulls' sexual performance is not improved by raising them with females or exposing them to females at the time of puberty.[230,1159] Rearing bulls in individual pens does suppress sexual performance.[1162] Bovine male courtship behavior is becoming an unobserved rarity in modern dairy farming with the advent and spread of AI. Bulls are now selected for (among other things) their willingness and ability to mount dummy cows, other bulls, or steers and to ejaculate into an artificial vagina. One might wonder how this willingness to leave behavioral interactions with the opposite sex up to random genetic drift eventually will affect the species. The courtship sequence is actually a series of reciprocal interactions between male and female. The female behavior has already been described.

Courtship Behavior

Starting late in the cow's proestrus, the bull will begin to graze beside the cow, guarding her from any other cattle. His attempts to mount will be repulsed by the cow. During proestrus, most females are attractive to the male and attracted to him but not yet receptive. The bull may attempt to drive the cow away from the herd.

Periodically, the bull will smell and lick the cow's vulva, often followed by the flehmen response (Fig. 4.6). Some, but not all, dairy bulls flehmen to estrous urine, and they do so more often than to mucus or to nonestrous urine.[694] Flehmen is followed by an increase in LH.[907]

As estrus approaches, guarding becomes more marked as both other bulls and nonestrous cows are kept away. When the cow is in full estrus, the bull will have a partial erection while guarding her, and accessory gland fluid or precoital discharge will drip from the penis.[791] The bull frequently nudges the female's flanks and either maintains head-to-head contact or, because of mutual genital sniffing and licking, stands in a reverse parallel (bigeminal) position with the cow. This position is common to the courtship sequence of all ungulates. The bull may rest his head across the cow's back while they stand in a T-position.[505] He makes several mounting attempts with a partial erection before the female will stand for him. When she is ready, the cow remains immobile and the bull mounts immediately. He fixes his forelegs just cranial to the pelvis of the female as he straddles her. Ejaculation occurs within seconds of intromission and is noted by a marked, generalized muscular contraction. The bull's rear legs may be brought off the ground during this spasm (Fig. 4.7). Dismounting and retraction of the penis follow rapidly. When bulls are used for hand breeding or for AI, lack of the stimulatory effects of the prolonged courtship may result in poor semen quality or poor reproductive behavior.

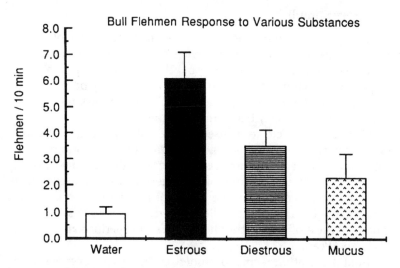

FIG. 4.6. The mean (± SEM) response of bulls to estrous and nonestrous urine, mucus, and water[694] (copyright 1989, with permission of Butterworth Publishers).

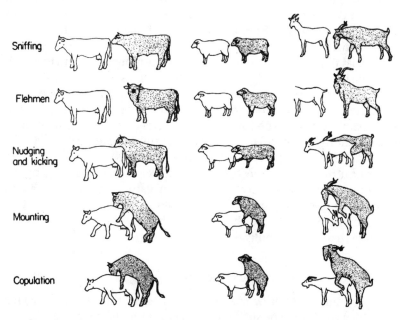

FIG. 4.7. Sexual patterns in cattle, sheep, and goats[586] (copyright 1974, with permission of Lea & Febiger; and Y. Rouger, unpublished data).

Sensory Stimuli

Experimental manipulations have shown that a variety of sensory stimuli are needed to elicit male sexual behavior. Tape recordings of cows will sexually stimulate a bull,[366] but these sounds are probably of a nonspecific arousal nature. Bulls used for AI apparently become classically conditioned (see Chapter 7, Learning) to the sounds in the collection arena, and these may help stimulate these animals.

Olfactory cues seem to be used by the bull during early estrus to monitor the stage of the reproductive cycle of the cow, however, no studies have been able to show sexual excitation caused by olfactory stimuli alone. Although volatile compounds have been isolated from estrous urine, these compounds do not elicit interest from bulls. Urine from a bull used as a teaser did elicit flehmen and other signs of interest, but none of these olfactory stimuli affected sexual performance of dairy bulls in an AI center.[1148]

Visual deprivation seems to hinder sexual response generally, but most markedly when the blinded individual is presented with a female in a novel situation.[592] The inverted U-shape seems to be the visual stimulus most likely to stimulate sexual behavior, whether this be in the form of a standing cow, a dummy, or an unfortunate person bending over. Without special conditioning, a wild bovid will only mount an estrous conspecific female. One of the aims of domestication has been to obtain "easy" breeders, that is, bulls that respond to a less specific series of stimuli than those offered by a female and that do so with little sexual foreplay.

Sexual responsiveness is influenced by the range of stimuli to which the male is exposed during ontogeny. Free-ranging or wild bovids gradually restrict their sexual behavior to interactions with females. Bulls not exposed to females do not establish this more narrowed range of sexual stimuli; hence, the ability of some rather bizarre stimuli to instigate mounting and ejaculation. Dairy bulls are also rarely raised with females, and so they can be stimulated by other bulls.

Malnutrition rarely affects sexual performance, as evidenced by the continued reproductive success of starving cattle in parts of Asia and Africa. This phenomenon has been experimentally demonstrated by Wierzbowski[1485] who found that underfed bulls actually were quicker to copulate than well-fed bulls.

Genetic and Experiential Factors

Individual bulls show marked differences in their levels of sexual behavior as measured by such parameters as mounts or ejaculation, the number of ejaculations per unit time, the latency to ejaculation, the number of ejaculations required for satiation, and the length of time to recover from postsatiation refractoriness. Although it is impossible to separate completely the genetic from the learned components of these differences, there is strong evidence that the differences between individuals as measured by such parameters are strongly genetic in origin. Work with monozygotic twins and triplets and comparisons between sires and sons show a high degree of similarity between these groups of related animals[132,588] (Fig. 4.8). European breeds seem more likely to mount an inappropriate sexual object (a male or anestrous female) than the Zebu breed, indication perhaps of a lower threshold in the European breeds. It seems unwise, therefore, to continue breeding a male that shows limited interest in mounting females, as doing so will only perpetuate the defect. Early experience and management, in addition to

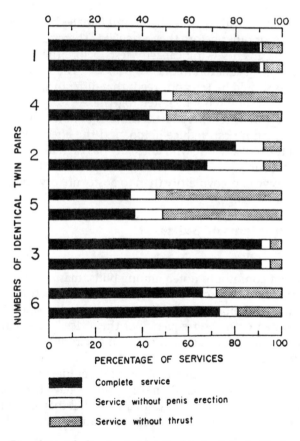

FIG. 4.8.   Ejaculatory behavior in identical twin dairy bulls showing percentage distribution of complete copulations, mounts without penis erections, and mounts without thrust[588] (copyright 1975, with permission of W.B. Saunders Co.).

genetic factors, are important determinants of sexual behavior. For instance, Zebu bulls raised in small groups had a faster reaction time (time to the first mount) than those raised on the range in large herds.[915]

Serving-Capacity Tests

Blockey[211-213] has developed a procedure for testing serving capacity (libido and copulation competence) in potential beef cattle sires whereby several bulls are tested for 20 minutes with heifers restrained in stanchions. The relative performance in this short test is well correlated with sexual performance in a 19-day pasture breeding test. Because dominance hierarchies are not strong in bulls under 2 years of age, they do not fight in the group breeding situation, but they can interfere with one another.[536]

Hereford bulls will consistently mount females by 9 months of age, but are a year or older before they ejaculate. Therefore, Hereford bulls should not be given serving capacity tests before they are 18-months old, when their ejaculation frequency peaks.[1160] The sex ratio in serving capacity tests should be 1:1, and breeds that differ in aggressivity (Hereford and Angus) should not be tested together.[1161]

Problems can arise with short-term tests of libido. A high environmental temperature can reduce libido, and the effect may not be the same in all the bulls, so one who normally has high libido may be more affected than one with low libido.[294] Other problems are that bulls are more responsive to preovulatory than to postovulatory cows,[536] so cows should be at that stage.

Clinical Problems of Bulls

Masturbation

Although commonly noted among bulls, masturbation causes no known reproductive problem. In general, masturbation is frowned upon on an anthropomorphical basis. Sperm quality or counts do not seem correlated with the frequency of this "vice." The bull performs pelvic thrusts, with his back arched, with a partially erect penis. Thus the penis moves in and out of the preputial sheath until ejaculation occurs. All bulls masturbate, especially at times of inactivity.[695] As shown in the following sections, all domestic animals have been observed to masturbate, and so it would be difficult to label this an abnormal behavior. See Fig. 4.9 for patterns of masturbation in dairy bulls.

Impotence

Loss of libido must be approached as a clinical problem with a rather large differential diagnosis in mind. The problem may be sec-

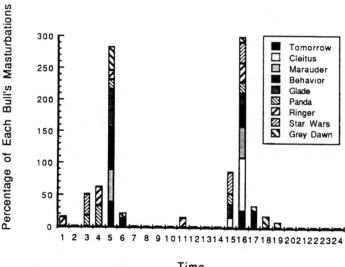

FIG. 4.9. The occurrence of masturbation by time of day. The masturbations observed are plotted by the hour of the day in which they occurred as the percentage of each bull's masturbations. Some bulls were observed longer than others, so it was necessary to use percentages rather than absolute numbers of masturbations[695] (copyright 1989, Elsevier Science Publishers).

ondary to almost any other organic disease. Bulls used for AI seem especially susceptible to musculoskeletal diseases, rendering them lame and unable or unwilling to mount. Obesity may be considered a pathological condition, and it caused many problems in the early development of the AI program. Obese bulls were difficult or impossible to arouse. Now the diets of bulls are more closely controlled; however, obesity should still be considered in assessing a case of loss of libido. Balanoposthitis or injury to the penis are specific conditions of the genital urinary tract that might be confused with a libido problem.[1203] Testicular atrophy involving both the Leydig cells and the seminiferous tubules may occur and be recognized by the reduced size and soft consistency of the testis. This is one case in which androgen administration might be expected to lead to a return of normal sexual behavior. A sperm count is recommended, however, as a hypospermatic animal will still be an unsuitable breeder.

Management practices may influence male libido. A large bull may be unwilling to mount a teaser steer or a cow in an icy corral or on wet concrete floors. One or two slips may be sufficient to condition him to ignore the stimulus of the teaser. Distractions of other animals or human observers should be kept to a minimum; this may be more of a problem in some bulls than others. By 2.5 years of age, most bulls are dominant over cows; before this, a dominant cow may prevent a young bull from mounting. Restraint of the cow may be of value if another, older bull cannot be substituted. Young bulls also lack experience, and their tentative approach while courting may stimulate aggression in a cow. If AI is not possible, and a proposed breeding is between animals far enough apart that one must be transported to the other, it is advisable to bring the cow to the bull. Males seem more sensitive to environmental factors; a new or different surrounding may trigger a reluctance to mount. Fortunately, it is also usually easier to truck a cow than a large, aggressive bull.

For reasons that must be classified as psychological, a bull trained to mount another male and ejaculate into an artificial vagina eventually may lose interest and refuse to mount, especially if he has been used too frequently. This inhibition may be overcome by changing the sexual stimulus to a new steer or bull. The stimulation of sexual behavior caused by a new female is called the Coolidge effect. A new teaser, preferably a cow and even better an estrous cow, or a moderate change in the environment or location of the mating arena may be sufficient to stimulate sexual behavior, but because of the risk of disease, cows are not allowed at AI centers. Other bulls are found to be more attractive teasers than steers. Although psychological in its origin, the Coolidge effect has a physiological limit. The maximum number of ejaculations will be the same as that obtained by electroejaculation.

Restraining the bull from the teaser or allowing him to watch another bull being collected also may arouse a previously disinterested male.[212,791,920] A caution regarding the latter technique must be made: a submissive or young bull being allowed to observe an animal dominant to him may be further conditioned against servicing through intimidation. In most ungulate species, copulations are restricted to the dominant males. The lack of sexual activity in the remainder of the male population has been termed psychological castration, although the effect is usually and quickly reversed if the dominant individual is eliminated.

The Buller Steer

Although most problems of sexual behavior in food-producing animals are those of insufficiency, manifestation of sexual behavior in steers is also a problem. Approximately 2% of feedlot steers are buller steers that are mounted by other steers. This occurs more often in steers that are implanted with stilbestrol or estrogen, but the level of hormone actually may be lower in the buller's plasma than it is in that of normal steers. There is a large component of dominance-related aggression in the buller syndrome. Penile erection and anal intromission rarely occur.

More aggressive animals mount, and the rate of mounting increases dramatically, when new steers are added.[813] The syndrome is seen most frequently when groups of animals are mixed, especially in crowded conditions. The economic losses resulting from the syndrome are due to the increased activity of the mounting steers and the harassment of the buller steer. None of the animals will gain weight as they should.[736] The usual means of treating bulling behavior is to remove the steers involved; however, electrified wire placed above the pens so that a steer that reared to mount would be shocked can be used to reduce the incidence of the behavior.[789]

SHEEP

The Ewe

Estrous Cycle

Sheep are short-light breeders, that is, they usually cycle in the fall of the year as the light phase of the photoperiod decreases. The ewe is polyestrous and will cycle several times during one breeding season if not bred; the average cycle for the ewe is 16 days (range, 14–20). The actual period of estrous receptivity is 30 to 36 hours in the ewe. Signoret[1322] has shown that breed differences in the duration of estrus in sheep exist even in estrogen-induced estrus in ovariectomized ewes. The duration of estrus in lambs and in the first estrus of the season, is shorter than that of the normal estrus. Puberty is dependent on the time of birth. Females born early enough in the year will cycle their first fall season indicating that puberty may occur as early as 4 months of age.

There are other environmental influences on the estrous cycle of sheep. Domestic sheep raised in the tropics and subtropics are nonseasonal breeders;[713] as with many other life forms in these areas, reproductive activity occurs all year. Breeds originating at higher latitudes and higher altitudes, such as the border Leicester, Scottish blackface, and Welsh mountain breeds, show a shorter breeding season than their more tropical or lowland cousins.[585] Thus, environmental input seems to temper genetic constraints.

Feral sheep display an even shorter breeding season than most of the domestic breeds, and it is likely that both predation pressures and the necessity of foraging over large areas for limited resources established estrous synchrony as an evolutionary stable strategy in the ancestral forms. In feral sheep, the adult rams form a flock separate from that of the ewes and lambs and only join the females during the breeding season. Flocking tendencies are strong in sheep, and a ewe lambing after the remainder of the flock would be unable to keep up with the movements of the flock.

The Ram Effect

Most sheep breeding is still done naturally, with the rams pastured with the ewes all year or introduced in the late summer. The introduction of the ram tends to synchronize estrus in a high proportion of the ewes 15 to 17 days later.[713,1387] Continuous exposure to a ram tends to increase the incidence of estrus in the normally anestrous period.[1198] The presence of a ram also shortens the duration of estrus, but this depends on direct physical contact with the ram and probably on mounting of the ewe by the ram, rather than on pheromones.[1112] Rams that court more vigorously cause more ewes to ovulate than less sexually active rams.[1125]

Estradiol-treated wethers can stimulate estrus in anestrous ewes, indicating that testosterone must be aromatized to produce the ram effect. The pheromone is present in the wool, wax, and antiorbital gland secretion of the ram.[821] The pheromone probably consists of more than one component.[304] Notice that in sheep the odor of wool rather than of urine, which is so important in rodents, appears to be involved in both sexual and maternal recognition. There is some effect across species. Although ram wool does not stimulate resumption of cyclic behavior in does, buck hair increased LH levels and thus induced ovulation in ewes.[1096] Farm animals do not appear to be completely dependent on odor, however; for example, the ram effect can occur even in ewes without the sense of smell.[305]

Olfactory cues are not necessary, but they may be sufficient to stimulate estrus, because the synchronization of the first estrus of the breeding season in sheep can occur in the absence of direct contact or visual cues.[27] Synchronization of estrus is a desirable property of the ewe cycle for management purposes, as breeding may be accomplished in a short time and the lamb crop will be born synchronously and tend to reach market size simultaneously.

Synchronization may also be accomplished by using progesterone compounds, and this procedure is used more as the practice of AI in sheep increases. Estrus occurs 48 hours after withdrawal of progesterone.[1404] Synchronization is more successful using these compounds if a male is added to the flock following the treatment.

Courtship Behavior

The estrous ewe can be identified because she follows or seeks out a ram, turns her head as the ram approaches, may circle to nuzzle his flank, squats and urinates, especially if he nudges her, fans or wags her tail, and stands to be mounted. The ewe may actively seek out the male and sniff the male's body and genitals and then thrust her head against his flanks. Estrous ewes will choose a ram on a two-choice test and will do so even on the basis of a photograph.[782,1398] She may call frequently with nonspecific bleats and be more active.[260] Standing occurs when the female is receptive, and she will look over her shoulder at the ram as he investigates and nudges her. Like cows, most ewes are in standing estrus at night.

The active role of the ewe in seeking the ram was demonstrated in an experiment in which two thirds of a flock of ewes were inseminated even though the rams were tethered.[892,895] The ewes apparently use olfactory cues to locate the ram because anosmic ewes were unable to find tethered rams.[477] Even when a ram and ewe are both free, ewes initiate half of the contacts between the sexes.[1092] Ewes can be attracted to a male in the next field or to an infertile male, so ram seeking does not guarantee pregnancy. Ram-seeking behavior seems correlated with estrogen levels,[894] but does not occur without the visual and olfactory stimuli of the male.

Competition involving agonistic behavior has been observed between estrous ewes over access to a ram. Older ewes are generally more successful in gaining access to the ram; otherwise, experience does not seem to be an important factor in ewe sexual behavior, but maiden ewes should be kept in separate flocks from experienced ewes or the latter will outcompete the former for the ram's attention.[910] Ewes seem to prefer rams of their own breed.[863]

The Ram

Sexual Behavior of Free-Ranging Sheep

The sexual behavior of the primitive and free-ranging Soay sheep may give some clue as to the optimal manner of raising breeding rams. Ram lambs remain with their dams from late spring when they are born until the following fall when they begin to chase ewes. They will be rejected by most adult ewes until they are larger. Meanwhile, the adult rams have been living in separate all-male flocks on separate feeding sites. Before the ewes are in estrus, the rams begin to move toward the ewe flocks. They will chase away strange rams but remain on friendly terms with their own flock mates. As the season progresses, however, they will begin to fight with one another. The fights consist of head clashings in these horned sheep and nudging behavior similar to that seen in courtship. All this takes place before the ewes are receptive, although young rams will have harassed the ewes already. When the ewes come into heat, the sexual contests between the males have been settled and the males can turn their attention to breeding. The winner of the most contests between males breeds the most ewes, but he does so sequentially by tending each ewe for a day, or at least half a day, grazing with her and performing the courtship activities described earlier in this chapter for domestic sheep.[577] The rams graze and rest less while courting and fighting and may lose weight. These observations indicate that the best way to rear breeding rams would be to leave them with females for as long as possible, certainly until puberty, and then put them in an all-male group.

The reproductive capacity of the ram is not seasonally limited as is that of the ewe. Estrous ewes may be satisfactorily fertilized at any season, although semen quality may decline somewhat during the spring.[1122,1123] Thus, the breeding season in domestic breeds is primarily determined by the environmental input to the female's hypothalamic–hypophyseal tract. Although the ewe may seek out the ram, courtship is more elaborate in the male than in the female.

Courtship Behavior. Like the female, the male spends a great deal of time sniffing the other's genitalia and urine; the flehmen response may be noted (see Fig. 4.10). As stated previously, the flehmen response

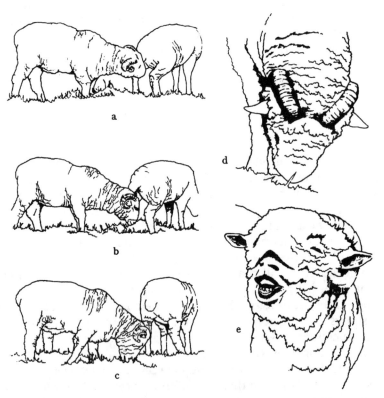

FIG. 4.10. Response of the ram to urine of an estrous ewe. *(A)* Urination by ewe; *(B–D)* ram nosing urine on the ground; *(E)* the Flehmen response[133] (copyright 1964, with permission of E.J. Brill).

has no visual communicative properties but may be a method of introducing material into the vomeronasal organ. Rams flehmen less frequently when ewes are in estrus, apparently because estrous ewes urinate less.[206] Urination appears to be a sign that the ewe is not in estrous. The male may also lick the female's genitalia, a form of tactile stimulation that may also be part of the testing procedure. Normally, rams can discriminate estrous from nonestrous ewe urine using olfaction.[210] Bulbectomized (no olfactory bulb) rams in a range situation show some difficulty in identifying estrous females; they are nonselective in the ewes they approach and test but are able to identify estrous ewes, presumably via visual and auditory cues.[890] In fact, confined rams may not use any sensory cue to detect estrus; they may rely entirely on the willingness of the ewe to stand. When given a choice of restrained ovariectomized ewes, rams sniffed, nudged, mounted, and copulated with ewes with equal frequency whether or not they had received estrogen (all had received progesterone).[1324] Ewes do differ in their sexual attractiveness to rams. This attractiveness is stable from estrus to estrus and depends, at least in part, on wool. Wooly ewes are preferred to shorn ones.[1397]

A ritualized kicking with a foreleg is performed as the ram orients himself behind the ewe; the leg is raised and lowered in a stiff-legged striking manner (see Fig. 4.11). Tongue flicking accompanies the nudging. Nudging is also stereotyped: the head is tilted and lowered while the shoulder is brought into contact with or oriented toward the flank of the ewe; simultaneously, the ram utters low-pitched vocalizations or courting grunts. Six nudgings of the ewe per minute are observed at the peak of the ram's sexual interest.[1385] Several abortive mounts may be made with pelvic thrusting but without intromission. When the tip of the glans penis contacts the vulvar mucosa, a strong pelvic thrust accomplishes intromission and ejaculation occurs immediately. After dismount, the ram may sniff the ewe's vulva, appear depressed, and stand near the female with head down; he may urinate.

The latency to ejaculation increases with the number of observed ejaculations during an observation period, but it is subject to large individual variation. The number of ejaculations that occur when a ram is presented with a large number of estrous ewes is similar to the number that occur in response to an electroejaculator, indicating that physical capability rather than libido is limiting. Males show a reluctance to remate a recently mated female, whether or not the test ram was the male who inseminated her originally.

If rams are mated to sexual exhaustion with one ewe (usually three to six matings),[716] rapid recovery may be obtained by introducing a new ewe, while a much poorer recovery rate is obtained by removing and reintroducing the original ewe.[161] This effect of novelty already has been described here as the Coolidge effect. Another method of demonstrat-

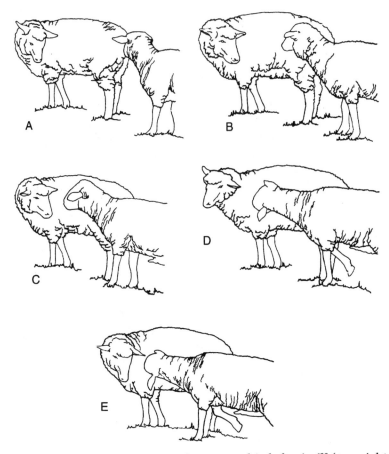

FIG. 4.11. Nudging sequence in sheep courtship behavior[133] (copyright 1964, with permission of E.J. Brill).

ing this phenomenon is to present four receptive ewes to a ram; he mates three times as much as he does when only one ewe is in estrus.[27] There is some evidence that habituation to a sexual partner may last several weeks after a mating as measured by the degree of recovery in the sex drive following satiation. These effects make evolutionary sense to the species and are fortunate for the rancher. A ram is capable of something less than 20 ejaculations daily. If many of these copulations were repeat breedings to especially persistent ewes, the fertility rate for the herd probably would be low. This breeding efficiency allows a rancher to introduce only a few rams to service a flock. In wild or feral sheep, a dominant male can spread his genes over a much larger proportion of the population than if the mechanisms for reduced repeat breedings by either himself or subordinate males did not exist.

Serving-Capacity Tests

Rams do not discriminate between ewes in natural estrous and ovariectomized ewes treated with progesterone and estrogen,[1149] so either may be used for serving capacity tests. Rams show similar rates of ejaculation and mounting whether ewes are restrained or free to move in a pen; however, if copulation is prevented (by covering the perineum), there are many more mountings and foreleg kicks as well as buttings.[1151] Foreleg kicking and sniffing the anogenital area of the ewe are considered courtship behavior, whereas attempts to mount, mounting, and ejaculation are considered copulatory behavior. The courtship behaviors are a good measure of libido because when copulation or mounting is prohibited males sniff and kick more if they have higher serving capacity (that is, ejaculations).[1150]

Prenatal Effect

There is an influence of the intrauterine environment on subsequent ram sexual behavior. Rams born cotwin to a ram have a higher serving capacity than those born cotwin to a ewe. The larger the litter size, the higher the serving capacity of its male members.[475]

Rearing Effects

It seems to be important for lambs to have heterosexual experience in the period between weaning and their first birthday in order for them to prefer ewes to rams as sexual partners. Ram lambs at 6 months will begin to exhibit adult sexual behavior sooner if they are exposed to estrous ewes for as little as 30 minutes per day for 4 days. By 8 months, exposure to estrous ewes does not result in an earlier onset of adult sexual behavior. This is important because the young ram lambs could be used for breeding before they are a year old. In addition to the immediate effects of exposure to females, there can be delayed effects. Exposure to estrous ewes for several hours at 10 months of age enhances sexual performance at 2 years in rams that were otherwise kept in all-male groups.[1152]

Dominance Effect

When more than one ram is being used to service a flock, dominance effects may influence the percentage of the flock each ram inseminates. Mature rams may dominate other mature rams and almost always will dominate yearling rams. Dominance orders, as determined by the number of butts each animal delivers when the animals are confined together, also exist and may be more marked between yearling rams. Rank in a food competition hierarchy corresponds to rank in the sexual hierarchy.[491] In a confined ewe flock, a dominant ram may copulate 12 to 15 times daily over the breeding season, whereas the subordinate ram(s) may average only two to five copulations daily.[712] The subordinate rams may service ewes; but this occurs mostly at times when pregnancy is not likely to result, that is, long after ovulation. In a feral population or a multiram pasture breeding situation, a number of rams may copulate with a ewe. The ram who copulates between the 9th to 15th hour of estrus is most likely to sire the lambs because that is when the ewe is most fertile.[756] In less confined conditions, the monopolization of mating by the dominant ram may be restricted only to the harem of ewes he is able to hold.[947]

Hulet[712] suggests using uniform age groups and allowing adequate space when groups of rams are to be used together in a mass-mating system. Three rams per 100 ewes or 20 to 50 ewes per ram has been suggested for a range breeding situation. It is unwise to use a pair of rams to breed a group of ewes, as agonistic activity may take precedence over mating behavior. Lindsay and Robinson[895,896] showed that three rams served 50% more ewes than a single ram over a 4-day period. It is unfortunate that dominance or sexual aggressiveness does not necessarily correlate with fertility; the combination of dominance with infertility in a flock ram could be disastrous to a rancher.

In pen breeding, a teaser, or vasectomized, ram is used to mark ewes by means of a harness and crayon attached to his brisket. A ewe that is in estrus will be mounted and, therefore, marked.[891] The marked ewes are then added to the breeding ram's pen. Problems can arise if the breeding ram is submissive to rams in nearby pens. He may be inhibited and not mount. Another problem is seen when young and mature ewes are added at once. Rams prefer or are monopolized by the older ewes. For most effective pen breeding, ewes should be checked for heat at 12-hour intervals and the breeding rams should be free of disturbance from rams or ewes in adjacent pens.[806]

The effects of the presence of the ram on the reproductive status of the ewe already has been discussed here. The ram is influenced in his turn by the presence of the ewe. Testosterone levels, testis size, and levels of aggressive behavior are all higher in rams in pens next to ewes than in isolated rams.[728] Rams, like ewes, prefer mates of their own or their dam's breed.[793]

The effects of castration in prepubertal rams are extremely variable;[133] there may be some decline in, or the complete elimination of, sexual activity. Postpubertal castration causes decreased sexual activity in rams,[27] but erection and intromission may persist for years postoperatively, albeit with a marked decrease in frequency.

Clinical Problems of Rams

Rams are often reared in large monosexual groups or individually, and Zenchak and Anderson[1533] have shown that the isolated males show some aberrations in behavior when first exposed to estrous females but most are eventually able to mate. The previously isolated rams were more proficient than rams raised in all-male groups.[258] Some of the rams raised in monosexual groups are sexually inhibited or show homosexual preferences and do so for long periods or permanently after maturity. In one study, after 9 days' exposure to estrous ewes, some rams (17%) still had not copulated.[715] The frequency of mounting in an all-male group cannot be used to predict sexual behavior in a mixed sex group.[1154] Continuous exposure to ewes eventually stimulates sexual behavior in most rams.

Zenchak and Anderson[1533] argue that some individuals in the male groups learn to associate courtship and mounting activities with agonistic behavior. When presented with a conflict, they may have used courtship and mounting instead of fighting to assert and maintain a social position. The repeated association of sexual and agonistic behaviors in maintaining a social rank may make it difficult for the ram to respond to the cues provided by the estrous ewe. Another explanation is that the inhibited rams learn to associate sexual behaviors with the odors of rams only. It seems reasonable, therefore, to suggest not raising rams in monosexual groups in order to avoid sexual inhibition or the development of homosexual preferences. Mixed groups are probably ideal, with isolated rearing less preferable but acceptable.[774]

GOATS

Free-Ranging Goats

During the rut, feral bucks join female groups and fight for access to the estrous does. The most successful goats are 5 to 6 years old. Older males select groups in which they are more likely to be successful, that is, groups with few other males whereas young males select the groups with the largest number of does.[408]

Collias[311] noted the close relationship between agonistic and sexual behaviors in male goats.

A detailed study of feral goats indicated that does permitted large, horned bucks to mount, but small, hornless goats were virtually excluded from breeding. When pursuing a doe the buck would gobble, a sound produced by moaning and rapidly thrusting his tongue in and out. Courtship activity would cease when the buck exhibited the flehmen response to a nonestrous doe's urine. Although large males were usually able to chase off rivals, occasionally many males would mount a female in quick succession without regard to their position in the dominance hierarchy.[1303]

The Doe

Goats, like sheep, are short-day breeders. Does in estrus show an increase in tail wagging, vocalization, urination, and mounting of other females. Does are in estrus for 39 hours once every 20–21 days.

The Buck

In general, billy goat or buck behavior is similar to ram behavior. The buck holds his tail straight up during courtship. A component of the mating sequence unique to the goat is the urination by bucks on their own forelegs and beards during courtship. This behavior is termed enurination. Occasionally, mouthing of the penis also occurs. Although various functions have been attributed to enurination, including increasing the intensity of the buck's odor and advertising his nutritional fitness, the behavior occurs most frequently in a situation of sexual frustration when the buck is restrained from mating. See Figure 4.7 for normal caprine sexual behavior.

Goats reared in a group of males from the time of birth have fewer agonistic interactions with other bucks at puberty. The dominant goats initiate homosexual behavior toward the subordinate ones. Homosexual behavior does not interfere with their ability to ejaculate into an artificial vagina in response to the stimulation of an estrous female.[1094]

Postpubertal castration causes a decline in sexual behavior in goats.[620]

Clinical Problems

The Doe

Silent Heat.   More silent heats occur in the spring.[291]

The Buck

Not all goats that mount homosexually in an all-male group will fail to mount heterosexually; those that do probably have a strong preference for one partner who happens to be male. The bucks that are mounted by other males are less apt to restrict their sexual activity to males. White bucks appear to be mounted more often than colored ones, which may be a function of the calm disposition of white Saanen goats rather than of color preference by the mounters.[1156]

HORSES

Free-Ranging Horses

Courting Behavior

Free-ranging mares will often seek out a stallion during estrus and display the normal behavioral signs of estrus before him.[462,1418] It is unusual for free-ranging 2-year-old mares to breed (1%) and even uncommon for 3 year olds to breed (13%).[1418] Stallions generally do not exhibit much interest in the sexual displays of young mares. Yearling mares may show very exaggerated signs of estrus and may attract males from bachelor herds. It has not been established whether the yearlings are in true or psychic estrus (see later here) because it has been most often observed in free-ranging horses. It is rare, but not unknown, for 2-year-old feral horse mares to deliver a foal, indicating that they can conceive as yearlings. It is hypothesized that the exaggerated signs of estrus in young mares may be a means of attracting stallions from a distance (that is, unrelated stallions). Incest is unusual in free-ranging bands; mares either leave the band when sexually mature, are "stolen" by a stallion forming a new harem, or the herd stallion is replaced by a young stallion. Those that remain in their sire's herd have a much lower foaling rate than those who join an unrelated male's herd. Those who join a half-brother herd have an intermediate rate.[779] Stallions base their avoidance of incest on familiarity rather than on kin recognition.[191]

Free-ranging stallions rarely acquire a harem until they are 5 or 6 years old. The years between the time they leave their natal herd at 2 and acquire their own mares is spent in bachelor herds. The stallion is the sire of foals born in his band in 85% of the cases. In the other cases, the mare has been bred by bachelor stallions or the stallion of another band.[773] In multimale bands, the dominant horse sires most of the foals based on electrophoresis of blood proteins. Dominant mares have dominant colts who sire more foals than foals of subordinate mares.[460]

The Mare

Estrous Cycle

Mares show the opposite seasonality in estrous pattern to that of sheep and goats; that is, they are long-light breeders and cycle in the spring. Foaling season is late winter and early spring, and the 11- to 11.5-month gestation period has dictated a rapid recycling and rebreeding if foals are to be born to coincide with the spring grasses. Most free-ranging horses are probably bred on the first, or at latest the second, heat after foaling.[1418] If breeding does not take place, estrus recurs approximately every 3 weeks. The average length of receptivity is 5 to 6 days. Breed differences in the length of estrus in different localities have been noted but are of uncertain significance owing to the probable interplay of genetic and environmental factors, methods of testing for the presence of estrus, and statistical analysis.[1454]

Modern management usually isolates the stallion from the mares. Teaser animals are used to determine the receptivity of the mare. The teasers are usually stallions that are introduced to a mare over some form of barrier; the mare is restrained and sometimes hobbled. A nonreceptive mare will react to the teaser's advances by squealing, striking, kicking, and moving away. During full estrus, the mare indicates receptivity by her immobility and by permitting the teaser to nibble her rump and withers. The mare exhibits a characteristic breeding expression, in which her ears are turned backward but not flattened and her lips are held loosely (see Fig. 1.6). She adopts a basewide squatting stance, urinates frequently, and rhythmically exposes the clitoris with a series of labial muscle contractions known as winking[85,92] (Fig. 4.12). Rectal palpation may be helpful to ensure that follicular development coincides with the behavioral aspects of estrus. The optimal breeding time is usually the 2nd or 3rd day of estrus.

Mares show individual preferences for stallions, and the preferences are influenced by the stallion's vocal behavior. The more the stallion

FIG. 4.12.   *(A)* The posture of the estrous mare. The clitoris is everted and the tail deviated. *(B)* The prance, or piaffe, of the courting stallion.[679]

neighs the more likely the mare is to approach him.[1139] The importance of vocal stimulation is demonstrated by playing a recording of a courting stallion's vocalizations and/or manipulating their genitalia.[954,1430] This makes it possible to detect behavioral estrus in the absence of a stallion. Exposure to male odor, a procedure that is effective in sows, elicits the signs of estrus in only half of the mares tested.

Foal Heat

Foal heat following parturition is quite predictable and ranges from 5 to 18 days, occurring usually 9 days after foaling. Regular estrous cycles usually continue thereafter at approximately 21- to 22-day intervals, although a few mares appear to have a lactational diestrus due to maintenance of the corpus luteum following the foaling heat. Other mares are cycling but do not show estrous behavior in the presence of a stallion. Instead, the mares protect their foals. Habituation to the stallion might result in normal estrous behavior. Many mares are bred on the foal heat because it is predictable and easily detected and planned for, although the endometrial epithelium is rarely restored or complete at foaling heat and usually does not accomplish restoration until days 13 to 25 following the heat. The 30-day heat following foaling is a safer time to breed; there is a higher conception rate.

Clinical Problems of Mares

Most of the sexual problems of mares occur at either end of the breeding season. Mares are seasonal breeders that come into estrus as daylight increases. Feral horses in the United States have a restricted breeding season. The foals are born in early summer, and the mares are bred on foal heat. No doubt, if all domestic mares were pasture bred during the late spring and summer, they would be increasing in number at the same rate as the wild horses. Mares normally are anestrus from October to February; however, mares can and do show estrus and conceive all year around, especially if artificial light is used to simulate long days and the mares are well fed and well sheltered. Despite our ability to manipulate the equine reproductive cycle, seasonality is expressed in the large numbers of abnormal heats observed in mares.

Split and Prolonged Estrus

It is at the transitional periods, the beginning and end of the breeding season, that such sexual abnormalities as prolonged estrus and split estrus are usually seen. The length of estrus can be prolonged from 5 to 7 days to up to 90 days. This may be accompanied by aggressive behavior. Other mares may show split estrus. Split estrus may consist of 1 to 2 days of "shallow" estrus in which the mare does not react to the stallion with vigorous squatting, tail deviation, and urination, but rather tolerates him and occasionally exhibits signs of estrus. A few days later, she may show strong signs of sexual receptivity. Both prolonged estrus and split estrus may be accompanied by active ovaries in which follicles are present but do not mature. In the fall, follicles may remain on the ovaries rather than regress, and this condition may be accompanied by complete anestrus or abnormal estrus. Mares that have either prolonged estrus or split estrus may begin to have normal cycles as the breeding season progresses and daylight length increases.

Anestrus

One of the most common reproductive problems of mares is anestrus. Mares may be either physiologically or behaviorally anestrous. The former includes mares with persistent corpora lutea and those with inactive ovaries. Older mares may have inactive ovaries due to degenerative changes in the endometrium and/or degenerative changes in the ovary itself. There may be senile changes in the ovaries, including the formation of germinal inclusion cysts. The age at onset of these senile changes varies from the midteens to well into the twenties.

A developmental cause of anestrus in mares is hypoplasia of the gonads and the reproductive tract. This condition, which is relatively rare, is similar to Turner's syndrome in humans because it is associated with an XO rather than the normal XX pair of sex chromosomes.

Although most pregnant mares do not show estrous behavior,[86] a few may. Hayes and Ginther[631] have shown that mares that show estrus while pregnant are carrying female foals. The sexual behavior is most likely to occur between days 35 and 40 of pregnancy, when accessory follicles are formed. The behavior of the pregnant mare can be distinguished from that of the nonpregnant estrous mare by the rapid tail lashing or wringing of the former as compared with the deviated tail of the latter. The pregnant mare will not usually stand for the stallion; she is not truly receptive. Her urine will be clear rather than cloudy, as is that of the estrous mare.

The common sexual behavior problems of mares are silent heat, psychic estrus, and excessive sexual behavior.

Silent Heat

Behavioral anestrus is also known as silent heat. Palpation of the ovaries may reveal normal follicles. Ovulation will take place normally. There is physiological, but not behavioral, estrus because the mare will not accept the stallion. One reason for silent heat may be related to the fact that mares do show mate preferences. These preferences should be considered in teasing a mare in silent heat. More than one stallion should be used. Environmental factors can influence the behavior of the mare just as they influence the behavior of the stallion. A mare may not show estrus if she has just been trailered to a strange stud farm or handled by a strange person. If the mare does not show signs of estrus even in a familiar environment and with exposure to several stallions, she may still conceive if she is artificially inseminated or tranquilized, restrained, and forcibly mated by the stallion.

Psychic Heat

Not all sexual abnormalities of mares are caused by a deficiency of sexual behavior; some result from an excess of sexual behavior. Some mares, for example, show estrous behavior without the normal physiological correlates of estrus. This abnormality is known as psychic heat. It may occur when any horse is brought into the environment of a solitary mare. In that case, it is usually a relatively short-lived phenomenon. Much more serious is psychic heat of performing mares. A mare that stops frequently to urinate and is attracted to stallions, geldings, or mares will not be a good competitive trail horse or show ring performer. Despite posture and behavior, the mare in psychic heat may not tolerate mounting by a stallion.

Progestins such as altrenogest 0.02 mL/kg orally or progesterone 0.4 mg/kg i.m. daily have been used successfully to treat mares that show severe psychic heat. The progestins are believed to act on those neurons in the brain, probably in the hypothalamus, that control sexual behavior. The independence of psychic heat from endogenous hormonal control is demonstrated by the failure of ovariectomy to relieve psychic heat in some mares. If psychic estrus persists in an ovariectomized mare, dexamethasone may be used in an attempt to suppress endogenous adrenal steroids that might be inducing estrous behavior, but there are serious side effects.

Excessive Sexual Behavior

There are two pathological conditions of the ovaries that can cause excessive sexual behavior in the mare. The abnormalities are granulosa cell tumors and persistent follicles. These conditions should be differentiated because one, granulosa cell tumor, should be treated by ovariectomy; the other, persistent follicles, usually resolves itself. The excessive estrouslike behavior of mares with persistent follicles is sometimes called nymphomania. The cysts will regress in time and do not need to be manipulated as follicular cysts of cows do. The persistent follicles may also be treated with gonadotropins, LH, or an increase in artificial day length to 16 hours or more. Mares suffering from granulosa cell tumor can show signs of constant estrus, but they are often more aggressive than mares in normal estrus.[521] Some tumors are virilizing, in which case the mares will attempt to mount other mares and will attack other horses of either sex. Other mares with a granulosa cell tumor may show anestrus or a mixture of changing signs as already described here. The affected ovary should be removed.

The Stallion

Stallions exhibit libido throughout the year, but show peak sexual behavior in the spring. Seasonal changes are also seen in sperm number and testosterone levels.[1140] Stallions with a harem have a higher level of testosterone than those in a bachelor herd, which in turn have a higher level than stalled stallions.[969] This indicates the importance of the social situation.

Sexual behavior may be seen in 2- to 3-month-old colts, with full penile erection during resting, play fighting, or mutual grooming;[1418] but the age at the first successful copulation varied from 15 months to 3 years in the same study. Tyler[1418] also noted that stallions actively prevented young males from mounting females.

Courtship Behavior

Courtship behavior will vary with the management practices involved. The following description assumes that the mare and stallion have free access to each other.

Driving, herding, or snaking with a distinctive head-down position is a behavior usually elicited by the presence of another stallion. Piaffe-like prancing is also a display to other stallions (see Fig. 4.12).

Intromission may only be achieved with a full erection, and this is correlated with the degree of sexual excitement. Thus, an adequate period of sexual foreplay is essential.[1454] Males may tend a female for several days before she is fully receptive. Nipping and nuzzling begins at the mare's head and proceeds gradually along the body of the mare to the perineal area. During this testing phase, he exhibits the flehmen response. As sexual excitement increases, the male calls with neighs and roars. The female allows the male to lick her around the rear legs and back. Full erection usually develops over several minutes in the mature stallion. Several mounts are usually made before intromission and ejaculation. During copulation the stallion rests his sternum on the mare's croup and may reach forward to bite her neck. Ejaculation occurs around 15 seconds after intromission and after approximately seven thrusts,[1400] and intromission lasts less than 45 seconds.

Postcopulatory tending was not noted by either Feist and McCullough[462] or Tyler[1418] in their study herds in Wyoming and England, respectively. The male may sniff the mare's genital area and exhibit the flehmen response, but the pair soon separates. Under test conditions, sexual satiation occurs after one to 10 ejaculations (average, 2.9).[196] Although stallions usually are limited to a few hand breedings per week, a 6-year-old Belgian stallion bred 20 mares in 9 days, with an 85% conception rate. Prostaglandin had been used to synchronize estrus, so 8 mares were in heat and were bred by the stallion on 1 day.[248]

Sensory Stimuli

Visual and sensory stimuli are probably vital to the display of sexual behavior; however, their importance may be modified through learning. Tyler[1418] believes that the visual stimuli of the mare's posture with raised tail are important for attraction and penile erection. This reaction may be generalized so that a dummy or phantom in the general shape of a mare will be mounted by a sexually experienced stallion. Inexperienced stallions will not mount the dummy.[1484] Experienced males will also mount a mare or dummy while blindfolded. The stallion is undoubtedly stimulated by olfactory information, but the stimulation may precede the copulation by several minutes to hours. Typically, a stallion will stop chasing an estrous mare to sniff and flehmen at the small volume of urine she has expelled. Odor stimulation of the vomeronasal organ may lead to an increase in LH and testosterone and consequently libido, so his behavior is synchronized with that of the receptive mare. Experienced males do not show any inhibition to mounting a mare or dummy when olfactory input is blocked, but inexperienced males will mount a dummy only when it has been sprinkled with urine from an estrous female.

Masturbation

Masturbation is normal behavior in a stallion. The stallion flips the erect penis against the ventral abdominal wall. Ejaculation rarely occurs. Stallions masturbate four times a day spending 38 minutes with an erect penis sometimes, but not always, accompanied by masturbation.[1399] This behavior usually occurs in the resting stallion, even one at pasture with mares available. Masturbation may occur in association with recumbency.[1487]

Clinical Problems of Stallions

Ten percent to 25% of stallions presented for breeding soundness examination have some behavioral problem. Those stallions most at risk are young and/or novice breeders, frequently bred stallions, those in a new environment, and those in transition from racing to breeding.[963] Young stallions appear to be particularly affected by exercise; as little as 30 minutes per day of lunging can decrease libido.[376]

Some common problems of sexual behavior in the male are as follows:

- stallions that show sexual interest in mares, but will not mount, or mount but do not ejaculate
- stallions that have low or no libido, that is, do not show interest in a sexually receptive mare
- stallions that will mount mares only when another specific horse is present
- stallions that injure, or "savage," mares or handlers
- stallions that self-mutilate
- stallions that mount but do not intromit
- geldings that behave like stallions

Sexual Experience and Decreased Sexual Behavior

Too much serious sexual experience too early is very detrimental to normal libido. Many stallions overused as youngsters are presented with sexual behavior problems. The most common problems are impotence or low libido. Other stallions that are overworked as studs may bite the mares viciously or be uncontrollable by their attendants. It is not always clear whether the young stallions have had traumatic or unpleasant experiences. They may remember being kicked by a mare or they may simply remember that they were exhausted. The resultant loss of libido can persist indefinitely unless treated.[1141]

A typical case is that of a 4-year-old Arabian stallion that showed different behaviors to different mares. He mounted without erection and bit a Morgan mare with whom he had been housed as a 2 year old. His sexual behavior was normal toward another mare with whom he had had no previous contact. He had been noted to display snapping (see Chapters 1, Communication, and 6, Development of Behavior) to the Morgan mare as a youngster and presumably was subordinate to her. As an adult he showed aggressive behavior toward her, perhaps because of an approach–avoidance conflict. He was sexually stimulated, but because she was dominant, he did not want to mount. His displacement behavior was aggression.

Stallions that are used as teasers, that is, to detect estrus in mares, but are not used for breeding may eventually show a loss of libido. In addition, they may show stereotyped behavior, such as stall weaving. It has not been determined how often a teaser stallion should be allowed to copulate to prevent these abnormal behaviors, and there are, no doubt, individual differences in response to use as a teaser.

Stallions can learn to inhibit sexual behavior as easily as they learn to express it. Stallions may be fitted with stallion rings, devices placed on the penis that cause discomfort if erection occurs. They are used on stallions that are in training or in other circumstances in which sexual behavior would be inappropriate. Many stallions apparently can be fitted with these devices and learn not to respond sexually when wearing them and yet respond normally when the rings are removed. Other stallions have learned too well and are impotent even when the ring is removed. Stallion rings and belly brushes are also used to prevent masturbation, which, as already noted, should not be considered either abnormal behavior or a cause of infertility. Ejaculation rarely occurs, so the behavior is unlikely to lead to a drop of fertility; attempts to punish masturbation, however, do cause libido problems.

Physical Impairment

The treatment of any behavior problem must begin with elimination, or at least identification, of any physical problem. The two most common physical problems associated with breeding are genital injury or limb injury. Any lameness or limb injury will inhibit the stallion's ability to mount, so he may exhibit normal libido and penile erection but will not mount. An older stallion with navicular disease or chronic arthritis is a typical example of the effect of organic limb disease on sexual behavior. When pain is the cause of libido problems, nonsteroidal anti-inflammatory agents such as phenylbutazone and flunixin meglumine may be of value. Another factor that may be responsible for failure to mount is improper flooring. If the flooring is slippery, for instance, the stallion will be reluctant to mount, especially if he has fallen when trying to mount a mare on other occasions. He is much more likely to mount if taken to an environment with a different substrate such as grass or tanbark. A stallion with a mild locomotor problem may mount but be reluctant to ejaculate in cold weather. In warm weather, when he is pain-free, he will be normal.

Injury to the genitalia can be a cause of breeding difficulties. Naturally, a stallion will avoid intromission if his penis is painful. Stallions may be reluctant to copulate long after the injury is apparently healed because they may not have learned that copulation will no longer be painful. Any impairment of blood flow to the penis may produce behavior problems such as ejaculatory failure.

Stallions with physical impairment of the legs or back should be mounted on secure mounts, that is, sturdy mares. The flooring should be adequate, and the mare should be the correct size so that the sternum of the stallion rests on her croup. Anatomical fit is important because even normal stallions may lose their balance and slide off a mare that is too small or too large. Physically impaired stallions can be trained to an artificial dummy mount that is secure when AI is permitted for the breed. At first, an estrous mare may have to be held next to the dummy, but stallions, like bulls, can be conditioned to ejaculate in the absence of the normal stimulus of the mare.

Breeding Environment

The total breeding environment must be considered because another cause of injury to the stallion can be a low roof or an overhang. A stallion rearing on his hind legs to mount a mare is considerably taller than when he is standing on all fours. If a stallion strikes his head while mounting he may not only sustain serious injury, but may also be inhibited from mounting on subsequent occasions.

Perhaps the most important aspects of the breeding environment are handlers. Handlers must be familiar with the routine of breeding and experienced in controlling horses. Some handlers are better able to calm stallions than others with equal experience, and the calmer the stallion the less likely are accidents. Such simple arrangements as placing all attendants on the same side of the mare and stallion can facilitate communication between them and prevent difficulties. Most breeding injuries and accidents occur when an inexperienced or highly nervous mare or nonreceptive mare is bred without adequate restraint or judgment. Because errors in detection of estrus can be made and because the situation is unnatural, mares should be hobbled before hand breeding. Distractions in the form of extraneous people or animals should be avoided.

There is no single treatment for all abnormal sexual behavior in stallions. Patience and time are necessary with almost all cases. It is advisable to take advantage of the stallion's seasonal breeding pattern and institute behavioral therapy during the spring and summer. Advantage should also be taken of the stimulatory effect of the presence of other stallions. It already has been noted here that wild stallions are most likely to copulate when other stallions are present; the same appears to be true of domestic stallions. Some stallions have been known to breed mares only if another horse, even another mare, is present. The stallion may regard the second horse as a competitor—another stallion—or there may be another, unknown reason.

The antianxiety drug, diazepam (0.05 mg/kg slowly i.v.), has been used successfully to overcome impotence caused by pain associated with breeding and for the loss of libido shown in a novel environment.[967,968] Imipramine (500 mg i.v.) has been used to treat stallions that will mount and intromit but not ejaculate.[964] Gonadotropin-releasing hormone may act directly on the brain of horses to stimulate sexual behavior[965] and could have clinical application.

Finally, some stallions have definite mate preferences, and they should be allowed to exercise these preferences while recovering from loss of libido or impotence. Tease the stallion with several mares and use the one to which he is most responsive for further treatment. A quiet mare is necessary for a stallion that has been injured by another mare. A stallion that will not mount a mare may ejaculate into an artificial vagina. He may gain confidence and overcome his fear by this process and can later be induced to mount a mare.

Vicious behavior toward the mare and the attendants, like most other abnormal sexual behavior, is most apt to occur when stallions are used for breeding outside the normal breeding season. Therefore, stallions may be unmanageable in January, but well-mannered by May.

Overuse and Rough Handling

Overuse and rough handling are often the cause of misbehavior of stallions during breeding. They may bite the mare or be generally intractable. Attempts to improve the horse's breeding manners should be delayed until normal libido and copulation have been reestablished. Punishment of a horse with sexual abnormalities will retard its progress. If the stallion's viciousness is not attenuated as his libido improves, various physical devices, such as a muzzle and breeding bridle for him and a withers protector for the mare, may be used.

Self-Mutilation

Self-mutilation is a very common behavior problem. Although it occurs in horses of both sexes, it is much more common in stallions.[383] The behavior consists of biting at or actually biting the flanks or, more rarely, the chest. The horse usually squeals and kicks out at the same time. The signs mimic those of acute colic, but can be differentiated because self-mutilation does not progress to rolling or depression and is chronic. The cause of the behavior is unknown, but because it usually responds to a change in the social environment, it is probably caused by sociosexual deprivation. Most breeding stallions lead deprived lives in that they are kept in stalls in isolation from other horses, particularly from other mares; however, most do not self-mutilate. The question arises as to whether stallions that self-mutilate should be used for breeding. Castration usually, but not always, stops self-mutilation. Preventing the behavior with the use of cradles and side poles does not remove the cause; the stallion will continue to vocalize and kick, so although he can no longer injure himself, he can still injure a bystander. Sometimes providing a stall companion such as a donkey will reduce the incidence of self-mutilation. Allowing the stallion to live on pasture with a mare will eliminate the problem in most cases. The chances that the stallion will be injured by the mare are less than the chances that he will injure himself or someone else by self-mutilating. Opiate antagonists will prevent self-mutilation.[385] Unfortunately, naloxone, the antagonist now available, is metabolized very quickly by horses and is quite expensive. (See Cribbing under Clinical Problems in Chapter 8 [Ingestive Behavior: Food and Water Intake] for a discussion of the involvement of endogenous opiates in equine "vices.") Simply reducing the grain in a stallion's diet and increasing his exercise and roughage can eliminate self-mutilation.[958]

Effects of Castration

A horse that exhibits stallionlike behavior could be either a cryptorchid from whom the undescended testicle was not removed at castration or a gelding in which sexual behavior persists. A negligible plasma testosterone will distinguish the gelding from the cryptorchid stallion.[533] The testosterone response to gonadotropin administration is the best test for castration. Sexual behavior persists in over one third of geldings.[897] The sexual behavior may be as innocuous as exhibition of flehmen or as extreme as mounting and intromission. The sexual behavior itself is usually not a problem, but aggression directed toward other geldings by the one who is acting like a harem stallion is. Another unwelcome stallionlike behavior is attacking foals, particularly newborn foals. Management can be used to prevent these problems. A gelding that acts like a stallion should be stalled alone or pastured only with other geldings and should not have access to foals.

Geldings may also self-mutilate. These are usually geldings that are displaying other stallion behaviors but unlike intact males, they self-mutilate in the presence of mares. Stall confinement or pasturing without visual contact with mares usually reduces the incidence of self-mutilation. If not, progestins may be used.

PIGS

The Sow

Like the cow, the sow is a nonseasonal breeder. Once regular cycling commences, the sow will cycle every 18 to 24 days (mean, 21 days) until bred. Puberty occurs at 5 to 8 months. The presence of a boar leads to the occurrence of estrus at an earlier age and in more gilts.[1388] Puberty is accelerated in gilts older than 160 days exposed to a strange male for 20 or more minutes per day.

Estrous Cycle

Like females of other species, the sow shows an increase in activity as estrus approaches.[42] Urination is frequent, as is calling to the male. The call is a soft, rhythmic grunt. An estrous female approaches the boar and sniffs him around the head and genitals. Estrous sows attempt to mount other estrous females. The increased motor activity eventually takes the form of searching behavior, which seems vital to the initial uniting of an estrous sow with a boar.[1325] Olfactory stimuli alone will instigate this searching; anesthetized boars readily attract estrous sows. Olfactory bulbectomy drastically impairs the ability of the sow to discriminate between males and females.[27] Signoret and Mauleon[1326] have reported that bulbectomy also eliminates sexual behavior and prevents normal ovulation and estrus. This has not been confirmed by Meese and Baldwin,[991] who found that bulbectomized females mated, conceived, and reared their litters, although there were deficits in maternal recognition (see Chapter 5, Maternal Behavior).

This proceptive behavior can be used to detect estrus in sows using electronic monitoring to detect which sows visit a boar housed in a pen adjacent to the sow. Boars and sows show individual differences in mate selection; each animal has its favorite or favorites.[1379] Boars differ in the degree to which they attract sows, but this attractivity is not related to their libido.[649]

Searching behavior appears to be under endogenous control and requires estrogen during behavioral ontogeny for full development. Gilts reared in isolation will show this behavior upon reaching puberty.[1325] The immobility response of the fully receptive sow, however, seems to require both tactile and olfactory stimuli. The specific auditory stimulus is the courting song of the boar (see later in this chapter).[1325] The immobility reaction, like searching behavior, does not seem susceptible to learned modifications. The olfactory stimuli to which the sow responds are pheromones present in both the saliva and preputial secretions of boars. The chemicals involved are metabolites of androgens and have been identified as 5α-androst(16-ene)3-one.[999] These compounds have been used experimentally[1186] and are available commercially to elicit the immobility reactions.

Olfactory and auditory stimuli from adult boars, supplied by an aerosol spray and a tape recorder, will increase the proceptivity of gilts toward young boars.[711] The presence of a boar has a slight effect on the rate of conception and number of piglets or the size of the litter.[647]

Estrus is less apt to be detected in sows with less than 1 m^2 (11 ft^2) of pen space and those living in pairs. It is also easier to detect estrus if the sows are housed across an aisle from a boar rather than in an adjoining pen. This may be because the sow has had close olfactory contact with the boar when the attendants were not present; she has shown the immobility response, but no person was there to notice.[644,646,647,650]

Clinical Problems of Sows

Perhaps, because of the somewhat rigid genetic control of sexual behaviors, aberrations are unusual. Breed differences in the length of estrus are seen,[1323] but these are minor and unimportant clinically. Although not quantified, some sows seem to have decided mate preferences and display strong aversions to specific males.[1325]

Failure to Reproduce in Confinement

The most important problem is failure of reproduction in confined gilts. Confinement and the social environment appear to play a role in inhibiting estrus in young gilts.[485] Puberty is delayed in regrouped or crowded pigs or roughly handled pigs,[138,300,644] but accelerated by gentle handling. The stress of trailering can also stimulate the onset of estrus. It is interesting that chronic stress of overcrowding delays puberty, but the acute stress of transport accelerates it.

The Boar

Courtship Behavior

Once contact with an estrous female has been made, the boar will pursue the female attempting to nose her sides, flanks, and vulva (Fig. 4.13). Unique to the pig is the boar's "courting song," which is used during this phase of courtship. This is a series of soft, gutteral grunts, about six to eight per second.[1325] Tactile stimulation of the female continues and increases in intensity as the boar's sexual excitement increases. The boar usually emits urine rhythmically; pheromones in the urine may further increase the female's willingness to stand. Several mounting attempts may be made until the female becomes immobile, after which mounting and intromission follow rapidly. The boar's ejaculatory time approaches that of the dog, although no copulatory lock occurs in swine. Ejaculation occurs within 3 to 20 minutes, with an average of 4 to 5 minutes.[1325] Consort behavior continues for a short time following copulation. Burger[263] observed that boars mated an estrous female around 10 times over a 2- to 3-day estrous period. Although domestic pigs are not considered seasonal breeders, libido and testosterone levels increase earlier in prepubertal boars if the day is artificially lengthened to 15 hours.[668] Boars tend to have greater libido in a separate mating pen than in their home stalls.[648]

Olfactory Stimuli

Olfactory cues seem unimportant in stimulating a boar to mount a female or a dummy. Olfactory bulbectomy, for example, does not pre-

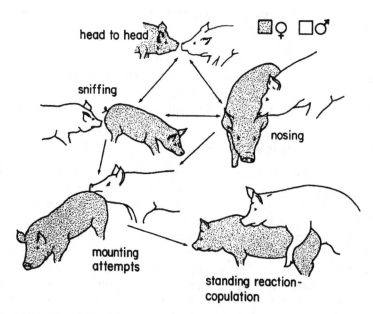

FIG. 4.13. The courtship sequence of pigs[1324] (copyright 1975, with permission of W.B. Saunders Co.).

vent normal sexual behavior.[222] Some, but not all boars can distinguish between estrous and anestrous females from a distance, that is, on the basis of olfactory information.[979] The initial contact with the sow, however, triggers a behavioral response from her: immobility, reflecting the degree of her sexual receptivity. It is this tendency toward immobility that seems to arouse the boar. Like the female response, this reaction is under fairly strict genetic control and not subject to much learned modification. Thus, what is noted by human observers as homosexual mounting or aberrant mounting of artificial stimuli is explained as being a normal response to immobile objects of approximately the correct size and shape. Although some breeds are easier to train for semen collection than others (for example, Yorkshires easier than Durocs[27]), training of a young boar to mount a dummy will usually be successful on the first few attempts.

Early Socialization

In the boar, as in most species, the early social environment is important to later sexual behavior. Boars raised in isolation from 3 weeks of age copulated less often with estrous sows than boars raised in groups. Boars with visual and olfactory contact with other boars were not nearly as inhibited as the isolates, indicating that contact with other pigs, even male pigs, is important.[647] Later, contact with females is also important because boars with visual and olfactory contact with sows copulated more often and ejaculated longer than boars kept in isolation or with visual and olfactory contact with other boars.[645,651] This is probably a hormonal effect; testosterone and corticosteroid levels are higher in boars that are in contact with sows.[898]

Clinical Problems of Boars

As with the other domestic animals, differences in the level of "sex drive" appear to be larger between individuals than between breeds. Low libido, however, has been associated with a high plane of nutrition; and, at least in the United Kingdom, is seen more frequently in Landrace than in boars of the large white breed.[56] Libido may be impaired through mismanagement of a young boar. A young male turned in with a group of gilts may be frustrated by excessive curiosity or bullied by the gilts. This incompetence or fear may become conditioned and a permanent problem. Supervision of early matings is recommended. A quiet sow or one recently serviced by a mature boar should be used for the first mating.[56] Another common problem is aggression by the boar toward humans. This is usually resolved by culling the boar, which removes the danger and prevents an aggressive animal from reproducing.

DOGS

The Bitch

Estrous Cycle

The domestic dog, unlike most of its canid relatives, is a nonseasonal breeder. The length of each estrous cycle is extremely variable from individual to individual and sometimes from one heat to the next in the same bitch. From one to four cycles yearly may be seen, with two being most usual. The basenji is an exception; one seasonal breeding per year is seen in the early fall.[526] Basenji-cocker spaniel crosses show both monocyclic and polycyclic activity, indicating genetic control of this aspect of the reproductive cycle. The onset of puberty also varies widely among individuals. No strict correlation of age at puberty may be made with either body size or conformation;[497] generally, however, the smaller breeds reach puberty earlier than the larger breeds. It would be very rare for a St. Bernard to reach puberty by 6 months, for example, but not unusual for a miniature poodle to do so. Thus, puberty onset for all dogs ranges from 6 to 15 months, with 7 to 10 months being the usual for the "average" dog.

Courtship Behavior

The correlation of hormonal levels and sexual behavior in the dog are illustrated in Figure 4.2. The first proestrus and estrus of a bitch's life is shorter than subsequent ones and the levels of LH and estradiol are lower.[288] She is less attractive to the male and less proceptive.[547] Courtship behavior is marked by play behavior in the proestrous part of the cycle, but this play behavior decreases during estrus. The female will run with the male, approach him using the typical play bow of puppies, and even whimper submissively. She will sniff and lick the male's body and genitalia. Urination becomes more frequent as estrus approaches and the posture used will frequently be the squat-raise (see Fig. 1.11). She may stand before the male momentarily during proestrus, but turns before the male can mount, often with a bark or growl.[295] Attraction of males and proceptivity appear in proestrus, but receptivity occurs later, during estrus.[154] During estrus, she stands more quietly to allow male investigation and eventually intromission toward the end of estrus. When the male touches her vulva, she will flex her body laterally;[609] while he is thrusting, she will move her perineum from side to side and ventrally, a motion that increases the probability of intromission. After the lock or the copulatory tie has been established, she may roll or twist and turn (the copulatory lock is discussed in the next section). Contractions of the constrictor vestibuli muscles and the anus occur as an afterreaction.

If the male does not mount, the female will "present" her hindquarters to him and even back into him and deviate her tail. An older and more experienced bitch may mount a young male and execute pelvic thrusts.[497]

Other social behaviors, in addition to sexual ones, are influenced by the reproductive condition of the bitch. Dominance relationships between females may shift, especially during metestrus. Males may defer to females in food competitions not because they are chivalrous males, but because they are more motivated to mount than to eat.

Courtship in Free-Ranging Dogs

Stray bitches avoid their male littermates but can be bred by a persistent brother. Estrous females attract two to seven males; they will show less proceptive behavior in the presence of many males and also less active rejection of nonpreferred males.

Clinical Problems of Bitches

Owners unfamiliar with canine courtship may be upset because the bitch appears to tease the male by soliciting and then threatening him if he mounts, but this is normal. Females may refuse males for any of several reasons. A bitch may display dominance over a male by not allowing the male to "stand over" her (a normal canine signal of dominance) or to approach from behind. Dominance relationships are learned, but can be established rapidly. Thus, dominance relationships probably help inhibit mother–son matings and may prevent certain sib–sib matings,[497] but may also develop quickly if an aggressive bitch is placed with a more submissive dog. Le Boeuf[860] and Beach and Le Boeuf[159] demonstrated definite female preferences for certain males. Refusal can range from avoiding a particular male to actively chasing and biting him. Not all the females rejected a male to the same degree, and Beach[151] showed that dominant males were not necessarily chosen as preferred mates. Beach also found that sexual preference could not be correlated with social affinity outside the mating period.[151]

The Dog

Sexual behaviors may appear in 5-week-old male pups, and mounting behavior becomes an important part of the male's social repertoire as it matures. As with many other mammals, mounting is used as a sign of dominance; a submissive animal will stand for a more dominant male, but standing over is not tolerated by the dominant animal. Fox and Bekoff[497] point out that most dogs are sexually mature physiologically long before they copulate for the first time. Perhaps the lack of dominance in young dogs inhibits early mating. Social contact is vital in the ontogeny of normal sexual behavior. Dogs raised in social isolation showed abnormal mounting orientation that persisted for longer than it did in dogs with similarly limited sexual experience but more social experience.[149]

Courtship Behavior

Male dogs are attracted to estrous bitches.[155] Urine of the estrous bitch appears to be more attractive to the dog than vaginal secretions,[393,406] but a component of the vaginal secretions, methyl *p*-hydroxybenzoate, has been shown to induce male sexual behavior when applied to the vulva of an anestrous bitch.[563] The pheromone may be considered a "releaser" of sexual behavior in the male. Mammalian behavior is not as stereotyped as that of fish and birds, so although sexual arousal may occur in all male dogs exposed to the pheromone, the expression of that arousal may vary considerably; therefore, male courtship behavior is extremely variable. Males may show extreme interest or indifference to females, although mating may occur successfully in either case. Play behavior may be marked or absent. The male sniffs the female's head and vulva; he may lick her ears. While canids do not show the classic flehmen response of ungulates, it is possible that the "tonguing" response seen during this olfactory investigation accomplishes transport of pheromones to the vomeronasal organ in a manner similar to that postulated for ungulates. The length of the play activity and olfactory investigation probably varies with the past experience of the male and female, perceived degrees of dominance within the pair, stage of estrus, and sexual satiation of either partner. Bitches can be forced to accept copulation by a strong and aggressive male who chases her until she is exhausted and holds her with his teeth by the neck or with his paw on her back. Only half of the stray male dogs are able to copulate, and those under a year of age rarely do so.[547]

The male mounts in response to female immobility; he grasps her with his forelegs just cranial to her pelvis. He thrusts with his pelvis and when intromission has been achieved, the rate of thrusting increases. Engorgement of the bulbus glandis and contraction of the vaginal muscles following intromission result in the copulatory lock or tie, a phenomenon most closely associated with canids, but not restricted to them. The male will usually dismount and turn around so that male and female are facing opposite directions while ejaculation occurs (Figs. 4.14 and 4.15). The lock may last 10 to 30 minutes (mean, 14 minutes[608]), after which, the bulbus decreases in size and the pair separates. Following copulation, recovery from sexual refractoriness may be rapid. Fox and Bekoff[497] report records of up to five copulations by a male dog in 1 day.

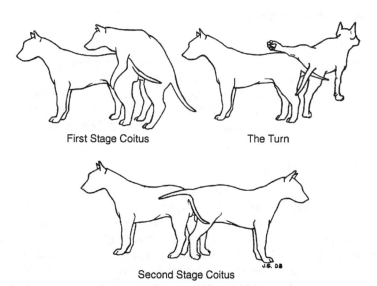

First Stage Coitus   The Turn

Second Stage Coitus

FIG. 4.14.   Coital positions of the dog[566] (copyright 1972, with permission of *Vet. Rec.*).

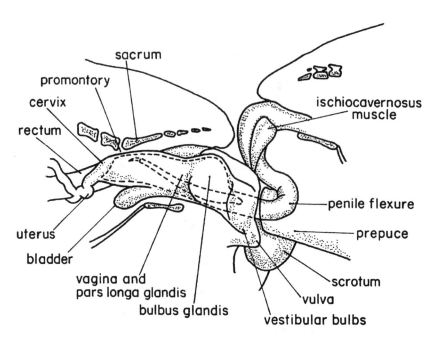

FIG. 4.15.   Relationship of the male and female genitals during the copulatory lock of the dog[566] (copyright 1972, with permission of *Vet. Rec.*).

Clinical Problems of Dogs

Impotence

Male impotence or loss of libido can be the result of organic disease. Most commonly this would be musculoskeletal disease such as hip dysplasia, arthritis, or trauma-induced pain in the hindquarters. Balanoposthitis is generally a mild disease in dogs and unlikely to affect sexual performance, although a severe form could conceivably do so.

Lack of Socialization

The lack of sufficient social contacts as a puppy may inhibit successful copulation. This becomes a very real problem, not only an experimental one, when a pup is ordered from a pet shop. Pups are weaned as early as possible (4–5 weeks) and shipped shortly thereafter to pet shops. The pup, if he survives transport, is kept isolated in the new owner's home, protecting him from infectious disease. When the owners finally do try to use their sheltered pet for breeding, they have great difficulty persuading the dog to perform or find it impossible to do so. Their dog has been essentially isolated from social contacts with conspecifics. Not only have the motor patterns of sexual behavior not been perfected, but they have never been placed in a proper social context. Some dogs may overcome this void in socialization, but their sexual and other behaviors may never be normal in direction or quantity. An example of poor libido is a pointer who was kept with his sister. Both the sister and his owners reprimanded the dog for sexual interest in his sister and when presented with an unrelated estrous bitch he had no libido and seemed frightened. With increasing age and exposure to other bitches, his libido increased.

Timidity

Timidity, especially in poodles and German shepherds, may be both learned and genetic in nature. Affected dogs may be inhibited to the point of impotence. Leaving the male and female together for several days, rather than allowing only a short breeding period, has been suggested. This prolonged period of socialization could, however, exacerbate a potential problem if the female is dominant.

A conditioned fear of any phase of the breeding program may be inadvertently instilled in a stud dog. Analysis of breeding techniques and reversal of the conditioning may alleviate the problem successfully. Thus, a young dog forced to court a very aggressive female may associate the rough treatment he received with the breeding process in general, with a specific breeding location, or with a specific color or type of female. Young dogs may make some clumsy mounting attempts and may otherwise prolong courtship, but should successfully mount a bitch within the first few exposures to a female. If the dog is a persistently timid breeder, however, he should be dropped from the breeding program. Artificial insemination is recommended if the timidity is suspected to be learned, or the result of the particular dominance relationship in the attempted mating. Most dogs will breed with most other dogs; therefore, mating problems with behavioral etiologies are unusual and require serious consideration.

Environmental Disturbances

Like other male domestic animals, dogs are sensitive to environmental disturbances in the breeding process. If one member of a breeding pair must be transported to the other, the female should be brought to the male. Noise and other disruptions in the breeding area should be minimized. The rather curious insistence of some breeders on helping a male dog mount and copulate might actually cause more difficulties than it is thought to prevent. Besides the physical disruption of the observer-helper, the breeder will likely be dominant to the dog and, thus, be somewhat inhibitory to the male's sexual performance. A timid dog may require the owner's presence if his dominance over a female is doubtful; but, as already discussed here, the continued use of a dog this timid would be unwise. Slippery floors, such as waxed linoleum, may prevent mounting, but should be a problem easily avoided.

Masturbation

Masturbation is not an unusual problem in house dogs. Semen quality or value as a breeder is not affected, but the habit becomes an embarrassing or annoying one for the owner. Masturbation using inanimate objects is probably seen in most puppies, but will become an insignificant behavior in the normally socialized adult. Although mounting can be an aspect of sexual behavior, it is also a sign of dominance, so mounting of people should be discouraged. To resolve the problem, the owners should teach the dog submissive behavior by counter-conditioning the dog to down stay when it attempts to mount, as well as consistently punishing mounting. Castration may eliminate or decrease the problem (see the next section). Owners reporting homosexual behavior in their dog should also be informed that mounting of one male dog by another is probably a sign of dominance.

Sexual Behavior After Castration

Prepubertal castration greatly reduces sexual interests. As mounting is an integral part of the dog's behavioral ontogeny and is used in agonistic interactions, all sexual behavior will probably not be eliminated. Hopkins et al.[674] studied the postoperative effects of castration on 42 postpubertal dogs in normal home situations (Fig. 4.16). They found that 90% of the dogs castrated to control roaming showed a rapid or gradual decline in this behavior. Intermale aggression was reduced noticeably in only 60% of the dogs and urine marking in the home, only 50%. Sixty-seven percent of the dogs showed a decrease in mounting behavior following surgery; mounting of people was reduced in seven of eight dogs castrated specifically for that problem. Mounting of other dogs was reduced in only one of four dogs castrated for that reason. The authors do not report any changes in the owners' handling or attitude toward their dogs following surgery, however, which might have been responsible for some of the behavioral changes noted. Age at castration was not correlated with the noted effects. Although in no cases did the occurrence of an objectionable behavior return to its preoperative level, some behaviors appeared intermittently for long periods following castration. Hart[608] reported that castrated dogs may retain sexual mount-

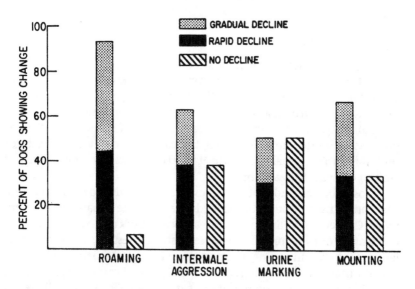

FIG. 4.16. Percentage of dogs experiencing rapid decline, gradual decline, or no change in four behaviors after castration[674] (copyright 1976, with permission of the School of Veterinary Medicine, University of California at Davis, and *J. Am. Vet. Med. Assoc.*).

ing behavior, with intromission, lock, and ejaculation for several years postcastration or indefinitely, although the frequency of the behaviors decreased.

CATS

Free-Ranging Cats

In a feral situation, one estrous female will be surrounded by a group of males. Tomcats will knock one another off an estrous queen or mount the tom that is mounting the female, but there is little overt aggression during courting. There is a male dominance hierarchy, based on size and age, that gives the dominant tom priority of access to females. Dominant males tend to be closer to estrous females than are subordinate males, but they do not mount more frequently.[1067] Fifteen copulations will occur in 24 hours.[356] In groups of cats living on farms, there seems to be only one sexually active tom. Although pair-bonding does not occur in the domestic cat, short-term consort behavior occurs normally. A male and female may remain together for several hours or days, mating many times. Juvenile males tend to move away from their birth area before their 3rd year.[875]

The Queen

Estrous Cycle

The queen is seasonally polyestrous and most cats will cycle at least twice yearly if not bred. Although population peaks occur in mid-January to March and May to June in the Northern Hemisphere, individual cats may be in estrus at any season. The nadir for reproductive output of a population is late fall, making the availability of kittens as Christmas presents very unreliable. If unbred, the cat will cycle every 3 weeks for several months. Actual estrus lasts 9 to 10 days without copulation and around 4 days if the cat is bred.

Most felids, including the domestic cat, are induced ovulators, and, thus, breeding may be accomplished whenever the female shows receptivity. Owners unwilling or unable to have a pet cat neutered may use this feature of the reproductive cycle to shorten estrus; artificial stimulation of the vagina using a cotton-tipped applicator stick will induce ovulation and shorten the receptive time. This is an especially useful procedure in terminating repeated heats.[496] Females will usually reach puberty at 6 to 10 months, but females born in April may not cycle until the following year.[496] Eckstein and Zuckerman[421] report that free-ranging cats may not reach puberty until 15 to 18 months, although "barn" cats born in May to July in Ithaca, New York, routinely give birth to their first litter 12 months later.

There appears to be avoidance of incestuous mating in that estrous females will travel farther from home if the closest male is related.[875]

Catnip. Nepetalactone, a volatile terpenoid found in the catnip plant (*Nepeta cataria*),[1448] and estrus elicit similar behaviors, and there has been continued debate as to whether this is a release of sexual behaviors or a nonspecific pleasure inducer.[611,626,664,738,872,1106,1403] As Hatch[626] points out, however, estrous behavior is similar to, but not identical with, catnip-induced behavior. Catnip does not cause vulvar presentation, vocalization, or foot treading, and cats in estrus do not head shake as do cats exposed to catnip. Also, male cats respond in an identical manner. Todd[1403] found that the body-rolling and head-rubbing behaviors characteristic of both the estrous and catnip-induced states could be induced in males and females by an extract of tomcat urine. As Hart[611] concludes, nepetalactone may be mimicking one of the compounds in male urine to which an estrous female may be especially primed, but to which most or all cats are sensitive. The response to catnip depends on the main olfactory system, not on the vomeronasal organ.[621]

Courtship Behavior

An estrous female will call and purr. She is restless and shows increased general motor activity. If she is a house cat she may run from one room to the next, stopping to call at each door or window. She may be very affectionate toward the owners. Urination occurs frequently, and she may spray. She rubs her head and flanks on furniture; glands in these areas may produce pheromones that contain information announcing the presence of an estrous female. She crouches, elevates her perineal region, and treads with her back legs.[1469] This will usually be accompanied by a rhythmic opening and closing of the claws of the front feet. Rolling, squirming, and stretching are seen. This activity occurs whether or not a male is present but becomes synchronized with male behavior when the female is interacting with the male. During proestrus she will roll and solicit the male's attention but act aggressively if he mounts.[1011] This may be termed postural acceptance and affective rejection. When fully receptive, she becomes immobile and stands crouched in lumbar lordosis and with her head held on the ground between her forelegs (Fig. 4.17). Her tail is deviated to one side, and she allows the male to mount. This latter sequence may be stimulated in an estrous female by scratching her over the dorsal tail base.

An estrous female may show a darting behavior in the presence of several tomcats. She will repeatedly run a short distance from the toms, and this may be her means of assessing the relative strength of the males as they chase her and try to displace one another.

Clinical Problems of Queens

Female–Female

Female–female mounting behavior is seen only rarely and usually in colony situations. Two estrous cats may try to squirm underneath each other when presenting to an inaccessible male; one may mount the other and perform malelike pelvic thrusts. Leyhausen[874] reports definite examples of female mate preferences in colony cats. He does not offer a clue as to the mechanisms of this selection but does observe that the dominant members of the colony were sometimes bypassed as mates.

Effects of Ovariohysterectomy

Spaying (ovariohysterectomy, or less commonly, ovariectomy or tubal ligation) is performed quite commonly on the domestic house cat. Spaying eliminates sexual behavior, but may affect other behaviors. Care must be taken not to ascribe changes in behavior to surgery, when

FIG. 4.17. Sexual behavior of the female cat. *(A, B)* The typical posture adopted by the queen in full estrus. Note leg flexion, lordosis, and deflection of the tail. *(C)* Tom holding queen with neck grip during intromission. *(D)* Postcoital rolling by the queen [1286a](copyright 1970 and 1987, with permission of Lea and Febiger).

the changes might be caused by maturational changes in the cat or changes in the cat's environment. A common behavior problem following spaying is an increase in aggression between cats in a multicat household, resulting either from the lowered progesterone levels of the spayed cat and/or the change in her odor that precipitated attacks by the other cat. Maternal behavior has also been observed in newly spayed cats, perhaps triggered by the fall in ovarian hormone levels, which is similar to the fall that occurs at parturition, and the presence of neonatal kittens.

The Tom

Courtship Behavior

The male probably locates an estrous female via olfactory cues deposited as pheromones in the urine and by some sebaceous gland secretions. A male placed with a female in a mating arena will spend some time investigating and marking the area with urine and anal gland secretions before mating. The cat shows a flehmen response, or gape, similar to that of ungulates (see Chapter 1, Communication, Figure 1.13). He calls to the female, circles her, and sniffs her genitalia. A nonreceptive female will actively, even violently, rebuff a male. When a female is receptive, the male approaches her from the side and behind and grips her neck in his mouth. He then mounts with the front legs, then the hind, and rubs her with his forepaws. Intromission follows a forward stepping with arched back and pelvic thrusts.[496] Ejaculation occurs seconds after intromission, and intromission usually lasts less than 10 seconds. The penis is covered with numerous small spines that apparently cause an intense stimulation as evidenced by the loud copulatory cry of the female with intromission. With retraction of the penis, the female rolls and claws at the male. The male will often lick his penis after copulation. Copulation may occur every 10 to 15 minutes for several hours. A seasonal variation in sexual readiness with a decline in the fall is seen in the male cat under experimental conditions.[82]

Clinical Problems of Toms

Reluctance of the male to breed a female is usually the result of the female's being nonreceptive and, thus, aggressive toward the male's advances. Inexperienced males may be especially intimidated by the aggressive responses of a proestrous female. In a laboratory, only one tom in three will consistently copulate with fully receptive queens. It is not surprising, therefore, that many visits to the tom are necessary before successful breeding takes place. Estrous females may indicate a mate preference by actively rebuffing one male or staying near another.

Effects of Castration

Castration is a widely accepted procedure for the pet cat. Prepubertal (6–8 months) castration generally eliminates sexual behavior. Fox[496] points out that although androgens are secreted by testes by 4 months, mating behavior does not develop until 8 to 9 months. Castration is usually physically difficult until the testes are reasonably well developed at 6 to 8 months. Some owners object to a feminine-appearing male and so delay castration until 12 to 14 months, or the first serious fight abscess. Rosenblatt and Aronson[1221,1222] and Rosenblatt[1219] point out that the effectiveness of castration in eliminating mating behavior depends on the previous level of sexual experience. Thus, owners should be advised to restrict the access of their cat to females until after surgery unless they do not mind the cat's continued sexual interests.

Hart and Barrett[614] studied the effects of postpubertal castration on fighting, roaming, and spraying. Castration seems much more effective in reducing or eliminating these behaviors in the cat than it is in the dog. Eighty-eight percent of cat owners interviewed 23 months after having their cats castrated reported a rapid or gradual decline in fighting, 92% reported the same for roaming, and 87% responded favorably for spraying. The failure of surgery to eliminate these behaviors completely probably is due to the learned components of the behaviors. Mounting behavior, either mounting inanimate objects, other cats, or the owner, occurs in 25% of castrated male cats. See Chapter 1 (Communication) for treatment of spraying by neutered cats. It is interesting that the copulatory or consummatory aspects of sexual behavior, but not the appetitive activities such as roaming, persist.

CHAPTER 5

MATERNAL BEHAVIOR

Maternal behavior is influenced by hereditary, experiential, and hormonal factors. Primiparous females are most likely to neglect or attack their offspring. During a sensitive period soon after birth, the mother of a single or small number of young forms a bond with her offspring. The offspring may take somewhat longer to recognize their mother. Litter-bearing animals such as pigs, cats, and dogs are not as exclusive in their bonds. Nest building occurs in pigs and to a lesser degree in dogs and cats. Nursing is less frequent in hider species such as cattle and goats than in followers such as horses and sheep. Weaning time depends on the number of offspring and the availability of food. Artificial weaning is almost always earlier than that seen in free-ranging animals.

INTRODUCTION: GENERAL PRINCIPLES OF MATERNAL BEHAVIOR

Internal Factors That Elicit Maternal Behavior

Hormonal and Neural Controls

Maternal behavior is characterized by sudden onset. One day we own a single cat who spends her day eating, sleeping, grooming herself, and hunting or playing. The next day, we own five cats, four of whom are kittens, and the original cat now spends almost all her time feeding and grooming the kittens. This behavior will gradually subside but, unlike other behaviors, is remarkably persistent. Aggression, for example, may be sudden in onset but does not persist very long. Other adult behaviors seem to have been rehearsed by the developing animal in play. Play includes elements of aggressive and sexual behavior, chasing, and fleeing, but not of maternal behavior.

What then is the basis of maternal behavior? It is certainly an innate behavior pattern, although experience does play a role, as is discussed later in this chapter. The genetic basis of maternal behavior is probably responsible for breed differences. For example, Targhee sheep do not show as strong maternal behavior as other breeds.[1334] Even if a behavior is innate, that is, a genetically programmed response to a certain set of stimuli, the behaving animal must be physiologically prepared to respond to the appropriate stimuli. In trying to determine the biological basis of maternal behavior, we investigate the hormonal basis of maternal behavior and the learned aspects, as well as those features of the neonate that may serve to release maternal behavior.

The combination of the proper hormonal milieu and the stimulus for maternal behavior, the neonate, plus prior experience of being a mother can elicit maternal behavior. The stimulation of maternal behavior appears to be under both hormonal and neural control. Estrogen rises and progesterone falls at parturition in sheep. Estrogen appears to facilitate, and progesterone to inhibit, maternal behavior in this species.[1319] In regard to neural control, one of the sequelae of parturition is cervical stimulation. Cervical stimulation will result in the reflex that stimulates oxytocin release. Oxytocin is released not only from the posterior pituitary into the bloodstream, but also from the terminals of cells whose cell bodies lie in the periventricular area of the hypothalamus whose axons can stimulate the neural mechanism underlying maternal activities in other parts of the brain. Brain oxytocin levels increase at parturition, at suckling (see Appendix 3), and when the vagina is stimulated,[787] and increasing oxytocin in the cerebrospinal fluid can stimulate maternal behavior.[786] Six weeks of treatment with intravaginal progesterone and estradiol, plus cervical stimulation at the time of introduction of the lamb, stimulated normal maternal behavior in anestrous ewes, indicating the importance of both hormonal and neural factors. Cervical stimulation will also cause a ewe that is already selectively maternal toward one lamb to be maternal toward another, alien, lamb.[792]

Hormonal priming by estrogen and progesterone, plus vaginocervical stimulation, is necessary in order to reduce aggression toward, or withdrawal from, alien lambs by ewes. Experience is also necessary for full expression of maternal behavior because only multiparous ewes would show positive maternal behavior—licking, sniffing, and low-pitched bleating—after the combination of hormonal and vaginocervical stimulation.[785]

The fact that primiparous ewes routinely reject their lambs if they have been delivered by Caesarian section also indicates the importance of neural stimulation by the passage of the lamb through the vaginal canal. The fact that multiparous ewes will readily accept their lambs even if they have been delivered by Caesarian section indicates the importance of prior experience in ovine maternal behavior.[32]

Learning

The evidence for the role of learning in maternal behavior is found mostly in higher primates. Monkeys that had been artificially reared made very poor mothers and very reluctant sex partners.[602] Apparently, a monkey must have been mothered in order to be a good mother spontaneously. It is interesting that monkeys that neglected or even killed their first offspring exhibited normal maternal behavior after the second pregnancy. This aspect of maternal behavior has not been well investigated in domestic animals; it is worth noting, however, that most problems in maternal behavior are seen in primiparous animals. Ovine maternal behavior, in particular, seems to be more independent of physiological changes after the ewe has mothered one lamb. The quality of maternal behavior in artificially reared cats, dogs, or sheep has not been documented. When maternal behavior in beef and dairy cattle is compared, the beef cattle exhibit more maternal behavior. These animals, at least on the range, raise their calves, and adequate maternal behavior is necessary for their calves' survival. Artificial rearing of dairy calves has been practiced on many generations of cows; consequently, few dairy cows have had much experience at mothering or at being mothered.

Concaveation. The presence of neonates can induce maternal behavior in virgin females and even in males. This phenomenon is called concaveation. When exposed to rat pups daily for 7 days, virgin female rats, and even male rats, will begin to retrieve the young, lick them, and even huddle over them in the typical nursing position. Mice will show similar behavior with no latency whatsoever as long as the pups presented are only 1 or 2 days old. Maternal behavior can, therefore, be induced in these rodents in the absence of hormonal stimulation, although hormonal stimulation accelerates the appearance of maternal behavior. The phenomenon of concaveation is used to force acceptance of alien (not the female's own) or rejected young. It is used to treat lamb and foal rejection.

External Factors That Elicit Maternal Behavior

What stimuli emanating from the neonate are important in maternal behavior? Some of these stimuli are presumably olfactory, such as the smell of a small conspecific wet with amniotic fluid. The appearance of the newborn may also serve as a visual stimulus, for Lorenz[902] has hypothesized that the short forehead, cheeks swollen by sucking fat pads, and the erratic gait of the neonate elicit maternal behavior in a number of species, including humans.

Summary

To summarize our somewhat sketchy knowledge of the biological basis of maternal behavior, we conclude that hormonal priming can lower the threshold for the initiation of maternal behavior that can be brought on and maintained as a response to the stimuli characteristic of the neonate even in the absence of the appropriate rise and fall of gonadal and pituitary hormones and cervical stimulation. Animals, especially those believed to be higher on the phylogenetic scale, can learn to be good mothers both by having been mothered themselves as infants and by having been mothers previously. Vaginal stimulation and oxytocin release appear to be important in sheep; otherwise the hormonal and central nervous system control of maternal behavior in domestic animals is virtually unknown and is a field that demands more attention from biological scientists. Figure 5.1 summarizes the factors involved in maternal behavior.

FIG. 5.1.   Factors that influence the expression of maternal behavior. The horse is used as an example, but there are similar influences on maternal behavior in all domestic animals[696] (copyright 1978, with permission of Veterinary Practice Publishing).

PIGS

The Free-Ranging Sow

Maternal behavior in sows can be divided into two prepartum behaviors, nest site seeking and nest building, as well as the postpartum behavior of nursing. One day before farrowing free-ranging sows will leave their herd and their normal home range, traveling as little as 50 m (153 ft) or as far as 7 km (4 mi). They will build several rudimentary nests before selecting a final site at which they dig a hole, which typically is 10 cm (4 in.) deep and 1.5 m (4.5 ft) wide. They will bring grass and sometimes sticks to the nest.[751] The sow will spend more and more of her time building her nest during the day of parturition. Usually, 3 to 7 hours elapse between the onset of nest building and farrowing. The sow stays with the piglets for the first 2 days and then leaves to forage for short periods. The nest will be defended against the sow's juvenile offspring and against other adults. Failure to defend the nest results in crushing of the piglets and a consequent threefold to fourfold increase in mortality.[1072] Nursing takes place every 45 minutes. Piglets do not all suckle at the same time at first, but later their behavior becomes more synchronous. For the first 2 days, the sow initiates all nursing, but after that the piglets initiate at least half of the nursing bouts. On day 7, the piglets leave the nest and the sow rejoins the herd. By day 9, the piglets sleep in the herd's communal nest. The pigs are weaned at 14 to 17 weeks.[749]

The Group-Housed Sow

Sows housed in group pens become more aggressive as parturition approaches.[69] Gilts prefer an enclosed area in which to farrow.[1137] Although multiparous sows do not show that preference, it is probably an innate behavior related to the behavior of the free-ranging sow who leaves her herd to farrow.[1348]

The Confined Sow

Farrowing Crates

Modern husbandry practices have all but eliminated most porcine maternal behavior except nursing. Sows are placed in farrowing crates that prevent them from turning around or touching the sides of the pen, and, consequently, the piglets are protected from crushing. Sows can be kept in an ellipsoid crate that allows them to turn around but does not result in any more crushing of piglets than a traditional crate.[904]

Because crushing of piglets by the sow is such a common cause of piglet mortality, it is interesting that sows do not respond to the feel or sight of a piglet under them; they do respond to the sound of a piglet squeal, although it must be loud.[725,726] Sows are most responsive to piglet squeals on the first 2 days postpartum, the time when piglets are in the most danger of being crushed if the sow lies on them.[723] Crushing of piglets is reduced by farrowing crates but not eliminated. Losses can be reduced further using a device that shocks the sow's belly when a piglet screams; however, the sow may also be shocked when extraneous noises trigger the device.[523]

Nest Building

The use of farrowing crates for nursing sows has resulted in much lower death rates for piglets, as the sow rarely can crush or cannibalize her young, but farrowing crates prevent sows from building the elaborate nests used by wild and feral swine; all that remains of the nest-building behavior is a futile pawing at the floor of the pen. The restlessness, which increases linearly during the last 48 hours prepartum, probably represents attempts at nest seeking and nest building. Nest building may be initiated by a rise in plasma prolactin.[285] Nest-building behavior consists of two distinct factors, gathering nest material and arranging it by rooting and nosing. Sows gather more material when there is no artificial shelter available.[753] When given access to an earth floor, sows will excavate a nest 8 hours before farrowing and farrow in it.[724] Sows in a pen provided with material (straw is preferred) build nests the day before and for several days after parturition.[600,852,1477] Nest building is triggered by endogenous factors, initially, which stimulate nosing and rooting, but external stimuli, that is, nest material, is necessary for pawing, carrying and arranging, that is, pigs do not engage in carrying nothing to their nest. If the pig has a preformed nest, nest building is increased, not decreased.[750] For example, if a hollow in the sand filled with 23 kg (50.7 lb) of straw is available, sows begin to nest build earlier and root more but do not carry as much straw to the nests.[68] Provision of sawdust to preparturient sows increases nesting behavior, shortens their labor, and may result in fewer piglet deaths.[328]

Parturition

Once labor begins, most sows lie down in lateral recumbency. The sow will swish her tail violently as abdominal straining takes place. Parturition usually takes 3 to 4 hours, but varies considerably with litter size and the condition of the gilt. If parturition is interrupted by moving the pig to a new pen after one piglet is born, there is a delay of hours until the next piglet is born. Opiates released in response to the stress of moving inhibit oxytocin release; however, exogenous oxytocin will reinitiate labor.[857] Most farrowings take place in the afternoon or night.[1324]

Behavior of the Sow Toward the Neonate.   When not confined, the sow will eat the placenta. The function of placentophagia remains unknown. It may be a recycling of nutrients or a form of defense against predators by removing odors. Kristal et al.[835] have found that placentophagia enhances analgesia in rats. The question of whether this occurs in domestic animals as well deserves investigation.

Sows do little licking of their newborn even when not confined in a farrowing pen. Therefore, human attendance at parturition is recommended. Although most piglets begin to breathe and quickly struggle free from the fetal membranes, a few will not. The removal of membranes, clearing of the airway, and stimulation of respiration can save a piglet that would otherwise die.

Behavior of the Neonatal Piglet Toward the Sow.   Piglets make a most startling transition from fetal to independent existence. They may be apneic for 5 to 10 seconds after birth. Then they give a few gasps before beginning to breathe regularly. Their eyes and ears are open, and they are able to walk immediately, although their gait is staggering for the first few hours. The firstborn may be slow to find the udder, but later-born pigs apparently respond to the voices of their littermates and quickly begin to seek the udder. Most piglets are nursing within 30 minutes of birth.[652] During farrowing and for some time afterward, piglets can suckle continuously, presumably because oxytocin levels are high; thus, they are rewarded for each suckle in the correct place, that is, on a teat.

Piglets are attracted to soft, warm surfaces,[1461] pig vocalizations, and the sow's odors, and they move in the direction of the sow's hair growth.[1208] Washing the udder with an organic solvent delays nipple location, as does blocking the piglet's sense of smell, indicating the importance of odors.[1049] The firstborn pigs appear to use thermal, tactile, and olfactory cues to find the udder, whereas the later born probably respond to the suckling sounds of their older littermates and walk straight to it because social facilitation is strong in pigs at birth. Suckling attempts are probably stimulated by tactile contact with a protuberance (the teat). Piglets rarely attempt to suckle on a haired portion of the sow; they will suck on the snout or the tip of the sow's vulva. Piglets nose the udder and intersperse nosing with gapes, the behavior in which the piglet opens its mouth as if to grasp a teat. Larger pigs do more gaping, which may account for their success in reaching teats. The nosing behavior of lighter pigs declines more rapidly than that of heavier pigs.[1209]

The piglet may find the udder, give a few inept sucks at a teat, and then make another circuit or two of the sow before it settles down to nursing.

Experiments using artificial sows have revealed that the piglets are attracted to the voice of the sow and to either end of the udder, but they avoid the middle and quickly abandon teats that give no milk.[754] Competition during formation of the teat order is intense and only one third of pigs end up on the teat they initially chose. Once a teat has been chosen and "won" by competition with other piglets, it is recognized by odor rather than visual cues.[755]

Nursing

Nursing causes release of opiates so that sows are less reactive to painful stimuli at that time. The opiates stimulate prolactin and somatotropin release.[1247] Approximately 10 hours after the birth of the first pig, nursing becomes cyclic.[871] Nursing bouts occur approximately every 40 minutes. The interval between nursing is longer at night than during the day. Small litters suckle less frequently than large.[1495] The sow ordinarily calls the piglets to suckle with a low-pitched rhythmic grunting. As the piglets begin to massage the udder with their snouts, the frequency of the sow's grunts increase from 1 per second to a peak of 10 per second. The more pigs massaging the udder, the faster the sow grunts and the less time until the release of oxytocin, which occurs at the peak of grunting followed in 25 seconds by milk letdown.[37] Sows may stretch a foreleg and rotate it toward the udder while the piglets are massaging the udder (foreleg rowing). Rubbing of the udder of a lactating sow can induce her to lie down and begin to give the nursing call. Stimulation of the anterior half of the udder and, especially, rubbing of the nipples in that area can increase the grunting rate.[507,511,512] Rubbing of the belly has a calming effect on even immature or male pigs and can be used to great advantage in handling swine.

A suckling bout is divided into four phases: an initial massaging of the udder for 1 minute; a quiet phase during which the piglets' ears go back and they stop massaging, which may correspond to the peak of the sow's grunts; true suckling for approximately 14 seconds while the milk is ejected, during which the piglets' ears are back, their tails are tightly curled, and their front legs are in rigid extension; and a final massage phase that is quite variable in length, 2 to 15 minutes.[552] The less weight a piglet gains, the more it will massage the udder after suckling, indicating that hunger drives this behavior.[1342] Young piglets often fall asleep on the nipple or curled beside the udder, whereas older pigs will nose the udder and pull the teats for some time.

Not all nursing bouts are successful. In 22% of the nursing bouts, the sow may call the piglets, who approach and massage the udder, but no milk is ejected. Unsuccessful nursing usually occurs less than 40 minutes after a successful bout. The piglets leave the udder as they do after a successful nursing but return much sooner.[513] The proportion of unsuccessful bouts increases if sows are moved to an unfamiliar pen.[1248] Apparently, unsuccessful nursing bouts occur in free-ranging sows.[752] Unsuccessful nursings can also occur when the sow terminates the bout by changing position. She may be responding to the vocalization of piglets fighting for a teat.[729]

The strong social facilitation and dependence on vocal communication exhibited by pigs can be used to practical advantage. If one sow in the farrowing house calls her litter to nurse, soon all the litters will be nursing. Nursing rates and weight gain can be increased by playing tape recordings of nursing noises to the sows at more frequent intervals than they normally nurse.[1361] A talented manager can imitate the noises and accomplish the same thing.[623] Piglets can also initiate suckling by their calls and persistent nudging at the sow. If half of a litter has been fasted, the hungry piglets will induce the sow to lie down, and then all of the piglets will nurse, although the nonfasted piglets will consume less.[688]

Despite the apparent low level of maternal activity in sows, piglets separated from their mother for even a short time (a few hours) exhibit considerable distress. They vocalize with either squeals or closed-mouth grunts[509] up to 21 times per minute. The vocalizations increase with the length of isolation. The vocalization changes to a higher frequency, quacking, vocalization when the piglets can hear their mother's voice, which they can discriminate from that of another sow.[1311] If the piglets are in a strange pen, they will make persistent efforts to escape, and they often urinate. The vocalizations are reduced if the piglets are isolated as a litter rather than individually. The effect of the presence of littermates is additive with that of the sow, so a litter placed in a strange pen with their mother gives only closed-mouth grunts and a few squeals. The olfactory cues are not sufficient to prevent vocalizations because the presence of the sow's bedding has no attenuating effect on vocalizations.[509] If the separated litter is closely confined and provided with a heat lamp, they are much quieter and weight losses, especially due to urination, are reduced. Similarly, cuddling of a piglet will reduce the number and volume of its squeals.

Mutual Recognition

Sows and piglets apparently use olfaction to identify one another but need more than 1 day and possibly as long as a week to learn. The piglets can identify their dam's feces, milk and urine odors,[1048] and vocalizations.[676] Sows respond to playbacks of piglet separation calls by vocalizing,[1457] but cannot discriminate their own from other piglets on the basis of their voices. They can identify their own piglets by the time they are a week old on the basis of olfaction.[677] Piglets can easily be fostered onto another sow when they and the sow's litter are less than a day old. After that, the fostered piglets are reluctant to suckle, walk around, and vocalize, perhaps because they had already formed a bond to their dam.[1152]

Sows will reject strange piglets older than 2 days. The rejection is based on olfaction, for Meese and Baldwin[991] found that anosmic sows would accept strange piglets.

Defensive Reaction. Sows normally exhibit strong defensive reactions when their piglets are threatened. They give a crescendo of barks, open their mouths, and attack. Only when pigs are defending their young are they really dangerous. The use of the farrowing crate has had a definite effect on the maternal behavior of sows. The sows can do nothing if their piglets are handled or hurt despite the piglets' loud distress calls. In herds where the piglets are handled often and by many people, as in university or research institutions, the sows become accustomed to the distress calls of their pigs. A sow may even continue to sit on a piglet that is screaming loudly and eventually smother it.

Weaning

Weaning begins at 5 weeks when the sow begins to aggress against the piglets; however, the piglets continue to suckle for 80 days. Under modern management techniques, piglets are weaned at 5 to 6 weeks or even younger. Because piglets eat little solid food at 2 weeks of age, weaning at this time is more stressful and is associated with greater inhibition of growth than weaning at 4 weeks.[1009]

There have been few studies on the effect of isolation on piglet behavior, although many piglets are raised under these conditions[242] in attempts to make artificial rearing of piglets commercially feasible, or in specific pathogen–free (SPF) pig production,[413] or segregated early weaning (SEW). To prevent piglets from sucking on one another and to prevent spread of gastroenteric diseases that plague artificially raised piglets, the piglets are housed separately. Baldwin[110] noted that pigs reared without the sow would defecate in the nesting area, whereas normally raised pigs do not. Piglets raised in germ-free isolators give distress calls almost continuously during handling and feeding; conventionally raised pigs give distress calls only when hurt.[1080]

Early weaning (at 3 to 4 weeks old) of piglets is often practiced in order to decrease the interlitter time.[1145] Early weaned pigs massage and nibble on one another, yet spend less time rooting or nibbling on other objects. If placed in cages, early weaned pigs dog-sit (on their haunches) seven times more frequently than piglets in straw-bedded pens,[1427] which indicates that flooring type, as well as age at weaning, influences behavior. Aggression is high when 3-week-old piglets are first weaned, but it decreases with time. The piglets spend approximately 70% of the daylight hours lying down, 13% exploring, and 9% feeding.[1513]

Clinical Problems

Cannibalism

Cannibalism occasionally occurs in sows; nervous primiparous gilts are the most likely offenders. Cannibalism is responsible for 4% of piglet deaths and occurs in 18% of litters.[329] The most common occurrence is immediately after parturition. In fact, many sows will bark at the first piglet that walks by their heads after parturition.[1176] Farrowing crates successfully prevent cannibalism unless an unwary piglet walks right in front of the sow. The tranquilizer, azaperone 2.2 mg/kg, has been used to treat cannibalistic sows.

Refusal to Nurse

A more common problem is seen in sows suffering from mastitis; a sow that is normally a good mother will attack her litter whenever they attempt to nurse. This early behavioral sign warns of disease before many physical signs can be detected. This type of behavior is seen in sows that become afflicted with mastitis late in lactation, after the farrowing crates have been removed. Sows that have the mastitis metritis syndrome shortly after farrowing are usually too ill to protest when the piglets suckle. The failure of the sow to eat and of the piglets to gain weight is the best clinical evidence of the latter syndrome.

Early Weaning

Although sows in a seminatural situation do not wean their piglets until they are 14 weeks or older, in a loose housing situation where sows can leave their piglets and socialize with other sows, the piglets were weaned by 5 weeks.[215-217] Sows spend less and less time with their piglets during their 1st month. Some sows that have the opportunity to leave their piglets may do so, but confining sows with their piglets reinstates normal maternal behavior.[215] This indicates that contact, particularly visual contact, with the piglet is necessary to sustain the maternal behavior.

SHEEP

Maternal behavior in sheep has an important clinical aspect, as most lamb mortality occurs within the 1st week of life in range-reared sheep. Mortality rates are from 5% to 15%, rising to as high as 50% in huge flocks in bad weather, even in the absence of predators.[1040] Abnormal or weak maternal behavior accounts for parts of these high losses and perhaps for some of the losses attributed to coyote predation. Most maternal rejection or simply poor mothering without absolute rejection occurs in young ewes and in those that had difficulty at parturition.[1447]

Parturition

Lambs may be born at any time of the day or night, with peak frequencies being noted at 9 to 12 AM and at 3 to 6 PM.[889] A few days before parturition, the ewe withdraws from the flock, if on the range, and seeks some sort of shelter. Shelter seeking by the ewe improves the environment into which the lamb is born so that its chances of survival are greater; however, the ewe is responding primarily to her own thermoregulatory needs whereby shorn, but not unshorn, sheep seek shelter.[909] Allelomimetic behavior is so strong in sheep that some of the flock may follow her. In a pen, the ewe will withdraw from social contact and seek a corner.[243] She will show restlessness, circle, vocalize, rub her head on her flanks, lick herself, and paw at her bedding 60 to 90 minutes before parturition.[420] Grazing and ruminating ceases. The older the ewe, the shorter the lapse of time between the onset of restlessness and the onset of labor. The interval between onset of labor and the appearance of the lamb can vary, but is usually 30 to 60 minutes.

Even before parturition, 20% of ewes show maternal behavior toward other lambs.[77] This prepartum maternal behavior results in lamb stealing (discussed later in this section). The amniotic fluid dripping from the vagina to the ground attracts the ewe. She will sniff and lick at bedding contaminated with amniotic fluid. This attraction to amniotic fluid can be used to predict parturition because only ewes close to parturition will eat food mixed with amniotic fluid. Ovine or caprine,[870] but not bovine, amniotic fluid is accepted,[81] indicating that there is some species specificity. The attraction of the ewe to this fluid may serve to keep the ewe in the area where the birth will take place, ensuring that the lamb will not be abandoned before it can get to its feet.[1332]

Behavior of the Ewe Toward the Newborn Lamb

Licking

When the lamb is born, the ewe begins to lick it, simultaneously emitting a special parturition call that is a very low-pitched gurgle or rumble, a call that also may be given before parturition. The call is heard only at parturition in domestic sheep, but it persists in the feral Soay sheep as a close contact call.[1318] The lamb's behavior influences that of the ewe. If the lamb is inactive, the ewe will cease licking. Licking of the lamb can be very important in cold or windy weather because it serves to dry the neonate; it additionally serves to stimulate the lamb. While the lamb is recumbent, the ewe licks its head, even restraining the lamb with a front leg to prevent it from standing. Licking of the perineal area stimulates the lamb to rise. Once the lamb is standing, usually within 30 minutes,[1447] the ewe continues to lick it, but mostly on the hindquarters. If the ewe stops licking, the lamb gives distress calls.[136] Finally, licking of the lamb by the ewe establishes the maternal–offspring bond, for the ewe will be able to identify her lamb by smell and taste. Usually, the fetal membranes are licked off the lamb and ingested, but the placenta is not eaten (Fig. 5.2).

Although mothers of twin lambs spend more time licking their offspring than do mothers of singletons, the increase is not double; so, twin lambs, especially the second born, are licked less than singletons.[1085] The lack of licking is reflected in the longer interval to successful suckling in twins.

Although lambs are able to stand within an hour of birth, it may

FIG. 5.2. Ewe licking the head of her newborn lamb (courtesy of Dr. Martin Siegel, Annandale, N.J.)

take 2 to 3 hours before they find the udder.[34,1447] The ewe plays a part in the search. She may either facilitate or inhibit teat seeking.

Suckling

The ewe is most attracted by the head of the lamb. As she circles to maintain head-to-head contact, she moves her hindquarters and udder away from the lamb, which hinders the lamb's attempt to find the udder. The innate pattern to which lambs appear to respond is the curved underline of the ewe. The newborn lamb moves toward the ewe's head and toward her udder—both areas appear to be attractive (Fig. 5.3). Once the lamb has established contact with an underline, odor, texture, and temperature probably serve to guide him.[1436] The warmest surface of the ewe is her woolless inguinal area; furthermore, the lamb is attracted by the odor and resilience of the inguinal wax.[198,199,1438,1439] Contact with the face of the lamb stimulates him to push his head up and forward. Contact with the lips causes him to open his mouth and protrude the tongue. Contact with his tongue causes the lamb to curl the tongue into the suckling position.[1439] This series of innate responses serves to bring the lamb to the ewe, then to the udder, then to the teat, and finally, to suckle the teat.

FIG. 5.3.   Initial orientation of the lamb to the ewe's underline, but in the axilla rather than the udder (courtesy of Dr. Martin Siegel, Annandale, N.J.).

One can determine whether a lamb has successfully sucked or not. Before he has suckled, a lamb will lift his nose and open his mouth and make directional movements with its tongue or lips when touched on the forehead or nose.[910,1437]

Lambs whose dams are stanchioned take longer to locate the udder,[34] indicating the active role the ewe usually plays. The drive to suckle is inhibited, but not eliminated, by intragastric loads of milk; consequently, hunger is not the lamb's only motivation.[35] Once the udder has been located, the lamb uses visual cues to relocate it for subsequent sucklings.[135]

Advantage can be taken of the features to which a lamb responds to build a colostrum feeder from which the lamb will suckle without human aid. The colostrum feeder consists of a fleece-covered horizontal ledge through which soft rubber teats protrude at a 45° angle, 50 cm (20 in.) off the ground.[516]

Once suckling has begun, it occurs with great frequency; twin lambs suckle 22 times during 16 hours of daylight, and single lambs, 6 to 14 times.[390,1056] Newborn lambs may nurse for as long as 3 minutes in one bout; later the duration falls to 20 to 40 seconds. Frequency of suckling and suckling duration both decrease with age. By the end of the 1st week, lambs suckle hourly and every 3 hours by 9 weeks.

Triplets nurse less often and for shorter duration than singletons or twins.[442,447] Probably because they are not receiving adequate nourishment from their dams, they are most likely to try to suckle an alien ewe. During the first 2 weeks, the ewe will allow one twin to suckle without the other. Later, the ewe will walk away when the lamb nudges her in the inguinal area and will refuse to let one twin nurse until the other is also present. If the ewe is lying down, the lambs will not only nudge her but will also jump on her back and paw at her in an attempt to make her stand. The ewes will call their 5-week-old lambs to them and then refuse to let them nurse. Such behavior encourages the lambs to stay in close contact.[443] As suckling decreases, grazing by the lambs increases. The ewe, meanwhile, will graze farther and farther from the lamb. Although the ewe stays within 10 meters of the lamb the first few days, she will soon increase her distance from it. During the 1st month of her lamb's life, the ewe will leave the flock to seek out her lamb if it has strayed. Thereafter, she will bleat, but remain in the flock.

Acceptance of the Lamb

There are three important time periods: (1) the time after parturition that a dam will accept a neonate, (2) the time necessary for the dam to form a bond to the neonate, and (3) the time the bond will persist if the neonate is removed after the bond is formed. The "critical period"

during which a ewe will accept a lamb is the first several hours after parturition.[311,1333] Normally, a ewe will stay within 2 meters of its lamb for
most of the 1st day.[909] If a ewe's lamb is removed immediately after birth
and before she has licked it, the ewe will accept any lamb presented to
her. Once the ewe has spent 20 to 30 minutes licking a lamb, her own or
a substitute, she will not accept another. If the lamb is removed 4 hours
after birth, the ewe will continue to exhibit maternal behavior if it is returned within 24 hours.[867] The importance of olfactory cues in the establishment of the bond is demonstrated by the fact that ewes will temporarily accept strange lambs that have been rubbed with the ewe's
placenta. Ewes can also be induced to follow their placentas.[311] Primiparous ewes will not accept a lamb whose wool has been washed free of
amniotic fluid, whereas multiparous ewes will,[869] indicating both the
importance of amniotic fluid and the importance of prior experience at
mothering. Olfaction is necessary for normal maternal behavior in
primiparous, but not multiparous, ewes.[868] If the ewes could not smell
the lambs, they bleated less and licked the lamb less; the lambs took
longer to suckle.

In hilly country, newborn lambs may roll down the hill away from
the site of their birth. The ewe may neglect the lamb because the odor
at the birth site is more attractive than the lamb itself. In this situation,
a weak lamb, or the weaker of a pair of twins, is most likely to roll away
and not be licked. A more vigorous lamb will survive and seek out the
ewe. If too long a time elapses between parturition and the presentation
of the lamb, the lamb may be rejected.

Mutual Recognition by the Ewe and Lamb

Recognition of the Lamb by the Ewe.   Recognition of lambs depends on at least three senses: olfaction, audition, and vision. It is the
wool of the lamb that contains the odor used by the ewe to identify her
own lamb.[26,28] It seems apparent that ewes base their recognition of
their lambs at a distance on vision and at a close range, 0.25 m (10 in.)
or less, on smell,[26,893] recognition cues being reinforced each time that
the lamb suckles. During nursing, the ewe sniffs at the tail and perianal
area of the lamb. Olfactory bulbectomy eliminates the preparturient lip
licking observed in Soay sheep (a feral sheep of ancient origin), as well
as the licking of the newborn and normal lamb recognition. The bulbectomized ewe will accept other lambs indiscriminately.[127]

At least two senses must be impaired before ewes are unable to find
their lambs.[1041] Visual cues may be most important because ewes had
more trouble finding a hidden lamb than a silent one.[24] By changing
the appearance of various portions of the lamb's body, it was found that

maternal recognition was most impaired by altering the appearance of the head.[25] Although it represents only 12% of the body surface, the head is apparently the area that the ewe uses to identify the lamb visually. There is also evidence that ewes use the color of their lambs to identify them. They reject their own lambs when they are dyed, and, if they do reaccept them, they will choose lambs of the same color when their own is not available.[26] Laboratory experiments (Chapter 1, Communication) indicate that sheep can perceive color; the studies on lamb recognition indicate that sheep use color vision.

The strong individual recognition of her lamb by a ewe, however, can break down. Multiple birth lambings of three, four, or more lambs are not uncommon, especially in Finnish Landrace sheep. As sheep are bred to produce litters, there may be changes in maternal behavior. The ewes rearing three lambs bleated less and approached a solitary lamb less often then ewes with singletons or twins.[1143] If several ewes and their "litters" are penned together, the ewes may not distinguish their own lambs from the others; communal suckling of all lambs by all ewes results.

Recognition of the Ewe by the Lamb.   Lambs detect their own mothers faster and more accurately with age. During their first several days of life, lambs are not able to discriminate their mother from other ewes very well except at very close range. A lamb separated from its mother will rush up to the nearest ewe and attempt to suckle, only to be butted aside. Lambs can distinguish their dam from an alien ewe at 24 hours of age by approaching her pen but if tested again will approach the same pen whether or not their dam is there. That indicates that once they have learned where to find the dam, they have difficulty finding her in a different location. By 3 days, they are able to recognize their dam from a distance, but they do not appear to use olfaction to recognize their dams even at close range. Vision and hearing are more important.[1079] Given a choice between their own mothers, similar (same breed), or dissimilar ewes (different breed), they will continue to be more attracted to similar rather than dissimilar ewes.[23,1316]

The importance of hearing was demonstrated by Arnold et al.,[73] who found that lambs were more apt to approach a ewe that was not its mother when the voices of the ewes were muffled. Technical advances in reproduction have allowed advances in understanding of auditory recognition. Most Dalesbred and Jacob lambs born after embryo transfer to Dalesbred ewes could identify the ewe on the basis of her voice, whereas most of those born to Jacob ewes could not.[1315] Sonographic analysis indicated that there were more intersheep differences among bleats of Dalesbred ewes than among Jacob bleats.[1312,1314]

As lambs mature, visual cues become more important. A lamb less than a week old is not affected by a change in his dam's coat, such as shearing or blackening, but a 2-week-old lamb may hesitate to join a visually altered dam.[21] Shillito[1317] also found that covering the pens in which the ewes were restrained slowed the approach of their lambs.

A critical period within the first few hours after parturition may exist for acceptance of lambs by ewes; however, the lamb is not restricted in time as far as his social attachments are concerned. The tendency to follow any large moving object is most marked during the first 3 days of life; for the next 3 days fear responses predominate, but from 6 days to 2 months lambs will continue to follow even an artificial sheep model.[1496] Lambs tend to follow large moving objects, but imprinting in the avian sense does not occur, for a lamb's attachment can be quite impermanent. Lambs easily can become attached to a nanny goat or to a human who feeds them. Lambs that had been normally reared with ewes quickly formed attachments to dogs when one of each species was penned together. Within 8 weeks, the lamb would follow the dog, vocalize if the dog was removed, and even run a maze to be reunited.[267] After living in a normal situation for 4 months, the lambs no longer preferred dogs to sheep. Therefore, social attachments in lambs seem to be relatively easily formed and equally easily dissolved. This phenomenon of attachment can be used to bond sheep to cattle. The cattle deter coyotes from attacking the sheep. See Chapter 2 (Aggression and Social Structure) under Guard Dogs for Predator Control.

Clinical Problems

Poor maternal behavior is often seen in ewes that have been in labor more than 30 minutes. The corticosteroid levels of such ewes are elevated, indicating that they are stressed.[243] Poor maternal behavior may vary; the ewe may reject the lamb outright, but more frequently she will lick it in a desultory fashion only and nervously avoid the lamb's attempt to nurse.

There are great breed differences in the frequency of abandonment of one of a pair of twin lambs. Merino sheep are much more likely to abandon one of their twin lambs than are Dorsets or Romney.[33]

Cross-Fostering

The problem of cross-fostering of lambs is a common one. In general, older ewes will more readily accept lambs than younger ones.[1333] Fortunately, if a ewe is exposed to young of its own species long enough, maternal behavior will occur. The process (concaveation) may take weeks, and the ewe should be stanchioned or somehow restrained so that the lamb will not be badly butted in the interval before maternal behavior appears. Tranquilization of the ewes with pherphenazine will facilitate acceptance[1068] but does not facilitate fostering of alien lambs onto ewes with their own lamb present.[1405] Diazepam administered after parturition will facilitate acceptance of an alien lamb by a ewe and could be used to facilitate cross-fostering as well.[469] A variety of methods have been used to facilitate cross-fostering. The time-honored method is to tie the skin of the ewe's own lamb over the lamb to be fostered. There are, however, quicker and easier methods of providing olfactory cues from the ewe's own lamb that do not depend on skinning a dead lamb. These include pouring amniotic fluid on the alien lamb, washing the lamb,[30] and putting a garment worn by the ewe's own lamb inside out on the alien.[29,31,1155] This method of transferring the familiar scent to the alien lamb also is successful in fostering a second lamb onto a ewe.

Visual cues should also be altered. Either the lamb can be tied so that it cannot stand, thus mimicking the attempts of a neonate to stand, or visual cues can be eliminated. Stanchioning the ewe is effective because the ewe cannot move away from the lamb or butt it, and if her view of the lamb is also blocked (eliminating visual cues), fostering is facilitated.[22,1156] Advantage should also be taken of cervical stimulation, as already discussed here, to facilitate fostering.[782] The technique of "slime grafting," in which the vaginal fluid of the ewe is rubbed on the lamb to be fostered, probably owes its success to the consequent vaginal stimulation in addition to, or instead of, the transfer of her odor to the lamb.

Mismothering

Mismothering, that is, maternal behavior directed toward a lamb that is not the ewe's own, is a common problem, especially in large flocks in which ewes are not penned separately for parturition. Up to 15% of lambs may be raised by ewes that are not their mothers. There are many combinations. A ewe's lamb may be born dead, and the ewe will steal another ewe's lamb. She may steal a lamb before parturition, have one of her own, and be credited with twins. Although "stolen" lambs may survive, it is impossible to make accurate statements about productivity of a given ewe under these circumstances. A shepherd might cull a productive ewe and keep one that never produces twins but often "acquires" them.[1462] The opportunities to mismother are increased when sheep are confined at parturition; however, if cubicles are provided, the incidence of mismothering is considerably lower in those ewes that choose to lamb in them.[561]

Oral Vices of Artificially Reared Lambs

Artificially reared lambs suck one another's navel or scrotum and eat feces. Such behavior may cause injury or interfere with weight gain.[1356]

GOATS

As parturition approaches, does, especially multiparous ones, leave the herd and seek a sheltered place, almost always near a vertical object, to kid. The does will defend this area and the kid both before and for the 1st day after the kid is born. Parturition is most likely to occur during the day at a time when goats are generally inactive. As parturition approaches, they grunt, paw the ground, kick, and lick their backs.[1173] After parturition, the kid will be licked for 2 to 4 hours. The doe will vocalize frequently, using a low pitched bleat similar to the rumble of periparturient sheep. Vaginocervical stimulation can be used to induce a recently parturient doe to accept an alien kid.[1211] There seems to be a critical period of an hour for acceptance of the kid; kids removed at birth and presented to the doe an hour later may be rejected.[580,1210] The doe must have contact with the kid for more than 5 minutes to become not only maternally responsive, but also selective in that response, that is, accepting only her own kids.[229,581,1212] Olfaction seems to be important; anosmic goats accept all kids, rather than only their own.[819,1210] The small ruminants seem unable to distinguish between species, as lambs can easily be cross-fostered onto goats and vice versa.

Intensive maternal behavior is short lived in goats because within a day, the kid will have left the doe to hide and the doe will rejoin the herd or stay nearby if she can obtain adequate forage there.[877,1081] She will approach the hidden kid several times a day and call to it. The kid will answer and emerge to suckle as infrequently as twice a day. In the absence of a proper hide, a dim area with vertical sides and a roof, hiding behavior may not be recognized, but it has persisted in domestic as well as feral goats.[876,1242] Goat kids should stand within 20 minutes and suckle within an hour.[38,660,878] The kid seeks the udder and usually searches the axilla first because the doe is turned toward the kid licking it. The kids of does with udders transplanted to the neck region located the udder as quickly as the kids of normal does.[1358] Two-day-old kids can identify their mothers visually, apparently they use their dam's pelage for recognition.[880,1243] Kids are more apt to be farther from their mothers than the nonhiding lamb that develops visual recognition only slowly.

Clinical Problems

Kid Rejection

The importance of olfactory identification of the young is emphasized by the following case: a week-old kid was castrated; an open castration technique was used in which incisions were made and the testes removed. When the kid was returned to the doe, she took one sniff and rejected it. The kid had to be hand raised (Dr. Mary Smith, personal observation). The smell of the fresh wound was probably responsible for the rejection response. Closed castration techniques should probably be used in suckling kids.

CATTLE

Some of the signs of imminent parturition are relaxation of the sacrosciatic ligament, slackening of the tissue of the perineum and vulva, distention of the udder and teats, and mucous discharge from the vulva. Unless the afternoon body temperature is below 39°C (102°F), parturition is unlikely even in the presence of all the other signs.[441] The normally lower body temperature of cattle in the morning interferes with the predictive value of temperature.[401,403]

Parturition

The majority of cattle do not leave the herd to calve. This is probably an example of flexibility of behavior that takes advantage of geography. When there are trees or rocks available, the cow would leave the herd and hide, but in an open pasture the risk of predation is less if she stays with the herd.[886]

Cows choose dry, elevated areas with shelter available, so those are the areas where one should search for lost neonatal calves. A periparturient cow will sniff and lick other calves especially if she is within 24 hours of parturition and the other calf has just been delivered, whereas after parturition all her activities are directed toward her own calf. Licking of alien calves does not cause rejection of the cow's own calf nor does suckling of an alien mother cause a calf to fail to suckle its own dam.[730] Other cows will sometimes push or butt a newborn calf. In a study of Hereford cattle, George and Barger[546] found that 82% of all parturitions take place between noon and midnight. Parturition times are distributed throughout the 24 hours, but dystocias occur mostly at midday.[1100,1523] A greater proportion of births will take place during the day if cows are fed late at night.[905] Arching of the back and an elevated tail occur for 1 to 3 hours before the chorioallantoic membrane ruptures. When the membranes rupture, the cow often licks the fluid and tends to stay near the spot, now attractive to the cow, where the fluid fell. Ninety-five percent of all cows are recumbent at the actual time of delivery.[1293] Approximately 100 minutes elapse from the rupture of the membranes to the birth of the calf. The placenta is eaten by 82% of the cattle. Parturition is longer in cows that give birth to large calves and in nervous heifers. In fact, labor may cease if nervous heifers are disturbed.[402] Handling of the cows at the time of calving appears to result in improved behavior at milking.[643]

Bonding

Heritability of maternal behavior is low in cattle, but some breeds are more maternal than others. For example, Angus are more maternal toward, and more defensive of, their calves than are Herefords and Charolais.[262] Contact between the cow and her calf for as brief a period as 5 minutes postpartum results in the formation of a strong specific maternal bond. Cows groom their calves during the early postpartum period, concentrating on the back and abdomen. Licking the calf occupies up to half the cow's time during the first hour postpartum; heifers lick less.[424] If contact between cows and their calves is delayed for 5 hours postpartum, 50% of the calves will be rejected; therefore, the critical period for formation of the cow–calf bond must be the first few hours postpartum. When the calf is removed after a brief initial contact, the cow vocalizes and is restless; however, after 24 hours she can no longer distinguish her own calf.[709]

Cows do not show kinship recognition of their calves. When twins were created by transferring an embryo into the uterus of already pregnant cows, the unrelated calf was treated just like her own calf.[1449] Calves can recognize their dams, but make many errors when trying to identify their dam from 20 other cows. They tend to choose a cow of the same coat color as their mother.[1058]

Suckling

The newborn calf shakes its head, snuffles, and sneezes. This behavior may begin during parturition as soon as the calf's shoulders are free of the mother's vulva. Some calves will remain motionless for up to 30 minutes after birth, but within an hour most calves can stand. It may take 30 minutes to an hour before the teats are located, and the cow's conformation may not provide the higher recess that the calf appears to seek. Passive transfer of immunity to calves is poor in cows that have had dystocias, presumably because the calves did not suckle as much or as often.[389] Most calves suckle within 3 hours, but up to a third of calves may not suckle within 6 hours of birth. This is particularly apt to be the case when the cow has a pendulous udder.[423] Once the teat has been located for the first time, the calf will be able to locate it much more quickly at subsequent nursings. As the calf suckles, the cow will lick the perineum, stimulating urination and defecation by her calf. Calves that have not suckled for the first 6 days of life cannot learn to suckle.[471] Younger calves or those that have had suckling experience can learn to suckle from another cow.

When suckling, calves assume a particular stance with spread legs so that their shoulders are lowered, allowing them to butt upward at the udder. The butting appears to function in the stimulation of milk flow.[587] Like other young ruminants, they nuzzle and lick along the cow, especially in high recesses like the axilla and groin and will mouth any hairless protuberance as they seek the udder.[1294] They appear to be confused when they encounter a hairless teatlike object that does not supply milk. They wag their tails while suckling, although not at as high a rate as lambs. Newborn calves normally suckle five to seven times a day. Usually, the number of suckling bouts decreases with age, but beef calves may actually suckle more frequently with age, possibly because the milk supply of the beef cow is small.[587,1089] The most regular suckling time is at daybreak, with other bouts occurring between 9 AM and noon, 3 and 6 PM, and 10:30 PM and 1 AM.[1446] Most suckling takes place during the day.[1266] Suckling bouts are long, approximately 6–12 minutes, and do not seem to vary with frequency of suckling.[543,589,1131,1340]

It is now possible to produce twins in beef cattle by embryo transfer. These twins, like triplet lambs, suckle more often and are more apt to suckle from an alien cow. They usually approach the udder from the rear rather than in the normal anteparallel position. It is the smaller of the twins that is most likely to suckle from an alien cow, and the cow suckled usually has a single calf.[1158] The twins are groomed less than single calves.[1155]

When calves are raised artificially on a nippled feeder, they show similar rates of nursing when the milk is similar in concentration to cows' milk, but nursing increases in frequency when the milk is diluted. The calves often stand touching the wall while sucking from the feeder, just as they would touch the side of the cow if they were nursing.

Weaning

Cows do not break the bond with their yearling calf when their next calf is born; they may even allow the yearling to suckle, although they are more aggressive toward the yearling than they were before the birth of the younger calf.[1432]

In *Bos indicus,* bull calves are weaned at 11 months, but heifer calves are weaned much earlier, at 8 months.[1188] The cow apparently invests more of her resources in a son that can produce many offspring per year than in a daughter than can produce only one.

Although most modern dairy calves are weaned at 1 or 2 days of age, there are alternatives. Cows that are allowed to suckle their own calves for 10 weeks produce more milk than cows whose calves are removed. Although the suckled cows produce a large supply of milk, some of the total goes to the calf, and they do not make up on marketable milk for all that the calf consumes. Calves allowed to nurse only twice a day gain more weight than those fed from buckets and those with continuous access to their dams.[440]

Clinical Problems

Sucking Problems

Nonnutritional sucking is a very frequent problem when calves are raised in groups, especially if they are pail, rather than nipple, fed. Nonnutritional sucking can occur 78 to 300 times a day.[589] Skin irritation or even hernias can result from prolonged sucking by one calf on the umbilicus or sheath of another. Calves that engage in nonnutritive sucking often fail to thrive.[1353] The incidence of intersucking on British dairy farms is 13%. If the problem exists on a farm, as many as 30% of the calves and 11% of the adult cows may be affected; there is apparently social facilitation of the behavior.[1509] Most dairy farmers solve the problem by penning calves separately; when this is not feasible, however, there are other procedures that may alleviate the problem, such as muzzles, provision of nipples for the calves to suck, or application of unpalatable substances to the part sucked, and changing to dry food rather than milk. Most nonnutritive sucking occurs immediately after feeding so provisions of nipples in the feeding area will decrease self-, or auto-, sucking and allo-sucking (sucking on others).

Calves weaned after 6 days are more apt to suck one another than calves weaned earlier. A related problem is that of self-sucking. Various harnesses and even surgical procedures, such as splitting of the tongue, have been devised to deal with the problem.

Cross-Fostering

As noted previously, cross-fostering can be accomplished by draping the calf with the skin of the cow's dead calf, if that is available. A more difficult problem is convincing a cow to accept a foster calf in addition to her own offspring. Fostering calves onto dairy cows is fairly easy because they have not been selected for maternal behavior that includes rejection of alien calves. Beef cows have been selected for these traits, and, hence, it is much more difficult to foster additional calves onto them. There are practical reasons for wishing to do so because a well-fed beef cow has a milk supply large enough for two calves. Dairy calves can be fostered onto these beef cows, and when raised in this manner will suffer far less from maternal deprivation and from the respiratory and enteric diseases to which artificially reared calves are susceptible.[708,797]

There are difficulties in persuading the cow to accept the dairy calf. Although the cow–calf bond is presumably formed when the cow first encounters the fetal fluids and the newborn calf, a beef cow may reject a dairy calf even when its own calf has been removed immediately after birth and when the foster calf is rubbed with fresh amniotic fluid. The cow apparently still can discriminate between a newborn and the older, larger, more active foster calf, probably on the basis of visual cues; blindfolding the cow may help.[797] It is possible to substitute one calf for another by removing the cow's own calf 48 hours postpartum, leaving the cow with no calf for 3 days, and then returning her own calf plus the alien calf. The bond to the original calf has been broken, but maternal responsiveness persists.[790] Placing a jacket worn by her own calf inside out on the calf to be fostered is helpful; the same technique is used in sheep.[658]

Much obviously remains to be learned about the basis of maternal bonding in cattle. Some cows will grudgingly foster the dairy calf and mother their own calf. Others will become promiscuous mothers that allow four or more calves to suckle.[801] The maternal bond has been broken in these cases and replaced by nondiscriminative tolerance. The promiscuous mother may suffer teat or udder damage when too many calves suckle. There is also the danger that mastitis may be passed from nurse cow to nurse cow by the calves. Cows that live in a group with their foster calves continuously present are more likely to be maternally selective than those cows that are exposed to the calves only for nursing periods twice a day.[1124]

HORSES

Parturition

The onset of parturition in mares is heralded by "waxing" of the udder, but the length of time between the appearance of udder waxing and the appearance of the foal may be quite variable, up to 21 days. The calcium level of the milk increases as foaling approaches and can be used as a predictor. Body temperature is lower the day prior to parturition.[1306] The mare will walk more and stand less the evening of parturition. Berger[190] has found that only primiparous mares leave the herd to foal; multiparous mares remain with the herd. In the first stage of labor, which lasts for about 4 hours, the mare is restless and will crouch, straddle, and urinate. The smell of fetal fluids is attractive to parturient mares. The mare will exhibit the flehmen response in response to amniotic fluid that is expelled. Sweat will appear on her elbows and flanks.[506]

During the second stage of labor, the mare will lie in lateral recumbency. This second stage is very violent and very short in horses, lasting less than a half hour. For that reason, it is important to have a veterinarian in attendance when complications are expected. There will not be time for professional help to reach the mare if problems develop in the course of labor.

Mares are notorious for their ability to thwart observation of their parturition. Although most mares foal at night, some will wait until they are released from their stalls in the morning in order to foal in the solitude of the pasture. Thus, parturition appears to be under some type of voluntary control in such mares, but the evidence is all anecdotal.

Postparturient Behavior

Ordinarily, the foal is delivered in such a way that the mare need only to turn her head to meet her foal muzzle to muzzle. The establishment of the maternal–offspring bond is still uninvestigated in horses. It may be based on olfaction, as the mare licks the fetal membranes and then the newborn; licking behavior usually is confined to a few hours after parturition.

Licking, as well as sniffing, is concentrated first on the head of the foal and later on the hindquarters, particularly the perianal area. The rate of licking decreases markedly during the 1st hour postpartum. Although the period for bond formation has not been identified in horses, the first hour is probably critical for the mare to learn to recognize her foal selectively. The foal appears to take much longer, perhaps as long as a week to recognize the mare. The foal will follow any large, moving object. The mare is usually very aggressive toward other horses and sometimes toward people for the 1st day or 2 after foaling. This behavior serves to keep away other horses that the foal otherwise might follow.

Suckling

Standing and suckling occur within the 1st hour after birth for pony foals and within the first 2 hours for Thoroughbred and saddlebred foals.[272,1228,1452] Foals suckle four times per hour at 1 week of age and gradually decrease the frequency to once per hour by 5 months[279,280,331,1418] (Fig. 5.4). Mares spend approximately two minutes nursing, during which the foal spends less than half the time actually suckling. The rest of the time is spent in nuzzling the teat and butting the udder.[134] The mare usually flexes her hind leg on the side opposite the foal, possibly conserving energy by shifting her weight to the stay apparatus[1494] (Fig. 5.5). Subordinate mares have shorter nursing bouts, because dominant mares aggressively disrupt nursing and nursing attempts.[1251] Nursing also occurs after any separation, even a very brief one, or after the foal has been frightened. When a foal approaches its dam to nurse, it often shakes its head and nickers or crosses in front of the mare. Foals turn their heads sideways to nurse, especially as they grow larger.

Colts may suckle more frequently than fillies when food is a limiting factor because sex differences in suckling have been found only in poor environments.[190,412] Nursing is often terminated by the mare, not by aggression, but by simply walking away from the foal. This behavior occurs most frequently during the first month of the foal's life.[331] The

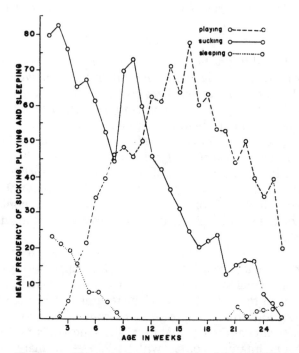

FIG. 5.4. Changes with age in frequency of sucking *(solid line)*; playing *(dashed line)*; and sleeping *(dotted line)* of free-ranging pony foals[1418] (copyright 1972, with permission of Academic Press).

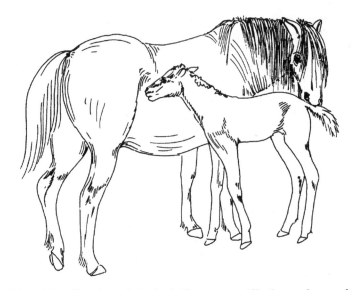

FIG. 5.5.   Identification of the foal. The mare sniffs the anal area of the foal. The hind leg on the side opposite the foal is flexed to facilitate nursing.[679]

foal may be forced to practice following the mare as a result of her behavior. The more skilled the foal is at identifying and following his dam, the more likely he is to survive. Mares do aggress toward their foals during nursing, but the aggression appears to be a response to bunting of the udder and does not prevent or shorten suckling bouts.

Weaning occurs at 40 weeks when the mare is about to foal but is prolonged beyond a year if the mother is not pregnant.[412] Under domestic conditions of abundant food, foals continue to suckle for 3 or 4 years, even when they are larger than their mothers. Orphaned or newly weaned foals will often attempt to nurse from nonlactating mares, suckle the sheaths of geldings, and investigate the inguinal area of any horse.

Mares seldom venture far from their foals throughout the first few months. Foals spend several hours per day lying down. During these periods, the mare remains with the foal, either grazing in circles around it or standing next to it. This behavior, the recumbency response, which probably functions to protect the foal both from predators and from becoming lost, wanes as the foal matures. When the foal is awake, it is responsible for maintaining contact with its dam.[332] The mare is within 5 yards of the foal 94% of the time during the 1st week and 52% of the time the 5th month of the foal's life. The typical equine family group will travel in the following order: mare, most recent foal, yearling foal, and then the other offspring in the order of increasing age.[1265]

Weaning

There are at least five different methods of artificial weaning: (1) removal of the foal from the mare and confinement of the foal by himself; (2) removal of the foal from the mare and confinement of the foal with another foal or foals; (3) interval weaning, in which the mare alone is removed from the pasture while the foal remains with the other younger foals and their dams. The other mares will be removed gradually in order of their foals' ages; (4) separation of mares and foals into adjacent corrals for 1 week and subsequent removal of the mares; and (5) feeding mares and foals separately and gradually increasing the duration of separation. This method is particularly valuable for the owner of a single mare–foal pair. Although pony foals seem less stressed when weaned as pairs rather than singly,[684] Thoroughbred foals are more stressed apparently because there is more aggression between members of the pairs. Feeding concentrates before weaning appeared to reduce stress.[671] In addition, separation of the two foals may also be stressful. Weaning by the fourth method, in which the foals can see and hear their mother but cannot make direct contact (or suckle) appears to be less stressful than methods in which the mother cannot be seen, probably because weaning is more gradual.[956] Weaning from the mother as a food source occurs before weaning from the mother as a social companion.

Mutual Recognition

The roles of the three senses, vision, audition, and olfaction, in the mare–foal bond is complex. The neighs (or whinnies) of the separated mare and foal are impressive, and horses make use of these calls to locate one another. The more frequently a mare neighs, the more quickly her foal will find her. The mare's neighs are not specifically recognized by the foal, but the mare neighs more often to her own foal.[1507] Changing a mare's or a foal's appearance by hooding, blanketing, and bandaging it does not interfere with recognition, yet visual cues must be involved because both foals and mares have difficulty finding each other when one is in a closed stall. They orient toward whinnies but need visual confirmation of the dam's or foal's presence.

Olfaction is also important. Mare and foal sniff each other's heads, and the mare sniffs the anal region of the foal (see Fig. 5.5). Masking olfactory cues with a strong odor greatly retards location of mares by foals, especially in the absence of visual cues.[1507] Thus, olfactory and possibly vocal cues (nickers) are used for close-range identification, whereas other vocal cues (whinnies) and visual cues are important for more distant communication of identity or presence (Fig. 5.6). Visual cues are probably not of vital significance, as many blind mares have successfully raised foals even in seminaturalistic conditions.

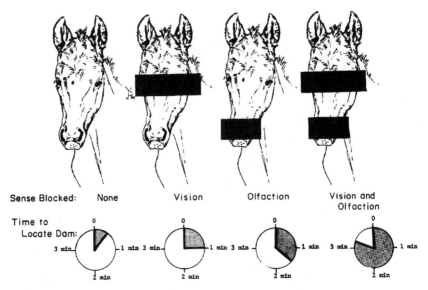

FIG. 5.6.   The effect of masking vision or olfaction or both on the time
that a foal takes to locate its dam[696] (copyright 1979, Veterinary Practice
Publishing).

Clinical Problems

Mismothering

Mismothering can occur in equids, although it is much more rare
than in sheep. A female mule adopted and successfully raised a foal;
only examination of the foal's karyotype revealed that the foal was a
Shetland pony, one of a pair of twins born to a pony mare in the same
pasture as the mule.[426] Mules may be particularly prone to this behavior;
in another case, a mule stole a calf from its mother and raised it.[1305]
Mules have successfully raised Thoroughbred foals born to them after
embryo transplant, but a donkey foal was rejected.[1305]

Foal Rejection

There are three types of foal rejection that occur immediately after
foaling: (1) rejection of suckling, (2) fear of the foal, and (3) attacking
the foal. All three forms are more common in the primiparous mare, but
the third form may occur again and again.

In the first type, many mares lick the foal and appear attracted to it,
but will not tolerate suckling. They will kick the foal if it persists. These
mares can usually be treated successfully either by tranquilization
and/or by milking the mare while holding the foal next to the mare and

feeding the milk to the foal from a bottle held in the inguinal area. The newborn foal normally nurses every 15 minutes, so this exercise should be repeated many times in order to nourish the foal, transfer colostral antibodies, and teach the mare that milking relieves tension on the udder. Gradually, the foal should be encouraged to suckle the teats and less and less restraint should be applied to the mare until the pair can be left alone safely.

In the second type of foal rejection, the fearful mare tries to escape from her foal and may injure him by running over him. She will kick when the foal approaches her. There are several behavioral methods that can be used to stimulate maternal behavior. Turning the mare and foal out in a paddock so that both the mare and the foal can avoid each other may lessen the mare's fear. Adding another horse may stimulate the aggression periparturient mares normally show, and this may be followed by acceptance of the foal. A large dog has been used for the same purpose, that is, stimulating maternal defensiveness.

In the third, and most dangerous, type of foal rejection, the mare actively attacks her foal, usually biting him in the withers and throwing him across the stall. She will not have licked the fetal membranes or the foal. She will also kick if the foal approaches her. These mares act toward a foal as foal-killing stallions do. Tranquilization with acepromazine and passive restraint for as long as 3 weeks, immediate punishment of aggression, and administration of oxytocin and progestins can result in acceptance. The mare may then be aggressive toward anyone who approaches the foal. This problem can be seen in any breed of mares but is most common in Arabians, indicating a genetic component. Passive restraint is best obtained by placing a pole across a box stall so that the mare cannot move sideways, forward or backward. She can still bite, but the foal can escape and soon learns to avoid her head. She can kick forward, cowkick, but a persistent foal can suckle. Usually the mare will accept the foal after a week or 2 of restraint. They tend to be merely tolerant of their foals, and some may relapse after a few weeks. Mares that reject their foals have lower levels of progesterone before parturition than normal mares.

Pain, such as that associated with passing of the placenta, may result in aggression toward the foal. Removing the source of pain will eliminate the aggression. Maternal rejection may occur after foaling. Too much disturbance of the mare and foal has been implicated, as has changing the odor or appearance of the foal. A common clinical problem is rejection of the foal that has had to be separated from his mother for treatment of a medical problem. The foal is changed in odor and may have had his appearance altered by clipping and bandaging. Allowing the mare to have visual contact with the ill foal may help to prevent rejection even if the foal is too weak to suckle for many days.

Redirected aggression also occurs. A mare may be aggressing against another horse and then bite or kick her foal. More frequently, the foal is simply kicked accidentally during a fight. Occasionally a mare may aggress against her own foal when an alien approaches—a failure of recognition.

Some mares do not reject their foals but do not respond vocally to them when they are separated. This can lead to injury to the foal because he will approach other mares and be attacked or will try to jump a fence or gate to return to the stall where it last saw the mare. Although this would be presented clinically as accidental trauma to the foal, it is the result of poor maternal behavior.

Mares can be vicious when protecting their foals, but they can also be vicious when weaning them. A mare may bite not only the foal, but also a nearby person when the foal attempts to nurse. This behavior also may be caused by mastitis or injury to the udder or may be of unknown cause.

CATS

Parturition

Gestation is 63 to 66 days in cats. There is a seasonal distribution of births, with the greatest number of litters being born in the summer and the least in autumn and early winter.[1147] Cats rarely deign to use the boxes carefully provided for parturition by their owners. Most cats will choose a cavelike place, such as a closet or a linen cabinet. A cat is attracted to the smell of the amniotic fluid at the birth site, so moving the kittens to a more suitable location will not entice her from her chosen spot. Parturition in the cat is characterized by a great deal of licking by the queen: self-licking mostly directed at the belly and genital area, licking of the fetal fluids from her body or the floor, and licking the kittens. The queen is responding to the fluids, rather than to the kittens, at this time.

The queen is typically very restless, and a normal protocol will list lying down, sitting up, licking of the vulva, squatting, bracing lordosis, circling, walking around the cage, lying down again, rolling, licking of a kitten, and so on. There is not a consistent pattern, and the behavior of the queen varies with the endogenous stimuli (from the uterus) and exogenous stimuli (from the birth fluids and kittens). The uterine contractions of labor can be distinguished from fetal movements because the raised hind legs of the queen will usually flex when she is in labor.[1271]

Most cats prefer solitude, though some highly socialized ones seem to be content only when the owner is present. Cats, with the exception of Siamese, are usually quiet during parturition. The restless behavior of the queen serves to stretch the umbilical cord of the newly delivered kitten. When the placenta is delivered, the queen will eat it and part of the cord with the same tilt of the head and pronounced chewing motions that she shows when consuming prey. In the process of eating the placenta, the queen stretches the umbilical cord so that there is little bleeding when the cord is severed. It is rare for the eating to extend to cannibalism of the kitten. The interval between kitten births can be as long as 1 hour even in normal births but is usually much shorter. As already mentioned here, the queen seems unaware of the kittens despite her bouts of licking them, as demonstrated by her inadvertent stepping on them in the course of her pacing and her ignoring of their cries. The bursts of activity are interspersed with periods of fatigue.

When the last kitten has been delivered, the queen directs her attention to her litter. She lies down with an encircling motion, positioning her legs in such a way as to form a U around the kittens. For the next 12 to 24 hours, she rarely leaves the newborns and then only for brief intervals to eat, drink, and eliminate. Each time she returns, she arouses the kittens by licking them, after which she encircles and then nurses them.

Suckling

Finding the Teat

Most kittens are suckling within an hour or 2 of birth. The cues used by the neonate to find the mammary glands are unknown in this species as well as in the other domestic species. The kittens probably use temperature cues but also the mobility and responsiveness of the adult cat to locate her. Kittens who are blind and deaf at birth apparently use smell and, probably to a greater extent, tactile sensations to locate the nipple. Kittens with anesthetized tongues can find the nipple but cannot suckle; kittens with anesthetized lips cannot locate the nipple but can suckle.[1321] Olfactory bulbectomy also eliminates the ability of kittens to find the nipple,[829] but damage to the olfactory bulb eliminates more than the sense of smell alone.

Kittens locomote by pulling themselves along with their front legs while paddling with the weaker hind legs. As they crawl forward, they turn their heads from side to side. When they encounter the nipple, they pull their heads back and lunge forward with open mouths. Eventually, the nipple is secured in the mouth. The position of the mother facilitates locating the mammary region. The responses of the kittens to the areola and nipple appear to be innate, almost reflex in nature.[1402]

Experiments conducted on artificially reared kittens revealed that before their eyes were open, they followed a path produced by their own body odors to find the nipple of the brooder. The kittens could also make tactile discriminations under these experimental conditions, learning to choose a nipple with bumps on it over one with concentric ridges when the former was associated with milk reinforcement.[1271]

Prolonged experience with suckling from an artificial teat (1 to 3 weeks) delays, but does not abolish, the kitten's ability to initiate natural suckling.[1218] Milk reinforcement is not necessary for suckling, as intragastrically fed kittens will initiate suckling on the teats of a nonlactating cat as rapidly as on a lactating cat, even on repeated trials.[823] The ability to initiate suckling disappears after 22 days of age if the kittens have been fed intragastrically.

Teat Order

By the 2nd day, kittens have a teat order determined, which is usually but not always followed. The largest kittens, however, do not appear to be those that acquire the best producing glands, in contrast to the teat order in piglets.[450] In some litters, no regular teat order is formed. Once the teat order is established, the kitten can use the presence of his sibling on either side to help guide him to the proper nipple, though sometimes the kittens obstruct more than facilitate one another's progress. Kittens massage the udder with treading motions of the paws. Treading persists in the adult cat, presumably as a pleasurable activity or in pleasurable situations.

The feline nursing period has been divided into three stages: stage 1, 1 to 14 days—mother initiates nursing; stage 2, 14 to 21 days—both mother and kittens initiate nursing; stage 3, 22 to 35 days—kittens initiate nursing. In the hormonally primed queen, the suckling of kittens is probably tactilely pleasant. However, as kittens grow older and their feeding demands become more persistent, the female develops an approach-avoidance type of behavior toward her kittens; she discourages their attempts to nurse from the 3rd week onward by moving away. Although in a home situation she could easily escape from the kittens, in a cage she can only jump to a shelf to which the kittens can soon gain access as their locomotor skills improve. Another ploy by the mother is to lick the kittens vigorously, thus preventing them from nursing. Figure 5.7 illustrates the change with time in the queen's relationship with her kittens. In a two-kitten litter, the time of weaning occurs later.[941] In the single kitten, these stages are not as discrete. The queen spends less time with a single kitten initially (30% versus 70% of her time), but she continues to allow it to nurse long after a larger litter would have been weaned. Apparently, a single kitten is less attractive than several kittens but also less aggravating. From the 4th week on, "barn" cats will begin to bring live or dead prey to their kittens (see Chapter 6, Development of Behavior); however, these females sometimes allow intermittent nursing until the next litter is born.

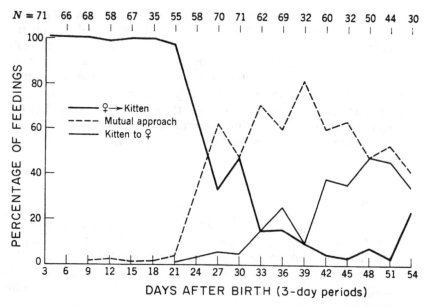

FIG. 5.7. Approach of the queen to her kittens and kittens to the queen during the first 8 weeks of life[1271] (copyright 1963, with permission of John Wiley & Sons).

Homing and Retrieval

Even before the opening of their eyes, kittens can return to their nest from several feet away. Apparently, they use olfactory cues, for washing the floor between the kittens and the nest prevents them from finding their way back.[1271]

Feline retrieval behavior is quite different from the canine form. Queens retrieve their kittens in response to auditory, not visual, signals. The more the kitten vocalizes, the more apt the queen is to retrieve him.[624] The queen usually picks the kitten up by the scruff of the neck, though occasionally she grasps the skin of the back of the head or even the kitten's whole head. Queens are able to lift and even jump several feet carrying large kittens. In fact, the peak of kitten carrying occurs when they are 3 weeks old. Picking others up by the scruff is an effective means of establishing dominance even over an adult cat, probably because the cat is being treated like a kitten.

Communal Nests

Cats frequently move their kittens to new nests, moving four to nine times before the kittens are weaned, a phenomenon many cat owners have observed. Various hypotheses have been put forth such as avoidance of predators or infanticidal males (if indeed there are any such males). In a seminatural situation the nests are moved closer to the food source.[464] A queen may share a nest with another lactating female where the kittens are suckled communally and where one cat can remain with the litters while the other hunts. The cats that share litters are members of the same group, but might not be related. Communal nesting has advantages to the kittens; they leave the nest earlier (20 days) than kittens in single-litter nests (30 days). Kittens in communal nests are moved to different nest sites more often than those in a one-litter nest.

Grooming

Grooming plays as important a part in feline maternal behavior as in most other species. Queens lick their kittens frequently and in particular lick the perineum to stimulate urination and defecation for the first 2 to 3 weeks of life. In common with most carnivores, female cats ingest the kittens' urine and feces for several weeks postpartum, thereby keeping the nest clean.

Acceptance of Kittens

Cats will accept alien kittens that are not too much older than their own at the time of parturition. Maternal behavior persists much longer in cats than in ungulates, so a queen whose kittens were removed at birth will accept one kitten weeks later and encircle it. Three kittens will be avoided or actually attacked under the same circumstances.[1271] In general, species that produce litters are more willing to accept foster young than are those that produce singlets or twins, probably because the mothers of litters do not discriminate between individual offspring.

Clinical Problems

There are few clinical problems of maternal behavior in cats. In fact, the efficiency of feline reproduction is much more of a problem. Occasionally a queen may reject her litter, but this happens less frequently than in other species.

Infanticide

Tomcats may kill their kittens, an abnormality of paternal, not maternal, behavior. Infanticide is rare in cats; there is only one authenticated case. There is no particular reproductive advantage to the male from infanticide in feline society, where many males can mate with each female[1066] (see Chapter 4, Sexual Behavior). Cannibalism by the queen, although infrequent, does occur, usually at parturition or shortly thereafter.

Mismothering

Cats sometimes care for one another's kittens or communally nurse.[914] An interesting variation on this occurred in a newly spayed cat that stole the kittens of another cat in the household. The problem was easily solved by shutting the natural mother in a room with the kittens. This case indicates that maternal behavior is independent of ovarian hormones and/or is stimulated by a dramatic decline in estrogen and progesterone levels as occurs at spaying and at parturition.

The nutritional, but not the social, development of orphan kittens will be adequate if they are artificially fed. Because cats socialize with one another almost exclusively as kittens, orphan kittens should probably be placed with the litter of another lactating queen if possible. Sometimes older kittens will show sucking abnormalities. They may suck on one another, on a human finger, on cloth, and on wool. The latter is a particular favorite of Siamese (see Chapter 8, Ingestive Behavior: Food and Water Intake). Occasionally, this behavior can be related to early weaning, but more commonly there is no immediately obvious explanation. The sucking is usually not injurious to the kittens' health and will gradually diminish in frequency.

DOGS

Parturition

The pregnant bitch gives little indication of her condition for the first 30 days. Late in pregnancy, her activity will decrease, and her appetite will increase. In the last weeks, she may wish to eat small, but frequent, meals as abdominal pressure increases. She may grunt each time she sits, especially if she is carrying a large litter. The slow development of the fetuses during the first half of pregnancy allows the bitch in the natural state to hunt and forage as well as ever. She is encumbered for only the last 2 weeks of the 60- to 63-day gestation period. At that time, she may even urinate or defecate in the house unless she is taken out frequently.

In contrast to the more independent feline behavior, when a nest box is provided, bitches usually will use it. If nesting materials, such as strips of rags or paper, are provided, the bitch will make a nest. Like the sow, she will make digging attempts in the absence of nesting material. Restlessness, inappetence, and a drop in body temperature are the cardinal signs of impending parturition. Once labor begins, the bitch usually lies in lateral recumbency. As labor progresses, her hind legs will twitch, and she will shiver; several strong abdominal contractions are visible just before the puppy is expelled. The fetal fluids appear to be attractive to the bitch. She will lick herself, any soiled bedding, and, almost incidentally, the puppy. The pup's head, umbilicus, and perineum are the areas most frequently licked. The bitch cuts the umbilical cord with her molars and then licks the stump. She may recut the cord if it is too long. If the placenta has not been delivered, she may extract it by pulling on the cord. The placenta is then nearly always eaten. She is usually silent as labor progresses, unless there is dystocia. Adult male dogs, if present, often whine.[209]

Pups are usually born at 30-minute intervals, but delays of a few hours are not pathological. Labor can be interrupted if the bitch is disturbed; the disturbance can be as mild as the entrance of another human observer. If she is resting between deliveries, the birth interval will be lengthened; if she is in the middle of labor, it may stop. Depending upon the temperament of the dog, 15 minutes to an hour are required for normal parturition to proceed.[209]

Although the bitch licks the neonatal puppy dry, she pays little more attention to it until all of the puppies are born. In fact, she may step on the puppies and ignore their cries at this time. Once all the puppies are born, she will lie quietly and allow them to nurse. She helps to orient the puppies to her by licking.

Suckling

The puppies move forward by paddling and turn their heads from side to side, as they cannot lift their bodies from the floor yet. If they encounter a wall or any cold object, they will change their direction. Once they encounter the mother's body, they nuzzle through the fur, usually attempting to burrow beneath her. High-chested breeds like Shetland sheep dogs present much more of a challenge to the pup seeking a nipple than do flat-chested breeds like cocker spaniels. When the pup locates a nipple, he does not immediately grasp it but rather noses it from beneath. Unless he opens his mouth at the same time as the nipple moves past, he will not succeed in grasping it. Puppies improve markedly in the first 2 or 3 days of life in their ability to locate nipples. Once the nipple is located, the pup jerks up with his head, pushes at the mammary gland with his front feet, and arranges his hind feet to support itself against the mother.[1196] All of this is not proceeding in a social vacuum; the other pups are also struggling for positions. The efforts of one will dislodge the other, and whining and scrambling will ensue. The supposedly serene maternal scene is, in fact, noisy and tumultuous.

Puppies do not appear to have as definitive a teat order as cats or pigs. The inguinal mammary gland is the preferred one and is the one approached by a single pup.[495]

Puppies have two vocal signals, the whine and the grunt. The whine is emitted whenever the puppy is cold, hungry, or separated from its litter or mother. Whining will stop immediately if the puppy's head and neck are covered with a warm towel or if it is again placed with its litter. Puppies show a distinct preference for soft surfaces; they will spend more time on a cloth-covered versus a wire-covered artificial mother.[727]

Textural, as well as olfactory, cues may help puppies locate their mother and her mammary glands. The grunt is apparently a pleasure communication that occurs when sought-after warmth or reunion is obtained. Despite the puppy's loud vocal response to separation, the bitch appears to notice that a puppy is missing when she sees it rather than when she hears it.[209]

Licking

Bitches lick their puppies a great deal. Licking serves three functions in puppies; two are common to other species, and one is unique to dogs. The licking arouses the pups to eat, and, when it is directed at the anogenital area, it stimulates urination and defecation that would otherwise not occur spontaneously. The bitch keeps the nest area clean by consuming the urine and feces of her puppies for at least the first 3 weeks of their lives. The third function of licking is retrieval. Bitches seldom carry their puppies. Instead they lead them back to the nest by licking the pup's head. The pup will orient toward the bitch and move toward her. The bitch will back toward the nest, while continuing to lick the pup that follows, until the nest is reached. Licking can be reinstated by substituting young (2- to 3-day-old) puppies for older ones.[828]

Weaning

During the first few days, the bitch spends most of her time with the pups. Nursing reaches a peak at the end of the 1st week. The amount of time that she spends with them decreases with time (Fig. 5.8). The undisturbed litter will be weaned gradually by the bitch; weaning usually will be complete by 60 days. During early lactation, the bitch always approaches the puppies to initiate a nursing bout. By 3 weeks, the puppies have opened their eyes and can locomote well. They then approach the bitch and initiate most of the nursing bouts. Bitches rarely punish their puppies until the 3rd week. Even during the weaning process, the punishment is mild enough that it may momentarily deter the pups' at-

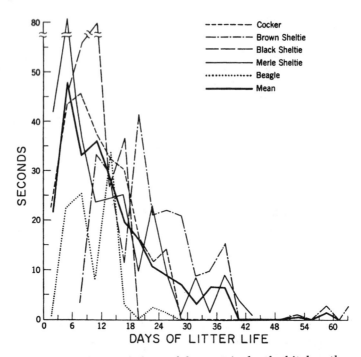

FIG. 5.8.   The decrease in licking of the puppies by the bitch as the puppies mature[1325] (copyright 1963, with permission of John Wiley & Sons).

tempts to nurse, but it will not inhibit them from trying again. The punishment may consist of a growl, a snarl, or an inhibited bite. The level of aggression of the bitch toward her puppies increases and the number of nursing bouts per hour decreases after the puppies' 3rd week.[1493]

Some bitches may regurgitate food for their pups during the weaning process at 4 to 6 weeks.[926,927,944] This behavior, commonly seen in wild canids, helps to maintain adequate nutrition in the young during the transition from a milk diet to a raw meat diet, when their powers of mastication and digestion may not have matured enough to enable them to survive on the raw meat.[495] Wolf pups beg for regurgitated food by licking at the mouth of the adult. This behavior is occasionally seen in domestic dogs and may be the basis of face licking and licking intention movements directed toward people by adult dogs.

Clinical Problems

Pseudopregnancy

Pseudopregnancy in the intact, nonpregnant bitch during the luteal phase of the canine reproductive cycle is so common that it may be normal behavior. Bitches do vary, however, in the severity of the behavioral signs. Some dogs show only slight enlargement of the mammary glands, whereas others go through pseudoparturition at 49 days postestrus and actually lactate. The pseudopregnant bitch becomes less active, mimicking the slowing down of the truly pregnant animal. She may make a nest, usually either in her own bed or in some cavelike environment, under a table, or in a dark corner. She may adopt a toy, a leash, a shoe, or some other object to mother. She will not only place it in her nest and assume a nursing posture next to it, but she will often defend it. Serious problems with aggression can arise in the pseudopregnant bitch. She may be generally aggressive, but, more commonly, she will only attack when her nest is invaded or her "offspring" are threatened. Mibolerone, an androgenic anabolic steroid, will relieve the signs of pseudopregnancy, but ovariohysterectomy should be recommended. The bitch who shows recurrent cycles of pseudopregnancy is prone to uterine infections, pyometra in particular. It is advisable to wait until after the behavioral signs of pseudopregnancy have passed before performing the ovariohysterectomy, as protective behavior persists if the animal is operated upon while pseudopregnant (B.L. Hart, personal communication). Pseudopregnant bitches can adopt and successfully raise foster puppies.[492] A bitch spayed during late metestrus may also exhibit material behavior including maternal aggression for a few weeks.

Maternal Rejection

Maternal rejection also occurs in bitches. It is most common in dogs that have undergone cesarean section and have been anesthetized during the time they would normally be licking and smelling the neonatal puppies. It rarely happens if the bitch has delivered a puppy normally before surgical intervention was necessary. The bitch may tolerate the first nursing better if some milk is expressed from the engorged glands before the pups suckle. She can be restrained while they nurse for the first time and sedated lightly if necessary. Attacking of puppies at some time after parturition is less common but does occur.

Moving Pups

A nervous bitch, especially one housed with other dogs or in an area with too much commotion, may repeatedly move her puppies. They will not be nursed often enough. Providing a nest box in a quiet room with no other animals and minimal visitors can solve the problem.

A related problem is continued carrying of pups in circles and without much other maternal behavior. Isolation and tranquilization may be helpful.

CHAPTER 6

DEVELOPMENT OF BEHAVIOR

Ontogeny in all the domestic species involves a decrease in time spent in close proximity to the dam and in sleeping. Especially in ungulates, time spent foraging for food or grazing increases. Time spent with peers increases, and much of the time is spent in play. There are sex differences in play; males spend more time play fighting and mounting than females. In cats and dogs, a sensitive period for socialization has been identified. These periods probably exist in the more precocial species as well, but occur within the 1st days rather than within the 1st months of life. Temperament tests have been devised for dogs, cats, horses, and pigs.

INTRODUCTION

One of the most pleasant aspects of owning animals is watching the young develop. Even in a world overpopulated with cats, kittens have not lost their attraction. The gangly foal and the playful kid are also very appealing. The veterinarian or animal scientist will find that there will be more questions about normal developmental behavior—when to take a puppy home and when to start various types of training—than are asked about adult behavior. To answer these questions correctly, the clinician should be familiar with the neurological development and behavioral maturation of various species because owners spend more time observing infant than adult animals.

DOGS

Critical or Sensitive Periods

During the past 20 years, the behavioral concept of critical periods has emerged, a concept that has had a strong impact on practical dog handling. Recently, the term *sensitive period* has replaced *critical period*. A sensitive period is a time in the life of an animal when a small amount of experience (or a total lack of experience) will have a large effect on

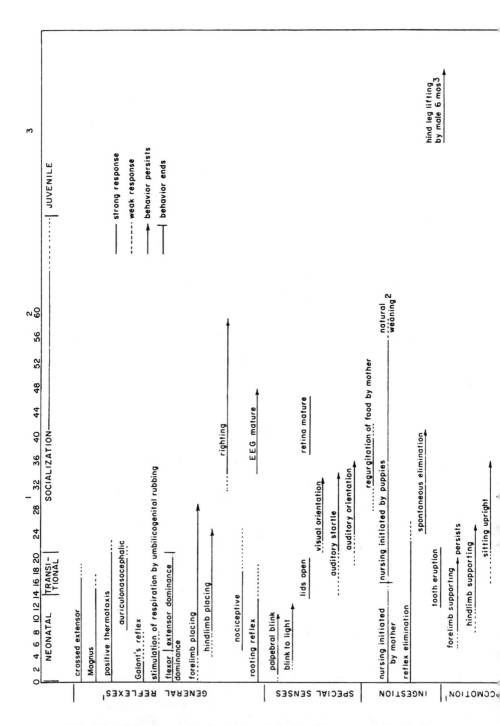

FIG. 6.1. Behavioral development of the dog. The *superscript numbers* refer to the references: (1) 494, 497; (2) 1196; (3) 188; (4) 209; (5) 1284; and (6) 500, 501.

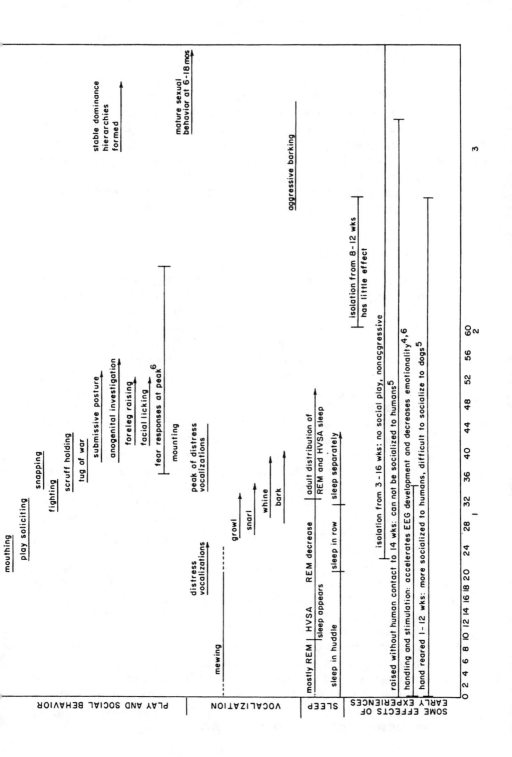

213

later behavior. The critical periods for dogs are defined as the neonatal period (1–2 weeks), the transitional period (3 weeks), the period of socialization (4–10 weeks), and the juvenile period (10 weeks to sexual maturity).[518,1283,1285] Nevertheless, socialization can occur in older animals, albeit with difficulty. For that reason, Bateson[146] has suggested that the adjective *sensitive*, rather than *critical*, be used with these periods. The sensitive periods are not sharply defined and may vary among breeds that are fast or slow to mature. Cocker spaniel puppies appear to mature more slowly than basenjis, for example. There are marked neurological changes during development that have been well described by Fox.[494] See Figure 6.1 for a canine development chart.

Neonatal Period

The neonatal period, during which the puppy is deaf and blind and able to find the nipple only through tactile and olfactory cues, was described in Chapter 5 (Maternal Behavior). Most of the puppy's time is spent eating and sleeping. The sleep is characterized by a high proportion of rapid eye movement (REM) or stage IV sleep. Urination and defecation do not occur spontaneously but can be elicited by stimulation of the anogenital area; usually such stimulation is supplied by the mother's licking. Puppies locomote with their front legs, pulling their hind legs along.

During this period, several reflexes are present that will gradually disappear as the central nervous system matures. One can use the presence or absence of these reflexes to determine the age of a normal puppy and to assess development if pathology is suspected in a puppy of known age. For the first 3 days, puppies show flexor dominance; that is, when picked up by the scruff of the neck they will flex their legs. From 3 days until the 4th week of life, there is extensor dominance; that is, the puppy will extend its legs when it is picked up. Gradually, normotonia appears.

The Magnus reflex is present for the first 2 weeks. The reflex is demonstrated by turning the pup's head to one side. Both front legs and hind legs on the side toward which the head is turned are extended; the legs on the opposite side are flexed. The crossed extensor reflex is also demonstrable for the first 2 weeks. The reflex is elicited by pinching on a hind foot; that foot is withdrawn while the opposite leg is extended. The rooting reflex is best demonstrated after the puppy is a few days old. The puppy will push its face into a cupped hand and crawl forward. This is the reflex utilized by the mother to retrieve a puppy (see Chapter 5, Maternal Behavior). Like the Magnus and crossed extensor reflex, the rooting reflex will wane by the 4th week.

Transitional Period

During the transitional period, the puppy begins to be bombarded by many more stimuli as his sensory organs develop. The eyes open between 10 and 16 days, although visual acuity is poor and puppies will not follow visual stimuli when the eyes first open. As vision improves, the puppy will no longer swing its head from side to side as it locomotes. The ears open and a startle response to auditory stimuli can be elicited at 14 to 18 days. By day 16, sound can be localized.[88] The crossed extensor reflex disappears from the front legs first, as does the Magnus reflex. Urination and defecation occur spontaneously; the bitch continues to ingest the excreta for several weeks. When kept in a kennel, the puppies will begin to leave the nest to eliminate and will use the same area as that of the bitch. If the bitch is paper trained this would be the ideal way to train a puppy, long before it can follow its mother outdoors. The puppy can support its weight on all four legs by 12 to 14 days, although normal adult sitting and standing will not be seen until 28 days. Tooth eruption begins to take place during the transition period, and the pups will chew on one another and begin to play clumsily and growl.

Socialization

The third period, socialization, is the most important from a behavioral viewpoint. During this time from the 4th to the 14th week, pups learn about their environment, about their littermates and mother, and about humans. Play begins and has its highest frequency in the socialization period. Canine play is discussed in more detail later in this chapter. Dominance hierarchies are formed. Strong avoidance behavior develops, and by 8 weeks, fear reactions are seen.

This period is also most important from a clinical standpoint. Puppies weaned and removed from the company of other dogs before the period of socialization will, as adults, often be difficult to handle in the presence of other dogs. They will either be frightened of other dogs or, less commonly, too aggressive. They will not be able to play with other dogs and will be difficult to breed. A dog that has not had the opportunity to interact with other dogs will be too human oriented. Male dogs may direct their sexual attentions toward humans.

Normally, pups are not weaned before 4 weeks of age, but they may be weaned even earlier if the bitch has died and the puppies must be hand fed. The nutritional requirements of the orphan puppy can be met, but many of the tactile and social requirements may not be. Hand-fed puppies usually will suck more on fingers and other objects than will normal puppies, indicating that the need to suckle has not been met, despite scheduled bottle feedings.[1224] It would be best to foster a litter of orphan pups onto another lactating bitch. If this is not possible, the litter should be kept together and separated only for an hour after feeding when suckling on one another is most apt to occur. A similar situation may arise if the bitch has lactation tetany, in which case the litter may have to be weaned quite early. Again, the litter should be kept together at least until the puppies are 6 weeks old.

The importance of social contact to the puppy during this period is demonstrated by the emotional reaction to separation from the litter. Six-week-old beagle puppies yelp 1400 times per 10 minutes when placed in a strange pen. Older and younger dogs are less disturbed[430] and, therefore, less vocal (Fig. 6.2). It is surprising that human contact

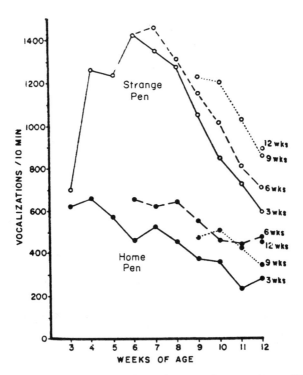

FIG. 6.2.   The average number of vocalizations by puppies at different ages and in two environments. Note that vocalization when alone in the home pen is uniformly lower than that in a strange pen. Puppies whose tests were begun when they were older than 3 weeks showed a slightly higher rate of vocalization, but the curves are parallel[430] (copyright 1961, with permission of the Helen Dwight Reid Educational Foundation, published by Heldref Publication, 4000 Albemarle St., N.W., Washington, D.C. 20016).

is more effective than canine contact in alleviating separation distress in 4- to 8-week-old puppies.[1132,1227] By 7 to 8 weeks, a fear posture, tail tucking, is first seen in beagle puppies.[361]

Socialization to humans is equally important. A dog that has had little contact with humans until 14 weeks rarely becomes a good pet.[1284] This is, of course, typical of kennel-raised, rather than family-raised, dogs and is sometimes called kennelitis. Such a dog is well socialized to other dogs but has had limited experience with humans. The dog, depending on his genetic background, may be overtimid or most difficult to dominate. Although normal dogs find contact with a human rewarding,[1351] the dog that has not been socialized with humans will not; it will, therefore, be a much more difficult animal to train.

There is a substantial amount of experimental evidence to support the sensitive period hypothesis. Most impressive is the effect of early isolation.[14] If puppies are completely isolated from the 3rd to the 20th week of life, they are markedly disturbed. Their learning ability is impaired.[1000,1389,1390] They are socialized to neither humans nor dogs, and their response to either species is fear. They are hyperactive and difficult to train. Beagles react most fearfully, and Scottish terriers are more hyperactive and show an impaired (higher) threshold to pain.[1001] Partial isolation from 3 to 16 weeks has different effects on different breeds. Beagles become less active; terriers more active.[527] Freedman[517] found that the early environment of puppies had no effect on the reaction of some breeds to mild punishment (saying "no" and hitting with a newspaper). Basenjis ignored the punishment; Shetland sheepdogs were always inhibited by the punishment. Beagles and wirehaired fox terriers were inhibited only if they had not been punished in early life. Even a week in isolation will produce changes in the canine electroencephalogram (EEG),[494] although the changes are transient.

Complete isolation is rarely, if ever, imposed on puppies except for experimental purposes, but exclusive human, dog, or cat contacts do occur. Dogs raised with cats prefer the company of cats to that of dogs and fail to recognize a mirror image.[494] Hand-raised or early-weaned (by 3.5 weeks) puppies will approach a human much more quickly than will dogs that were weaned at 8 weeks with little human contact before that time.

Not only the presence or absence of human contact but also the quality of that contact will affect the puppy's later behavior. It has been shown in many species that early handling can influence emotionality in later life; the dog is no exception. Dogs were subjected to varied stimulation (exposure to cold, vestibular stimulation on a tilting board, exposure to flashing lights, and auditory stimulation) from birth until 5 weeks of age. The stimulated pups differed from controls in several physiological and behavioral parameters. They showed earlier maturation of the EEG, larger adrenal glands, and lowered emotionality—

which enhanced problem-solving ability in novel situations—than the nonstimulated pups. Most interesting was the finding that they were dominant over nonstimulated controls in a competitive situation.[494] Some effects of early stimulation are not permanent. Development is accelerated, but the nonstimulated animal eventually catches up. Not only social deprivation but also food deprivation during early life can affect later canine behavior. Elliot and King[429] found that food-restricted puppies were more attached to their handlers during the period of deprivation and that they later showed increased eating rates and an increased intake of highly palatable food.

Juvenile Period

During the juvenile period, a dog increases in size and in competency at adult activities. Puppies begin to show adult sexual behavior at 4 to 6 months when they begin to show greater attraction to estrous bitches than to spayed ones. This attraction increases with age until they reach 2 years old, at which point dogs are fully mature.[153] Although considered to be adult at puberty, most dogs do not mature socially until 18 months or later.

Conclusions

The conclusions to be drawn from the experimental studies and clinical observations of the critical periods of development of dogs is that dogs should be exposed to both dogs and humans during the period of socialization. The exposure to both species should be pleasant because fear responses are also strong during this period. The effects of isolation are most pronounced in dogs that have been isolated during the period of socialization; however, isolation, or even partial isolation, in a boarding kennel for several weeks during any portion of its 1st year can reduce the sociability of a dog and increase its fearfulness. The importance of human socialization to canine training is exemplified by the study of Pfaffenberger and Scott,[1134] in which 90% of the dogs (mostly German shepherds) that were home raised from the 12th to the 52nd week of life were trainable as guide dogs for the blind. Dogs that remained in the kennel for the same period failed the training program. Properly socialized dogs will be much more willing to work for a reward as simple as verbal praise or even reunion with the human handler.

Neurological Development

There is an old adage that "One can't teach an old dog new tricks," but it is equally difficult to teach a very young one. As discussed in Chapter 7 (Learning), 6-week-old puppies could not solve a barrier problem, nor could 4-week-old puppies (even after 13 days of training) remember the location of hidden food for more than 10 seconds, although 12-week-old puppies could remember for 50 seconds. Puppies less than 21 days old cannot learn to pull food into their cage with a ribbon,[13] and 5-week-old puppies took twice as long to learn a visual discrimination as 12-week-old ones.[494] Neonatal puppies, however, can learn to avoid an aversive stimulus.[1350]

The poor performance of the young puppy is not surprising in light of the stage of development of its nervous system. There is only 10% dry matter in the brain at birth. The adult percentage (19%) is not reached until the 4th week of life.[494] Myelin is almost completely absent from the newborn puppy's brain and appears gradually over the next 4 weeks. Conduction speed along nerves is related to the presence of myelin and the more rapid reactions of the month-old puppy attest to the myelinization of its central nervous system. At birth, the length and width of the canine brain are nearly equal. The increase in length of the brain with age is due to an increase of the frontal and occipital areas. A great increase in the complexity of the gyri and sulci also occurs.[494]

Placing measures the ability of a dog to put its paws onto a table when held up to it, usually without visual contact. Placing reactions mature in the following order: chin placing (if the dog's chin makes contact with a surface, he will reach for it with his paws), visual placing (if the dog sees a surface, he will put his paw out), contact forelimb placing (if the dorsal surface of the paw is touched to the lower surface of the table, the dog will place the paw on the upper surface), contact hind limb placing and tail placing (if a dog's tail makes contact with a surface, the dog will reach toward that surface with its hind legs). By 6 to 9 weeks, all these reflexes are functional.[344]

Sleep

Sleep shows many changes in duration, type, and posture with development. The newborn puppy spends most of its time (96%) sleeping except for brief nursing bouts. Most of the sleep in the neonate is REM or stage IV sleep. Only 1% is slow wave sleep (SWS). Rapid eye movement sleep is associated with dreaming in the adult human, but one wonders what the newborn puppy dreams of, given that its experience is limited to intrauterine life. Owners are often concerned about the twitching exhibited by puppies during the neonatal period, but this is normal.

Newborn pups sleep in a heap, which may serve to prevent heat loss. A puppy removed from its littermates will waken and whine until it is either returned to the litter or placed on a soft warm surface. Holding a puppy will often calm it, probably because of the warm body contact. As puppies mature, the percentage of time spent in REM sleep drops from 85% at 7 days to 7% at 35 days. Meanwhile, the percentage of time that the dog is awake has increased to 62% by day 35, and SWS occupies the other 31% of the 24 hours.[499] By 3.5 weeks, puppies sleep in a row with side contact only. Later, they will sleep apart but may sleep against a wall for contact.[1196] Even adult dogs often try to maintain contact while sleeping by curling against their owner or simply lying on the owner's foot. Much of the nocturnal distress of the newly weaned or separated puppy can be alleviated by providing it with a warm "companion" such as a hot water bottle or even an old, and preferably dirty, sweater with lots of olfactory stimuli.

Play

Puppies, kittens, lambs, and even foals are attractive both because their foreshortened faces and awkward gaits inspire maternal, or at least protective, attitudes in humans[902] and also because they play. Play remains an enigma not because we do not know how animals play, but because we do not know why they play. Play appears to be important in the development of the social organization of animals, but that does not explain solitary play. Play may be important simply as a form of exercise or perhaps as a means of practicing and perfecting the skills necessary for the hunt, in the case of carnivores, or the escape, in the case of herbivores. None of the above explanations explain adult play and why it persists more in some species and in some individuals than in others. Finally, play is presumably pleasurable and may, therefore, be its own reward whatever the ultimate value to the organism may be.

Play in puppies begins when they are about 3 weeks old, with mouthing of one another. The mouthing is concentrated on the head region of the opponent. This should not be surprising because it is the cranial nerves that are most myelinated in the suckling animal. The biter and bitten will get maximal sensory input from play that involves the puppies' heads. As the puppies' strength improves and as their teeth erupt, the mouthings become genuine nips. Four-week-old pups may nip painfully, but the violent reaction of their littermates and, in particular, their mother to painful bites soon teaches them to inhibit the force of the bites. The early weaned or orphaned pup will not learn to inhibit its bites; it is up to the owner to punish, albeit mildly, the painful nip. The worst disfavor one can do a puppy is to wear heavy gloves when playing with it; the dog will not learn to play gently or to be submissive to humans. Safe chew toys should be substituted for human hands.

Play Fighting

By 4 to 5 weeks, play fighting becomes more skilled as the puppies' motor and perceptual skills improve. Scruff holding and shaking or worrying appear. Pouncing, snapping, and growling occur in the course of play. The facial expressions of the adult dog replace the masklike expression of the younger pup. Tug-of-war is a favorite game with littermates. Wrestling bouts occur with the puppies alternating the standing-over and lying-on-the-back positions.

Sexual Play

Elements of sexual behavior appear at 6 weeks. The puppies will mount, clasp, and perform pelvic thrusts without regard to the sex of the partner. Male puppies, in particular, exhibit this behavior. Dogs deprived of all play experience and social contact as puppies can mate but are often misoriented when they mount and, consequently, achieve fewer intromissions.[149] The poor sexual performance of socially isolated dogs indicates the importance of play in puppyhood to normal adult behavior.

Characteristics of Play Behavior

It is important to the participants as well as the observer that play be distinguished from serious behavior. This is particularly true of fighting behavior. By 3.5 weeks, puppies can effectively signal that "what follows is play." The signal most often used is the play bow, in which the dog lowers its forequarters and often paws at its own face while wagging its tail (see Fig. 1.10C). The play bow is an innate, not a learned display, for it occurs in hand-raised puppies.[178] The play face is distinguished by an open mouth and erect ears. Other signals are the exaggerated approach, repeated barking, approach and withdrawal, and pouncing and leaping. A submissive dog is more successful in soliciting play than is a dominant one. Perhaps this is because the dominant dog usually is taken seriously by its subordinates.[177]

Play in the dog, as in all species, is characterized by actions from various contexts (aggressive, sexual, and so forth) incorporated into unpredictable sequences in which the actions are repeated and performed in an exaggerated manner. A typical sequence would begin with a play bow, followed by an exaggerated approach, veering off, a chase, general biting, head shaking while biting, rolling and wrestling, reciprocal chasing, more wrestling, inhibited biting, rearing, and pushing with forepaws. Typical bouts last 5 to 15 minutes in puppies between the ages of 3 and 7 weeks. The more play exhibited by young canids, the less true aggression is manifested, as shown in a comparison of the pups of three canid species—dogs were the most playful, coyotes the least, and wolves were intermediate. An analysis of play and fighting can be used to identify canids of unknown genotype.[179]

Play is not only valuable for the development of normal behavior; it is also diagnostically useful. Play occurs most often in the warm, well-fed, healthy puppy. The absence of play behavior in 3- to 9-week-old puppies is an indication of pathology. Social play is the most common form in dogs, but solitary play does occur. The dog's pouncing upon and carrying a stick is an example. Tail chasing occurs in the absence of another puppy to chase. Games of fetch between owner and dog are the outgrowth of chasing play. If the game is not initiated during the period of socialization it is very difficult to teach,[1284] especially to a dog that is not genetically a retriever.

Exploratory behavior may be classified as a thrill-seeking type of solitary play. Exploratory behavior increases with age, not decreasing after 10 weeks of age as social play does. By 6 weeks of age, the puppy has mastered many of its social skills. It can signal play and aggression. It approaches another dog and investigates the inguinal area. It is beginning to form a dominance hierarchy. It eliminates in the same area as its mother and littermates do. It can eat food and sleep alone. In the next week or 2, it should be ready to become socialized with humans.

Toys are important to adult dogs, and they will spend 24% of their time using toys, but this can actually result in a decrease in time spent playing with other dogs.[706] Nylon bones are used longer than rawhides, although the latter are preferred initially.

Temperament Tests

Puppies that are exploratory at 6 weeks may not be at 12 weeks and vice versa. The same is true of social dominance.[1517] These findings indicate that tests of puppies at 7 weeks, popular as a means of predicting adult behavior, are unlikely to be valid. The tests involve scoring the puppies' reactions to handling and their willingness to approach or follow people. Activity is a better predictor of later behavior than is the puppy's response to handling at 7 weeks.[164]

Some dogs are surrendered to shelters because they have behavior problems. Because the former owner may not have been forthright about the problem, a temperament test has been developed to identify behavior problems in adult dogs. Barking, separation anxiety, and aggression toward cats, joggers, or other dogs could be predicted, but aggression toward the owner—dominance aggression—was not.[1423] A very extensive temperament test has been developed by Netto and Planta[1071] to evaluate dogs for aggressive tendencies.

CATS

See Figure 6.3 for a feline development chart.

Critical or Sensitive Periods

Feline sensitive periods, occurring at 2–7 weeks, are earlier than in dogs.[1416] A litter of kittens isolated for the 1st month will be reluctant to approach people even if they are genetically friendly (see later here). Handling for 15 minutes per day from 2 to 6 weeks will result in friendly kittens.

The most detailed study on the role of early experience in adult feline behavior was that of Seitz,[1292] who separated kittens from their mothers at 2, 6, or 12 weeks. Kittens weaned at 12 weeks did not cry upon separation, even though they had been living on their mother's milk alone. Kittens weaned at 6 weeks cried for a day or 2. Those weaned at 2 days and fed by dropper cried for 1 week. As adults, the early weaned kittens showed the most random activity, such as trying to escape from a carrying cage, and were most disturbed by novel stimuli. When tested with food, they were most persistent in trying to obtain food secured under a wire cover but least successful in competing with other cats for food. The early weaned cats were also the slowest group to learn to associate the sight of a light with food.

Konrad and Bagshaw[827] also weaned kittens at 2 days of age. The kittens were fed by nipple and handled as little as possible. When tested in an unfamiliar room, the cats raised in the restricted environment explored, played, and approached less than conventionally reared cats. Kittens raised in isolation from 40 days of age spent more time close to another cat than did kittens raised communally, but all the cats spent more time with another cat as they grew older.[274]

Handling kittens each day for the 1st month accelerates eye opening and EEG synchronization. Such cats are more active and aggressive when confined and are quieter in a novel environment.[995] In another study, Meier and Stuart[996] found that cats that had been handled and raised in a stimulating environment made fewer errors in a visual discrimination task. Kittens raised with their mothers and without handling were very slow to approach humans. In another study, handling of the kittens appeared to impair their ability to learn some tasks several weeks later.[1492] It appears to be difficult to slow a kitten's development. Neither limitation of food nor severe hypoxic episodes affected kittens' development, although treatment with a goitrogen did delay development of solid food ingestion and locomotory skills, as well as slow physical development.[193]

The physiological basis of the behavioral abnormalities seen in early weaned kittens may be inferred from the changes in the function of the visual pathways observed in cats reared either in the dark or in an environment in which they had no, or very limited, visual stimuli, such as horizontal or vertical lines. Both behavioral and neurophysiological evidence demonstrated that the visual system, especially the cortical components, does not develop normally; cats exposed only to horizontal lines show little response to vertical lines.[205] Kittens raised without opportunity to see their front paws because they were either in darkness or wearing Elizabethan collars have difficulty in visual placing. They extend their paws appropriately, but may miss a small target that nor-

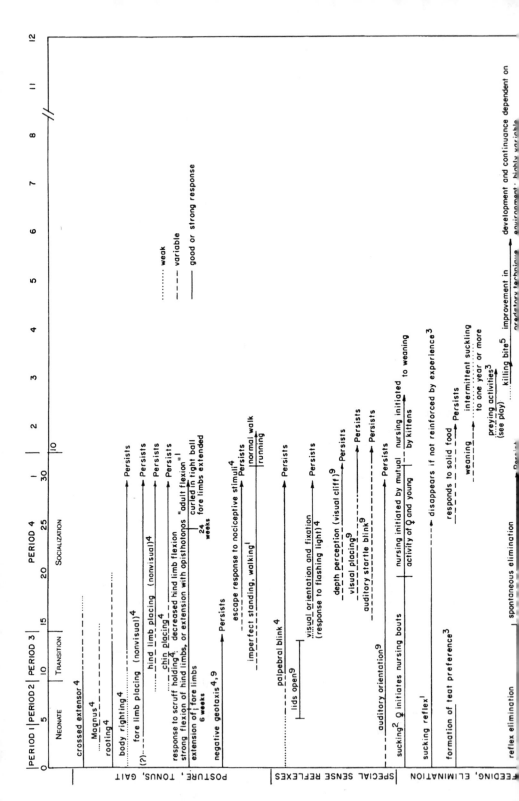

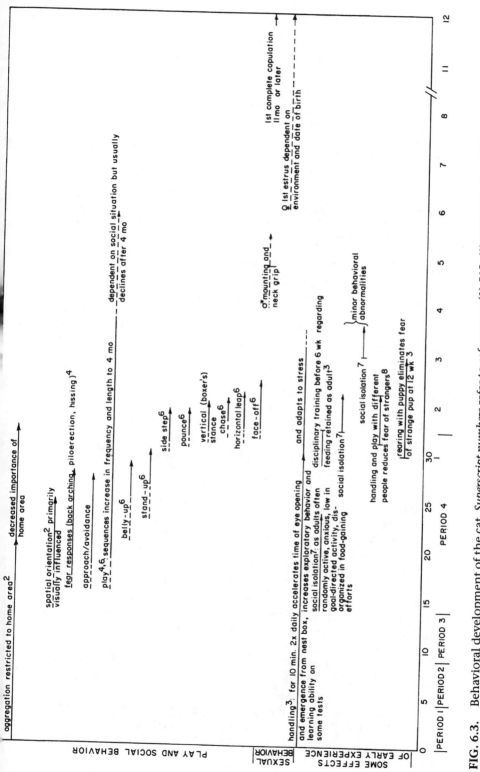

FIG. 6.3. Behavioral development of the cat. *Superscript numbers* refer to references: (1) 815, (2) 1223, (3) 496, (4) 493, (5) 874, (6) 1467, (7) 1292, (8) 310, (9) 1435, and (10) 1416.

225

mally reared kittens reach 95% of the time.[637] Kittens must learn to match paw position to target position. Similar changes occur in more complex behaviors: a cat that never had the opportunity to play as a kitten does not respond to the appropriate play signals as an adult. Kittens have adequate genetic capabilities to form the neuronal connections necessary for normal vision or social behavior, but the complex connections between cortical neurons form with visual or play experience during a sensitive or critical period.

Neurological Development

The neurological development of the cat has not been studied as systematically as that of the dog. The kitten shows a dominance of flexor tone for the first 2 weeks of life and then a dominance of extensor tone for the second 2 weeks. The motor cortex involved in forelimb movement develops during those first 2 weeks and cortical control of the hind limbs in the second 2 weeks. This is reflected in the locomotion of the kitten. It drags itself by its forelegs at first, but later the pushing movement of the hind legs grows stronger.[496,1197] The eyes open at 7 days (range, 6–10 days) and orienting responses to auditory stimuli develop a day or 2 before.[488] Between the 3rd and 6th week, cats develop the ability to land on their feet (air righting).

Visual acuity improves 16-fold between 2 and 10 weeks of age. The development of the cytoarchitecture of the sensory cortex is interesting in that the cortical layers of the kitten brain are arranged in an orderly fashion with few dendrites linking the cells. The adult cat brain possesses disordered layers with many dendritic processes on the cells, which apparently pull the cells out of the original orderly alignment. It is hypothesized that the interconnecting dendrites form with increasing sensory experience.

Adult cats and dogs will respond to a silhouette of their own species as they would to a real animal. Five-week-old kittens do not even orient themselves to a cat silhouette, but 6-week-old kittens will, and the frequency approaches the adult level by 8 weeks. Adult cats are apparently threatened by silhouettes and will show piloerection toward a silhouette on its first presentation. Five-week-old kittens show no piloerection and 6-week-old kittens show very little, but 8-week-old kittens show the adult response to silhouettes.[824] Hypothalamic (see Appendix 3) stimulation does not elicit adultlike affective response with piloerection and enlarged pupils until 3 weeks, although sensory motor responses, such as arching and jaw movements, can be elicited at 4 hours.[814]

Adult cats show a unique expression, the gape, to conspecific urine (see Fig. 1.13B). This response is not seen in kittens less than 5 weeks old and is essentially similar in frequency and performance to adult gaping at 7 weeks.[824]

Kittens can make ultrasonic vocalizations. In general, the frequency limit and range fall with age. Deafened kittens produce vocalizations similar to those of normal kittens, indicating that learning is not important; however, their calls are louder than those of normal kittens, indicating that feedback through the auditory system normally occurs.

As kittens mature, they become more proficient at finding their way back to their home area. They also vocalize less when placed on a cold surface (30 cries per minute at 1 day of age as compared with 17 cries per minute at 15 days). Once the eyes have opened, the kittens use visual cues to find their nest; prior to that time they use olfaction. Very young kittens will become less active and less vocal when placed on a warm rather than cool surface, but the calming effect of thermal stimuli is lost after the 1st week.[519,520,1220] Isolation produces most vocalizations (four cries per minute) at 3 weeks of age; younger and older kittens vocalize less. Response to restraint remains high and unchanged (five cries per minute) throughout development.[625]

Sleep

Sleep in kittens also shows a developmental pattern. For the first 3 weeks, the EEG cannot be correlated with the other behavior defining the different sleep stages, such as eye movements and muscle quiescence. Although the percentage of time that kittens are awake remains constant, the percentage of active REM sleep decreases and that of quiet sleep increases. Muscle twitching, which is characteristic of REM sleep, also decreases with age. The sleep cycles are also much shorter than those of the adult cat. Kittens also pass directly from the awake state to REM sleep; adult cats almost always pass through SWS sleep before entering REM sleep.[973] It is not until 3 months of age that forebrain maturation and environmental influences mediate a mature sleep–wake cycle.[675]

Play

Play in kittens is first seen at the beginning of the 3rd week at the same time that the queen begins the process of weaning by repulsing the kittens' attempts to nurse (see Chapter 5, Maternal Behavior). Play in cats has been most thoroughly studied by West,[1467,1468] Caro[276-278] and Martin and Bateson.[942] Although severe malnutrition leads to a suppression of play, less severe restriction of the food of a lactating queen leads to more play, especially more contact play, by her kittens. This is presumably in response to milk deprivation of the kittens and may be a form of early weaning—preparing the kitten to hunt for its own food.[147] Perhaps, some of the kittens whose play is too exuberant for their new owners were deprived while suckling.

Play in kittens begins with gentle pawing at one another. As kittens improve in coordination, biting, chasing, and rolling replace simple pawing. One kitten is usually in the belly-up position (kitten lies on its back with all four legs held in a semivertical position). Social play increases from 4 to 11 weeks and then declines relatively rapidly (Fig. 6.4). At first, three or more kittens may play together, but by 8 weeks, almost all play is between pairs of kittens. A reliable sign of play is the arched back and tail, but a definite play signal has not been defined in cats, although tail position and movement have been suggested.[1467]

Play Periods

There are usually four play periods per day. Almost an hour a day is spent in play at 9 weeks of age. Most kitten play bouts begin with a pounce and end with a chase. In between, the kittens frequently face off, hunching forward with tails arched out and down. They may bat at one another. Kittens also assume a vertical stance in play, rearing back on their hind legs, sometimes standing up by extending the legs. Various leaps are seen also. Kittens are much more apt than puppies to paw rather than bite at one another. The prevalence of pouncing, stalking, and chasing in feline play may be evidence that it is practice for hunting. During play bouts there may be one chase per minute. Play may occupy 9% of the kitten's total time and only 4%–9% of its energy expenditure, indicating that play may be important, but it is not calorically costly.[940]

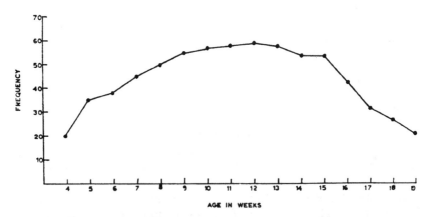

FIG. 6.4.   The change in playing behavior of kittens with age. Social play reaches a peak at 12 weeks and then declines. Frequency refers to the percent of the daily 90-minute observation period in which play was observed[1467] (copyright 1974, with permission of *Am. Zool.*).

Predatory Play

The mother plays an active role in the development of her kittens' predatory behavior. Not only do the mothers attack and eat prey in front of their kittens, but they also vocalize to attract the kittens' attention to it. These behaviors occur when the kittens are 4 to 8 weeks old. After that, the mother defers to the kittens in that she rarely kills and almost never eats the prey. The kittens are more apt to interact with the prey if the mother has just been interacting with it than if a littermate has, indicating that the mother has a greater influence.[277,278] Kittens also learn other tasks better from watching their mother than from watching another cat.[293]

There is a very definite increase in predatory activity around 8 weeks. At that time, most kittens will kill and eat mice, and more of their behavior is directed toward prey than toward playing with one another. Once the prey is dead and eaten, the kittens return to playing with one another, indicating that the motivation to play is still present but is overridden by the motivation to hunt. Social play and predatory play are not correlated and probably are controlled by different systems.[276] By 2 months of age, those kittens that will be frightened rather than aggressive toward prey and other cats can be identified; these same kittens are reluctant to explore and to relax with people in a new environment.[6] This is unfortunate because some people prefer that their cat does not hunt, and many wish their cats to be less aggressive toward other cats; but almost all owners want their cat to be friendly, even in a novel environment.

Sexual Play

Elements of sexual behavior are not seen in kitten play, but there is one sex difference in feline play. Males show more object contact than females; females with male littermates play with objects more than do females with no male littermates.[143] Play may be more important for intraspecies socialization in cats than it is in other species because the ancestral species are solitary for much of their adult lives.

Solitary play in kittens also begins to decline at 4 months, but the decline is much more gradual.[1467] Kittens will chase small rolling objects or even a moving string. They particularly like to bat at suspended objects, such as window shade pulls or tassels. Many of the pounces and face-offs of social play may be performed by solitary kittens with "imaginary" playmates, a mirror, or their own shadow. Solitary play persists in many adult cats. Playfulness is a factor for which breeders should select because it enhances the pleasure a cat gives to his owner as well as to himself. Social play may also occur between species. Cats will often play with dogs with which they are familiar. Interspecies play consists mostly of chases by the dog and pounces by the cat.

Several factors may contribute to the decline of play in kittens. Subadult cats begin to sleep more during the day. Older cats tend to spend more time sitting quietly but alertly. Male kittens show sexual activity by 4.5 months and attempt to mount and bite the scruff of females, who will reject these attempts until they reach sexual maturity a few months later. Young feral cats may also devote more time to finding their own prey. When canine and feline play are compared, dogs are found to chase (especially in a group), mouth, wrestle, shake, and indulge in solitary play more than cats. Cats stalk and ambush more frequently.[19]

Relationships With Humans

Handling of kittens during their sensitive period of socialization from 2 to 7 weeks is important.[772] Handling kittens for less than 30 minutes twice a week from the time the kittens are 5 weeks until they are 8 weeks does not increase their friendliness.[1190] More handling beginning at an earlier age does, especially if the kittens are genetically inclined to be friendly.[961]

Feline Temperament Test

There have been several attempts to categorize feline temperament. One can (1) record behavior in the home environment, whether that is the barnyard, the living room, or the laboratory; (2) record behavior in a structured test situation, usually while exposing the cat to a familiar and/or a strange person and a novel stimulus; (3) rate various characteristics (active, aggressive, agile, curious, excitable, playful, solitary, tense, vocal, voracious, watchful), reactions to other cats (equable, hostile, fearful, or sociable), and reactions to people (equable, hostile, fearful, or sociable); or (4) accept owner reports.[1002] The personality of cats seems to be composed of three characteristics: alertness, sociability, and equability.[458]

A laboratory test of sociability is that of Adamec,[5] who showed that as early at 10 weeks, cats differed in their reactions to strange people, to rats, and to the aggressive vocalization of adult cats. The cats that demonstrated fearful behavior in those three situations were classified as defensive and represented 25% of the population. These cats are the ones least likely to accommodate to a multicat household. Another kitten temperament test[772] reveals that cats that are vocal as kittens are vocal as adults, that very active kittens were less likely to spend time with people as adults, and that some kittens were timid.

Clinical Problems

Two common clinical behavior problems in cats are aggression between two or more cats in a household and rejection of the tom by the estrous queen. Both may be related to failure to socialize adequately to other cats as kittens. Kittens usually are removed from the mother at 6 weeks, long before the peak of playful interactions at 11 weeks. Cats that have remained with other kittens longer than 6 weeks may be more tolerant of other cats, including courting toms, as adults.

Playful behavior itself can be a behavior problem, particularly if it occurs in the middle of the night.[165] This is most apt to occur when the kitten has been alone, and probably asleep, most of the day and has not had much opportunity to play. Punishment may inhibit the kitten's play, but it is more likely simply to move out of range and to continue racing about and knocking over objects. A scheduled play period in late evening is the best treatment. Cat toys also help. Adding a second kitten might help because two kittens usually play with each other, but they may become incompatible as adults.

HORSES

The Foal's First Day

First Hour

The perinatal behavior of foals has been described by Waring,[1453] who studied American saddlebreds, and by Rossdale,[1228] who studied thoroughbreds and ponies. The foal can move its head and legs immediately after birth. The suckling reflex appears within the first few minutes. The suckling reflex is elicited by anything put into the foal's mouth. Righting itself to sternal recumbency and the first attempts to stand occur within the first 15 minutes, but the foal will fail in a dozen attempts, so it may be an hour before the foal stands. Pony foals can stand at a younger age than those of the long-legged breeds.

The foal begins to use all of his senses within the first hour. He experiences tactile stimulation from his mother's licking and will begin to respond and orient to visual and auditory stimuli within the first hour. Within the 1st hour, he will begin to communicate by nickers to his mother and by snapping (Fig. 6.10) at any fearful object. The foal can walk soon after he can stand, although he will not be well coordinated for another few hours. As soon as the foal can walk, he begins to search for the udder. He may attempt to suckle from the walls of the stall or from inappropriate parts of his mother, as well as suck when no oral contact has been made (vacuum suckling). The feature that the foal innately seeks is an underline, so he will attempt to nurse from the axilla as readily as the inguinal region. Defecation occurs within the 1st hour also.

The Rest of the First Day

Successful suckling is the major event of the foal's 2nd hour of life. Although pony foals suckle within the second half hour of life, a further 30 minutes is necessary before saddlebred and thoroughbred foals are able to suckle. By the 2nd hour, the foal has also begun to follow his dam or any other large moving object. Lying down is another difficult task for the foal to master but is usually accomplished within the 2nd or 3rd hour. The foal will then sleep; a few foals will sleep standing up if they have not been able to lie down but will fall down if they go into REM sleep. By the 3rd hour, the foal can also groom himself and gallop. Within the 1st day, the foal can play, urinate, flehmen, and graze as well as communicate, suckle, and locomote; in other words, he is already a well-coordinated, functional horse. See Figure 6.5 for the behavioral development of foal.

The First Year

The ontogeny of the foal has been extensively studied in Welsh ponies,[332,338,340] New Forest ponies,[1418] Camargue ponies,[238] thoroughbreds,[846,848,849] and Belgians.[134] There is little difference in the time budgets and rate of development of these four types of horses even though the environments varied considerably in the degree of confinement and amount of forage available. Thoroughbreds and other horse breeds are more apt to be managed intensively; therefore, differences in time budgets are more apt to reflect artificial feeding and stall restraint rather than breed differences in ontogeny. For example, Kusunose and Sawazaki[849] found that thoroughbred foals lay down more often while stalled at night than during the day on pasture.

The Mare–Foal Bond

The distance between a mare and her foal is proportional to the age of the foal, that is, young foals are closer to their dams than are older foals. It is the foal that is responsible for this proximity in most circumstances. He follows the mother; this changes when the foal lies down. Then the mare remains close to the foal either stand resting or grazing in circles around the foal.[332] This behavior wanes as the foal matures, but one can almost guess the age of the foal by how close the mare remains while the foal sleeps. The other circumstance in which the mare follows the foal is when the young foal ventures more than 10 m (33 ft) away.

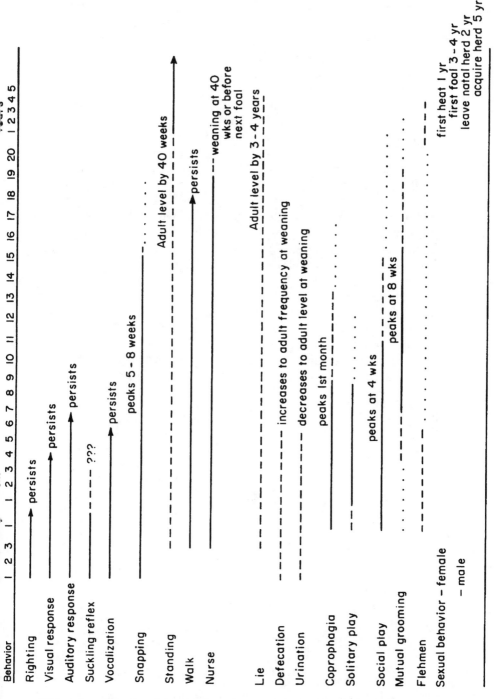

FIG. 6.5. Behavioral development of the horse. References: 190, 336, 337, 338, 340, 341, 409, 1452.

Grazing

Foals at first must spread and flex their legs in order to graze, especially if the grass is short; later, their necks lengthen in relation to their legs and they can graze more comfortably. Foals gradually increase the length of time that they spend grazing from 4 to 16 minutes per hour during the period from birth to 4 months. Thereafter, the increase is more rapid, reaching adult levels (60%–70% of the time) at natural weaning (40 weeks).

The development of feeding behavior is interesting because social facilitation plays such an important part. Foals graze only when their mothers are grazing[340] (Fig. 6.6). This illustrates the importance of providing creep feed for a foal in a location from which the foal can watch his mother eat. Drinking may not occur in foals on lush pasture. They obtain all their water needs from their dams' milk and moist grass, and although they follow their dams to water, they do not drink. In arid areas, foals do drink.[239]

Adult horses normally will avoid feces,[1102] but coprophagy is a normal ingestive behavior of foals, although its function is unknown. There may be both negative and positive consequences: the foal may ingest ova of parasites, but he may also be ingesting bacteria and protozoa that inoculate his gastrointestinal tract with the proper flora. The dam's feces are consumed in preference to those of another horse.[335,336,502]

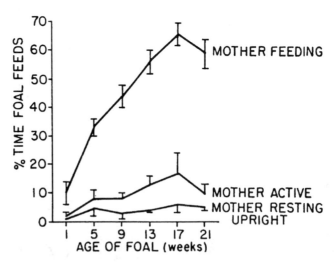

FIG. 6.6.   The mean percentage of time the foals spent feeding when their mother was feeding, active, or resting upright[340] (copyright 1985, with permission of *J. Anim. Sci.*).

Sleep

Foals rest either standing or lying. As foals develop, they spend less time lying and more time resting upright.[333] They spend a great deal more time lying than adult horses. The percentage of time spent in lateral recumbency decreases with the age of the foal from 15% (1st month) to 2% after weaning.[238] Resting in sternal recumbency does not change very much throughout the foal's first 6 months (averaging about 15% of its day) and is still higher than adult levels in 2 and 3 year olds.

Standing and resting also occurs in foals. The foal usually stands beside the mother, often facing in the opposite direction in order to take advantage of her tail to ward off flies.

Play

As the foal matures, he spends less time resting; he suckles less and grazes more. In between these activities, he plays. For the first 2 weeks, play is solitary. Foals gallop away from and toward their mothers, which may be a form of exploration or even thrill-seeking behavior.[19] Play in foals is one of the best examples of play as exercise. Seventy percent of locomotion in foals is in a play context.[453] Foals at first play with their mothers by nibbling at their legs and mane. Later, this will become true allogrooming. Social play with other foals gradually increases with age, and solitary play declines; by 8 weeks, solitary play rarely is seen (this, of course, would not be true of foals that do not have companions available; see Figure 6.7). In lone foals, solitary play persists and social play may include dogs and humans.[1265] Foals may also play with inanimate objects, such as twigs, tossing them into the air.

Although at 20 weeks of age foals still spend more than half of their time within 5 m (16.4 ft) of their mothers, they are most apt to leave her to play.[332] There are definite sex differences in play: colts mount and fight; fillies chase and mutually groom one another. Play in foals often centers about the head. Nipping of the head and mane, including gripping of the crest, accounts for the greatest number of play sequences. Rearing up and mounting is frequently seen, especially in colts. Properly oriented mounting is seen even in very young colts. Chases are a common play sequence.[1272] Side-by-side nipping can progress to circle fighting in which each foal attempts to bite the tail and legs of another. Figure 6.8 illustrates the types of play seen in pony foals on pasture.

When colts do mutually groom, they tend to groom fillies rather than other colts.[339] Grooming the mare is part of the courtship behavior of the stallion. These sex differences in play may prepare the animals for their adult roles.

Foals have a dominance hierarchy that is related to their age and

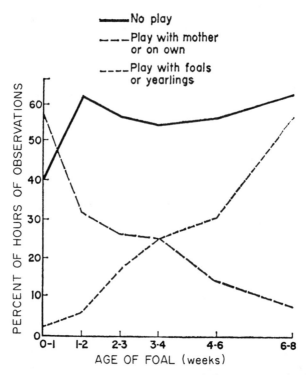

FIG. 6.7. Changes with age in foal's choice of play partners. As play with the mother decreases, play with other foals increases[1418] (copyright 1972, with permission of Academic Press).

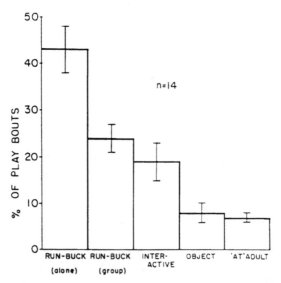

FIG. 6.8. Relative frequency of various types of play. The mean percent of total play bouts by 14 foals in which the type of play was running and bucking alone, running and bucking as a group, interactive, manipulation of an object, or play at an adult. Standard error of the mean is shown as a *vertical bar*[341] (Copyright 1987, wtih permission).

their dams' rank. Colts receive more aggression than fillies. Fillies are more likely to kick, and colts, to bite.[61]

Flehmen and Snapping

Flehmen behavior is also much more frequent in colts than in fillies. It is probably investigatory behavior. Flehmen peaks during the colt's 1st month, possibly because his dam will be in estrus then and/or because he is still somewhat precocially masculinized owing to his prenatal hormonal environment[340] (Fig. 6.9).

The facial expression of snapping (also known as tooth clapping or champing) occurs almost exclusively in foals and subadult horses (Fig. 6.10). This expression persists in zebra and donkeys as yawing, the mouth movements associated with estrus. In fact, the facial expression of snapping and that of yawing occur in the same circumstances, that

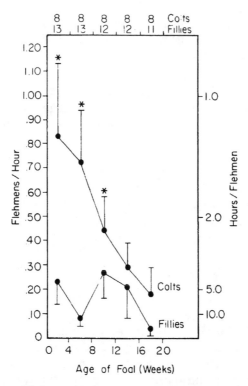

FIG. 6.9.   Rate of flehmen by foals of various ages. Colts exhibit flehmen more than fillies only during the first few weeks. *Asterisks* indicate significant difference[337] (copyright 1985, with permission of Academic Press).

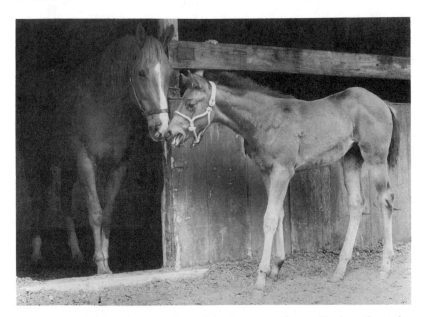

FIG. 6.10. Submissive snapping of the immature horse. Foal on the right approaches its dam while snapping[1507] (copyright 1980, with permission of Elsevier Science Publishers).

is, an approach-avoidance situation. A colt is most apt to snap when approaching a stallion to whom he is apparently attracted, but who is also frightening. The estrous donkey and zebra mares and, rarely, a submissive, estrous mare[1516] may be attracted to the stallion but may also be frightened. Snapping in foals is likely to occur when a stallion is courting the mare. The rate of snapping falls rapidly with age from one every 3 hours during the 1st month to one every 20 hours during the 6th month. The peak of snapping is during the foal's 2nd month of life.[338]

The Juvenile Period

Play and activity in general decrease with age; 2 and 3 year olds are more active than adults. Ninety-seven percent of colts and 81% of fillies leave their natal band.[1252] The age at which colts leave depends on whether there are colts of the same age in the band; if there are, the colts stay longer. The colts join other males, sometimes their older brothers, to form bachelor herds.[777] A great deal of the young bachelor's time is spent play fighting. Those colts that remain in their natal herd do not have this experience and appear to be slower to mature. The peak of colt play occurs at 3 years. At 5 years, he will begin to show adult male behaviors such as marking of urine and feces, true aggression toward other stallions, and driving of mares.[670] At 5, he is able to take over his own herd either by defeating an infirm harem stallion, replacing a dead one, or competing successfully against other bachelors for a filly that has left her natal herd.

PIGS

Pigs are intermediate in their development at birth. They can walk, albeit unsteadily, within a few minutes of birth; they can see and hear. Their brain development is not complete at birth, however, and some homeostatic mechanisms, such as temperature regulation, are not yet mature.

The neurological development of the pig has not been extensively studied despite the considerable clinical application such a study would have. Piglet mortality is very high—approaching 20%. Some of this mortality is due to infectious disease, but a considerable number of deaths are "accidental." The piglets may wander out of their pen and drown in a gutter or simply become chilled and die from exposure, or they may wander in with older pigs that will maul them to death. Other piglets are crushed by the sow even though she is in a farrowing crate. Normal piglets do not wander; they stay close to their littermates, the heat source, and the sow's udder. Piglets, especially those among the last delivered, may suffer various degrees of brain damage as a result of hypoxia during birth. If these brain-damaged pigs could be identified and hand reared or otherwise given extra protection, piglet mortality would fall.

When piglets are castrated, they stand and suckle less and lie more.[980] This response occurs in piglets castrated at any age from day 1 of life, indicating that they are able to perceive the pain.

Piglets have some unique physiological problems that affect their behavior. Piglets are born almost hairless and with little subcutaneous fat for insulation. Their small size, lack of insulation, and low energy reserves leave piglets very vulnerable to hypothermia.[1052] Piglets solve the problem of heat conservation behaviorally rather than metabolically. They huddle with their littermates, thus decreasing their surface area and, in effect, make one large animal from 12 small ones. If a heat source is provided, they will lie next to it. Unless the environmental temperature is over 26°C (79°F), a heat lamp should be provided.[1401] Piglets that do not huddle or that persist in wandering away from the sow and the heat source have very likely been brain damaged at birth. A quick inspection of the farrowing house can enable one to determine whether the temperature conditions are correct: if piglets are sprawled on the floor in extension, they are warm enough; if they are crouched on their sterna with their legs drawn under them, they are too cold.

Sleep

As with most newborns, piglets tend only to eat and sleep. Piglets sleep for 16 minutes of every hour. Like other young animals they spend a lot of time in REM, when they assume a crouching position in sternal recumbency rather than lying in lateral recumbency as they do in non-REM sleep. The number of REM bouts decreases with age, but each bout remains of similar length.[842]

Teat Order

Nursing behavior has been described in Chapter 5 (Maternal Behavior). The teat order is formed on the 1st day. The peak of aggression occurs 1 hour after birth. By day 6, only 10% of the piglets change teats.[652] To reduce injuries to the sow's udder or to the piglets that may occur during the formation of the teat order, it is common to clip the incisors and canines of day-old pigs.[510]

Feeding

Piglets begin to eat creep feed around day 12; their intake is less than 5 g/d, but within a week they are eating 10 times as much.[1104] Intake can be increased by providing more feeders so that advantage can be taken of social facilitation.[58] Piglets who start consuming solid food earlier will consume more.[59,218] Piglets that had access to the most productive teats gained less weight when first weaned. They had not learned to eat solid food as soon as their hungrier litter mates.[36] Piglets are more apt to eat solid food if the food is sweet, as suckling piglets possess a well-developed sweet taste preference.[686]

Piglets rest for most of the day during the 1st week but are active most of the time by their 8th week. In a seminaturalistic environment, piglets begin to graze by 4 weeks, and the percentage of time spent in this activity rises from 7% to 42% by 8 weeks.[1129]

Play

In a Seminatural Environment

When sows leave the maternal nest and rejoin the herd, the piglets are gradually integrated into the social life of the older animals.[1128] Piglets will choose to investigate a novel object and other forms of play, such as scampering, increase in the presence of novel objects.[1514] Play in piglets develops in the 2nd week of life and wanes by 5 weeks (Fig. 6.11). Failure to play is of diagnostic value in determining the seriousness of neonatal pig disease.

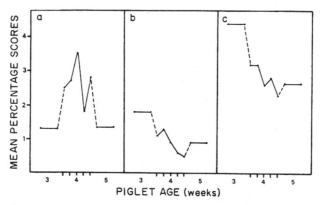

FIG. 6.11.   The change with age in play fighting and solitary play in piglets. The graphs give the mean percentage of minutes in which early weaned piglets were scored as *(A)* biting littermates; *(B)* scratching their bodies; and *(C)* scampering during the first 5 weeks of life, expressed as a percentage of the number of minutes in which the animals were scored as active. Piglets were weaned at the end of the 3rd week.

In a Confined Environment

Piglets begin to explore their environment by rooting, biting, and chewing within the 1st days postpartum. Male piglets play more frequently than females, and play occurs more frequently between same sex pairs than mixed sex pairs. Play in piglets is characterized by play fights. These fights are usually head-to-head confrontations in which each piglet chews and roots at the other's shoulders and neck. In older pigs (3–4 weeks old), chases and gamboling can be observed. It is at about the same age that the typical porcine startle reaction (a woof and freezing behavior) can first be elicited.

Exploratory behavior is very pronounced in piglets and consists of rooting and mouthing anything that is new in the environment. Pigs with an enriched environment, that with straw, logs and branches, spend less time manipulating the sow's udder and later spend less time nudging or tail biting.[1130] In the absence of manipulatable objects, piglets chew the floor and walls of their pens and one another. Therefore, toys should be provided. Pigs prefer hourglass rubber dog toys to chains, ropes or rubber hoses.[57]

Elimination

Pigs begin to eliminate only at the edges of their pens as they mature.[1470] In pet minipigs, this behavior can be used to facilitate housebreaking.

Relationships with Humans

An investigation into the effect of early handling revealed that piglets handled daily were not larger than nonhandled littermates but were more aggressive when penned with strange pigs.[1273] The effects of human handling depends on the quality of the handling and can have economic importance. For example, Hemsworth et al.[643] found that pigs treated pleasantly (stroked) grew more rapidly and had better feed conversion than those treated unpleasantly (pigs were shocked if they approached the handler). Handling of pigs from 0 to 3 weeks or from 9 to 12 weeks reduces their fear of humans.[642] Pigs living in enriched environments from birth to farrowing are more difficult to drive, possibly because they are less fearful.[162]

Hemsworth et al.[644] have found that gently handled pigs not only were less afraid of humans than harshly handled ones, they also came into estrus sooner. More intensive handling by an individual resulted in pigs spending more time with humans, avoiding them less and being easier to catch.[1378] Unfortunately, few farmers would be able to spend 10 minutes a day with each pig.

Temperament Tests

Restraint, by holding the pig on his back, has been used to assess active and passive personality in pigs.[661] There appear to be three important traits: aggression, sociability, and exploration, as measured by aggression to an intruder, social dependence, and response to novelty.[486] Pigs raised in an enriched environment are less tractable, and they vocalized more in an open-field test.[162]

RUMINANTS

There have been very few behavioral studies of development in ruminants. Despite the detailed knowledge we possess about growth rates in these food-producing animals, next to nothing is known about the daily activity of young ruminants or how their relationship to their environment and their peers changes with maturity. The change in suckling patterns with age is discussed in Chapter 5 (Maternal Behavior). Some general statements can be made. Young ruminants are born in an advanced state of development; they are true precocial animals. They can stand and walk within a few hours of birth and can apparently see and hear.

Lambs

By 30 days of age, young lambs will spend 60% of their time with other lambs, but for the first few weeks they stay quite close to their dams. This behavior is to be contrasted with the behavior of the kid, who is much more apt to stray. The lamb may depend on its proximity to his mother for protection, whereas the kid uses its own behavioral pattern, that of freezing, to protect itself.

Play

By a month of age, play is well developed in lambs.[1040] Play in sheep begins with investigation of one another when the lambs are only a few days old. Play consists of intentional butting, vertical leaps, rearing up on the hindquarters—which may be a play signal—and twisting the forequarters and kicking. Play may be centered around rocks or mounds so that one lamb can butt others from above. The lambs push one another, lay their heads on one another, and mount each other. Male lambs mount much more than females and are the only ones to "nudge," raising a foreleg under the belly of another lamb while standing close behind it. The sex differences in lamb play are mediated by prenatal testosterone.[1093]

Lambs form groups by a few weeks of age and rest, graze, and play together. Disportive or solitary play also occurs in which lambs gallop and leap. Adults may join the lambs in play. Lambs are especially playful in the evening. By 4 months of age, play begins to wane.[576]

Weaning

Natural. Weaning does not exactly parallel decline in milk production. Milk production falls gradually, but suckling stops abruptly when milk production reaches a threshold.[79] A study of free-ranging Soay sheep found that the ewe–lamb bond ceased just before the estrous period. The ewe lambs continued to follow their dams, but the ram lambs did not. None of the lambs slept touching their dams as they had done previously. The young sheep associated with their peer groups, the ewes remained in the dam's home range groups, and the rams wandered off to join a ram group.[576]

Artificial. Weaning lambs at 2 days causes a rise in cortisol and immune suppression. Lambs weaned at 2 or 15 days do not gain weight as quickly as unweaned lambs.[1064] When weaned at 3 weeks, lambs will bleat, but the bleating rate is lower if they are paired with another lamb, especially if that lamb is its twin.[1146]

Feeding

Lambs tend to prefer foods their dams eat, rather than avoiding what she rejected.[1021] They will eat low-quality roughage diets if their dams have done so.[379]

When suckling lambs were offered creep feed at 3 weeks of age, their intake was variable and they did not consume the solid food on a daily basis until 2 weeks later.[454] Grazing times of lambs who had had a week of experience with pasture prior to weaning were twice as long as those of naive lambs. Lambs will eat less of a shrub avoided by their dams,[1020] but this may not apply to more palatable foods.[1392]

Exposure to a novel flavor (onion and garlic) from a month of age for 3 months resulted in only a temporary preference for those flavors.[1077]

Social Relationships

The social relationships of sheep are formed in the first few weeks and appear to persist for life in the undisturbed flock, but this may vary with breed and environment. Whether a yearling ewe associates with her mother may depend on environmental factors such as high population density. The yearling may gain more weight, and the ewe will incur no costs. The ewe is followed by her lamb and the lamb is submissive to her. Even as adults, daughters will follow the mother, and their lambs will be close at the heels of their respective dams.[1278,1281]

Sexual Behavior

Sexual behavior develops gradually in ram lambs. Despite the sexual elements of play, 6-month-old lambs rarely even investigate a ewe in estrus. At 10 months, they investigate ewes in estrus, and at 13 months will mount; but only at 17 months is the complete mating sequence, including copulation, seen in all rams.[1497]

Sleep and Activity Patterns

Sleep can occupy up to 40% of the lamb's day, but only 15% of the adult's.[1241] Grooming behavior, scratching with the hooves or teeth, occupies as much as 9% of the lamb's time, but far less of the adult's.[576]

Relationship With Humans

Restraining lambs in a head gate (stanchion) and handling them did not increase their tendency to approach people; allowing the lambs to approach on their own did.[946]

Temperament Tests

Temperament tests in sheep consist of isolating the sheep, measuring its approach to a novel object and to a human. Ewes are more fearful than rams, and ewes given testosterone are less fearful.[1428]

Kids

Kids appear to be much livelier than lambs. Even in the confined environment of a pen, they will play a game that can only be described as "king of the mountain," as each kid leaps onto the highest available horizontal surface, which may be an overturned bucket or even another animal.

Lickiter[879] has described changes in activity patterns of goat kids. Lying decreases markedly around 4 weeks from 60% to 70% of the time to 30% for the next few months. Ruminating begins at about 4 weeks as they begin to graze. It is apparent that the change to a herbivorous ruminant accounts for the biggest change in behavior during development. Presumably, this is true of all ruminants and contrasts with the more gradual changes that can be seen in the foal. Standing increases somewhat at the time that lying decreases. By 15 weeks, grazing is the predominant activity, and behavior of the kid is synchronous with that of its mother.

Play

Other types of play are short bursts of running (less than 15 seconds), leaping into the air, kicking, butting and mounting (more in males than females), mouthing objects, and standing on the hind legs with the forelegs against a vertical object. Play occurs most at dawn and dusk in 1-hour periods separated by a period of rest. Restraint in a pen or other forms of play deprivation are followed by a rebound of play in which play may last for 3 hours.[292]

Vocalizations

The call of the kid changes during development. The fundamental frequency falls from 600 kHz on the 1st day to 250 to 350 kHz on the 5th day. By 4 weeks, there are sex differences in that the male voice is three octaves below the female voice. Call lengths vary. The orienting call is 1 second in duration, the distress call, 1.5 seconds.[865]

Social Relationships

The pattern of older animal dominant over younger is true of a stable herd of goats, but not if strange adult animals are mixed.[1360] The dominance that an adult nanny shows over an alien kid is apparently never challenged by the kid even when it is itself adult.

Temperament

Goat kids can be divided into timid and bold animals on the basis of their approach to humans. The behavior of the mother, and in the case of timid kids, that of even strange goats, affected the willingness of the kids to approach humans.[912]

Calves

The development of calves has not been studied in great detail. In particular, beef calves have not been studied, so most of the descriptions of play and ontogeny are based on Brahma or Masai cattle *(Bos indicus)* or on the primitive Maremma cattle *(Bos primigenius taurus).*

Calves, as with other ruminants, must change in a few weeks from simple-stomached milk-drinking animals to grazing ruminants. Rumination increases to occupy 7 hours per day by 7 weeks of age in calves kept on pasture. Sleep time declines to 4 hours per day. Calves tend to sleep in groups or "kindergartens." This attraction of calves for one another results in the calves' spending more than half of their time 15 m (48 ft) or more away from the mother. The mothers will leave their usual sleeping areas to rest near their calves.[1188] When calves that had been suckling were separated from their mothers, they spent more time with other calves. This effect is due to social rather than nutritional needs; if the dams remained, but nursing was prevented by cloth placed around the udder, the calves did not shift their social contact to their peers but instead maintained close contact with their mothers.[1433]

Play

Play in calves has been studied by Brownlee.[257] Calves, as well as older cows, have a special play vocalization, the baa-ock. Calves play in a variety of ways, often trotting or galloping with the tail elevated. They often buck, kicking up and to the side with both hind legs, but these kicks are not aimed at anything in particular. Directed kicking, which is playful rather than aggressive, may be aimed at a stationary or moving target. Calves also play by making noise with inanimate objects, such as buckets or latches. They may be increasing their auditory stimulation level just as children increase their vestibular stimulation level on slides, swings, and merry-go-rounds.[19] At 7 weeks, about 3 minutes per day is spent in play.[1231] This type of play, solitary play, is more commonly seen in young calves. Social play, especially frontal butting and mounting, predominate in the older calf. Play tends to occur during grazing bouts.[1440] Calves head butt with each other or with inanimate objects. This is the same action used by adult cows in dominance interactions. They may prance, paw the ground, or gore as adult bulls do, and even threaten human attendants. Soft snorting noises may accompany play. Mounting behavior is also seen in calves. Mounting and pushing play decrease with age, whereas butting increases.

There are sex differences in bovine play. Male calves play more than females, the same pattern that is seen in most animal play. Exploration peaks later than play; yearlings are more apt to investigate a novel object than older or younger cattle.[1057] Males mount, push, and exhibit the flehmen response more than females, but butting and social licking are seen equally in both sexes. Mounting and pushing by bull calves are usually directed toward other males, but the flehmen response is directed toward females.[1187]

As in other species, play can be used as a diagnostic criterion, for calves play more when well fed and healthy than when malnourished or ill. They also play more often in fine weather than in foul. Play is stimulated in calves by any change in the environment. They play most when let loose from confinement, after gaining access to new terrain, or even when new bedding is placed in their stalls. A new pen mate, the arrival of a human attendant, or even the stimulation of scratching their backs may set off a play bout in calves.

Activity Patterns

Weaned calves raised in individual pens spend their time in the following manner: standing, 40%; ruminating, 28%; feeding, 22%; grooming, 5%; and drinking, 2%. Such calves quickly learn to anticipate feeding times and become restless at those times. Many calves are able to make contact despite being penned separately, by making tongue contact through the opening where they are fed.[807] Calves raised in single pens are most likely to associate with the calves in adjacent pens when they are released in pasture. The singly raised calves rarely associate with group-raised calves. There is no difference in weight gain between the groups.[252] Dominance hierarchies are not stable even though artificially fed calves may compete fiercely for access to their feed.[273] Calves raised in isolation for the first 10 weeks of their lives had higher cortisol values when stressed as yearlings than calves raised with the opportunity to interact with other calves,[324] indicating the long-term affects of stress during development.

Social and Sexual Behavior

Bulls mount females by 9 months of age but do not usually achieve ejaculation until they are a year or older. Aggressive behavior increases markedly among bulls from 9 to 18 months, but stable hierarchies have not formed by 2 years.[1160]

Relationships With Humans

Handling influences cattle more if the breed has not already been selected for docility.[220] Handling in the first 10 days, or for 10 days when the calves are 6 weeks old, or just before or after weaning at 8 months all seem to eliminate aggressive behavior. Handling 6 weeks after birth and after weaning appears to be most effective.[221] Calves raised in isolation are more friendly to humans and learn more quickly; they are not handicapped in achieving dominance.[1166]

Temperament Tests

There are individual differences in the fearfulness of cattle that can be measured by their response to a novel object, environment, feed in an unfamiliar place, or a startling stimulus.[219] One cattle temperament test consists of leading, restraining in a corner, and stroking. There were genetic effects, a heritability factor of 0.22 for docility, and environmental effects as well. Cattle kept indoors were more docile than those kept outside.[862] Grandin et al.[568] has found a correlation of position of hair whorls and agitation in a chute. A whorl above the eyes may predict agitation.[567]

Behavior Problems

One problem of raising calves in groups is that they will suckle on one another, particularly on one another's mouth and ears and the scrotum, and only rarely, on the prepuce. The suckling occurs most frequently in the 15 minutes after a milk meal;[365] weaning from milk to grain reduces the frequency of cross suckling.[887]

Nonnutritive suckling is observed in calves following a milk meal drunk from a bucket. The motivation to suckle after drinking milk lasts for less than an hour. Calves who are hungrier before they drink the milk will suckle more afterward.[1245]

CONCLUSION

The drastic changes that are taking place in management of young ruminants have resulted in a much higher incidence of infectious disease. The role of behavioral stress in the etiology of neonatal pneumonias and diarrheas remains unknown. The economic value of these animals would dictate that more information is needed on their behavior in both naturalistic and highly artificial environments in order to best advise those who undertake lamb or veal operations.

CHAPTER 7

LEARNING

Learning occurs in all animals, but it is particularly important for the usefulness of dogs and horses. Testing relative intelligence among species and breeds is controversial, but all species tested can be operantly and classically conditioned and can form taste aversions. Housebreaking is the one essential task a pet dog must learn, and several methods are presented. The distinction between negative reinforcement and punishment is important because most horse training depends on the former, not the latter.

INTRODUCTION

Types of Learning

Learning in animals can be classified into various types. Horses will be used for most examples.

Habituation

Habituation, considered the simplest type of learning, is the long-term, stimulus-specific waning of a response, or learning not to respond to stimuli that tend to be without significance in the life of the animal.[1395] Horses habituate to the feel of a halter on their heads just as we habituate to the feel of glasses on ours. We make use of habituation when we try to desensitize horses to the sound of crowds at a horse show. Another example of habituation is a horse's response to traffic on a road beside its pasture. When first put in the field, he will react to traffic on the nearby road; later he will not. Other examples of habituation are found in pigs that soon ignore a sparkler over their feed trough[362] or dogs that are not repelled after a few exposures to a dog repellent.

Classical Conditioning

Classical conditioning or signal learning was first demonstrated in dogs by Pavlov. An unconditioned stimulus (UCS), such as the sight of

251

meat, which produces a response (R) by the animal, such as salivation, is paired with a conditioned stimulus (CS), for example, the sound of a bell. The stimuli are paired repeatedly until the conditioned stimulus alone elicits the response; the dog salivates when the bell is rung. Because horses do not salivate at the sight of food (the food must be in their mouths before salivation is stimulated), it is interesting to speculate on what might have happened to the field of psychology if Pavlov had used horses rather than dogs.

Perhaps the best illustration of classical conditioning is the release of oxytocin in response to the jangling of milk equipment; oxytocin causes contraction of the myoepithelial cells of the mammary gland, or milk "letdown." Cows normally release oxytocin in response to suckling on the teats by the calf or squeezing of the teats by the negative pressure of the milking machine or the hands of the milker. When a cow has been milked in the same environment a number of times, the sounds of the approaching machinery and milk cans will have been paired with the milking process, and those noises alone will elicit oxytocin release.[432] This type of classical, or Pavlovian, conditioning reduces the time required to milk a herd of cows.

A pet animal's fear reaction to the smell of a veterinary hospital or the sight of a person in a white coat is often a classically conditioned response. The dog or cat responds to a painful stimulus (UCS) with the fear or escape response. The hospital and the professional staff are the conditioned stimulus. It does not take many pairings of these stimuli to produce a fear response whenever the animal encounters the conditioned stimuli. For those who become veterinarians because they genuinely wish to help animals, it is somewhat disheartening to discover that many of their patients turn tail and hide whenever they approach, sometimes even outside the veterinary clinic. The behaviorally oriented clinician will make every effort to reduce the painful and frightening incidents, especially in a young animal's first visit, so that a conditioned fear response will not develop and hinder future patient–veterinarian (not to mention client–veterinarian) relationships. Figure 7.1 shows classical conditioning of the goat.

Operant Conditioning

The third type of learning is called operant, or instrumental, learning. Operant conditioning was first demonstrated by Thorndike[1393] using cats as experimental animals. Hungry cats were placed in slatted boxes. Food was available outside the box within sight and smell of the cats. At first, the cats struggled vigorously to reach the food. Eventually, some of them, by chance, pulled a latch string that opened the box door. The cats were then free to consume their reward. Each time a cat was replaced in the box, he took a shorter time to escape, making fewer

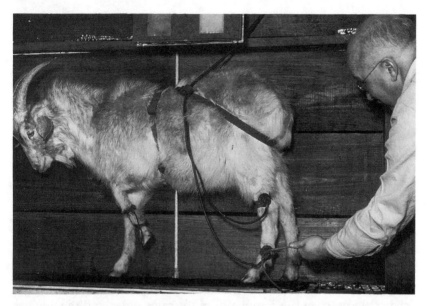

FIG. 7.1. Classical conditioning. A goat is conditioned to lift its leg when a metronome ticks, by pairing the sound of the metronome with shock to the foreleg. The late Professor Liddell, who compared the rate at which various species of farm animals learned a classical conditioning, is shown.

and fewer extraneous motions, until eventually he pulled the latch string immediately upon being placed in the box.

Operant conditioning is also called instrumental learning because the behavior is the instrument by which the reinforcement is obtained. A laboratory example of instrumental learning is a rat in a "Skinner box," named for the psychologist who popularized the technique,[1329] in which an animal presses a bar to obtain food, water, or electrical stimulation of his brain. Figure 7.2 demonstrates a dog using a Skinner box to feed herself.

Operant conditioning is also used on the farm. Swine use electronic feeders. Horses use automatic waterers and cows can be taught to enter an area to be milked by a robot, that is, the cows milk themselves.[1503] Operant conditioning devices that dispense food when a cat or dog presses a switch are sold as novelty items.

Chaining

A fourth type of learning is chaining. Chaining is the performance of a series of operant response in sequence. A good example are goats who were trained to jump three hurdles, walk on a raised walkway, pass through two barrels, and press a lever 10 times.[1309] Many dog owners in-

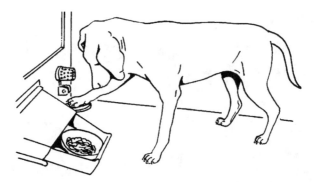

FIG. 7.2.   Operant conditioning. The dog pushes a pedal (operates on the environment) to obtain a dish of food[683] (copyright 1978, with permission of Veterinary Practice Publishing).

advertently chain obedience commands, with the result that the dog sits, shakes, lies down, and rolls over when the owner says, "sit." The owner always gives the commands in the same order, and the dog has chained the responses.

Discrimination Learning

A fifth type of learning is discrimination learning. Animals can learn to discriminate between various visual, auditory, or tactile cues. One of the simplest tests of visual discrimination in horses, cattle, and sheep was done by Gardner.[537-541] Her studies revealed that all three species could learn to choose a feedbox covered with black cloth instead of two uncovered feedboxes. With increasing trials, the number of errors decreased. The animals retained the discrimination when tested over a year later.

Conceptual Learning

The highest type of learning, that is, the one that demands the most intelligence, is conceptual learning. The simplest form of conceptual learning is the ability to respond to a common quality or characteristic shared by a number of different specific stimuli. We are most familiar with this from the children's shows on TV in which toddlers are exhorted to tell, "Which of these is not like the others?" In that case, difference is the concept.

Apparently, horses can form concepts. The experiment was an operant conditioning task. The horse simply had to push one of two hinged panels. The correct panel was unlocked, allowing the horse access to a bowl of grain. The incorrect panel was locked, but there was a bowl of grain behind it to ensure the horse did not choose the panel because he could smell the grain. The correct choice was not always on the same side. The first problem was a simple discrimination between a black panel and a white panel. Next, he had to discriminate between a cross and a circle. The third problem was to distinguish a triangle from a rectangle and the next to discriminate triangles from half circles and various other patterns. In the true test of conceptual learning, the horse had to choose between two shapes he had never seen before, one triangular and the other nontriangular.[1260]

There appear to be other types of learning that are difficult to classify, such as taste aversion, imprinting, and imitation.

Imprinting

As defined by Lorenz,[903] imprinting is a special process that (1) can only occur during a definite and short period of the animal's life; (2) is irreversible; (3) involves an attachment to an object that will later evoke adult behavior patterns, including sexual behavior; and (4) involves reactions to a particular object that can be generalized to all objects in that class, for instance, all humans or all duck decoys. Subsequent laboratory studies have revealed that in ducks, imprinting does not appear to be irreversible or to influence any adult patterns of social behavior.[1029]

Imprinting is often misused and confused with socialization. Imprinting occurs most commonly in birds and involves a following response. Ducklings will follow their mother because she is the first moving object they see. If a red ball is the first moving object they see, they will follow that. Imprinting probably occurs in horses, too. Neonatal foals will follow any large moving object. That is the reason they may follow a human if the mare has not yet risen after foaling. Imprinting occurs more rapidly in frightened animals, which explains why a newborn foal may follow a horse that has been biting him. Mares may have developed the tendency to guard their foal against any animal or person who approaches, not only to defend the foal from predators but also to prevent the foal from following the wrong animal.

The few studies done to date indicate that the following tendency is not irreversible. Cairns and Johnson[267] showed that normally reared lambs that presumably followed their mothers began to follow dogs when the two species were housed together, but they lost the response after returning to the company of other sheep. Scott[1281] hand raised a lamb that subsequently showed abnormal social and maternal behavior. The lamb's situation was quite comparable to that of the geese that were hand raised by Lorenz.[902]

Dr. Robert Miller[1017] has popularized "imprint training" in foals. His technique involves habituation and probably learned helplessness when practiced on a foal too young to stand or otherwise resist. The foal is not really imprinted on humans. Mal et al.[923] found that foals handled for 10 minutes twice a day during their 1st week and then weekly until weaning at 4 months of life were no more likely to approach people than were foals that received only routine and emergency veterinary care.

Sambraus and Sambraus[1256] have shown that in order for goats, pigs, and other domestic animals to direct sexual behavior toward humans, the animals must have been isolated from conspecifics and been in close association with humans.

Imitation

Animals can learn by imitation, by observing others. This form of learning has been most thoroughly studied in cats, cattle, and horses and is discussed in the section on learning in those species.

Conditioned Taste Aversion

Taste aversion, or bait shyness, is the process by which an animal learns to avoid a food not because it tastes bad, but because he associates it with illness, particularly gastrointestinal malaise. This form of learning has long been recognized by those attempting to rid farms of rats. When first used, a poison usually kills many rats; but after the first application, very few rats are killed. The animals that survive will no longer eat the bait. The same phenomenon occurs when rats are exposed to radiation and at the same time offered a novel food. They soon avoid the food that they associate with radiation sickness.[535]

There are three unique characteristics of taste aversion that differentiate it from classical and operant conditioning. One is that it appears to be specific for taste and olfaction; other stimuli like visual or auditory cues will not be avoided. Rats will learn to avoid the taste of saccharin, which they normally like, but not a blue solution. On the other hand, birds readily learn to avoid novel-colored foods; avian species, which possess few taste buds, apparently depend more on sight than on taste for food identification. Second, the illness must be of internal origin, a general or gastrointestinal malaise. External injury, as from electric shock to the feet, is not a sufficient stimulus. Third, the novel taste and the illness can be widely separated in time, and learning will still take place. This is in contrast to both operant and classical conditioning, in which the stimulus and the response must be close together in time for learning to take place. Sheep and cattle can form taste aversions to a particular food even when the illness follows the ingestion by as much as 8 hours.[264,839]

Taste aversions can be formed even if the animal is anesthetized during the time illness was present. Anesthesia blocks other forms of learning, so taste aversion may be noncognitive learning.[1165] It is always difficult to know how an animals feels particularly when the sensation is something subtle, such as nausea in an animal that cannot vomit. Whereas anesthesia does not prevent taste aversion, an antiemetic does, indicating the critical importance of nausea.[1165] The higher the doses of toxin and, presumably, the sicker the animal, the stronger the conditioned taste aversion. The concentration of the flavor is not important, but sheep would generalize from one food to another if the flavor was the same.[853]

Taste aversion has three uses, one experimental and two practical. Experimentally, taste aversion can be used to determine what substances an animal can taste or perceive. An animal will show an aversion to a substance at concentrations far below those at which it would show preference or aversion if the taste had not been paired with illness. Practically, taste aversion has been used to teach coyotes to avoid lamb. Repeated pairing of lamb with injection of lithium chloride, which produces nausea and vomiting, resulted in a definite aversion to live or dead lambs by the coyote.[583] Taste-aversion techniques can be used on a large scale to reduce livestock predation by wild coyotes. The second practical use is to teach livestock to avoid a poisonous plant. Many poisonous plants, like larkspur, do not make the animal nauseated but instead are chronic poisons. The cattle or sheep can be taught using another emetic to avoid that plant before they encounter it on the range.

Formation and Strengthening of a Learned Task

Shaping

In teaching an animal an operant task, one can wait until the animal performs the desired activity and then reward it as Thorndike[1393] did, or one can speed up the process by shaping the behavior. If, when teaching a dog to heel, the trainer first rewards the dog for staying within a yard of his side, then a foot, and finally only when the animal walks quietly exactly beside the trainer, he is "shaping" the dog's behavior.

Circus horses are usually shaped by being rewarded for a simple task such as trotting around the ring, then for trotting close to another horse. Later, the horse is rewarded for rearing on command, perhaps with special urging with a whip to get the first rearing motions. At last, the horse can be induced to rear up and put his legs on the horse in front. Then the various movements are chained.

Animals that are trained to perform complicated and relatively unnatural tricks are usually reinforced with food rewards and reinforced for each correct response at first. The same techniques are used to teach chickens to play baseball or to teach pigs to put giant coins in a bank.[244]

Autoshaping

Autoshaping is a phenomenon whereby an animal makes a response directed toward a stimulus that precedes a reinforcement. If a light signals that food will be delivered in a few seconds, pigeons will peck at the light, and dogs will lick the response keys. An attempt to autoshape horses failed.[396]

Reinforcement Schedules

When operant conditioning is employed, a variety of schedules of reinforcement can be used. The animal can be rewarded, for example, after every response, after every 10, or after every 20 responses. These schedules are called fixed ratios or FR1, FR10, and FR20, respectively. This technical detail is important because the higher the fixed ratio, the faster the animal will respond; even more important, the longer it will take for the response to be extinguished or forgotten. The animal will go on responding for some time after he is no longer rewarded. There are many situations in which owners inadvertently put their animal on high FR schedules. Take, for instance, the dog that barks while its owners are eating. They may have given him food once or twice when he barked, until they became annoyed at the behavior. The dog continues to bark while the owners try to ignore him. Finally, they relent and give him some food; they have just increased the ratio. Dog and owner may adjust to this new level, but often the adjustment is temporary and the level of response (barking) needed for reward increases again. Dogs have been trained in the laboratory to bark 33 times for each small food reward and cats to meow 15 times.[1028,1255] The problem dog at the dinner table may continue to bark hundreds of times even though his owners do not give him any more food.

Another type of reinforcement is called fixed interval (FI). In this case, the animal is rewarded for a response that occurs after a certain period of time has elapsed since the last reward. Animals do have a good time sense, and their rate of responding will slow down after a reward and then increase sharply just before the end of the time interval. If animals can learn to respond in this manner, it is not surprising that they learn to expect their owners home at a given hour. Another variant of reinforcement ratios is the progressive ratio (PR) in which, for example, the animal must respond once for the first reward, twice for the second, four times for the third, and so on. The number of responses per reward increases progressively. This technique is used to measure the strength of preferences for food.

A more long-lasting response, that is, more difficult to extinguish, follows a variable interval–reward (VI) schedule. Here, the reward follows the first response after 1 minute has elapsed, then after 3 minutes, then after 5 minutes, and so on. The highest response rate follows variable ratio reward (VR). The owners of the barking dog may find themselves rewarding their dog on this type of schedule if they inconsistently reward the barking, depending, perhaps, on their own mood on any particular evening or on which family member gives in to the pet. A high rate of barking may contribute to the owners' giving in sooner, but the owners are probably not counting barks; they are merely "holding out" for as long as possible, a war of nerves that the dog invariably wins.

Obviously, however, a variable ratio–reward schedule is to be recommended in routine animal training; after a task has been learned using continuous reward, owners should supply verbal or food rewards sporadically during a training session rather than after every trick.

One of the most difficult tasks for the animal trainer is to get the animal to understand the experimenter's instructions.[554] Dogs, for example, have remarkable olfactory acuity and can be taught to detect gas leaks, hidden narcotics, and fatty acids at very low concentrations, but Becker et al.[170] found it extremely difficult to get dogs to learn to turn right in response to an olfactory cue. It is also important to know what is rewarding for an animal. Dogs learn faster when the reward is simple contact with a passive person than when the reward is stroking or picking up.[1351]

Rewards

Positive and Negative Reinforcement

An animal will learn for both positive and negative reinforcement. Positive reinforcement is a reward, usually food but sometimes social interaction, for performing a response. Negative reinforcement is something aversive applied until the animal makes the response. One pulls on the horse's mouth until he stops, or the rat is shocked until he moves to the other side of the cage. Many field dogs are trained using negative reinforcement varying from a pinch to a shock collar.

A practical application of negative reinforcement (shock avoidance) learning is now commercially available. Both dogs and goats learn that they will be shocked if they cross a buried wire or "invisible fence." An auditory signal on the transmitter collar they wear warns the animal that a shock is forthcoming if he approaches the buried wire.[457]

One of the most difficult concepts for owners to understand is the difference between negative reinforcement and punishment.[227] Punishment is something that occurs after an action as a consequence. The dog chews the slipper and the owner hits him. Timing is very important. Punishment will not decrease the frequency of the behavior unless it occurs when the animal is misbehaving or within a second or 2 of the termination of the behavior. Most owners feel that they can punish a dog hours after he has chewed a slipper or eliminated in the house, and they are surprised when the dog does not learn.

Learned Helplessness

A phenomenon that has considerable application to practical animal training has been discovered in dogs and cats. This phenomenon is learned helplessness.[922] Normal, naive dogs, when first placed in an active avoidance situation in which impending shock is signaled, at first escape the shock, once it has begun, and later avoid the shock by performing the necessary task, such as jumping over a barrier, during the signal before the shock begins. Dogs that previously have been exposed to unavoidable shock act in a quite different manner. They not only fail to learn to avoid as naive dogs do, but also fail to escape; they simply sit and take the shock. These experimental findings indicate that the same form of aversive stimulus should not be used first as inescapable punishment and then later as negative reinforcement that the dog should learn to avoid. Improper use of the popular shock collars or invisible fences may produce learned helplessness in dogs, and any form of inescapable punishment may inhibit later learning.

COMPARATIVE INTELLIGENCE

Which animal species are the smartest?[202] This is a question commonly asked by lay people. Although knowledge concerning the IQ of the pig may not contribute much to one's medical or husbandry skills, a well-informed discussion of the facts and the pitfalls involved in assaying relative intelligence will be appreciated by the questioner. Furthermore, in species that are commonly trained, such as dogs and horses, many of the behavior problems revolve around learned tasks or, more frequently, tasks not learned. A dog that refuses to be housebroken or a horse that runs out of a jump are good examples. Volumes have been written on training horses and dogs so only the underlying principles will be discussed here.

Methods of Measurement

Brain Weight to Body Weight Ratio

An anatomical approach to intelligence can be used. There may be a correlation between brain size and intelligence.[1194] The brain weight to body weight ratios in decreasing order are human 2%, cat 1%, mongrel

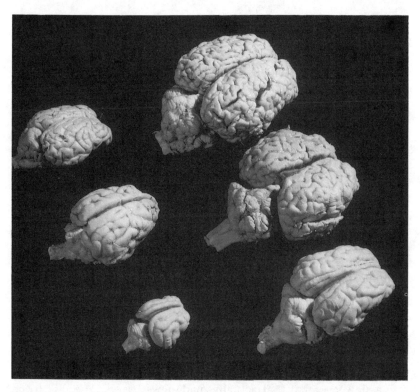

FIG. 7.3.   The relative brain size of various domestic animals. *Right row from top to bottom:* horse, cow, and pig; *left row:* dog, sheep, and cat.

dog 0.5%, rat 0.3%, goat 0.3%, horse 0.1%, and pig 0.05%.[359] Figure 7.3 illustrates the brains of domestic animals.

The brain to body weight ratios of various breeds indicate that increases in brain weight are not linearly related to increases in body size. The smaller the dog, the higher the brain to body weight ratio. The toy poodle and Pekinese have ratios of over 1%, whereas the Saint Bernard and Great Dane have ratios of 0.2%; the medium-sized breeds such as the cocker have a ratio of 0.6%.[251]

Pigs and most other animals have suffered by domestication in this regard. Pigs have been bred for larger bodies, so wild pigs tend to have larger brain weight to body weight ratios. The fallacy of using so labile a parameter as body weight to judge intelligence is manifest when one realizes that the malnourished pig, which shows definite intellectual impairment,[137] has a greater brain weight to body weight ratio than the well-nourished pig. This occurs because the brain appears to be spared when the rest of the body is stunted by malnutrition.[1476] It should be mentioned that the inferiority of the intelligence of women to that of men purported by many male scientists and thinkers of the 19th century was supported by brain weight to body weight ratios similar to those just mentioned. Gould[564] reviews the work of Broca and his disciples and shows how the same figures they used to prove male superiority could prove just the opposite, even if only such obvious factors as age at death, cause of death of the donors whose brains were examined, and the "sexual mass" (the differences in muscle mass and body fat) of the two sexes are evaluated. The value of these ratios in assessing comparative intelligence between species should thus be looked at skeptically.

Problems of Cross-Species Comparison

Learning Rates

Another way in which intelligence might be measured is to compare learning rates of various species on the same task. There are many confounding factors that may invalidate this approach also. The task must be physically possible for all species tested. A task requiring manipulation with the forelimbs, which a rat could perform with ease, would be nearly impossible for any ungulate. Scaling also presents a problem. Is a 60-ft maze appropriate for a cow if a 6-ft one is for a cat? Similarly, one must be careful that one is not measuring athletic ability rather than intelligence. A dog that can run fast may complete a task before a slower animal that actually made fewer errors. The task to be learned should also be within the normal behavioral repertoire of all species to be tested. A cat can easily be taught to pounce on an object; a cow rarely performs such actions.

Classical Conditioning

Liddell and Anderson[884] and Liddell et al.[885] used classical conditioning to measure comparative intelligence. They compared the number of trials necessary to produce leg flexion in response to the conditioned stimulus, the sound of a metronome. The unconditioned stimulus was a shock to the foreleg. Dogs were most easily conditioned. Pigs were the most easily conditioned of the farm animals, followed by goats, sheep, and rabbits.

Delayed Response Method.   Several experiments comparing intelligence in a variety of species were done early in this century.[970] For example, Hunter[718] used the length of time an animal could remember which of three boxes identified by a brief illumination held the food reward. This technique, called the delayed response method, revealed that a rat could delay its response for 10 seconds, a raccoon for 25 seconds. Children 2 years old could delay their responses for 25 minutes, and dogs, for 5 minutes. In other experiments, it was found that cats can remember for 6 minutes, adult dogs for 18 minutes, and goats for 30 minutes, although the goats had a more intensive signal to remember than the other species.[1339] Of two horses tested by Grzimek,[578] one could remember for only 15 seconds, the other for 60 seconds. The delayed response time is quite variable, however; the delayed reaction time for cats varied from 18 seconds to 16 hours, depending on the test and the investigator.[921]

Multiple-Choice Method. Hamilton[596] used a multiple-choice method to test comparative intelligence. The animals could escape from the apparatus through one of four doors; the correct or unlocked door was never the door that had been unlocked on the previous trial, the win and shift strategy. Humans were superior in this test, followed by monkeys, dogs, cats, and the one horse tested. The horse engaged in stereotypic behavior, making repeated attempts to escape through the door that had been unlocked on the previous trial, the win and stay strategy. Monkeys did as well as pigs in choosing doors when the correct response was the second door from the end, but an orangutan did poorly.[1524]

Avoidance-Response Method. Willham et al.[1489] have suggested another method to measure learning ability both among and within species. An avoidance response is taught using a shock for the unconditioned stimulus and a buzzer for the conditioned stimulus. The animals must cross a barrier either to avoid or to escape shock.

It was necessary for cats to undergo 12 trials to learn to avoid a shock by jumping on a shelf.[360] Dogs needed only 4 trials to learn to avoid a shock, pigs 10, and horses 8. This does not necessarily mean that dogs are more intelligent than horses and that horses are more intelligent than pigs. A much higher level of shock was used on the dogs and pigs than on the horses, and this may have increased or decreased the learning rates.[584,832,1338]

Even very young animals have been tested. Newborn kittens cannot learn to escape an aversive stimulus (an air blast),[93] whereas puppies can.[1350]

Maze Learning. Gardner[541] compared the learning ability of cattle, sheep, and horses and found that the horses and cows learned visual discrimination better than sheep. Maze learning has been used to assess species differences in learning ability. Karn and Malamud[770] found that dogs learned a double alteration maze better than cats. In a series of maze tests using the Hebb-Williams maze, in which different configurations are made, in some of which the animal can see the solution, whereas in others, it is hidden from view (see Fig. 7.4 A), children made the fewest errors; dogs, cows, goats, and sheep made approximately the same number of errors; pigs and cats made more. Horses made more errors than pigs or cats (see Fig. 7.4B).[955]

Object Permanence.   Cats and goats have been shown to comprehend object permanence. They will watch the place where an object disappeared from view and go to that place to find the object.[1386] Out of sight is not out of mind. This is the level of insight usually obtained by 12- to 18-month-old children.

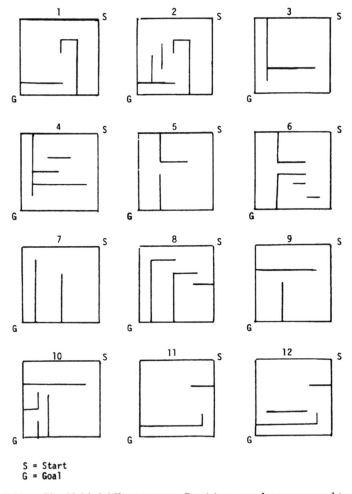

S = Start
G = Goal

FIG. 7.4A.   The Hebb-Williams maze. Partitions can be rearranged to form 12 or more barrier problems. Each day the animal has a new problem to solve, with 10 trials on each problem. The number of errors is used to compare with that of other species[955] (copyright 1981, with permission of J. Anim. Sci.).

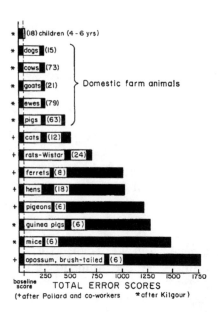

FIG. 7.4B. Comparative maze learning in various species of animals. Children made fewest errors and opossums the largest number of errors in a Hebb-Williams maze (courtesy Dr. Ron Kilgour, Ruakura Animal Research Station, Hamilton, New Zealand)[802] (after Pollard et al. 1971)[1144].

Summary

One might also answer the questioner on comparative intelligence with the observation that each of the domestic species has apparently had enough intelligence to survive for several million years on its own and several thousand years with people. Nevertheless, studies of comparative intelligence can, if performed properly, answer some questions as to the role of particular learning abilities in the survival of a species.

PIGS

Pavlov apparently felt that pigs could not be classically conditioned. Pigs salivate in response to a bell[1369] and increase heart rate in response to a metronome tick.[931,932,1035] Early work also indicated that they could run a maze for a food reward[1060] or choose one of a series of doors to gain access to food.[1525] Difficulty was encountered in teaching a concept such as "center" to the pigs; they could choose the middle of three, but not of five, doors.[1525]

Sex, Breed, and Age Differences

In the years since the pioneering studies already described here, pigs have been used in many types of learning situations to validate or cast doubt upon concepts developed using laboratory species. Pigs learn to make more correct responses when trials are spaced in time rather than grouped. As had been shown in other species, four trials a day for

10 days produces learning superior to that shown by pigs given 40 trials in 1 day.[767] Classical conditioning occurs more quickly if the interval between the unconditioned stimulus and the conditioned stimulus is 2 seconds.[1075] Although pigs usually require a food reward, they will swim a maze when the reward is reunion with their littermates. The speed and accuracy of solving a water maze is not correlated with the pig's ability to learn to avoid shock by jumping over a small barrier.[598]

On a given task, there are sex and breed differences.[1488] Durocs learn avoidance more quickly than Hampshires.[1488,1489] Palouse make fewer errors in visual discrimination than Pitman Moore miniature swine.[818] Artificially reared pigs make fewer errors in visual discrimination tasks than sow-raised pigs.[888] Yorkshires perform better in a T-maze than Poland Chinas; crossbreds are intermediate. Females perform better than males.[1479]

Although season of testing, body weight, and age of dam do not account for variation in avoidance learning,[1488,1489] pigs from large litters ran a maze more quickly than pigs from small litters.[1479] Apparently, when the reward for completion of the maze was return to the company of littermates, social isolation was more severe and the reward of reunion greater when the litter size was greater.

Operant Conditioning

Pigs will not perform an operant response for some types of sensory reinforcement, such as pig noises,[114] but will for others, such as light[123] and brain stimulation.[125] Older pigs (40–150 days) learn to avoid shock less well than younger pigs (3 weeks)[832] (Fig. 7.5).

Operant conditioning of pigs is used for a variety of purposes: on the farm, in animal acts, and in the laboratory. Feeders with hinged covers that the pig must open with its snout are a good example of a very simple operant response that pigs learn rapidly, and they can learn to use electronic feeding stations. Pigs are often found in European circus acts because of their trainability.

Consideration of the anatomy and normal behavior patterns of pigs indicates that it would be much easier to teach them to manipulate their environment with their snout than with their feet. Consequently, pigs have been taught to push a panel with their snouts for either a food reward[128] or for heat in a cold environment.[120]

To measure a pig's motivation for the reward, a progressive ratio technique can be used. (A progressive ratio is a schedule in which an animal must make one response for the first reinforcement, two for the second, four for the third, eight for the fourth, and so on, until the breakpoint, at which the animal stops responding). Kennedy and Baldwin[788] have used a progressive ratio technique to measure the strength of preferences for various sweet solutions.

FIG. 7.5. Avoidance learning in pigs. (A) To avoid a shock, the pig must jump the barrier when a buzzer sounds. The late Ulric Moore, who performed some of the early experiments on learning in pigs, is shown. (B) The ability to learn to avoid shock decreases with age in Duroc and Hampshire pigs.[832]

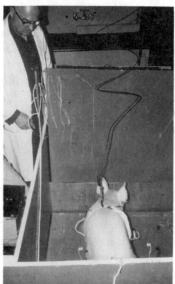

A

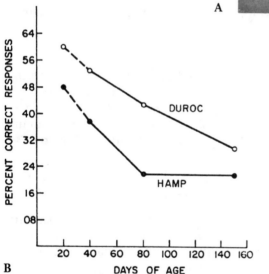

B

A conditioned anxiety reaction can be produced in pigs by training them to press a panel for a food reward and by then adding a tone that signals a shock if the animal presses the panel.[128] The pigs learn to inhibit panel pressing, although experimental neurosis is sometimes produced;[343] tranquilizers do not reduce the inhibition of responding.[353] Conditioned anxiety does lower heart rate[354] and increases endogenous levels of corticosteroids, but the rise in corticosteroids is not as great as when the pigs are exposed to cold or chased with a goad.[129]

Visual Discrimination

Although pigs are better at spatial (right versus left) than at visual discrimination tasks and must be taught to make visual discriminations before 20 weeks of age,[817] pigs do have color vision similar or slightly superior to humans and can discriminate between wavelengths of light differing by only 25 μm.[818] Pigs have difficulty in reversal learning of either a visual or spatial discrimination; they abandon the original correct response, but their performance remains at chance levels for many trials. Overtraining, training carried beyond criterion on the first trial, does improve reversal learning.[818,1466] Most visual discrimination tests involve pressing one of two panels for the reward. If the animal has learned the discrimination, it will get a reward after six responses on FR6, but, if it has not learned, it will take many more responses, an average of 18, to get a reward by chance. Pigs may fail to learn a visual discrimination not because they are unable to either learn or to see, but because they are willing to work very hard (respond many times) for one food reward. They will tolerate being rewarded at a chance level, whereas another species will expend the minimum amount of energy.

Effect of Drugs

Dantzer and his colleagues have studied the effects of various drugs on learning in pigs.[351-355,1042,1043] Pigs apparently do not suffer from learned helplessness as dogs do.[1043] Dantzer[351] has found that pigs treated with diazepam will press a panel more times for food than untreated pigs; diazepam stimulates feeding (see Chapter 8, Ingestive Behavior: Food and Water Intake).

Effect of Malnutrition on Learning

Avoidance learning has been used to study the effects of malnutrition on brain function. Barnes et al.[137] demonstrated that pigs previously malnourished but subsequently rehabilitated for several months show poorer avoidance learning than well-nourished pigs.

DOGS

Housebreaking

The first task that all house pets must learn is voluntary control of the anal and urinary bladder sphincters. There are probably as many methods for housebreaking dogs as there are books on dog training.

Both classical and operant conditioning methods have been recommended.

Immediately after a meal the gastrocolic reflex operates to increase motility of the large colon and rectum. As a result, filling of the rectum will stimulate relaxation of the smooth muscle of the internal anal sphincter and the striated muscle of the external anal sphincter. If a dog is taken outside after every meal, the conditioned stimulus of being outside will soon replace the unconditioned stimulus of the gastrocolic reflex. Some clinicians recommend the use of glycerin suppositories after a meal when the dog is taken outside. The principle is the same; the suppository is the unconditioned stimulus and is more reliable in action than postprandial defecation.

Operant conditioning is more useful in teaching voluntary control of urination for which there is no reliable unconditioned stimulus. Newspapers are spread all over the room where the puppy is confined. Gradually the area of newspaper is decreased; the puppy is placed on the newspaper when he squats to urinate and praised when he urinates in the proper place. The newspaper can then be laid outside the door to encourage the dog to go outside to urinate and to even whine to go outside. In other words, the puppy behavior is shaped by first being rewarded for a general action; later, the reward is contingent on more and more specific actions of the animals. It is often valuable to continue to reward the use of newspaper in case the dog must be left indoors for longer periods than one can expect him to retain a bladder full of urine. (Be consistent—do not punish the animal if he uses the daily paper that has been left inadvertently on the living room carpet.)

Still another method takes advantage of the innate reluctance of animals to soil their sleeping quarters. The puppy is put in a small cage or crate and kept there except for trips outside to eliminate every hour or 2. The premise is that the dog will not urinate or defecate in the cage, and once the animal can control his bowels and bladder for long periods, he will generalize the control to the whole house and will no longer have to be closely confined. It would be asking too much of a young puppy, whose control is not good enough to wait an entire night without eliminating; besides being somewhat cruel, this may only condition the pup to accept urine and feces on his bed and on himself. This technique has merit for those who can stay home with their puppies or for an older dog that is still poorly trained.

Whichever method is used, the trainer must use appropriate and consistent rewards and punishment. Verbal praise and a perfunctory pat are ample reward; in fact, as already mentioned, Bacon and Stanley[94] and Stanley and Elliot[135] found that dogs will learn to perform better when simple contact with a passive person is the reward than when stroking and picking up is the reward.

Punishment of the animal for misbehavior must come as soon as possible (a second) after the offense. The unconditioned stimulus is the punishment, and the response is avoiding the pain. Unless the conditioned stimulus, inappropriate urination or defecation, is closely paired in time with the unconditioned stimulus, the animal will not learn. If it is punished 10 minutes after it has performed its misdeed, it is more likely to be the trainer who becomes the conditioned stimulus than the act of elimination. Dogs who act "guilty" are those in whom this kind of conditioning has taken place. They have learned to associate the presence of a pile of feces in the house with a painful experience; they have not learned to associate their act of eliminating with the painful experience.

Although taking the puppy outside to eliminate is a good method of training, a common mistake that owners make is to put the dog outside for longer and longer periods. The dog does not know what he is to do outside and, if he should eliminate, the owner is not there to praise him. Praising the dog when he returns to the door is a good way to teach the dog to return home but not to housebreak him. Still another complication is that of the dog who is walked on a leash, but whose walks always end as soon as he eliminates. It learns to avoid eliminating in order to prolong the walks. Walk lengths should not be contingent on elimination. A dog may also be afraid to eliminate when the owner is present because he has been punished when "caught in the act." The dog has learned to avoid eliminating when the owner is present, rather than to avoid eliminating in the house.

The problems of housebreaking have been covered in great detail because they represent 20% of the behavioral complaints of dog owners.[1443] Owners will tolerate many defects in their animals, but few will tolerate house soiling, as a case of a Lhasa apso donated to a veterinary college illustrates. The dog suffered from lissencephaly (absence of normal gyri and sulci) (Fig. 7.6) and had many learning disabilities and visual deficits, but the owner's reason for donating the dog was its inability to learn bowel or bladder control.

From the clinician's point of view, house soiling in a previously well-trained animal indicates a medical problem. A dog with diarrhea will not be able to retain voluntary control of its irritated intestines. Ulcerative colitis must be treated as both a behavioral and a medical problem; stress should be reduced and the appropriate drugs prescribed. Dogs with chronic nephritis do not concentrate their urine, so they must excrete large quantities of dilute urine and will not be able to retain or limit their micturations as well as they could when healthy. There are many diseases which may cause polyuria. In addition to medical problems, old dogs may suffer from cognitive dysfunction syndrome (see below) similar to human senility. A common sign of cognitive dysfunction is housesoiling.

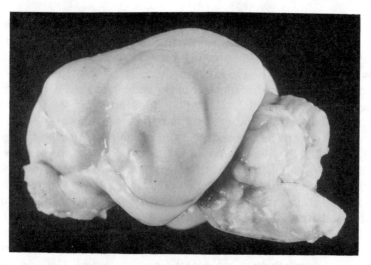

FIG. 7.6.   An example of a learning deficit associated with organic disease of the central nervous system. The brain of a Lhasa apso terrier that suffered from lissencephaly and could learn neither bowel nor bladder control.

Canine Geriatric Cognitive Dysfunction

Old dogs may have physical problems such as deafness, cataracts, heart disease, or arthritis, but many old dogs are euthanized for behavioral reasons. They forget to be housetrained or pace and whine at night. Older dogs have learning deficits that can be measured in objective laboratory tests such as delayed, nonmatching to sample. The dog is shown an object, such as a circle, and then after a delay of 5 seconds is given a choice between a circle and a rectangle. He must choose the rectangle (nonmatch to be rewarded). Beagles over 10 years old could not learn this.[1015] There are lesions in the hippocampus (see Appendix 3), the part of the brain essential for forming some types of memory similar to those found in people with Alzheimer's disease.[342]

Breed Differences

Intelligence among dog breeds is a perennial topic of conversation among owners. No one has tested a large enough number of dogs to evaluate them statistically, but Coren[319] has ranked dogs for both obedience and problem solving. The complete examination includes 12 tests: (1) giving the dog the usual cues for a walk—leash, keys, and so forth—but going to an unusual exit and determining if he follows; (2) showing the dog food and hiding it under a can and timing how long it takes him to find it; (3) rearranging a room and timing how long the dog explores;

(4) throwing a large towel over the dog and determining how long it takes him to extricate himself; (5) staring and then smiling at the dog to see if he will approach; (6) throwing a small towel over food and timing how long it takes the dog to extricate the food; (7) showing him the location of food and taking him out of the room and then back in to see if he goes directly to the food after having been out of the room for 5 seconds (short-term memory); (8) the same procedure as in (7), except extending the time he is out of the room to 5 minutes (long-term memory); (9) hiding food under a board so that the dog must use his paws to retrieve it; (10) calling the dog by the wrong name to see if he recognizes his own; (11) teaching the dog a new command such as sitting in front of you; and (12) arranging a barrier problem in which the dog has to go around a barrier—a longer route when a straight-line approach is blocked. The dogs are rated from 1 to 5 on these tests, with short times getting higher scores in the timed events. The overall best scores were made by, in alphabetical order: Doberman pinschers, German shepherds, Norwegian elkhounds, poodles, pulis, and Shetland sheepdogs. Obedience judges rank the breeds similarly: Border collies, poodles, German shepherds, golden retrievers, and Doberman pinchers. This indicates that there are breed differences in learning ability or, at least, in acquisition and performance of various tasks.

The most carefully controlled studies on breed differences in learning ability were carried out at Jackson Laboratory in Maine. Scott and Fuller[1284] reviewed the experiments, which compared learning ability in five breeds of dogs: cocker spaniels, beagles, wirehaired fox terriers, Shetland sheepdogs, and basenjis. These particular breeds were chosen because they did not differ much in body size, nor did any breed possess a breed-specific anatomical peculiarity, such as the achondroplasia of basset hounds. The five breeds were tested for their ability to learn three types of tasks: forced training, reward training, and problem solving.

The forced method of training was used to teach the dogs to sit still on a scale, to heel on a leash, and to stay and jump on command. In all three types of tasks, the cocker spaniels ranked highest in correct performance. The trainability of cockers in these situations is probably the result of selection within the breed for dogs that would crouch in response to a hand signal. Although cockers are not often used for hunting now, the behavioral predisposition remains. All the dogs learned to heel within the 10-day training period, but there were marked differences in the types of errors that dogs of the various breeds made in the early sessions. Basenjis fought the leash and often pulled ahead or lagged behind; Shetland sheepdogs interfered with the trainer, that is, tangled the leash around his legs. Beagles vocalized in protest (Fig. 7.7).

Reward training consisted of showing the dog a piece of fish in a box and then restraining the puppy behind a wire gate before allowing

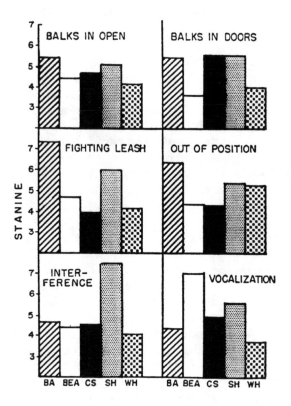

FIG. 7.7. Breed differences in response to leash training. BA = basenji; BEA = beagle; CS = cocker spaniel; WH = wirehaired fox terrier; SH = Shetland sheepdog. The higher the stanine score, the more often the dog exhibited the behavior[1284] (with permission of University of Chicago Press).

him to run to the box and eat the fish. The position of the box was changed to measure goal orientation versus habit formation. Basenjis performed best on this test, probably because they could run the fastest. With the additional trials, all the dogs reduced the time they took to reach the reward. When motor skills of the five breeds were compared, the basenji was also best.

Problem solving was also studied. The first type of problem the dogs were supposed to solve was a barrier or detour. The dogs were separated from a food reward by a wire barrier. When they learned to run around a short barrier, the barrier was extended and later formed into a U-shape to increase the difficulty of the problem. The dogs were only 6 weeks of age and had considerable difficulty in solving the barrier problem that an adult dog could master easily. The puppies often yelped, and it was noted that they never solved the problem while yelping but instead engaged in stereotypic activity. Few puppies did well on the barrier test, but basenjis, which are already active at 6 weeks, did best. Puppies of other breeds still tend to be fat and clumsy at 6 weeks and would react to failure by going to sleep.

Another type of problem was a manipulation test in which the dogs were tested for their ability to pull a dish of food from under a box by pawing or pulling it out with their noses. Later, the dish was positioned in such a way that it could be maneuvered out from under the box only by pulling on a dowel and string attached to the dish. Again, the basenjis were the most successful. Most interesting was the effect of repeated failures on the dogs' performances. Although all puppies at first scratched and nosed at the box containing the food dish, those that had failed often took one look at the box and simply sat down to await the end of the trial, thus precluding any success.

When maze learning was tested in RLLRRR (right, left, left, and so on) or LRLL mazes, beagles did best. They completed the maze most quickly and made the fewest errors.

A final test of problem-solving behavior, when the dogs were 22 weeks old, was a delayed response test in which the dogs were shown a visual cue that indicated which side of a T-maze led to freedom. Before the dog was allowed to run the maze, a delay from 1 to 240 seconds was imposed. As in all problem solving, there were great individual differences, but cocker spaniels could remember after the longest delay, and Shetland sheepdogs had the poorest memory.

The extensive studies at Jackson indicate that care must be taken in comparing intelligence, even within a species, since breeds of dogs differ markedly in their relative performance depending on the task to be learned. Fox and Spencer[498] have shown that dogs improve in their ability to make a delayed response. If puppies were trained for 13 days beginning at 4, 8, 12, or 16 weeks, all increased the interval over which they could remember during testing, but the 4-week-old puppies could never remember longer than 10 seconds, whereas the 12-week-old dogs could remember for 50 seconds. Sixteen-week-old dogs did not perform as well; they made many errors. Fox and Spencer[498] explain the poor performance at 16 weeks by postulating a lack of inhibition at that age. Stanley et al.[1349] also studied the ontogeny of learning in dogs. They found that puppies less than a week old could be conditioned to suck more from a nipple when milk was the reward and to inhibit sucking when a quinine solution was the aversive stimulus, although earlier studies had found that puppies less than 18 days old could not be conditioned.[528] Stanley et al.[1350] also found that puppies less than a week old can learn to escape from a cold stream of air and will even move from a comfortable carpeted surface to a hard cold one to do so (Fig. 7.8). Puppies less than 2 weeks old could also learn to choose a wire or cloth model that contained a nipple for a milk reward if given five tests a day, 2 hours apart.

It is much more difficult to train nervous dogs than calm ones.[418,1515] Angel et al.[55] found that administration of a tranquilizer facilitated op-

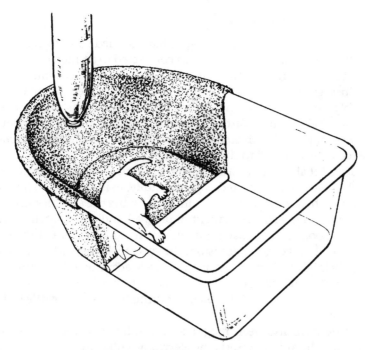

FIG. 7.8. Learning in neonatal dogs. The puppy learns to avoid the textured side, for which puppies have an innate preference, if that side is associated with a blast of air.[1350]

erant conditioning of a genetically nervous pointer. Petting has a physiologically demonstrable calming effect and can be used to facilitate learning. For example, dogs classically conditioned to expect a shock following a tone have a higher heart rate during the tone,[532] but heart rate declines if the dogs are petted during tone presentation.[911]

Training is easier if the innate responses of dogs to auditory signals are used. Dogs will increase their activity to high tones and inhibit it to low tones,[959] and this is reflected in the signals shepherds use to signal their dogs.[960]

Because social facilitation is strong in dogs, slow dogs tend to run faster with another dog than they do alone.[1441] This is in contrast to cats, who run slower for a food reward in the presence of another cat.

CATTLE

There have been few studies of learning ability in cattle, and most of those published described attempts to increase production or reduce labor on the farm.

Operant Conditioning

Kiley-Worthington and Savage[800] have trained cows to come in to be milked when an automobile horn connected to a timer and the electric fence was sounded. Albright et al.[17] trained cows to come into the barn in a given order, but the cattle reverted to their original order, probably based on dominance, as soon as the trainer was absent. Wieckert et al.[1480] also trained cattle to come to a feeding trough when an auditory stimulus was delivered to the cattle from a small timer-activated tape recorder attached to the cow's halter.

Offord et al.[1090] conditioned cattle to eat in response to an auditory stimulus and then attempted, unsuccessfully, to increase food intake in the cattle when they were free-feeding by playing the auditory stimulus. Dairy cows have also been taught to press a handle with their muzzles to obtain food and to make right and left handle discriminations. The most difficult task was to accustom the cows to lifting, rather than lowering, their heads for food because they were accustomed to eating from the floor.[1472] Cattle did not learn an operant task by observing other cattle performing that task.[1431] More Limousine than Aubrac cattle learned the operant task.

Cattle have also been taught to use individual feeders that open only when the animal inserts the electronic collar around its neck into the feeder.[771] These systems are now widely used because cattle can be group housed yet feed themselves individually. Other cattle have been taught to enter a feeding stall when one signal, a bell, is presented and to leave the stall when another signal, a buzzer, was presented.[1307]

Moore et al.[1036] taught seven Jersey cows to press a panel for a food reward. The cattle learned to respond to various schedules of reinforcement: a fixed ratio, in which a certain number of presses resulted in the food reward; a fixed interval, in which food was delivered at intervals and only one panel press at the appropriate time was necessary for the food reward; a variable ratio; and a variable interval. The variable ratio schedule produces the highest response rate in cattle as in other species. Cattle can learn both radial-arm mazes and parallel-arm mazes, that is, they learn that food will not be in a location they have previously visited even if visits were separated by as much as 4 hours, but not by longer intervals. Kilgour[802] found that Jersey cows learned 12 detour problems in the Hebb-Williams maze (Fig. 7.4A,B) and made few errors after four runs of a given detour test. There are marked genetic effects on maze learning ability.[63]

Conditioned Avoidance

Many farmers teach their cattle to defecate in the gutter behind their stanchions rather than on the stall floor by running an electrified wire that shocks the cow whenever it arches its back to eliminate in the wrong place. Because cattle defecate 17 times a day, the cows learn quickly and the average fecal output of 60 to 80 pounds is deposited in the gutter.

Cattle sometimes receive shocks through milking machines. Cows were taught to press a bar for food in order to determine how large a shock had to be before it disrupted a cow's feeding and other behaviors. They stopped bar pressing when a shock greater than 7 mA was applied to one teat or 6 mA to all four teats.[1474] This type of conditioning can be used to determine what is painful to cattle.

A practical application of bovine avoidance learning is to train cattle to respect electric fences by confining them in a pen with a sturdy fence beyond the electrified wire. The cattle will then respect an electric fence even in a new environment and will not be in danger of breaking through it.[962]

Taste Aversion

Cattle can form a taste aversion to alfalfa pellets and corn (Fig. 7.9). Cattle can be taught to avoid licorice-flavored alfalfa pellets, but the aversion is much weaker than that to beet pulp even when the same illness that is produced by lithium chloride is associated with the taste.[1169] This demonstrates the general rule of taste aversion learning: that it is harder to learn to avoid a very palatable food. Older cows are easier to teach a taste aversion that younger ones. Social facilitation can outweigh aversion learning in cattle as well as sheep.

Visual and Auditory Discrimination

One of the earliest studies of learning in cattle is one of the most extensive in that large numbers of cattle were tested. Gardner[537,539] studied the ability of cows to discriminate a box covered with black cloth, which contained feed, from two empty and uncovered boxes. The errors per trial fell as the trials proceeded. Guernseys, Brown Swiss, and Holsteins performed better than Jerseys, shorthorns, and Ayrshires. The cows had retained the learning when tested a year later. More errors were made when the cloth was moved below or, in particular, above the feed box. Calves have been taught to make shape and orientation discrimination.[113] See sheep and goats section for further discussion of visual discrimination in ruminants.

Effects of Age

When the ability to remember the location of a feeder was tested over a period of 5 days, heifers learned more quickly than older cattle, but cows after the second calving remembered the location best when

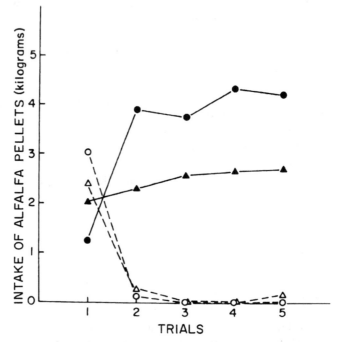

FIG. 7.9. Taste aversion learning in cattle. On day 1 (trial 1), the experimental cows *(open symbols)* were offered alfalfa pellets. After they had eaten the food, LiCl, which produces mild gastrointestinal malaise, was infused into the rumen through a fistula. The control cows *(closed symbols)* were infused with 0.9% NaCl. Four days later (trial 2), when all cows were offered alfalfa pellets, the cows that had been treated with LiCl ate almost nothing, whereas the controls ate more than they had initially. The experimental cows learned to avoid a feed associated with illness[1529] (copyright 1977, with permission of Baylor University Press).

tested 6 weeks later. In both tests, primiparous cows were intermediate in performance.[830]

Cattle can learn to shift from one arm of a maze to another (win and shift strategy) if they ate all the food in the initial arm on the previous trial.[678] It is interesting to compare the behavior of cattle in the very artificial maze situation to that on pasture. When tested in a pasture with high-quality forage, they display the win and shift strategy, but on a poorer pasture they display the win and stay.

From a practical point of view, cows can be "trained" to use cubicles to rest as adults by providing experience with cubicles as heifers.[1084] Cows can learn to activate a robotic milking machine for "on demand" relief of mammary pressure. They are variable in their requests to be milked but averaged three times a day.[1503] No milking was done from midnight to 6 AM, and the system was timed so that a cow could not be milked more frequently than every 4 hours.

SHEEP AND GOATS

Liddell et al.[885] have shown that sheep and goats can be classically conditioned to flex a front leg in response to a ticking metronome that had been paired with an electric shock, although the small ruminants did not learn as quickly as pigs or dogs. Liddell[883] has also demonstrated that sheep can learn simple mazes. Cairns and Johnson[267] have demonstrated that lambs will run a maze to be reunited with a cohabitant, whether the animal was another sheep or a dog.

Operant conditioning techniques have been used by Baldwin and Yates[131] to study thermoregulatory behavior. The sheep learned to activate a heat lamp by sticking their muzzles through a slit to break a photoelectric beam. Unshorn sheep did not learn to turn on the heat at 5°C (41°F), but shorn sheep, deprived of the thermal insulation of their fleece, did. Sheep have also learned to press a bar for reward in response to a tone signal and to refrain from pressing the bar for 30 seconds after each reinforcement.[1257-1259]

Goats can learn to press a bar for the reward of electrical stimulation of their brain[1126] and press over a thousand times in 45 minutes for a food reward.[1310] Sheep and goats can learn to make fairly fine visual discriminations between different shapes and different orientations of the same shape. For example, they can learn not only to discriminate between a circle and a square but also to discriminate between a triangle pointing right and one pointing left.[112,113] Their learning rate improves with additional discrimination tasks, indicating that these ruminants form a learning set, or can learn to learn.

Unfortunately, conditioned safety, taught by feeding increasing percentages of sagebrush in the diet before goats were released on the range, did not cause the goats to eat any more sagebrush than nonexperienced goats when they were free to select a diet on the range.[1199]

Sheep will learn to move faster through yards (chutes and alleys) with experience and will do better than inexperienced sheep months later, but training in moving a different way through the same yard is worse than no training.[721] Apparently, the sheep have to forget their previous experience. Giving the sheep a food reward at the end of a yard is an efficient way of speeding their movement through it but will not help if an unpleasant experience such as restraint is also associated with the same movement.[722]

The ability of sheep to perform operantly conditioned tasks has been used in experiments with sodium-deficient sheep. These sheep learn to press a bar for sodium bicarbonate in proportion to their sodium deficiency. Sheep can also learn to discriminate which of five feeders contain food.[1532] Sheep have also been studied by Gardner[541] and Liddell,[881,882] as was discussed under Comparative Intelligence.

HORSES

The horse, like the dog, is a species that is only useful to humans if trained. In fact, despite their aesthetic value, few horses are kept as pets unless they can be ridden or driven. Myriad volumes have been written on the training of horses, and it is appropriate to mention only a few of the basic principles.

It is easiest to teach a horse a natural response; consequently, horses can be trained to race at a very early age. Two-year-old horses on the race track are very common; a horse under age 5 in a dressage class is a rarity. Most horse training is based on negative reinforcement: applying an aversive stimulus until the horse performs the response. The best approach to horse training is to try to substitute conditioned stimuli (a voice command or subtle pressure from the rider's legs) for unconditioned stimuli (the painful flick of a whip). In this manner, neck reining can replace direct reining. When punishment must be given, it should be applied as soon as possible after the misdeed. A slap on the pony's muzzle 30 seconds after it has nipped will only serve to make it head shy; a blow 1 second after the nip may inhibit further aggression.

Perhaps the best example of intelligence in horses is Clever Hans. This 19th-century Arabian stallion could perform mathematical problems by tapping out the answer with his hooves. He was able to give the correct answer whether or not his trainer was present. It was finally discovered that someone had to be present who knew the answer to the problem. The horse was able to perceive some subtle change in the person when he reached the correct number. Clever Hans was more clever at interspecies communication than he was at mathematics.

See Types of Learning in the Introduction to this chapter for examples using horses.

Operant Conditioning

A single horse was taught by Myers and Mesker[1061] to make a typical operant response, pushing a lever with his muzzle for a half cup of grain. Once the horse had been shaped using continuous reinforcement, the schedule was changed to three and then to 11 responses for every reinforcement. The horse increased its rate of responding as the FR ratio (fixed ratio of responses per reward) increased. When a fixed interval schedule was imposed, the horse at first "sulked" by refusing to turn toward the lever, but later responded with the scalloped pattern typical of laboratory rodents on fixed interval schedules. The practical application of the study is that a horse will perform better when it is not rewarded for each performance.

Operant conditioning is used to evaluate drugs in equine pharmacology.[1511] The horses are usually conditioned to break a beam with their head for a food reward. The rate of responding will be lower if a depressant, such as acepromazine, is given or will be higher if a stimulant, such as methylphenidate (Ritalin), is administered. The action of an unknown drug can be tested by comparing its effect on responses to that of other, known, drugs. Using this technique, one can determine if the horse is being stimulated or depressed, that is, whether the horse's performance is being improved or worsened. Operant conditioning has also been used to measure environment preferences of horses.[687]

Visual Discrimination

Surprisingly few formal studies on learning in horses have been performed, despite the amount of training that is necessary to teach a Lippizaner to perform the courbette, or to teach a riderless cutting horse to select a steer and separate it from the herd. Gardner's studies[538,540] revealed that a horse could learn to choose a feedbox covered with black cloth and containing food instead of two empty feedboxes. With increasing trials, the number of errors decreased. The horses retained the discrimination when tested over a year later, but like cattle they found it difficult to choose the correct box when the cloth was moved to a position above or below the box. In general, the horses made slightly more errors than cows in a similar situation.[537,539] Using figures on the covers of feedboxes, Giebel[549] was able to teach ponies to discriminate between many pairs of figures to obtain feed. If the symbols used were too similar to be perceived as different by the ponies, they would exhibit weaving behavior, swaying from side to side over the boxes without making a choice. As in Gardner's study, the discriminations were well retained for long periods. Fat horses make more errors on a simple visual discrimination test than thin ones, which is probably a result of poorer motivation for the food reward rather than intellectual deficit.[953]

Maze Learning

Yearling quarter horses can easily learn a simple right or left turn maze in trials and can learn to turn in the opposite direction (reversal learning). Punishment did not improve performance.[833] Using a similar maze, Haag et al.[584] found that ponies who learned a maze reversal (to turn right rather than left) most quickly also learned to avoid a shock in the least number of trials. There was, however, no correlation between learning ability in either task and position in the dominance hierarchy.

The maze is shown in Figure 7.10. Horses can also learn to form taste aversions.[1530]

Observational Learning

There have been two attempts to demonstrate observational learning in horses. In both cases, the horses, young quarter horses, watched another choose one of two feed buckets. In neither case did the horses that observed perform any better than those that had not had the opportunity to observe.[98,106] This is particularly interesting because horse vices, in particular cribbing, are believed to be learned by observation. Tests of other types of learning might reveal that horses do learn by observation.

Influences on Learning

Handling

Mal et al.[924] and colleagues[923] have shown that early handling (imprint training) does not influence trainability but that extensive han-

FIG. 7.10.   Maze learning in horses. The horse enters the maze and must make a right turn in order to leave the maze and receive a food reward.

dling for the first 6 weeks is more beneficial than the same amount of handling later in the foal's life.

A moderate, but not an extensive, amount of handling improves a young horse's performance in a maze learning test.[638] Perhaps the most useful information is that handlers can predict trainability after working with a horse for 10 days, and although another interpretation might be that handlers determine performance, these scores are correlated with trainability under saddle as judged by different people. The more emotional a horse, the poorer its learning ability.[474,639,919]

Age

Weanling foals learn to make the correct choice in the maze shown in Figure 7.10 with fewer errors than adult mares, but the latter move faster, so despite entering the wrong side, they reach the food just as quickly as the foals. Orphan foals do not appear handicapped in their learning ability but move even more cautiously than normally reared foals.[693]

Breed

When quarter horses and thoroughbreds were compared on ability to learn a visual discrimination, the quarter horses did better, apparently because they were less distracted.[919]

Frequency of Training

In general, pauses between training bouts result in faster learning, apparently because the biochemical processes involved in learning do not happen instantaneously. Learning must consolidate a process that involves formation of new protein in the relevant part of the brain. The problem is that, although the horse spends less time in training to learn optimally, more days have elapsed. Nevertheless, as Rubin et al.[1235] demonstrated with avoidance learning, spacing training bouts is more efficient than crowding them into a few days.

How many repetitions of a task should there be in each training session? Sixteen seems to be the optimum when negative reinforcement is used.[957]

CATS

Whereas dogs were the animals used in the earliest experiments on classical conditioning,[113] cats were studied in the first experiment in operant conditioning.[1393] Cats learned to operate on their environment in order to escape from puzzle boxes. They also learned to pull strings to which a piece of food was attached, selecting from the one attached to the meat from among several others.[7] Manipulating strings is a task that dogs perform poorly (see earlier discussion); cats may be more anatomically than intellectually suited to the task. Cats can be classically conditioned,[603] and experimental neurosis can also be conditioned in cats, as in most species,[50] by requiring them to discriminate between two very similar stimuli (auditory, in this case).

Discrimination

The cat's ability to learn discrimination has been used to great advantage by psychophysicists in studying vision. For example, color vision can be studied by teaching cats to discriminate between two symbols and then to discriminate between the symbols when they differ only in hue. Cats can, in fact, make this discrimination but only after 1400 trials; they do have color vision, as both behavioral tests[1289] and electrophysiological studies[306,1200] attest. The color stimulus must be large (that is, a big object) before the cat is able to make use of the hue. Nonneman and Warren[1078] used a two-cue discrimination to measure the salience or importance of various sensory modalities to cats. The animals were taught to feed from one of two feeders, the one with a buzzing noise and flashing light associated with it. Later, one feeder flashed and the other buzzed; the cats went to the flashing feeder, indicating that auditory stimuli are less important to cats than visual ones when both are carefully equated for intensity.

Rewards

Unlike dogs, cats will not usually perform in order to be reunited with the experimenter. They will perform for food rewards. Feline finickiness can even interfere with the reward value of food, but in general, cats will work harder for food rewards if the experimenter is the one who feeds the cats in the home cages. It is even more difficult to teach a cat to press a bar for water; water must be withheld for a week.[105] Kittens will learn more quickly when the reward is freedom to explore a room than when the reward is food.[1014] Kittens learn to make a light–dark discrimination more slowly than 35-day-old cats.[1216]

Brain Stimulation

Cats can be classically conditioned when electrical stimulation of the cortex (see Appendix 3) rather than the sound of a bell or some other sensory cue serves as the conditioned stimulus.[394] Stimulation of the posterior hypothalamus is both rewarding and aversive to cats. A cat will learn to work to turn on and to turn off hypothalamic stimulation.[1204,1205]

Conceptual Learning

Cats are able to form learning sets, that is, to form concepts. Warren and Baron[1455] showed that cats could learn to solve a problem, such as choosing the object on the left when identical black squares were the stimulus, and would learn much more quickly on the next problem to choose the object on the left when white triangles were presented. After four problems, the cats' errors fell to 36% of the original errors, and only 58% of the number of trials originally necessary were needed to reach criterion. Cats seldom show insightful behavior; they do not learn to move a light box under a suspended piece of fish in order to reach the fish.[7] Captured feral cats learn a discrimination more quickly than cage-reared ones.[1486] These findings indicate that a varied environment or experience may lead to an increased learning ability in cats.

Imitation

Learning by observation or imitation takes place in cats. Cats watching a cat press a bar or jump a barrier to obtain food learned to press the bar or jump the barrier much faster than cats who did not observe a trained animal. Cats can also be misled. If the cats watched a cat that obtained food by simply approaching but not pressing the bar, they learned to bar press for food more slowly than nonobserving cats.[757] Kittens can also learn by observation, and they learn more readily by watching their mothers than by watching another adult cat.[293]

Both cats and dogs can be taught to make auditory discriminations and to lift the lid on a food pan when one pitch but not another is sounded. When the two sounds are too close to be discriminated (less than one-third or one-quarter tone apart), the animals exhibit experimental neurosis. The cats respond to all tones as positive, and most dogs refuse to respond to any.[416] Early visual experience can influence a cat's performance on visual discrimination tasks.[1527,1528] Old cats (10 to 23 years) do not learn as well as younger cats. They are most apt to fail to learn, even after 1000 trials, if the conditioned stimulus begins too far in advance of the unconditioned stimulus.[607]

A final note on performance of cats: whereas dogs tend to run faster when competing with other dogs, cats do not; in fact they may refuse to compete for food in a runway situation.[1502]

CHAPTER 8

INGESTIVE BEHAVIOR: FOOD AND WATER INTAKE

There are a variety of factors that influence feeding in animals. Some of these factors act in the short term, others act in the long term; still others are only operational in emergency situations. Taste can either stimulate or inhibit intake. Gastric factors, particularly gastric fill, can suppress feeding. Changes within the intestinal tract, such as an increase in osmotic pressure and release of the hormone cholecystokinin, can bring a meal to an end, thus inhibiting intake. These are all short-term factors. Eating in response to a lack of metabolizable glucose is an emergency mechanism that is probably not involved in meal-to-meal initiation of eating. Increasing blood levels of glucose does not suppress feeding. When food is freely available, animals, including carnivores, eat many meals a day. Under these circumstances, initiation of a meal is probably a response to waning of the satiety signals. The long-term controls of feeding in which intake is controlled as part of the regulation of body weight or body fats are obviously operational in pigs and other domestic animals. The feedback from fat cells to the brain is a protein, leptin, which inhibits intake. All these signals are integrated in the brain, in which a variety of neurotransmitters and anatomical sites are involved in feeding. The finding that depressants introduced into the brain stimulate feeding reinforces the concept that hunger occurs when satiety signals weaken.

Drinking is associated with eating or stimulated by an increase in osmolality or a decrease in blood volume. Salt intake is controlled by the angiotensin aldosterone system.

INTRODUCTION

General

Animals typically show a growth curve that includes a short, dynamic phase in which weight gain per day is large and a much longer static phase in which there are no major gains or losses in weight but rather oscillation around a mean or set point of body weight. Most domestic animals grow rapidly for several months or the 1st year following birth and then plateau at a mature body weight.

Animals treated by veterinarians fall into two general categories: either food- and/or fiber-producing animals or companion animals. The problems of ingestive behavior of animals differs from category to category. Animals used for human food rarely survive much beyond the dynamic phase of weight gain. The objective of the producers and the veterinarians who advise them is to maximize weight gain per unit of time. The dairy cow presents a special case. She must be a good producer of milk and, therefore, must increase her food intake, yet she should route that increased energy intake not to body fat, but to milk production.

A quite different problem is presented in some companion species, especially household pets and horses. In a seminaturalistic setting, like a pasture, a horse can be fed ad libitum, but when presented with an energy-rich concentrate diet that it would not have encountered in the wild, the horse may overeat. Acute problems such as colic or founder may result. Chronically, a simple shift upward in body weight, obesity, may result. Dogs and cats face a similar problem. They can maintain their body weights on relatively unpalatable, dry chow diets but may succumb to obesity or digestive upsets when offered a highly palatable diet ad libitum.

The physiology of the controls of food and fluid intake has been studied extensively by such diverse scientific groups as nutritionists, physiologists, animal scientists, and psychologists. Knowledge of the physiological mechanisms involved in hunger and satiety will enable us to stimulate intake for maximum yield or to control body weight without inducing hunger in an animal that tends toward obesity. The physiology of hunger is reviewed here, using the pig as an example; next to the laboratory rat, the domestic pig has been more thoroughly studied in this regard than any other animal. What is known about ingestive behavior in other species is then discussed, and the unique control of food intake in ruminants is described.

Consideration must also be given to the mechanism by which these influences are integrated in order that animals not only survive but maintain an optimal and nearly constant body weight. Until recently, there was a missing link between body fat stores and the brain. The question was how the animal "knows" that it is becoming fatter. Various mechanisms such as level of free fatty acids in the plasma had been hypothesized to be the signal from the fat stores to the central nervous system, but none could be proved. An obese strain of mice (obob) differing from lean controls (Ob–) in one gene was discovered. The product of the gene found in normal mice is formally termed ob protein, but popularly called leptin. Leptin is carried to the brain in the bloodstream and acts on receptors in the area lining the cerebral ventricles (see Appendix 3). When stimulated by this protein, these receptors initiate changes in feeding behavior (it decreases) and stimulation of the sympathetic nervous system to increase lipolysis via ß-adrenergic receptors. In this way, less food is taken in and more fat is broken down, thus decreasing fat stores. Advantage has already been taken of ß-adrenergic effects on fat stores. Pigs are fed ß-adrenergic drugs such as clenbuterol to decrease fat deposition, thus sparing calories for lean meat, that is, muscle growth (see Fig. 8.1).

One of the first signs of illness is lack of appetite. This lack of appetite or anorexia is caused by cytokines. Cytokines are polypeptides produced by one cell that influence other cell factors. There are numerous cytokines with numerous functions, but one of those functions is relevant to feeding behavior. The anorexia of illness is probably mediated by cytokines. For example, cytokines such as interleukin-1 can reduce food intake and rumination in goats without producing fever (see Fig. 8.2).[1424]

CONTROL OF FOOD INTAKE IN PIGS

Meal Patterns

Pigs are essentially diurnal animals; therefore, most feeding takes place during the day.[91] Feeding behavior is a circadian rhythm modified by environmental temperature. Pigs avoid eating when daily tempera-

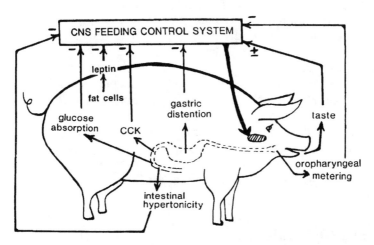

FIG. 8.1. Integration of physiological factors that stimulate and depress food intake in the brain (drawing by T. Richard Houpt).

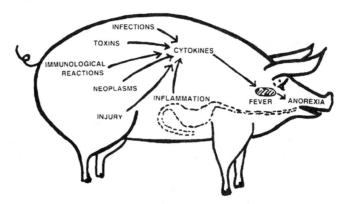

FIG. 8.2.   Integration of pathological factors that depress food intake (drawing by T. Richard Houpt)

tures are highest and, therefore, eat early in the morning and late in the evening.[459] When meals of individual pigs were recorded, the pigs were found to eat 8 to 12 meals per day, with the number of meals decreasing as the pig grew larger. Many "snacks" of less than 50 g are also eaten.[197] In contrast to anosmic rodents, anosmic pigs ate with the same meal pattern as intact pigs.[115]

Seventy-five percent of feeding takes place during the day in lactating sows. They consume eight meals a day.[397] If food is available from a ball that dispenses it only when the pig moves it, foraging in the bedding decreases. This means of obtaining food results in a more natural feeding behavior in which the appetite behavior, rooting, is combined with the consummatory behavior, ingestion of food.[1526]

When housed in groups, the pigs tend to eat at separate times probably to avoid competition at feeders, whereas individually housed pigs tend to eat at the same time as their neighbor.[560] Piglets housed in groups eat more if there are more feeders available because they are socially facilitated to eat but need not compete as much for a feeder.

Electronically controlled feeders have enabled animals, especially cattle and swine, to be kept in groups; yet, all have access to feed and their individual intake and meal patterns can be evaluated. There is no consistent rank order of entrance into electronic feeders, but groups of sows that have been in a pen longer eat before more recently added sows.[245]

Social Facilitation

Two animals housed together usually eat more than the sum of their intake when each is housed separately. This is true of most social animals. Social animals tend to do things as a group; therefore, when one pig goes to the feeder, all the pigs go to the feeder. Cole et al.[309] have demonstrated that group-penned pigs eat more than separately housed pigs. The same phenomenon, social facilitation, leads all pigs to attempt to eat from one set of feeders while ignoring other feeders. Social facilitation of eating begins early; all members of a litter nurse together.

Social facilitation may increase food intake, but this tendency can be offset by the opposing tendency of subordinate pigs to eat less in the presence of dominant pigs. Pigs may refrain from eating even in the absence of overt aggression by the dominant pig.[124]

Palatability

The adult set point of body weight (see Defense of Body Weight) applies to an animal on a given diet. An increase in palatability can result in a shift upward in body weight set point. More simply put, animals eat more and gain weight when their food tastes good. Pigs show a marked preference for sweet substances,[768,788] consuming up to 17 liters of sucrose solution per day. Advantage has been taken of this preference to increase intake.[18] The intake and weight gain of pigs are not affected by the addition of a substance that tastes very bitter to humans, indicating a species difference in taste perception.[204] Newly weaned pigs often show a drop in weight gain, and sweet pig starters are used in an attempt to stimulate intake. Although suckling neonatal pigs show nearly as strong a preference for the various sugars as do adults,[686] these attempts are not always successful. A more rewarding approach may be to add to the sow's feed a flavor that will appear in her milk and then to add the same flavor to the pig starter. Pigs will learn to associate a given flavor with a familiar food (in this case, sow's milk), and they will more rapidly ingest that flavor in solid food than they would a strange flavor.[268] As a result of testing 129 flavors, McLaughlin et al.[986] found that one, a cheesy flavor, would increase intake of newly weaned pigs.

The color of food is important. Sows avoid blue food.[725]

Environmental Temperature

Food intake is inversely related to environmental temperature; therefore, in hot weather animals eat less. The classic explanation is that "animals eat to keep warm and stop eating to prevent hyperthermia."[249] Certainly inhibition of food intake in hot weather reduces specific dynamic action and other metabolic heat as further heat loads to the animal. Under normal conditions, food intake increases or decreases in response to environmental rather than body temperature; but when body temperature rises to pathological levels, as in fever, food intake also decreases.

Food intake is also stimulated by cold environmental temperature. This thermostatic control of food intake is part of body temperature regulation. When more energy must be applied to maintain body temperature, more energy is taken in.[733] Those areas of the brain that are involved in temperature regulation, in particular the anterior hypothalamus (see Appendix 3), are also involved in the increase in food intake observed in the cold. Changes in temperatures of the brain are not correlated with the initiation and termination of meals; and thermostatic eating is a response to changes in ambient or environmental temperature, not to changes in body temperature within the physiological range (Fig. 8.3). If cold temperatures are too extreme, the animal will not be able to compensate for the energy lost by increasing its intake and will lose weight.

Gastrointestinal Factors

None of the many factors that stimulate or depress intake has been involved in the control of normal meal initiation and termination. Glucoprivic eating is an emergency mechanism. Body weight set point is maintained, but over a period of days, not hours. Environmental temperature also influences food intake over a period of days. The most likely candidates for the meal-to-meal controllers of intake are those factors that are closely associated with the act of eating. Following a meal, there are changes in the gastrointestinal and plasma level of various constituents, any one or several of which might influence food intake. Food intake ceases before intestinal absorption is complete. If it

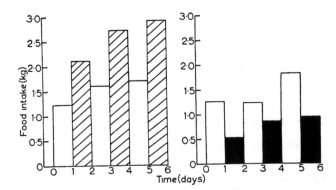

FIG. 8.3. Daily food intake of pigs fed ad libitum. Pig on *left* was subjected on alternating days to temperatures of 25°C *(white)* and 10°C *(hatched)*. Pig on *right* was subjected on alternating days to temperatures of 25°C *(white)* and 35°C *(black)*[733] (copyright 1974, with permission of Pergamon Press).

did not, animals would consistently overeat because food would continue to be ingested during the lag between ingestion of food and its absorption. Gastrointestinal hormones are released as soon as food is present in the upper gastrointestinal tract and may be satiety signals. Of the many gastrointestinal hormones, such as gastrin, secretin, and enterogastrone, cholecystokinin-pancreozymin (CCK) is the one that holds most promise as a satiety agent.[548] The hormone is released from the mucosa of the upper gastrointestinal tract. Cholecystokinin-pancreozymin has several physiological actions: it stimulates contraction of the gallbladder; it stimulates release of pancreatic enzymes; and it has been shown to inhibit food intake in hungry animals, including pigs.[689,701] Further, CCK acts on the stomach, presumably by slowing gastric emptying, to produce feelings of satiety. Cholecystokinin-pancreozymin is bound by two types of receptors, CCK-A and CCK-B; CCK-B receptors are found in the brain and mediate panic behavior, whereas CCK-A receptors are found peripherally in the gastrointestinal tract as well as in the brain stem. Agonists of CCK-A, but not CCK-B, suppressed feeding in pigs.[1109] Fat in the intestine suppresses intake out of proportion to the caloric value of the fat,[573] and a CCK antagonist prevents that suppression.[1184] Immunizing pigs, but not lambs, against CCK results in a significant increase in intake and weight gain.[1117]

Removing the gastric contents before a meal has little effect on pigs' meal size. There are gastric influences on feeding, but they do not act instantly. Satiety after meals high in protein or fat is probably mediated through the release of CCK, but another satiety mechanism may exist for high-carbohydrate foods. Suckling pigs show a depression of food intake following gastric loads of isotonic glucose solution, but not following gastric loads of isotonic sodium chloride solutions, indicating that there may be glucoreceptors in the gastrointestinal tract[688,1354] that produce satiety. Hypertonic solutions of either glucose or sodium chloride depress intake of both suckling and more-mature pigs.[688,702,704] As food is being digested, the osmotic pressure in the intestine rises and stimulation of osmoreceptors in the intestine may be part of the basis of preabsorptive satiety (Figs. 8.4 and 8.5).

Estrogen Levels

When female pigs are in estrus, food intake is depressed and activity levels rise. Gilts eat 4 kg (9 lb) less during a week in which they are in estrus than in weeks when they are in another stage of the reproductive cycle.[522] The increase in activity has been quantified: sows in heat walk 14,000 steps per day, whereas sows that are not in heat walk 5000 steps per day.[42] Observation of the food intake and general activity of a sow can assist the producer in identifying a female in estrus.

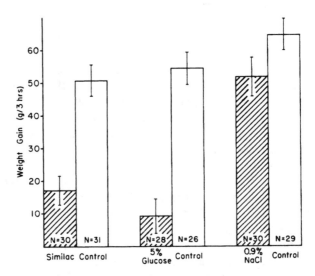

FIG. 8.4. Effect of various gastric loads on 3-hour intake of suckling pigs. Note that milk (Similac) and isotonic glucose, but not isotonic saline, depressed intake[686] (copyright 1976, with permission of Oxford University Press).

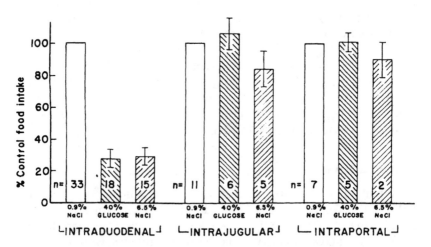

FIG. 8.5. Effect of hypertonic injections administered into the duodenum *(left)*, the jugular vein *(center)*, or the portal vein *(right)* on subsequent food intake of pigs. Only intraduodenal injections depressed intake relative to saline-injected controls.

Feeding at Parturition

Sows fed ad libitum during gestation have almost complete anorexia on the day of parturition, and although their food intake increases during lactation, it does not reach gestation levels for several weeks. In contrast, sows fed half of their ad libitum intake eat on the day of parturition and during lactation when feed was available ad libitum ate more than previously ad libitum–fed pigs.[1463] This could be a result of depleted fat stores and decreased leptin production in the restricted sows or increased insulin and reduced mobilization of nonesterified fatty acids.

Imbalance of Dietary Amino Acids

Animals generally show nutritional wisdom in that they select an adequate amount of protein. For example, growing pigs given a choice of a protein-free or an adequate-protein diet ingested sufficient protein for maximal weight gain.[372] In certain circumstances, however, nutritional wisdom is not shown and pigs will choose a nonprotein diet over a protein diet.[1206] This occurs when the protein contains an imbalance of essential amino acids. The explanation for the marked depression of food intake seen when the imbalanced diet is the only one offered, or for the selection of no protein rather than an imbalanced protein, is unknown. It is speculated that an imbalance in essential amino acids leads to an imbalance in central nervous system neurotransmitters. Accumulation of a neurotransmitter might depress intake.

When tryptophan is deficient in the diet, the concentration of serotonin for which tryptophan is a precursor is low and food intake is suppressed.[655] The area of the brain involved in this behavior may be the cerebral cortex (see Appendix 3), rather than the hypothalamus.[1076]

Glucose Utilization

Animals increase their food intake when the rate of glucose utilization in the brain falls. Experimentally, this phenomenon can be demonstrated by administering a competitive inhibitor of glucose, 2-deoxy-glucose (2 DG), which decreases glucose uptake in the brain.[1355] Similarly, food intake can be increased by administering an agent that markedly lowers plasma glucose (Fig. 8.6), such as insulin. Although eating in response to a lack of utilizable glucose (glucoprivation) can be readily demonstrated in a variety of species, it is probably not a practical method for stimulating intake because the dosage necessary to stimulate food intake is perilously close to the dosage that produces hypoglycemic convulsions. Eating in response to glucoprivation is an emergency mechanism the animal uses when its endogenous energy

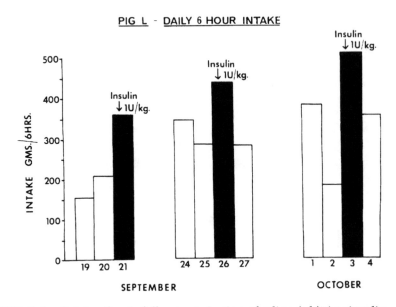

FIG. 8.6.   Intake of a pig following injection of saline *(white)* or insulin *(black).*

supply is approaching exhaustion. It probably is not involved in initiation and termination of meals in the free-feeding animals.

Defense of Body Weight

The constancy of body weight in the adult animal is believed to be the result of an innate set point for body weight or, perhaps and more likely, a set point for total body fat stores. The set point is most easily detected when body weight is artificially manipulated. For example, if an animal is starved for a few days it will, when food is again freely available, eat more than it did before the fast and rapidly regain the weight lost during the fast. In some cases, pigs show compensatory increases in intake after food restriction; in other cases, they do not increase their intake over that of nonrestricted controls, but their weight gain is greater, presumably owing to increased efficiency or decreased energy output.[1006] When diets are diluted with noncaloric bulk, intake rises so that caloric intake remains constant.[1016,1098] Conversely, if the animal is force-fed, it will gain weight but will decrease its voluntary intake while being force-fed.[115] Similarly, animals that have gained weight while being injected chronically with insulin will when treatment ceases, decrease their intake and lose weight until their weight is at the preinjection set point. The leptin system described earlier is probably the mechanism in pigs but has not been studied in that species. Set point is also maintained by adjustments in energy output that can be varied both by changes in motor activity and by metabolic changes in heat production.

The set point of body weight is more difficult to determine in the young, rapidly growing animal because it is being adjusted upward. The set point can be changed by manipulation of the neonate's nutrition. The starved piglet may remain stunted for the rest of its life, as conclusively shown by Widdowson.[1475] Similarly, overfeeding of piglets, as may occur in a small litter, may result in a permanently higher set point of body weight. If a young growing pig is fed 120% of its normal intake intragastrically, it will grow faster, indicating that appetite, rather than genetic growth potential or gastrointestinal fill, limits the rate of growth.[1116]

In response to changes in consumer preferences, pigs are now selected for leanness, so the set point for body fat is low and an obese pig is rarely seen on the farm. There are, however, genetically obese strains of pigs maintained for experimental purposes.[663,943] Obesity can also be produced by dietary manipulation in young meat-type pigs.[582] The number of fat cells did not increase in these pigs exposed to a palatable, high-carbohydrate diet, but the fat cells were larger, that is, they contained more fat.

In general, intake can be stimulated in the adult animal by any manipulation that lowers body weight or body fat stores beneath the set point. The neonatal piglet can respond to fasts of short duration with an increase in food intake, but longer fasts may produce irreversible hypoglycemia, coma, and death.

Integration of Factors That Stimulate and Inhibit Intake in the Central Nervous System

Central nervous system stimulants, in particular amphetamines, have been used extensively in human medicine to treat obesity. Less stimulating and less abused but less effective drugs, like phenylpropanolamine, are also used as over-the-counter aids to dieting. Chemical control of obesity is rarely used in domestic animals. Domestic animals' weight is controlled by the owner, and in rapidly growing meat-producing or lactating animals, depression of food intake is generally avoided. Amphetamines apparently have their effect on neurons or neurotransmitters that mediate satiety or determine set point. Stimulants increase satiety and lower body weight set point.

Unlike insulin, central nervous system depressants, in particular the benzodiazepine tranquilizers that affect the gamma amino butyric acid, or GABA, receptor, such as diazepam and elfazepam, can probably be used to stimulate food intake when given as food additives to pigs, cattle, and horses.[255,987] Elfazepam, for example, has been shown to increase pigs' mean intake from a control level of 220 g/h to 570 g/h.[987] A GABA-A receptor agonist increases feeding, and GABA receptor blockers decrease feeding.[118]

Many compounds when applied directly to the central nervous system stimulate food intake; for example, calcium or magnesium ions or pentobarbital injected into the cerebral ventricles stimulate food intake for an hour after injection.[119] Whether use can be made of the effects of these ions, or of barbiturates and tranquilizers, to increase food intake of food-producing animals economically and safely remains to be seen. All those interested in manipulating serotonin (5HT) function as a means of decreasing aggression should also be aware that agonists of the 5HT receptors increased spontaneous feeding in pigs.[116] Fluoxetine (Prozac) produces taste aversion in rats.

Kappa opiates increase feeding in pigs, and opiate blockers inhibit feeding.[117,126] It is somewhat confusing that there is a postfeeding opiate-mediated hypoanalgesia.[1246]

The controls of feeding in pigs have been reviewed by Houpt.[700] All the factors that have been discussed here—environmental temperature, rate of glucose utilization, palatability, social factors, estrus, gastrointestinal hormones, and the presence of glucose or hypertonic substances in the gastrointestinal tract—are operating at the same time. The information is integrated, presumably in the brain, and the animal eats more or stops eating accordingly. The role of the various neurotransmitters remains unclear.[739] Such factors as social facilitation and avoidance of protein-imbalanced diets are mediated through the cerebral cortex. Changes in food intake in response to temperature are mediated through the anterior hypothalamus.

Two areas have been identified as ones that may integrate at least some of the information carried to the central nervous system from the rest of the body: the lateral and ventromedial hypothalamus (see Appendix 3). Stimulation of the lateral hypothalamus causes animals to eat. Lesions in the lateral hypothalamus of pigs cause a decrease in food intake and body weight,[794] and destruction causes animals to stop eating, even to die of voluntary starvation. The ventromedial hypothalamus is involved in determination of body weight set point. Destruction of that area results in hyperphagia and a rapid gain of weight until a new set point is reached, at which time food intake drops and body weight is maintained at the new, much higher levels. Increase in food intake and body weight has been observed in pigs following ventromedial hypothalamic lesions.[794] The periventricular areas of the hypothalamus are also important in controls of feeding but have not been studied in domestic animals. Use of brain lesions in meat-producing animals is probably not practical because the increase in weight is nearly all due to an increase in fat.[90]

Clinical Problems

Polydipsia

Scheduled induced polydipsia has been produced experimentally in pigs, that is, when food is only available intermittently and in small quantities, the pig will overdrink.[1357] Sows in a farrowing crate may show polydipsia, perhaps because the waterer may be the only thing available for them to manipulate.[1244] Pigs that are fasted or on a severely limited ration may also increase their water intake (that is, show polydipsia).[1521,1522] Notice that in the free-feeding pig or one that is only food deprived for a few hours, drinking accompanies eating, but in a very hungry one drinking occurs as a compensation for lack of food.

Effects of Diet on Stereotypies

The role of fiber in swine feeding has not been assessed, but it does appear to reduce psychogenic polydipsia and stereotypic behavior in sows.[1202] High-fiber diets reduce rooting behavior, whereas low-protein diets increase rooting.[254] There is considerable evidence that feeding sows free choice will reduce or eliminate stereotypic behavior.[1382,1383]

CONTROL OF FOOD INTAKE IN DOGS

Meal Patterns

Dogs that have free access to foods 24 hours a day eat many small meals a day, mainly during daylight hours.[1054] This meal frequency indicates that dogs are diurnal, rather than nocturnal, and that once-a-day feeding is not "natural," or at least not preferred, by dogs. Beagles living in individual kennels with dry food freely available tend to eat three times a day, at dawn, at dusk, and whenever fresh food is given.[1181]

Social Facilitation

Many animal owners have noted anecdotally that the addition of another animal to the household increased the original pet's interest in food. At times, the increase can be a pathological hyperphagia or increased food intake.[492] Social facilitation of food intake has been quantitatively measured in puppies.[743]

Palatability

Taste preferences must be determined for each species; one cannot assume that because something tastes good to humans, it tastes good to animals. Dogs prefer meat to a high-protein, nonmeat diet, and they show preferences for one meat over another. These preferences, in order, are beef, pork, lamb, chicken, and horse meat.[685,901] Not only flavor, but also the form in which the food is offered is important in palatability. Dogs prefer canned or semimoist food to dry food.[812] They prefer canned meat to the same meat freshly cooked and prefer cooked to raw meat.[901] Dogs do not prefer familiar food; in fact, they prefer a novel flavor of canned food. Puppies also tend to prefer novel food, but palatability and maternal effects may outweigh these[470,1054] (Fig. 8.7). Dogs familiar with semimoist food eat only a little canned food at first, when both are available, but soon show a preference for the canned. Dogs also have an innate preference for sucrose in liquids or solid food.[565] They are unusual in preferring fructose to sucrose (humans and pigs prefer sucrose). Lactose, but not maltose, is also preferred.

Palatability does not depend on taste alone because elimination of olfaction (anosmia) eliminates preference for one meat over another, although anosmic dogs still preferred meat or a sucrose diet over a bland cereal diet.[685] Odor is important, but only for detecting minor differences between food, as in distinguishing lamb from pork. Such major preferences as that for meat over nonmeat and for sucrose are not af-

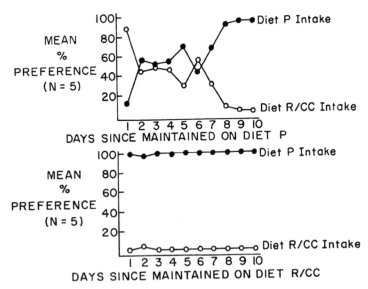

FIG. 8.7.   Effect of novelty on food preferences. Dogs normally prefer diet P to diet R/CC, but will show preference for either R/CC or P if it is a novel diet. Preference for P is sustained *(lower graph)*, but that for R/CC is reversed in a week *(upper graph)*[1054] (copyright 1977, with permission of Monell Chemical Senses Center and Academic Press).

fected. Odor is most important for locating food;[1207] taste is most important for identifying food.

The food preferences are reflected in dogs' attraction to the odors of meats. The odor of cooked meat is preferred to that of raw meat, and that of fresh meat to that of 3-day-old meat.[167] Despite their tendencies to roll on rotting organic matter, dogs showed little preference for aged dirty diapers or fish.

Taste is important in a long-term sense as well. Dogs who have been adequately nourished but deprived of taste will overeat when food is again available, indicating that deprivation both of the taste of food as well as of calories is important.[858] The practical application of this is that provision of a calorie-free, but tasty, food might be less stressful than fasting.

Environmental Temperature

Food intake is believed to be controlled, in part, by the regulation of body temperature. Consequently, animals tend to eat more in the cold and less in the heat. This has been demonstrated in environments as diverse as Alaska and Florida and with breeds as different as huskies and beagles. Food intake doubled in the winter as temperatures fell from a summer high of 20°C (68°F) to a winter low of −17°C (1°F).[415] Indoor dogs eat less than those housed outside. This reduction in intake is probably due to the warmer indoor environment.[841]

Gastrointestinal Factors

Oropharyngeal factors alone are not sufficient to induce satiety. Dogs surgically prepared with esophageal fistulas can swallow food, but if the fistula is open, the food will drop from the esophagus and not reach the stomach (sham feeding). In this case, dogs will sham feed for hours, stopping only to rest. It is interesting that placing food into the stomach at the same time that the dog is eating and swallowing food does inhibit further sham feeding.[745] Apparently, in dogs at least, both oropharyngeal and gastric stimuli are necessary to induce satiety. It has also been found that gastric loading immediately before a meal does not markedly inhibit food intake unless more than half of the normal meal is placed in the stomach.

Food placed in the stomach just prior to a meal does not inhibit food intake, but a gastric load given 20 minutes before a meal does.[745] The delayed effect of a gastric load indicates that food must either be absorbed or that some humoral factor must be released by the presence of the food in the upper gastrointestinal tract to inhibit food intake. There is evidence in various domestic animals that both the products of di-

gestion and humoral factors inhibit food intake. Therefore, glucoreceptors in the liver or portal system may be involved in satiety.

Food intake of dogs is suppressed by CCK and glucagon and is
markedly depressed by the opiate blocker naloxone.[866] Neither intraduodenal fat nor peripherally administered CCK suppresses sham
feeding in dogs, although intraventricularly administered CCK does.[1108]
The difference in the effect of CCK on sham feeding and normally feeding dogs indicates the importance of integration of gastrointestinal and
humoral events for satiety.

Estrogen Level

During estrus, bitches tend to eat less; conversely, removing the
source of estrogen stimulates food intake. Both cats and dogs have a
lower metabolic rate following castration or spaying, which also accounts for their tendency to gain weight.[1215] Therefore, one of the side
effects of ovariohysterectomy is a tendency for the spayed bitch to eat
more and gain weight (Fig. 8.8). Presumably, this is due to the removal
of the estrogen source. When a dry diet is available to dogs for only an
hour a day, neutered dogs do not gain weight—a strategy owners could
use.

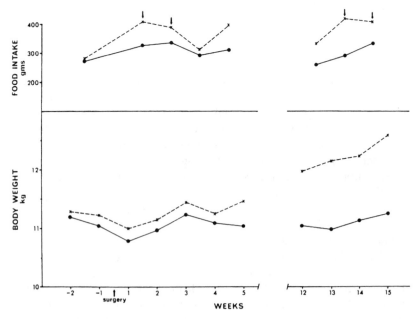

FIG. 8.8.   Effect of ovariohysterectomy on food intake and body weight
of beagle bitches. *Solid lines* = intact. *Dashes* = spayed bitches[682] (copyright
1979, with permission *J. Am. Vet. Med. Assoc.*).

Glucose Utilization

Glucoprivic eating is seen in dogs. Dogs treated chronically with insulin eat more and gain weight,[575] and eating in response to glucoprivation produced by an inhibitor of glucose utilization has been demonstrated in dogs.[703] A lack of glucose stimulates intake, and an excess usually does not. Neither intraportal nor intrajugular glucose suppresses intake.[185,186]

Defense of Body Weight

Dogs can defend their body weight against both decreases and increases. Dogs neither gain weight when force-fed nor lose weight when placed on calorically dilute diets. When dogs were given additional food each day through gastric fistulas, they decreased their oral intake, but not enough to prevent weight gain.[1304] The lack of caloric compensation may be due to the intragastric route of force-feeding; oropharyngeal signals, that is, taste and texture, were not elicited by the food, so satiety was not achieved.

Dogs can increase their volume of intake when their diet is diluted.[744] Dogs deprived of food will eat when food is again available and, thereby, defend their body weight. Different breeds of dogs defend different body weights and different degrees of adiposity. A casual comparison of bulldogs and salukis makes this clear. The problem of overeating and obesity in dogs is discussed here in the section about abnormalities of ingestive behavior (see Clinical Problems). Dogs can also select some dietary components to obtain sufficient nutrition. For example, dogs can regulate their protein and energy intake when offered diets differing in protein content. When offered such a choice, they chose a 30% protein diet.[1213]

Integration of Factors That Stimulate and Inhibit Intake in the Central Nervous System

The effect of central nervous system depressants on food intake in dogs has not been systematically studied, but there is anecdotal evidence that dogs recovering from anesthesia will eat ravenously before they are fully conscious. Because of the danger of choking in dogs recovering from anesthesia, food should not be available to them. There is also anecdotal evidence that treatment with progestational agents increases food intake. This side effect of progestins used to prevent ovulation should be investigated further.

Dogs lesioned in the ventromedial hypothalamus show hyperphagia and weight gain.[1233] They also show hypersecretion of gastric acid.[1063] The excess production of hydrochloric acid might be related to the increase in food intake that has been noted in laboratory rats and dogs with lateral hypothalamic or dorsomedial amygdala lesions (see Appendix 3).[479]

Clinical Problems

Obesity

Obesity is the most common behavioral problem involving ingestion. The cause is, most simply, an intake of energy that exceeds the output of energy. Therefore, obesity is most often observed in a nonworking animal fed a highly palatable diet.

Clinical reports[51,357,945] indicate evidence of obesity in 20% to 30% of the dogs. These studies and a considerable body of anecdotal information indicate that twice as many spayed as unspayed bitches are obese. The reasons for this have just been discussed in the section on controls of feeding in dogs.

There are differences in breed susceptibility to obesity. When a palatable food is made freely available to beagles, some get very obese; in a similar situation, terriers do not become obese[1054] (Fig. 8.9).

High dosages of amphetamines have been shown to depress food intake in dogs.[320] Their use as a treatment for canine obesity has not been evaluated carefully.

Fortunately, obesity in animals is easily controlled by food restriction and use of commercially available low-caloric dog foods. Owners may be concerned that their dogs will develop behavior problems if their food is restricted, but laboratory dogs on restricted intake do not become more aggressive, nor are they coprophagic. At first they are more active, but after a week or 2 they are less active than they were before food restriction, apparently in an effort to conserve energy.[334]

Anorexia

A much more difficult clinical problem is anorexia. Anorexia can occur in dogs. An example was a Siberian husky that would not eat. The dog's owner cooked steak, eggs, stews, almost anything, but the dog was chronically underweight and ate intermittently. The dog was examined thoroughly, and no organic cause for the ingestion problem was found. Discussion with the owner revealed that she had simply taught the dog not to eat, and that this was the principal way in which the dog secured the attention of its owner, who led a very busy life. This was the finicky eater gone to the extreme. Starting when the dog was 8 weeks old, the owner adopted the habit of sitting with the dog while it ate and coaxing it along. The owner, a nursing home nutritionist, was very concerned with eating. The first time the pup did not eat, it was whisked off to the

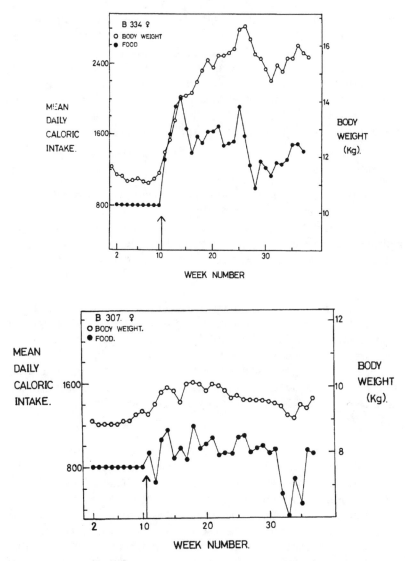

FIG. 8.9.   Food intake and body weight of two beagles. Ad libitum feeding began at *arrow*. *Lower graph* is of a dog whose intake and body weight remained normal. *Upper graph* is of a dog that overate and became obese[1054] (copyright 1977, with permission of Monell Chemical Senses Center and Academic Press).

veterinarian. The owner did not believe the advice she got from this veterinarian or several others, and instead of just waiting for the pup to get hungry, she went home and cooked a large meal of eggs for it. This pattern persisted, and finally, not even steak excited the dog; it lived on dog biscuits that were always available and ice cream once a day. The treatment was to put the dog on regularly scheduled meals with no treats and to wait. Within 3 weeks and after a 5-day stretch of not eating, the dog was eating one good meal a day and gaining weight.

Pica

Another type of abnormal ingestive behavior is pica, or eating materials that are not normally food. Puppies are notorious swallowers of inappropriate objects that must be surgically removed from the gastrointestinal tract. Occasionally, pica may be the sign of a nutritional deficiency, but more frequently, it is simply an extreme form of oral exploration.

Grass Eating. This behavior is so common that it should not be considered a true pica. It may be a means of obtaining roughage (see Wool Chewing and Other Picas in cats), provision of an emetic, because dogs may vomit afterward and because it occurs in dogs with gastritis and other upper gastrointestinal problems. Grass eating in some situations may not be ingestion, but rather tasting and smelling substances left on the grass. This mouthing of grass is particularly common after a rain.

Coprophagia. One of the most common and most disturbing forms of pica is coprophagia in dogs. Unless the feces contain viable parasite ova, coprophagia affects the owner's aesthetic values more than the dog's health. Sprinkling pepper or some other noxious substance on feces may inhibit the vice. A better approach is to inject hot sauce into the center of several fecal masses so that the dog cannot tell that they have been adulterated. A much more effective treatment is to administer pancreatic enzyme tablets orally. This probably imparts an objectionable taste to the dog's feces. A more drastic treatment would be to treat the animal with apomorphine (0.04 mg/kg [0.02 mg/lb]) immediately after it ingests the feces. The taste aversion that may result from associating the eating of feces with vomiting and nausea could break the habit.[583]

Psychogenic Polyphagia and Psychogenic Polydipsia

Both psychogenic polyphagia and psychogenic polydipsia have been reported in dogs.[492] Psychogenic polyphagia is frequently noted when a rival pet is introduced into a household. Feeding the animals in separate places sometimes alleviates the problem, as does extra attention to the original pet. Psychogenic polydipsia may occur, but care must be taken to differentiate it from the polydipsia secondary to polyuria. Water intake may be restricted in the case of psychogenic polydipsia but would be dangerous in the case of primary polyuria.

Adipsia and Hypernatremia

This syndrome is rarely encountered but has been found in Miniature Schnauzer dogs[323] and in cats. It is interesting because it demonstrates two principles of ingestive behavior: (1) that there are multiple controls of thirst and (2) that animals will increase their intake of diluted diet in order to keep their caloric intake constant. These animals do not drink, in response to the osmotic stimuli, so they present with hypernatremia. They will, however, respond to hypovolemia, so they drink when furosemide, a diuretic that lowers blood volume, is administered. To ensure that the animals consume enough water, they can be maintained on a slurry of canned food and water. The controls of food intake are normal, so they inadvertently ingest water while eating.

CONTROL OF FOOD INTAKE IN CATS

Meal Patterns

Cats, like dogs, eat many small meals (12) per day when given free access to food, but unlike dogs, cats eat both in the light and in the dark.[1054] One might argue that this intake pattern is not natural, yet the caloric intake per meal is approximately equal to that contained in one mouse. A feral cat with good hunting skills might easily catch 12 mice (or three rats) per day.

Social Facilitation

Cats have not been shown to increase their food intake when housed in groups rather than individually, nor does the sight of one cat eating stimulate other cats to eat when food is freely available. Nevertheless, many owners report that their cats eat only when they are nearby and that provision of food in the bedroom can prevent 5:00 AM meowing. Even when food is available in the kitchen, the hungry cat seems to need the owner's presence in order to eat.

Palatability

Cats are notoriously finicky. This reflects the strong influence of palatability on food intake in cats. There is one striking peculiarity of cats: they do not prefer sucrose as most animals do. Cats do not show a sucrose preference for aqueous solutions of sugar, but they do prefer sucrose in salt solutions. The taste of water may block the expression of a sweet preference in cats.[144] The feline preference for sucrose in saline solutions soon turns to aversion because vomiting and diarrhea result in the sucrase-deficient adult cat.

Cats will not eat diets containing medium-chain triglycerides or hydrogenated coconut oil. This may be because they break the triglycerides down to fatty acids in their mouths (cats possess lingual lipase) and are peculiarly sensitive to bitter tastes. They will avoid as little as 0.1% caprylic acid or 0.000005 M quinine. In comparison, rabbits and hamsters avoid quinine only at 0.002 M.

Cats prefer fish to meat and, unlike rodents, prefer novel diets to familiar ones.[636] Cat food manufacturers, no doubt, take advantage of both of these feline preferences. If the new diet is not more palatable than the familiar one, the cat will, after a few days, begin to choose the familiar food.[1054] While cats are drinking milk, their EEG is synchronized as it is when they are drowsy.[287]

When presented with a choice between a food frequently made available to them and one that is rarely offered, cats will choose the scarcer one, even if both foods are dry commercial cat food.[296]

Environmental Temperature

There has not been a complete study of the effect of temperature on food intake in cats, but one demonstration showed that cold environmental temperature, rather than changes in brain or body temperature, affects feeding. This study found that cats drinking milk not only show a decline in brain temperature, but also cease to eat.[12] If brain temperature was the factor affecting feeding, one would expect the cat to eat even more as his brain temperature fell.

Gastrointestinal Factors

Gastrointestinal depressants of feeding have been studied to some extent. Glucoreceptors that suppress feeding in cats appear to be present in the liver.[1249] Cholecystokinin-pancreozymin and bombesin, a peptide related to CCK, suppress food intake of cats.[96,97]

Hormonal Effects

Like dogs, cats increase their food intake when treated with oral progestins. The same hormone might be used to stimulate feeding in anorexic cats, but there are dangerous side effects.

Ovariohysterectomized and castrated cats have a lower metabolic rate,[1215] and if the cat is fed all it wants of a very palatable food and is not very active, obesity can result.[1264]

Glucose Utilization

In contrast to most other species, cats have not been shown to respond to glucoprivation with an increase in intake. The glucose analogue, 2-deoxy-D-glucose, failed to stimulate feeding in cats,[741] but the failure to stimulate may have been a result of species differences in the effective dose of the drug rather than to species differences in the control of food intake.

Defense of Body Weight

The most important determinant of body weight in cats appears to be cyclical in nature. Cats lose and gain body weight in cycles of several months' duration.[1178] For this reason, it has been difficult to demonstrate defense of body weight. There have been two studies in which it appeared that when their diet was diluted, cats did not eat more and, therefore, lost weight.[666,765] In both studies, the diet, a dry food, was diluted with a dry diluent, either kaolin or cellulose. The cats actually ate a smaller volume than they ate of the undiluted food, indicating that it was unpalatable. In contrast, when cats' food is diluted with water, they do compensate by increasing the volume of intake and maintaining caloric intake constant.[284,1054] Because the cats were consuming more of a diet when it was diluted with water, their water intake was increased. A watery cat food may be used to increase water intake when clinically indicated, that is, in cats with urolithiasis or hypernatremia.[282]

Integration of Factors That Stimulate and Inhibit Intake in the Central Nervous System

Cats have been the subject of many studies of the central nervous system. Stimulation of the hypothalamus electrically, for example, influences food intake. It has been shown that cats with electrolytic lesions in the ventromedial hypothalamus become obese,[45] whereas cats with lesions in the lateral hypothalamus are both aphagic and adipsic.[44] Lesions in the preoptic area may also lead to an increase in food intake,[1519] as well as depress sexual behavior and scent marking (see Appendix 3).[622]

Opiates

The role of endogenous opiates in feeding has not been well investigated, but the opiate blocker naloxone suppresses intake in cats.[489]

Clinical Problems

Obesity

The incidence of obesity in cats has increased in the past 20 years from 10% to 29%.[1264] Some of the risk factors for obesity in cats are neutering, confinement in an apartment, and free-choice feeding of a prescription diet.[1264]

Anorexia

Anorexia is seen most often in cats. A moderately sick cat may further compromise its health by refusing to eat.

Although an otherwise healthy cat may fast for several days and will then be willing to eat large amounts or even hunt prey,[3] anorexia can be a serious clinical problem. Anorexia is most commonly encountered in hospitalized cats or in cats that are initially healthy but are then placed in a boarding kennel. Although an animal can be maintained by intragastric feeding, it is far more beneficial to the animal to reinstate voluntary eating because food taken by mouth stimulates gastric and intestinal secretions much more than does food given by stomach tube.

Advantage can be taken of the fact that depressants stimulate intake. The benzodiazepines, in particular diazepam, have been used to stimulate intake in anorexic patients. A common approach to this form of anorexia is treatment with benzodiazepine tranquilizers, which stimulate food intake in laboratory animals[1504] and in horses.[255]

Plant Eating

Cats frequently eat grass. This is observed in free-ranging, prey-killing cats as well as in those eating canned or dry diets. It is not surprising, therefore, that cats may eat house plants. Plant eating can have serious consequences to the cat because so many houseplants are poisonous, but regardless of whether the plant or the cat are at risk, the behavior is undesirable. If owners observe their cat eating a plant, they can punish the cat or frighten it away, but the cat, like the destructive dog, is most apt to misbehave in the owner's absence. The best solution is to provide the green plants that the cat apparently needs and certainly desires. Plants that are safe for cats can be purchased at pet suppliers. While the cat is learning that it is permitted access to one plant, the decorative plants can be moved out of its reach. Later, the cat should be taught to discriminate the plants it must not eat from those that it may. A water gun can be used to punish the cat and aid in the discrimination process, but such a method, of course, depends on the owner's presence. The cat would have to be separated from the plants in the owner's absence. Another method is to spray the leaves of the plant with a hot pepper solution. Still another method that would prevent elimination in the pots of large plants, as well as eating or climbing the plants, is to place mousetraps in the pot upside down so that if the cat jostles the pot or the soil, the traps will snap shut, scaring the cat.

Wool Chewing and Other Picas

Wool chewing or sucking is a behavior problem that occurs with greater frequency in Siamese or Burmese cats than in other breeds. It should be differentiated from nonnutritive suckling that many early weaned kittens will perform. Wool chewing might be related to early weaning. In comparison to the age at which free-ranging cats are weaned (6 months), most domestic kittens are weaned early, but wool sucking is usually not presented as a clinical problem until the cat is an adult.

The behavior is characterized by chewing with the molars rather than sucking. The material chewed is usually wool, but in the absence of wool, the cat will generalize to other materials, including upholstery. The cats do not chew on raw wool but prefer knitted or loosely woven material. The behavior is sporadic, but large holes can be produced in a matter of minutes. The behavior seems to be related to feeding because (1) fasting stimulates the behavior and (2) access to plants or bones or even dry (as opposed to moist) food decreases the incidence of the behavior. There is no evidence of a nutritional deficiency, but it appears to be caused by a craving for fiber or indigestible roughage. For that reason, treatment should be aimed at supplying fiber. A higher-fiber diet should be fed, and safe plants should be made available. Offering scraps of sheepskin with the wool has been effective in some cats. If that does not eliminate the problem, give the cat woolen material, such as an old sweater or sock that it may eat. In order to teach the cat to differentiate between things it is allowed to eat and those it is not, treat a variety of other wool objects with a cologne and a solution of hot pepper sauce. The principle is that the cat will learn to associate the smell of the cologne with the unpleasant taste of the hot pepper sauce and will avoid objects that smell of the cologne. Articles of clothing, blankets, and so forth can then simply be sprayed with the cologne to deter the cat.

In severe cases in which the owner is considering euthanasia, feeding the cat a chicken wing each week is recommended as a source of chewable material. The danger of injury to the gastrointestinal tract from the bones is less than the danger of euthanasia (particularly considering that many cats kill and eat birds with impunity). Finally, the cat may be treated with a tricyclic antidepressant, such as clomipramine, which will help if the problem is a compulsive one.

Cats may chew a variety of other materials. Plastics, especially telephone or computer cords, plastic bags, articles of clothing, or wood may be chewed. In one case, the serotonin blocker and appetite stimulant cyproheptadine has been successful in inhibiting plastic eating (Dr. E. Schull personal communication).

CONTROL OF FOOD INTAKE IN HORSES

Meal Patterns

The normal feeding pattern of horses, as detailed in Chapter 3 (Biological Rhythms and Sleep), is to graze continuously for several hours and then to rest for longer or shorter periods depending on the weather conditions and distances that must be traveled to obtain water and sufficient forage. When offered a pelleted complete diet, a similar feeding pattern emerges: many long meals. It is interesting that whether the horse is grazing or eating pellets, the statistical definition of a meal (feeding with breaks no more than 10 minutes' duration) is the same.[855,948] The grazing horse chews at a rate of 30 to 50 bites per minute for 8 to 12 h/d. The frequency of oral sterotypies in stalled horses on low-roughage diets is probably a reflection of the oral activity seen in a natural environment.

Social Facilitation

Horses eat when other horses eat and eat more if they can see another horse eating.[1370] This is important when creep feeding a foal and when encouraging an anorexic horse to eat. Horses appear to prefer to eat from the floor and from shallow buckets or mangers. This enables the horse to see in all directions between its legs and may have evolved as an antipredator strategy. Grazing is, of course, from the ground rather than from the usual chest height of a manger. The problem that arises when feeding confined horses is to prevent ingestion of parasite ova while still encouraging intake.

Palatability

Randall et al.[1177] have reported the basic taste preferences of immature horses. They show a strong preference for sucrose, but no preferences for sour (hydrochloric acid), bitter (quinine), or even salt solutions. At high concentrations, all the solutions except sucrose were rejected. In a study of adult ponies, 9 of 10 showed a strong preference for sucrose.[628] Salt intake varies from 19 to 143 g/d.[1274]

Grazing

Grazing behavior is selective, that is, unless the pasture is composed of only one species of plant, the animal has the opportunity to be selective as to the species ingested. For example, in the Mediterranean climate of the Camargue, horses consume graminoids in the marshy areas, moving to the less preferred long grasses in the winter.[948] Similarly, meadow and shrubland were the vegetation types grazed by feral horses in the Great Basin of Nevada, but food preferences and nutritional needs had to be weighed against the dangers of exposure in the winter and the irritation of insects in the summer.[190]

There have been several studies of the plants that ponies and horses choose to eat on pasture. Tyler[1418] found that the New Forest Ponies of England eat eight different species of plants but avoid the poisonous Senecio. These ponies will eat acorns, an overconsumption of which leads to acorn poisoning. Horses have been reported to become ill repeatedly after eating buttercups *(Ranunculus)*.[1529] Even in the laboratory, ponies find it difficult to learn to avoid a feed that poisons them; the ponies fail to learn unless the illness occurs immediately following ingestion[698] (see Chapter 7, Learning). Archer[64] found that of 29 species of grass, horses and ponies preferred timothy, white clover (but not red clover), and perennial rye grass. Dandelions were the most preferred herbs.

Horses prehend at a rate of 25 bites per minute (prehending bites should be differentiated from chewing). Horses, sheep, and cattle each have different modes of prehending grass: horses use their lips; cattle, their tongues; and sheep, their teeth. Presumably the ability to be selective varies with the method of prehension as well as with the size of the muzzle. Sheep should be more selective than horses. The reason selectivity is important is that the grazer can choose a more nutritious plant or plant part or can avoid a toxic one.

Horses vary considerably in their selectivity, which may account for deaths of only a few horses on a pasture where poisonous plants grow.[934]

Gastrointestinal Factors

The horse appears to depend more on pregastric signals to satiety than other species. When horses sham feed, that is when most of what they have swallowed falls out of an esophageal fistula, they do not eat more than they do when the food reaches their stomachs.[1170] This is in contrast to dogs, which will eat very large meals under the same circumstances. Eventually, the sham-fed horses compensate for the lost feed by eating sooner than normal, but apparently the signal for satiety is some form of oropharyngeal metering, that is, 25 swallows means enough has been eaten.

Another indication that chemoreceptors in the gastrointestinal tract are not important controls is that when administered either intragastrically or intracecally, neither glucose nor cellulose, the main digestible constituent of grass, and its breakdown products, the volatile fatty acids solutions, depress intake until they have been absorbed. Intake is suppressed, but only after a long latency. There is a mechanism for controlling caloric intake, but it appears to be postabsorptive.[1171,1172]

The first sign of colic in most horses is anorexia. This clinical observation, as well as controlled studies, indicates that pathological distention of the gastrointestinal tract inhibits feeding in the horse as in other species. This anorexia is probably mediated through pain receptors that travel in the sympathetic nerves. Analgesics intended for use in horses are often tested by determining whether a horse with a dilated cecum will eat when treated with the drug. In this case, the drug is probably working peripherally, so pain signals are not relayed to the brain, whereas diazepam works by inhibiting a centrally acting inhibitor of appetite.

Central Nervous System Depressants

Brown et al.[255] in a preliminary study found that the central nervous system depressant, diazepam, stimulates food intake in horses and that the commonly used tranquilizer, promazine, had a similar effect. Horses, in common with other more thoroughly studied species, increase their food intake when the brain, presumably the areas involved in satiety, is depressed (Fig. 8.10).

Defense of Body Weight

Horses can regulate their energy balance by controlling the amount they ingest. In other words, horses do not eat a fixed amount or eat until they are ill, but they increase or decrease the amount they eat to compensate for caloric dilution or enrichment.[855] The pony gradually shifted to ad libitum feeding does not colic but does gain weight and is, therefore, in danger of laminitis; but the body weight reaches a plateau.

Clinical Problems

Equine stable problems, the so-called vices, can be divided into oral and locomotory behaviors. These problem may be classified as stereotypies because they are repetitive, serve no known function, and occupy a large part, >10%, of the animal's time. Cribbing and wood chewing are the common oral stereotypies, whereas stall walking, pawing, and

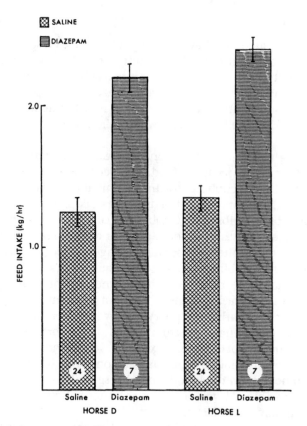

FIG. 8.10.   Intake of two horses following intravenous saline *(hatched)* or diazepam *(lined).*

weaving are the common locomotor behaviors. The latter are considered in Chapter 9 (Miscellaneous Behavioral Disorders). Although confinement is common to all these problems, cribbing and wood chewing appear to be influenced by diet. McGreevy et al.[981] found that the type of bedding and the type of feed, as well as the social contacts of thoroughbreds, influenced the incidence of stereotypic behaviors. Horses fed more roughage and roughage of more than one kind, horses bedded on straw, horses fed twice or more than three times a day, and horses with more visual contact with other horses had fewer stereotypic behaviors. Although toys may not help with oral problems, provision of a simple foraging devise that delivers food as the horse rolls it might help.[928,1501]

Cribbing

Cribbing is an oral behavior in which the horse grasps a horizontal surface, such as the rim of a bucket or the rail of a fence, with its incisors, flexes its neck, and aspirates air into its pharynx. Some horses aspirate air without grasping an object. This is called aerophagia or windsucking; the latter term can also be used to refer to pneumovaginitis. It was thought that the horse swallows the air, but this does not usually occur.[983]

Cribbing occurs in association with eating, in particular eating grain or other highly palatable food.[553,845] See Figure 8.11 for an illustration. The relation of cribbing to eating is similar to that of nonnutritive suckling in calves that occurs after drinking milk. Two-and-a-half to 5% of thoroughbreds crib.[949,1429] Cribbing occurs less frequently in endurance horses (3%) than in those used for dressage or eventing (8%), who are confined in their stall for much longer periods.[982] It is a clinical impression that cribbing occurs more frequently in confined horses, but once established, it may persist even when the horse is on pasture. It may be the result, rather than the cause, of gastrointestinal problems. Aspirating air or inflating the pharynx may be a pleasurable sensation to an animal experiencing gastrointestinal discomfort. The one undeniable consequence of cribbing is excessive wear of the incisor teeth. Although the opiate blockers such as naloxone will inhibit cribbing,[385] opiates do not rise in the blood when horses crib. In fact, horses that crib have lower blood levels of opiates than noncribbing horses, but the blood levels may not reflect brain levels.[553]

The simplest method used to prevent cribbing is to place a strap around the throat just behind the poll so that pressure is exerted when the horse arches its neck and even more pressure is exerted when the animal attempts to swallow. The horse is, in effect, punished for cribbing. If a plain strap does not suffice, a spiked strap or metal collar can be used. A common observation of a horse wearing a cribbing strap is that it continues to grasp horizontal objects with its teeth but does not swallow as much air. Many stables are designed or modified so that there are few horizontal surfaces available, but water and feed containers usually provide the horse some opportunity to crib. Shock has also been used to punish horses that crib.[108] Muzzles may also be used. These are wire baskets that permit the horse to eat and drink, but not grasp a horizontal surface. Some horses learn to grasp a stick, pull it into the muzzle vertically, and crib on that.

There have been a variety of surgical treatments for cribbing. These treatments include buccostomy; cutting the ventral branch of the spinal accessory nerve (9th cranial); myotomy of the ventral neck muscles; or a combination of partial myectomy of the omohyoideus, sternohyoideus, and sternothyrohyoideus and neurectomy of the ventral

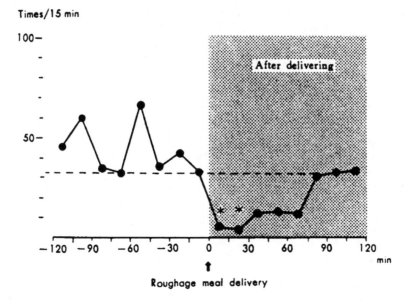

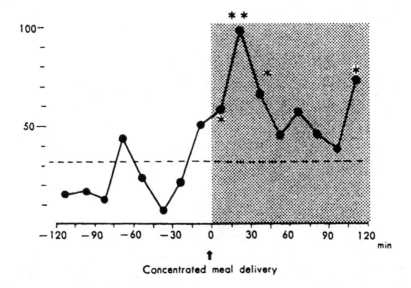

FIG. 8.11.   Mean frequency of cribbing before and after roughage meal *(upper graph)* and concentrated meal *(lower graph)* delivery. *Dotted line* indicates the mean level. * $P < .05$, ** $P < .01$[845] (with permission of *Jpn. J. Equine Sci.*).

branch of the spinal accessory nerve. The success rates of these treatments vary from 0% to 70%.[472,487,572,597,769,1099,1415]

Is it necessary to prevent cribbing? Unless the otherwise healthy horse is losing weight or suffering from colic as a result of swallowing air or flatulent colic, the behavior is not interfering with the horse's well-being. The noise of cribbing often annoys the owner, but that is not a good reason to subject an animal to the risk of surgery and the possible side effects of infection and disfigurement that may interfere much more with the animal's function than cribbing did. Cribbing is considered an unsoundness, but this may not be justified. Cribbing is also considered to be contagious, but this has never been proved. It is possible that the environment that causes one horse to crib causes the other horse in the same environment to do so;[981] young horses may be more likely to learn the habit from an adult than are other adults.

Wood Chewing

Both cribbing and wood-chewing (lignophagia) horses grasp horizontal surfaces with their teeth, but the wood-chewing horse actually ingests the wood, whereas the only damage the cribbing horse does is to mark the wood with its incisors. In contrast to cribbing, wood chewing appears to have a definite cause—a lack of roughage in the diet. Wood chewing is more common in dressage and eventing horses than in endurance horses who spend more time outside their stalls. Several investigators have noted that high concentrate diets or pelleted diets increase the incidence of wood chewing. Feral horses as well as well-fed pastured ponies have been observed to ingest trees and shrubs, so it would seem that there is some need or appetite for wood even when grasses are freely available. Farm managers are well aware that trees, especially young trees, must be protected from horses on pasture. Horses cannot digest wood; nevertheless, there may be some role for indigestible roughage in equine digestion. Jackson et al.[740] have found that wood chewing increases in cold, wet weather.

Eliminating edges, covering edges with metal or wire, and painting the surface with taste repellents are the traditional methods for preventing horses from wood chewing, but providing more roughage is a better practice both behaviorally and nutritionally. If roughage is provided, the horse's motivation to wood chew is reduced rather than thwarted. An increase in exercise reduces the rate of wood chewing.[840]

Polydipsia

Psychogenic polydipsia is also seen clinically in horses. The owner's main complaint is that the horse urinates frequently and copiously, so its stall is wet; in this case, the polyuria is secondary to the polydipsia. In horses, as in small animals, the condition must be differentiated from diabetes insipidus and other causes of secondary polydipsia.

CONTROL OF FOOD INTAKE IN CATTLE

Meal Patterns

The time that cattle spend grazing on pasture and eating in feedlots or loose housing has been reviewed in Chapter 3 (Biological Rhythms and Sleep). Cattle eat most of their meals during daylight but may graze at night if short days or hot weather preclude obtaining the nutrient requirements during the day. Grazing time is 400 to 650 minutes per day at a bite (prehension) rate of 60 per minute.[435]

Cattle kept in the confinement of a feedlot or barn eat approximately a dozen meals a day.[290] Cattle show individual cyclic patterns of intake.[1366] Rates of eating have also been measured. Eating rate varies with the physical form of the feed. Cattle can eat 2.72 kg (6 lb) of hay in an hour.[207] Meal is eaten more slowly than pellets, which in turn is eaten more slowly than a slurry. Dilution of the meal with water actually increases the amount consumed per minute. Silage is eaten more slowly than hay, probably because of illness produced by consuming the fermented product too quickly.

In some dairy parlors, the cows are fed only while they are being milked. An important consideration for dairy farmers, therefore, is the speed with which cows can consume concentrates. It takes a cow 1 to 2 minutes to eat a 0.5 kg (1 lb) of grain; slightly less time is required if the grain is cubed.[761] If a high-producing cow is milked in too short a time, she will not have time to consume the grain that she needs; another arrangement must be made to feed her. If cows are not fed grain in the milking parlor, however, they will not be as eager to enter. All these factors must be considered when planning a milking parlor and feeding system.

Rate of chewing in cattle has been studied by direct observation, by recording of jaw movement using either a balloon under the jaw or a strain gauge on the halter.[163,371] These studies have led to the conclusion that cattle who eat fast, chew fast and eat more and, therefore, produce more. Cows should be selected for rapid intake both to solve the problem of eating enough while being milked and to help reduce negative energy balance during early lactation.

Feeder design is important. Cattle that have to reach under a horizontal bar to reach feed may injure their necks. In other designs such as a post and rail, there may be wastage of feed.[1127]

Social Facilitation

Cattle will also eat more in groups than they will individually, and first calf heifers will eat more if grouped with older cows. Production may not be increased. These increases must be balanced against the negative effects of dominance disputes.

Another effect of the social hierarchy during group feeding of grain is that subordinate cows eat faster than dominant ones, probably because they have less time to eat before they are displaced by a dominant cow[483] and must compensate. Transponder-activated feeding devices are, theoretically, an excellent way to feed group-housed animals and still record individual intake, but several problems arise. Intake decreases as the number of cattle per feeding station increases, especially when there are more than five animals per station. When limited amounts of concentrates are fed in that manner, the cows' activities are disrupted as they test the station to determine whether they will be fed. Dominant cows can displace subordinate ones after the door has opened for the subordinate.

Palatability

Ruminants appear to have a definite sweet preference. What tastes sweet to people may not taste sweet to cattle. Based on electrophysiological recording from the chorda tympani and glossopharyngeal nerves, cattle can taste saccharin but do not respond to aspartamate.[641,1290]

Calves show a glucose preference that is similar within a pair of twins.[184] Cattle reject 20% sucrose.[557] Cattle reject acetic acid at 0.08% and quinine at concentrations greater than 2.5%.[558] Twin cattle show similar rejection responses, as well as similar preferences responses.[184] With regard to the more usual feed substances, factors in addition to taste and odor must be considered. Texture and ease of prehension may explain why cattle prefer pelleted to unpelleted feed. Unchopped silage is preferred to chopped silage,[400] in part because more bites are necessary to collect the chopped silage.

A good example of the complexities of preference testing is that dairy heifers will choose a mean ratio of 40% grass silage and 60% corn silage, but the individual preferences ranged from a 34% to a 75% preference for the corn silage.[1464]

Grazing and Selectivity

The preferences of ruminants for plant species vary and depend on a number of factors. These include (1) the growth stage of the plant—most ruminants prefer fast-growing succulent species; (2) the mixture of species—clover, for example, may not be eaten if it is growing in a pure stand but will be eaten if grasses are growing with it; (3) the season of the year—ruminants will consume species that are green during the winter although the same species will be rejected in the summer when other species are green; and (4) the relative abundance of herbage. These factors are particularly important in determining when a ruminant will ingest a poisonous plant. Many poisonous plants are bitter in taste or harsh in texture and are, therefore, avoided by ruminants; but when the poisonous plant is the first green plant to appear in the spring (hemlock or skunk cabbage) or the only forage available on a pasture, it will be ingested.

In addition to these factors, the location of the plant, in particular its distance from water (in arid regions) or shelter, will influence the amount of it consumed.[74] As well, cattle avoid fecally contaminated areas.

Various methods have been devised for determining preferences of grazing animals. The most accurate and most direct is to collect the food eaten by sheep or cattle with open esophageal fistulas. Another method is to observe the animals closely. This method is more accurate if the forages are planted in narrow pure stands from which the ruminants may choose. A third method is to sample the plant species and heights of the plants before and after the pasture has been grazed. Studies using these methods have revealed that cattle graze selectively; that is, they do not necessarily eat plants in proportion to the number of each plant species in a pasture.[657] For example, Cowlishaw and Alder[322] found that cattle most preferred meadow fescue and timothy; perennial rye grass and cocksfoot were the next most preferred plants; and Agnotis and red fescue, the least preferred.

Cattle wrap their tongues around grass and pull to prehend it. This method of forage limits them to plants higher than 10 mm. Sheep can graze much closer to the ground because they use their teeth to prehend, a fact that may have led to antagonism between cattle and sheep producers on a shared range.

Ruminants are less likely to graze on plants that contain old growth or cured material. Rapidly growing plants are the most palatable, provided that they are high enough in fiber.[542] Cattle prehend a gram or 2 at a time, taking in 30–50 g in a minute. As bite size increases, the cattle are both prehending and chewing with the same jaw movement.[851] Fasted cattle spend more time grazing that nonfasted ones, and although bite size was unaffected, they chewed less, swallowing larger particles.

Do animals forage optimally? Do they increase the time spent at patches of high-density food and spend longer at patches that were hard to obtain, that is, a longer walk from water or resting areas? These questions are only partially answered. Cattle apparently graze those patches that give them the largest bite weight and instantaneous intake. For example, cattle will select short, dense patches over short, sparse ones but will select against short, dense patches when tall, dense patches are available. There seems to be selecting for maximum bite weight and rate of intake.[378]

One must consider the effect of adding concentrates to supplement cattle on pasture or range. Supplemental food blocks are used frequently for delivering extra calories, vitamins, minerals, or medication for cattle. Placement of these blocks is important because cattle will not use blocks placed close to the watering site as much as those placed in grazing areas.[283]

Environmental Temperature

Cattle eat less when environmental temperatures are high. In addition to the direct effects of heat on intake, there are also effects on plant growth that may render them less palatable. One of the problems of raising cattle in the tropics or other hot environments is that food intake, and therefore production, falls. The use of tropical breeds and crosses with tropical breeds of ruminants helps alleviate the problem because these cattle eat more in the heat or eat more at night when the temperatures are lower.

Cold ambient temperature increases food intake in ruminants as in simple-stomached animals. This can be observed not only when environmental temperature falls, but also when the animals' heat loss is increased. (See Baile and Forbes[100] and Forbes[483] for reviews of all aspects of the controls of food intake in ruminants including thermostatic factors.)

Flies can also influence feeding. Cattle actually eat more when face flies are present in large numbers.[395]

Estrogen Level

Food intake falls in estrous cattle.[720] The decrease in intake can be used to identify cows that are ready for breeding. Stilbestrol is an estrogenic drug that has been used to fatten steers. In the doses given, it does not suppress intake and the anabolic effect of stilbestrol leads to weight gain.[640] After stilbestrol was banned, a mixture of estrogen and progesterone has been used to improve weight gain in feedlot cattle.

Rumen Fill and the Products of Rumen Fermentation

The anterior gastrointestinal tract of the ruminant (the rumen) is an anaerobic fermentation chamber in which bacteria-producing cellulase can release the energy that is utilized by the animal. This energy is otherwise unavailable to the animals and enables cattle, sheep, and other ruminants to survive on grass and roughage diets. In the case of grazing ruminants, the time necessary to collect food and fill the rumen is probably the limiting factor in food intake. For example, when cattle are grazing on the open range, they spend 56% of their time grazing and another 21% of the time ruminating[590] (see Chapter 3, Biological Rhythms and Sleep). In the case of cattle fed diets high in grain for fattening and in the case of lactating dairy cattle eating concentrates, other factors inhibit food intake because cattle fed a 75% concentrate diet spend only 2.5 hours a day eating.[669]

The end products of bacterial digestion in the rumen are the volatile fatty acids—acetic, butyric, and propionic. When acetic acid is added, the rumen food intake is depressed; similar amounts injected via the jugular vein have no effect on food intake.[103] It therefore seems likely that acetic acid may stimulate receptors located in the rumen epithelium or in the portal circulation. These receptors may, through central nervous system connections, inhibit food intake. Above and below approximately 50% forage in the diet, intake is reduced. On low-roughage, high-energy diets, production of volatile fatty acids is probably the most important satiety factor. On low-energy, high-roughage diets, rumen fill or abdominal fill is probably a satiety factor (Fig. 8.12). Intake falls in pregnant or fat cattle, presumably because of the increase in abdominal fill by the fetus or fat.[200,762]

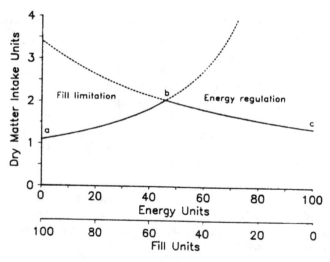

FIG. 8.12. Reciprocal relationship between dietary characteristics and dry matter intake. *Line a to b* represents intake limited by the fill effect of the diet. *Line b to c* represents intake limited by the energy demand of the animal. *Dashed lines* represent unattained intake predicted by extrapolating the theoretical equations[1007] (with permission of *J. Anim. Sci.*).

Humoral and Central Neural Factors

Few of the humoral or neurotransmitters have been studied in cattle. Most of the studies are performed on smaller ruminants and are discussed here under Sheep and under Goats. Intraportal propionate suppresses intake in cattle.[428] Cholecystokinin-pancreozymin may have a paracrine function, acting locally on the vagus nerve to suppress intake, although blood levels of CCK do not vary when the cow eats meals of concentrate or roughage.[530]

Corticosteroids increase food intake in cattle, but increase fat deposition, whereas ß-adrenergic agonists such as clenbuterol decrease fat deposition and increase muscle growth without increasing food intake.[483]

Defense of Body Weight

When dry cows are allowed to eat ad libitum, intake reaches a plateau in Jerseys, indicating a set point of body weight. Holsteins under similar conditions continue to gain for a longer period, probably indicating that a breed selected for high production has less inhibition of intake.[1031]

Although cattle can vary their intake with availability of feed, there are constraints. When concentrates are available 24 h/d, cows eat 80% more than when they are available for only 5 hours. Hay intake increases only 20%. This is strong evidence that rumen fill is not the only influence on satiety.

Cattle can, within limits, decrease their food intake as caloric density increases. Even young calves demonstrate this ability by decreasing their intake of milk replacer as the percentage of water in the replacer falls.[1133] The ability of cattle and other ruminants to respond to dietary dilution, that is, to increase their volume of intake as the nutrient density decreases, has been reviewed by Baile and Forbes.[100]

Low protein content of the diet suppresses food intake in ruminants as well as in simple-stomached animals, but because of the rumen microbial production of protein, the level of protein in the diet can be much lower before anorexia occurs. Furthermore, nonprotein nitrogen, particularly urea, can substitute for protein if there is enough carbohydrate available to the bacteria for protein synthesis. Offering concentrates can decrease forage intake, but the decrease varies with quality of the forage. Intake of low-quality hay will be less affected than intake of good-quality hay (see Fig. 8.12).

Parasitism

Gastrointestinal helminths stimulate the release of their host's gut hormones, including cholecystokinin. Therefore, the heavily parasitized ruminant loses weight not only because the parasites consume some of the food their host ingests, but also because the host ingests less.[1376]

Clinical Problems

The Fat Cow Syndrome

An abnormality of feeding behavior that is being seen with increasing frequency is the fat cow syndrome.[1046,1047] Under some types of management, dairy cows may be fed in a group that includes both lactating and dry cows. The dry cows may overeat, gain weight, mostly fat, and later at calving or during lactation become ill or even die. The pathogenesis of the syndrome remains unclear, but the initial problem is that the cow overeats. High-producing dairy cattle have been bred to eat large amounts in order to supply the metabolic fuel for lactation. When lactation ceases, the increased food intake may persist and the fat cow syndrome may result.

CONTROL OF FOOD INTAKE IN SHEEP

Although more studies have been conducted on the effects of rumen manipulations in cattle than in sheep, most of the other physiological studies of ruminant intake have been conducted on sheep, probably because of their size. Although it is fairly easy to make and maintain a bovine rumen fistula, the cost of feeding and housing sheep is much lower than that for cattle. Furthermore, it is much easier to anesthetize a sheep to prepare intracranial cannula and it is much less expensive to administer drugs to a 50-kg (110-lb) animal than a 400-kg (880-lb) one.

Meal Patterns

Sheep eat approximately a dozen meals a day in a laboratory. Sheep can eat at a rate of 15–30 g of pelleted feed per minute, but only 9 g of hay per minute. Sheep eat a kilogram more if feed is available free choice rather than for a limited period of 3 h/day.

Light stimulates growth and increased food intake in lambs. Even a flash of light at midnight—a skeleton day—can increase intake.

Social Facilitation

Confined, isolated lambs eat less than grouped ones.[1458] Sheep eat less when in metabolism cages, which may be due to a lack of social facilitation in a gregarious species, as well as to confinement.[481]

Palatability

Sheep do not appear to have a very marked sweet preference. Nevertheless, it was only 5% sucrose and glucose that were preferred; higher concentrations produced no preferences or aversion. Lactose was rejected at a concentration of 2.5%. Sheep also showed no preference for molasses for which cattle show a strong preference.[557]

Lambs prefer flavors paired with nutrients to flavored saccharin solutions, a form of nutritional wisdom.[265]

Lambs prefer soybean meal to commercial pellets, both of which were strongly preferred over fish meal, flocked maize, or whole oats. Rolled barley and sugar beet pulp were moderately well accepted.[358] Sheep prefer pelleted feeds to chopped feeds.[1425]

Sheep are much more sensitive to unpleasant taste. They reject sour solutions of hydrochloric acid, acetic acid, lactic acid, or butyric acid at concentrations lower than those found in the rumen, yet the volatile fatty acids must surely be tasted when the sheep ruminates. A bitter solution of quinine (0.004 M) was rejected, but not one of urea, even at concentrations high enough to produce fatal toxicity.[556] Experience also plays a role in that sheep that initially rejected quinine-treated hay eventually accepted it even when nonbitter hay was available.[482]

When given a choice of two feeds, one with a level of protein content above their requirements and one below, the sheep ate enough of each to maintain their requirements. Even when straw is fed, the sheep can select the most nutritious portion and leave the less-digestible parts behind.

Defense of Body Weight

Blaxter et al.[208] found that intake in ad libitum–feeding sheep remained relatively constant from 1 year of age onward. Sheep can compensate for considerable dilution of their diet with straw or even inert diluents.

It is not surprising that fasted sheep eat faster, taking larger bites than those who have been grazing at will, but it is interesting that their food preferences change, tending toward higher roughage feeds, such as grass, rather than the higher protein plant, clover.

Rumen Factors

Fiber decreases intake; and the smaller the particle size, the greater the intake of those diets in which rumen distention limits intake. There is additivity of rumenal acetate and intraportal propionate in suppressing intake of sheep, indicating that several factors acting in concert could be necessary for satiety. This is the more natural situation than an increase or decrease in a single metabolite or hormone. When given a choice, sheep do not eat the most calorically dense food but choose sufficient long fiber to optimize rumen function.

Intestinal Receptors

Gastrointestinal factors are as important in ruminants as in nonruminants. A high osmolality in the duodenum will suppress feeding and stimulate drinking of sheep.[281] Factors such as CCK and propionate act additively in suppressing ruminant intake,[456] but immunizing lambs against endogenous CCK did not result in a significant increase in intake.[1341,1412] If digesta enters the duodenum more rapidly, the sheep eats more, which suggests that abomasal fill may produce satiety.[925]

Grazing and Selectivity

It takes 0.34 seconds for a sheep to open and close her mouth in prehension bite; therefore, she prehends at a rate of 60–80 bites per minute.[289] Sheep masticate (chew) 60–70 times per minute. Sheep prehend grass by breaking it between their lower incisors and upper dental pad. That and their narrow muzzle enable them to be more selective than cattle. They select leaves rather than stems. When the protein content of a diet is insufficient, food intake may be depressed. This may occur on poor-quality pastures. Although a supplement such as urea and molasses may be provided and will increase intake, 19% of sheep will not consume it.[399]

The rate of food intake depends on the plants grazed. For example, sheep can prehend and masticate clover faster than grass.[1119] Sheep expend energy moving from place to place while grazing, and they increase their food intake 20% to 50% over that in confinement in compensation. Sheep have peaks of grazing activity every 8 hours. They graze most intensely just before sunset.

The factors affecting grazing are bite rate, bite size, time spent grazing, and species of food selected. Learning, especially learning from the dam's food choice; social facilitation of the beginning, but not of the end, of a grazing bout; and physiological state are all important. In general, the preferences are correlated with the dry matter and carbohydrate content of the grasses. Sheep select a diet higher in protein and lower in fiber content than that obtained from clipped pasture samples, indicating the advantage to the animal of selective grazing.[633,1363,1460] A practical consideration in pasture management is that sheep, as well as horses, avoid fecally contaminated grasses, although they seem to prefer grass subjected to urine contamination.[936]

It is unclear what senses are involved in forage preferences. Anosmic sheep do not avoid fecally contaminated feed as normal sheep do.[1411] Anosmic sheep and sheep with impaired vision (hooded with translucent eye coverings), however, showed preferences very similar to those of intact sheep, although the visually impaired sheep tended to graze at one level rather than grazing higher or lower on the plant to select the most succulent portions.[70] Even cutting of the gustatory nerves had little influence on forage preferences.[71] In the laboratory, anosmic sheep are no different than intact sheep in their meal patterns.[121]

Both food preferences and social factors influence food choices in sheep. Sheep in larger flocks graze for longer periods than sheep in smaller flocks.[1118] The satiety state of the grazer also influences plant selection. Previously fasted sheep eat more, as compared with nonfasted sheep, because the former take larger bites and choose fewer legumes than nonlegumes.[1073] When a fungus such as endophyte infests fescue, the sheep avoid it apparently because they are nauseated.[20]

Environmental Temperature

Sheep eat more when they are cold and less when they are fat. For example, sheared sheep eat 50% more after shearing as a result of the loss of insulation of their wool.[1384] At very low temperatures (–10°C, or 14°F), intake may be inhibited and the consequences of greater losses of energy of heat will be compounded by a decrease in energy intake.

Hormonal Factors

Ewes eat less during estrus. Infusion of the amount of estrogen equivalent to that which occurs at estrus also suppresses intake.[483] Food intake decreases during late pregnancy in ewes carrying more than one lamb. The decrease can be as large as 40% if the only available forage is low-quality roughage. This can result in ketosis or pregnancy toxemia of sheep. During lactation, thin sheep eat more than fat ones—a phenomenon seen in all ruminants studied thus far.

Glucose Utilization

Cross-circulation from hungry to satiated sheep stimulates intake of the satiated animal (and depresses intake of the hungry one), indicating the existence of humoral factors that influence feeding.[1298] Glucose or insulin levels may be among the humoral factors. Because virtually everything that the ruminant eats is exposed to rumen bacteria before reaching the intestine, very little dietary glucose becomes available to the animal. Instead, the ruminant depends on the volatile fatty acids for energy and produces glucose by the process of gluconeogenesis. Plasma glucose is low, approximately half that of simple-stomached animals. It was assumed, for these reasons, that ruminants were fairly independent of glucose utilization and would not be expected to eat in response to glucoprivation. It has been demonstrated in both sheep and goats, however, that food intake can be markedly increased by insulin and the glucose analogue 2-deoxy-D-glucose.[699]

Central Neural Mechanisms

The easiest generalization that can be made about central stimulants of intake is that factors that depress activity, such as anesthetics and opiates, increase feeding. For example, central nervous system depressants, such as calcium, barbiturates, and benzodiazepines, also stimulate intake in ruminants.[148,377,831,937,987,1296,1297]

As the number of known neurotransmitters and neuromodulators grows greater, the hope of complete understanding of the pharmacology of central neural controls of feeding and other behaviors grows dimmer. Cholecystokinin-pancreozymin, discussed earlier here as a peripheral satiety factor, also exists in the brain, where it can act to produce satiety in sheep.[102]

The opioid peptides are also involved in feeding behavior. The mu and kappa receptor agonists given intracranially appear to stimulate intake in sheep; gamma agonists depress intake.[101]

Although sheep increase their food intake when norepinephrine is injected via the cerebral ventricles (see Appendix 3), cattle do not show the same response; nor does a clear picture emerge when adrenergic agonists or antagonists are injected.[104] Cerebrospinal fluid from hungry sheep stimulates intake by satiated sheep, indicating that there are humoral factors within the brain (or the ventricles) that influence feeding.[943] The serotonin antagonist cyproheptadine increases food intake in sheep.[414]

There has not yet been any practical application of these findings to increasing the food intake of fattening cattle or lactating dairy cattle. It is not yet clear whether these drugs, which stimulate intake over the short term, would, if administered chronically, produce a long-term increase in food intake or meat and milk production.[1327] The problem of the effects of barbiturates and tranquilizers on human consumers also remains unknown. The latest approach is to stimulate production with bovine growth hormone or somatotropin and rely on the animal to increase intake to match the increased production (output).

Obesity can be produced in sheep by offering a pelleted diet free-choice. The sheep will consume three to six times their maintenance requirements. Once the sheep have gained weight, their intake falls and they actually eat more slowly than lean sheep. This indicates that the feedback from fat, presumably ob protein, is inhibiting intake. These sheep are more sensitive to the anorexic effects of opiate blockers.[15]

The hypothalamus of the ruminant appears to be as intimately involved in the control of food intake as the same brain area is in animals with simple stomachs. Lesions of the lateral hypothalamus produced hypophagia.[100] Kendrick and Baldwin[783,784] have shown that there are cells in the zona incerta of sheep that respond to the sight of food, especially a preferred food, and others that respond when food and/or liquid are in the mouth. There are cells in the hypothalamus that respond to the sight of food, the approach of food, and the ingestion of food.[918]

CONTROL OF FOOD INTAKE IN GOATS

Meal Patterns

Goats eat 12 meals a day (remarkably similar to cattle and sheep), 8 of them during the day. Goats are more selective feeders than sheep and have a longer vertical reach than sheep of the same weight.[483] Browsing behavior is a skill that must be learned. Goats learn to break twigs off the plants rather than to chew them off.[1095] Attempts to increase consumption of goats with plentiful, but not particularly palatable, foods such as sagebrush have not been successful.[1199] Goat selectivity is strongly influenced by the plants to which they were exposed in their 1st year of life and by their dam and peers.[201]

Goats readily consume glucose and sucrose solutions at concentrations as high as 40%.[181] Goats do not prefer bitter substances but will accept quinine-adulterated water at nearly 10 times the concentration that a rat will.[181,183] Field studies confirm that goats have a high tolerance for bitter taste.[321]

Angora goats are in danger of urolithiasis. It is possible to increase their water intake, and thereby decrease their risk of urethral blockage, by adding vinegar or orange flavor to the water.[316] Increasing the crude protein of concentrates fed to goats increases their food intake and their milk production.[95] Propionate and lactate suppress intake in goats, but the site of action is beyond the rumen, presumably in the liver. Food intake falls during estrus.[434]

Electrical stimulation of the lateral hypothalamus elicits eating in goats, and stimulation of the ventromedial hypothalamus depresses intake in the same species (see Appendix 3). Vasopressin suppresses intake in goats and may be the mechanism by which stress suppresses intake.

WATER INTAKE

At least three types of stimuli elicit thirst: a dry mouth, an increase in the osmotic pressure of the blood, and a decrease in the blood volume.

Increase in Osmotic Pressure

Hypertonic sodium chloride infused intravenously causes thirst in pigs[734], horses,[1368] and dogs.[1510] An increase in the osmotic pressure of the blood is believed to stimulate osmoreceptors located in the anterior hypothalamus.[1505] Goats drink copiously when hypertonic saline solutions (2% NaCl) are injected into the anterior hypothalamus through a cannula.[54] Similarly, a horse traveling in a trailer in hot weather with food, but no water, available will exhibit thirst as a result of an increase in the osmotic pressure of its blood and consequent stimulation of osmoreceptors. Free-ranging cattle drink once a day. Under these conditions, lactating cattle drink larger amounts but drink no more frequently.[1229]

Decrease in Blood Volume

Decrease in blood volume, for instance as a result of hemorrhage or peritoneal dialysis,[297] may stimulate thirst. A decrease in the blood volume would also be expected in a cow that produces 36 kg (80 lb) of milk (95% water) per day. A lactating cow drinks 45 kg (100 lb) more water a day than a dry cow for this reason.[590] The drug furosemide, frequently administered to race horses shortly before they perform, causes a decrease in plasma volume and thus stimulates thirst in horses[1368] and sheep.[1534] Water is lost when animals are heat stressed, whether they cool themselves primarily by panting (dogs), saliva spreading (cats), or sweating (horses). Therefore, it is obvious that water intake increases with environmental temperature.

Angiotensin

Recently, experimental interest has focused on still another type of drinking, that in response to the hormone angiotensin. Angiotensinogen is acted upon by the kidney hormone, renin, and then by a pulmonary converting enzyme to form the octapeptide angiotensin II, which has several actions. It releases another hormone, aldosterone, which is involved in sodium reabsorption by the kidney. As its name implies, high doses of angiotensin II can increase blood pressure. Most interesting is its action on water intake; angiotensin II is the most potent dipsogen known. Angiotensin II, or procedures known to release the hormone, stimulate water intake in a wide variety of animals, including dogs,[476] cats,[315,1367] sheep,[2] goats,[53] pigs,[130] and horses.[52] The role that angiotensin release may play in normal thirst remains to be determined.

Situations in which blood supply to the kidneys is compromised stimulate renin release and, therefore, angiotensin release that could be expected to lead to thirst. Indeed, dogs in congestive heart failure show increased water intake.[1174]

Dry Mouth

A desalivate animal takes frequent small drafts of water in order to swallow dry food. This type of drinking, prandial drinking, is also seen in intact pigs that take a small quantity of water into their mouths before swallowing a mouthful of grain.[659]

Multiple Causes of Thirst

Just as most food intake appears to occur without a major change in body energy balance, most drinking occurs without a change in body fluid balance. The osmotic and volume depletion stimuli to drinking are emergency mechanisms to restore body water under life-threatening conditions.

Most drinking in domestic animals is prandial, that is, in association with meals. For example, 75% of water intake in pigs is in association with meals. Twenty percent of the drinking occurs after feeding (within 10 minutes); 30%, as pauses in feeding; and 25% precedes meals. The drinking occurs before there are any changes in blood volume or osmolality. The animals appear to be drinking in anticipation of their needs, and the physiological mechanisms are the same as those that control the cephalic phase of gastric secretion.[705]

Thirst produced by overnight water deprivation is a combination of osmotic thirst and hypovolemic thirst. In contrast to the osmoreceptors that stimulate vasopressin secretion, those that stimulate thirst in the dog appear to be located on the blood side of the blood–brain barrier.[1396] Water intake of thirsty dogs is reduced 80% if water is injected intracarotidly so that the osmoreceptors of the brain are no longer stimulated. Intravenous administration of an equal volume of water has no effect. If thirsty dogs are injected with isotonic saline so that peripheral blood volume is returned to normal, their water intake is reduced 20%. If both intravenous isotonic saline and intracarotid water is administered, water-deprived dogs do not drink.[1175] Despite this evidence that thirst is inhibited when blood volume and osmotic pressure return to normal, dogs cease drinking before these parameters return to normal.[1510] There may be signals from the gastrointestinal tract that inhibit water intake, as there appear to be signals that inhibit food intake. The stomach has been implicated,[1410] but gastric fill alone does not suppress drinking in dogs;[1335] duodenal osmoreceptors are more likely candidates for the source of inhibition.

Water intake falls with environmental temperature, and this may account for the increased incidence of impaction colic during the winter when horses eat more and drink less. Providing warm water with meals increases water intake 40%.[836]

Cattle fed silage—a somewhat wet feed—and limited amounts of concentrates drink four times a day. The drinks ranged from 1 to 13 liters (1–14 quarts).[483] There may be central sodium receptors in cattle that when stimulated by an increase in NaCl in the cerebrospinal fluid increase water intake. Water intake increases in the heat and decreases in the cold, but when the environmental temperature falls below 0°C (32°F), water intake increases with the increased food intake necessary to control body temperature. Dogs may not be as thirsty in the cold. Their threshold to osmotic stimuli is higher, and their blood volume is elevated.[1336]

SPECIFIC HUNGERS AND SALT APPETITE

The concept of the nutritional wisdom of animals has not been validated. Animals apparently cannot innately choose a diet containing a vitamin or other nutrient in which the animal is deficient.[1232] Horses fed a diet deficient in calcium will not eat more of a calcium supplement to correct the deficiency.[1275] There is better evidence for a phosphorus-specific hunger. Cattle and other ruminants chew on bones, and some will consume phosphorus supplements. What animals, at least laboratory animals, can do is to learn to associate a particular diet with improvement in health. Conversely, animals can learn to avoid a diet they associate with a feeling of malaise. Given proper clues, such as a distinctive flavor, most domestic animals could probably learn to choose a diet that would correct a deficiency rather than choose a deficient diet.

There is, however, one nutrient that nearly all species select innately when they are deficient: sodium.[672] Salt hunger is a well-recognized phenomenon, especially in herbivores whose diet tends to be low in sodium. In most species, removal of the adrenal glands and the consequent hyponatremia are followed by life-saving ingestion of sodium chloride. Pigs, for instance, drink sodium chloride solutions and survive, following experimental adrenalectomy.[908] Ruminants can be made experimentally sodium deficient by creating a salivary fistula. The large loss of sodium bicarbonate in saliva produces a deficiency that sheep can correct by drinking precisely the quantity of sodium solution they need to bring their sodium levels up to normal. Treatment with a diuretic also stimulates sodium appetite in sheep.[1534]

Salt appetite is not stimulated during pregnancy and lactation in sheep despite the extra sodium demands.[1013] Sheep can be divided into those who excrete sodium predominantly in the feces, that is, who do not absorb excess sodium, and those who excrete sodium predominantly in the urine. The fecal excretors have a greater sodium preference.[1012]

Sodium-deficient cattle and sheep are able to learn an operant response to obtain a sodium reward.[1,1330] Sodium status is apparently assessed in the hypothalamus, where changes in the concentration of intracellular sodium act to initiate transcription and translation processes. The protein synthesized alters the ionic or membrane characteristics or increases the neurotransmitter capacity of the neurons.[370] The hormones that mediate salt hunger are angiotensin and aldosterone acting in synchrony. Because of the hormonal involvement, there is a latency for the appearance of the salt appetite.

Furosemide causes a loss of sodium as well as water in the urine. Sodium appetite is stimulated in horses treated with furosemide.[692]

CHAPTER 9

MISCELLANEOUS BEHAVIORAL DISORDERS

When an animal is presented for treatment of behavioral disorders, a complete history is necessary before a rational treatment can be prescribed. (See Appendix 2 for sample history forms.) Some behavior problems are the result of confinement and separation from group members and include destructive behavior in dogs and stereotypic behaviors such as stall walking and weaving in horses. Phobias usually develop to loud noises but can form to many things associated with pain or fear. Some trailer problems of horses are the result of phobias, and others, the result of poor trailer design. Behavioral and pharmacological treatments are suggested here. This chapter considers common, and some uncommon, behavior problems of domestic animals that do not fit into the classifications of aggression, maternal behavior, sexual behavior, or other problems that were covered in previous chapters.

THE INTERVIEW

The interview should be a pleasant experience for owner, clinician, and patient. Owners should be put at ease with a handshake, eye contact, and a compliment about the animal. House calls are the ideal, especially for feline cases, but are not always practical. Interviews should be conducted in an office or consulting room rather than in an examination room in order to put both the owner and the animal at ease. Schedule plenty of time. The average case of an aggressive dog takes 2 hours for history and instruction. It is very important to be nonjudgmental in posing questions. Ask where the dog sleeps, rather than saying "You do not let the dog sleep on the bed, do you?" When the owner first mentions that the dog runs free, do not criticize; wait until the history is taken to remind the owners of how bad they will feel if their aggressive dog bites a child while loose.

BEHAVIORAL HISTORY

The type of history form depends on the practitioner's personal preference. In a university clinic where many different people, including clinicians and owners, will be using the forms or when information is to be statistically analyzed, a computerized or multiple-choice form is best (see Appendix 2A), whereas those who interview all of their own patients may prefer an essay-type form. Some owners will be daunted when asked to fill out such a long form and will answer each question with a word or two. In contrast, they will be more than willing to answer the same questions at length verbally. With the increased use of facsimile machines, sending and receiving histories before the consultation is a convenient as well as time-saving strategy.

Before instituting treatment of any behavior problem, a detailed history should be carefully taken (see Appendix 2A, Canine Behavioral History form). Let us use a destructive dog as an example. It is important to determine just when and where the animal is destructive. The owners usually will have attempted to deal with the problem themselves, and the methods they have used should be recorded. It is always important to maintain a neutral attitude when taking a history because any criticism of the owners' methods at this time will probably inhibit the owners from volunteering any additional information as the interview continues.

Environment

The dog's environment should be described fully. In particular, the schedule of the entire family should be noted. The usual location of the dog when people are at home and when he is alone should be determined. The amount of exercise or opportunity to exercise given to the dog should be recorded. Perhaps the most important information in the dog's history is the owners' usual behavior just before they leave the dog and just after they return home. Excessively emotional or stimulating leave-takings and greetings by the owner have been implicated in canine destructive behavior. The actions of the owners when they discover that the dog has been destructive, in contrast to their actions when he has not been destructive, should be ascertained.

Early History

The dog's early history often gives a clue as to the cause of any misbehavior. For example, if the dog was obtained from a kennel at 6 months of age, he may never have been properly socialized to people during the socialization period of age. If the dog was obtained as a 3-

week-old puppy or was hand raised, he may be too dependent on people and not properly socialized to dogs. If the dog was obtained from a pound or an animal shelter, he may have been placed there because he was destructive in his original home. The owners should be able to supply this information and to indicate why the dog was obtained by them. For instance, if the dog was obtained to be a companion for a child during the summer, the dog may exhibit destructive tendencies when the child goes off to school (and the parents to work) in the fall.

Training

Along with the dog's early history, any training that the dog received should be noted. The relationship of the dog to the owner can best be elucidated by the type, or absence, of training given to the dog. If the dog could not be trained by the owner, the dog may either be dominant over the owner or he may be genuinely untrainable. It is not unusual to encounter cases in which dogs are destructive and also eliminate in the house. Occasionally, the dog has never been completely house-trained, but in most cases, the dog is trained and will not urinate or defecate in the house—if the owners are home. It is only in the absence of the owners that he eliminates in the middle of the living room, for instance. Many owners regard this as spiteful behavior, but it is much more likely to be a sign of emotional stress. Dogs may also eliminate in the house because they are scent marking with feces and/or urine. At least some "accidents," especially those that occur on beds or couches, may be scent marking in response to strong human odors.

Other Behavioral Problems

Finally, the clinician should carefully question the owner as to any other behavior problem exhibited by the dog. Most dogs that exhibit one problem will exhibit many, possibly in response to a single cause. It often helps to inquire specifically about behavior that may be abnormal, such as feeding or sexual behavior. The dog may be a finicky eater or he may overeat and be obese. It may bark or mount people's legs. There are two reasons why the veterinarian should be aware of these other behavior problems: a complete picture of the dog's behavior will emerge and the other misbehaviors may worsen as the animal becomes less destructive unless the cause of the destructive behavior has been removed.

Behavior history forms for cats (Appendix 2B) can be shorter than those for dogs because cats rarely are trained and do not require the attention that dogs do. Because cats are much less likely to demonstrate their misbehavior during an interview, videotapes of the cat, her litter-box, the place she soils, and her interactions with other cats and with people should be obtained. Equine histories are often incomplete because horses, particularly those with behavior problems, change hands many times. Race horses have a variable environment as they move from home farm to trainer to race track, and the personnel who handle them are rarely the owners. The veterinarians will be different in each location, so a complete medical history is difficult to obtain. A history for general equine behavior problems (Appendix 2C) or ones for specific problems such as foal rejection (Appendix 2D) and trailering (Appendix 2E) should be used.

Instructions

Instructions are the most important part of treating a behavior case because owner–animal interactions rather than drugs or surgery determine the outcome. If possible, demonstrate the treatment techniques and always give written instructions both to remind the owner and to inform any absent member of the family. Send a copy to the referring veterinarian. Along with any further diagnostic testing that he or she should perform and information of drugs prescribed, ask the client what they are going to do to be sure they understand and are willing to comply.

DOGS

Destructiveness

There are two major types of behavior problems of which dog owners complain: aggression and destructiveness. The first type is a serious threat to public safety and has been discussed in Chapter 2 (Aggression and Social Structure). The second type may occur with greater frequency and is less amenable to treatment than is aggression. A dog quickly ceases to be man's (or woman's) best friend when he has scratched the new rug to shreds, gnawed the antique desk, and strewn the contents of the garbage pail through the house. Destructiveness appears to be increasing in frequency, although a careful survey has not been performed. There are two recent sociological trends that may help to explain the greater incidence of destructiveness: more people, especially young people, own dogs; fewer women remain at home during the day. The net result of these two trends is that many more dogs spend 8 hours or more alone per day, and therefore, the opportunity for destructive behavior is present. The following examples illustrate varieties of destructiveness.

Clinical Cases

Case 1.   Ezra was a golden retriever adopted from a laboratory at age 4 months by a first-year veterinary student. At 6 months, he began to be destructive in the owner's absence. At first, most of the behavior seemed to be directed toward food because he would scratch at a cupboard door and drag out and consume the cat food as well as any garbage in the trash can. Treatment was initiated using an operant conditioning device to dispense food. When the fixed ratio was higher than 10, that is, he had to press a bar 10 times for each bit of food, he would not use it, and he continued to be destructive. Medication was begun using amitriptyline and caging during the owner's absence. If anything was left within reach of the airline crate, he would drag it into the cage and chew it even if a rawhide chew was already available. After a few months, the owner believed he was completely better and so she left him for a few minutes to visit a neighbor. He chewed the lining of the coat she left behind. After a year on medication, he can be left in the house without confinement or medication.

Case 2.   This case was much more dramatic. The dog, a beagle, had a history of convulsions and was maintained on an anticonvulsant. The convulsions increased in frequency, and at the same time, the dog became very destructive in the owner's absence. She ate through several wooden cupboard doors and finally, when she ate all the owner's Christmas presents, she was taken to the veterinary clinic. In the clinic, the dog responded well to a different anticonvulsant and on discharge was no longer destructive. The response to medication indicates the organic disease was present, but the dog showed the typical destructive behavior pattern of chewing only in the owner's absence.

Case 3.   An adult female German shepherd had been accustomed to being alone and had been well behaved until she had a litter of puppies. When she was temporarily separated from the puppies by a barricade, she began to chew, not the barricade but the furniture. It is interesting to note that the owner had separated her from the puppies because he did not think that the bitch should be eating the puppies' feces, a behavior that is a normal component of canine maternal behavior. The destructiveness ceased when she was again allowed free access to the puppies.

In this case, the destructiveness has recurred sporadically. It has usually been controlled by painting the chewed object with a noxious "hot" sauce, but only after the large dog and two of her now fully grown offspring have done considerable damage. They have, in separate incidents, ruined an antique table, spread plants and their soil over the house, and eaten several rungs of a banister. Although the bitch may have "missed" her puppies in the initial episode, each dog had two other dogs for companionship in the subsequent episodes. It has never been determined if all three dogs or only one or two are actually doing the damage.

Case 4. An adult spayed German shepherd had been given to the present owner because he could not be restrained in a pen or a cage. Its early history was unknown. The present owner tried to keep the dog in a fenced yard, but the dog jumped the fence. When the fence was made higher, the dog clawed at the back door or jumped against the windows of the house. If he was tied, he howled and struggled until he was in danger of choking himself. If he was kept inside, he would claw the door, jump through the window, and try to get out as desperately as he tried to get in. He could tolerate no restraint of any kind. He behaved in a similar fashion when he was restrained in a room with other dogs.

The owner finally put the dog in the basement while she was at work. The dog tore up the rug on the basement stairs and then began to eat the wood of the stairs themselves. He would also defecate all over the basement floor rather than on the papers provided. When a young dog was put in the basement with him, the young dog began to eliminate everywhere also, and the older dog continued to be destructive. Neither tranquilizers nor anticonvulsants attenuated the behavior. The owner had trained the dog fairly well, and he was obedient, but somewhat aggressive, when the owner was present.

The owner tried behavior modification for a few weeks, but there was no improvement, and the dog was euthanized. There were no abnormalities on either gross or microscopic examination of the tissues, including the central nervous system.

Etiology of Destructiveness

What are the causes of destructive behavior? Unfortunately, the definitive answer to this question has not been found. Some intelligent guesses can be made on the basis of the situations that lead to destructiveness, but the inadequacy of our knowledge is reflected in the lack of success of the treatments proposed for destructiveness. Once the causes of destructiveness are determined, it should be possible to find effective treatments. There may well be different causes for different types of destructiveness.

Breed and Other Predisposing Factors

In addition to environmental causes of destructive behavior, there may be a breed incidence factor as well.

The potential dog buyer should be warned of the possibility of destructiveness as a problem. According to a survey of veterinarians and obedience judges, West Highland white terriers, Irish setters, Airedale terriers, German shepherds, Siberian huskies, and fox terriers are the breeds most apt to be destructive.[618] The peak of destructiveness is 18 months of age. This is a problem that dogs appear to "grow out of," but the behavior can be chronic if mistreated or mishandled. Working couples, students, and apartment dwellers should be discouraged from selecting those breeds. These people should not be denied the pleasures of owning a dog, but they should choose their breed carefully. There have been facetious suggestions that urban dogs should be of a new breed that needs no exercise and can be trained to eliminate in a litter pan. Another characteristic should be added—ability to tolerate owner absences for 10 consecutive hours or so. Until such a dog appears, potential dog owners would do well to question the breeder as to the sire's and dam's behavior. One can ask where the dog is kept during the day or while the owner is away to elicit the information. Only an exceptionally honest breeder would volunteer that his or her dogs have behavior problems.

Puppies from pet shops and those left for long periods (6–8 hours) as puppies are predisposed to destructive behavior.[1299] Dogs from pet shops and breeders are more likely to be dominant aggressive. Dogs adopted as strays were more likely to be coprophagic.

Absence of the Owner from the Home. An examination of the environment in which destructiveness occurs reveals that the behavior almost always takes place in the owner's absence. It is not just that the owners punish the dog when they see the dog chew or scratch and that the dog has, consequently, learned not to do so in their presence; the owners have seldom seen the dog misbehaving in such a way. This circumstance, and the fact that the rise in incidence of destructive behavior appears to parallel the rise in the numbers of working couples, indicate that being alone must have something to do with the dog's misbehavior.

It is very difficult to pinpoint when a dog is destructive and what his behavior is just before and just after he chews and scratches because the behavior occurs only when no one is home. One way of observing a dog by himself is to use closed-circuit television. A small closed-circuit system, similar to those used in stores to prevent shoplifting, would be a good investment for practitioners genuinely interested in treating behavior problems. Those who have used this method have found that most dogs appear to become destructive shortly after the owner leaves, but a few become anxious at approximately the time that the owner is expected home. The dog then begins to chew and scratch. Dogs appear to have a fairly accurate sense of time and can, therefore, predict arrival times if the owners have definite schedules. Other dogs simply cannot tolerate being left alone. This is usually expressed by destructiveness, sometimes by barking, and sometimes by both.

Types of Destructive Behavior

Lack of Environmental Stimulation.   Boredom may be a cause or a contributing cause of destructive behavior. This is most apt to be the case if the dog is a large and/or young animal that does not receive adequate exercise. It may sound anthropomorphic to speak of bored dogs, but animals appear to need environmental stimulation just as humans do. Animals will work for light, to obtain access to another environment, or for brain stimulation. Dogs, therefore, may find chewing and scratching less boring than just lying around. In more scientific terms, they find such activity rewarding. This type of destructive behavior is random—not directed at either portals of egress or personal objects of the owner.

Separation Anxiety.   Dogs are social animals that, by nature, need the company of their "pack." This is what attracts us to dogs as pets. Yet, we expect them to accept long hours of solitude every day while we work or go to school; it is not surprising that some dogs exhibit anxiety when left alone. The causes of problem behavior might not be clear. Dogs that were previously well behaved when left alone may become anxious as a result of a change in the owner's schedule. After being boarded, many dogs will exhibit anxiety when left alone. Still others may grow increasingly anxious for no apparent reason.

The dog with separation anxiety will be destructive, vocalize, and/or eliminate. Many dogs do two or more of these behaviors. Separation anxiety in dogs can manifest many different ways. The dog may dig at the owner's spot on the sofa, chew on her shoes or clothing, or rip his own bed. Some dogs may be destructive, especially at points of exit such as doorways or windows.

Barrier Frustration.   Another type of destructive dog is one that cannot tolerate being enclosed. This behavior is called barrier frustration. It may result if the dog has been punished by being put into a closed room or into a fenced yard, but sometimes it occurs in the absence of such treatment. Barrier frustration is also caused by the presence of something very desirable on the other side of the barrier. An older dog may suddenly begin to chew the window sill after years of remaining quietly at home alone. Careful observation may reveal that another dog or a cat parades past the window frequently; the dog inside cannot give chase, but can only try to remove the barrier.

Another case was a dog who showed extreme barrier frustration of sudden origin, but probably of different etiology. This dog had previously lived in a house with other dogs, but when his owners, both of whom worked, got an apartment to themselves, the dog tried to accompany them to work. He would tear at the door or jump through a closed window and run to the restaurant where the husband worked. The dog would act in a similar manner if tied in the back of a pickup truck, but would be content to sit alone in the cab of the same truck. Many dogs are more content in cars than in houses possibly because the car is den-like and, no doubt, smells of both the owners and the dog. The German shepherd of case 4 is also an example of barrier frustration because he tried as hard to get into the house when outside as to get out when he was inside.

Prevention

Prevention of the development of misbehavior is always easier than eliminating established misbehavior, but because the exact cause of destructive behavior remains unknown, preventive techniques have not been established. There are, however, several things that the puppy owner should avoid so that destructiveness does not develop. Leaving puppies for long periods is associated with destructive behavior later in life. Puppies should not be given old shoes or a piece of rug to chew on because they will not be able to differentiate between the castoff shoe and the new, $40 pair of shoes, nor between an old scrap of rug and the living room carpet. Toys can be provided, but they should be of a type and texture that the dog can easily distinguish from forbidden objects. One toy given only when the owner leaves is better than many toys that are always available. Suitable toys are hollow rubber ones with food such as cheese inside or rawhides. Leave-takings and greetings should not be emotional. Playing tug-of-war games may also increase the puppy's tendencies to chew in play or self-stimulation.

Treatment

Substitution of Rewards. A behavior has to have some reward value to an animal for it to persist. It is easy enough to see the reward value of digging in the garbage; the reward is food. It is more difficult to imagine what the reward is for a dog that scratches, but escape from confinement is one possibility, especially if the scratching is directed at doors or windows. Chewing can have several rewards, escape or a form of play or self-stimulation. Some dogs run in order to play; others chew. Dogs that chew for stimulation should be able to transfer their chewing behavior to a more suitable object, such as a nylabone or rubber toy. If care is used in selecting a bone that cannot be splintered, bones can be provided because it is much easier to transfer chewing behavior to a naturally preferred object like a bone than to an artificial toy. Provision of hard chewable dog biscuits or chow should also help. If a dog chews just before the owner comes home, the owners should teach the dog to fetch and chew a bone or chew toy whenever they come home. The dog may then transfer his chewing from furniture to the toy.

Meanwhile, the objects that the dog should not chew should be made as undesirable to the dog as possible. This is attempting to substitute a negative consequence for a positive one, the reward value to the dog. As long as the negative aspects of chewing are greater than the positive ones, the dog will refrain from chewing. Most owners have thought of this solution, and in some instances (for example, case 3) painting the chewed object, such as a table leg or a plant, with a hot sauce of chili and cayenne pepper is effective.

In a variation of this technique, Campbell[270] suggests spraying at the dog with an aerosol deodorant while holding the spray can near the object chewed. The spray should not be directed at the dog's eyes, but at his muzzle. The object chewed should be sprayed also. Most dogs dislike aerosol sprays and will avoid the smell of an object when the smell is associated with the spray. There are also several commercial dog repellant sprays, but the efficacy of these products, particularly over a long period, is questionable.

The main disadvantage of making any one object or even several objects unpleasant to taste or unpleasant to smell is that the dog will simply transfer his chewing to another site. Another disadvantage is that some objects, curtains or rugs, for example, cannot be spread with a noxious concoction. These negative techniques are most effective for dogs that have just begun to be destructive. Like most bad habits, human or canine, destructive chewing is easier to break earlier in its course.

Companionship. Loneliness has been mentioned as a cause of destructive behavior, and in some cases, relieving the loneliness relieves the dog of its destructiveness.[492] Students often find that their dog can safely be left with their parents at holiday time because the parents also have dogs. A destructive dog may stop misbehaving if a cat or dog or even a turtle is provided for companionship. So far, no one has been able to predict which dogs would profit from having a companion. Young dogs who are destructive because of lack of environmental stimulation are most likely to be helped by a companion. It rarely helps a dog that shows barrier frustration (case 4). The one companion that does appear to inhibit destructive behavior in nearly all cases is the owner. Unfortunately, few owners can change their lifestyle so that the dog is not left alone for long periods. If the dog is a desirable pet except for his destructive tendencies, the veterinarian might suggest that a home be found in which the dog would not be left alone. A family with small children and a caretaker that does not work would be suitable, but any prospective owner must be warned that the dog can never be left alone in a place where he could do any damage.

Behavior Modification. A form of behavior modification can be used to treat destructiveness. Instead of the dog being punished for being destructive, he is rewarded for good behavior. The first step of the behavior modification program is for the owner to determine the longest possible period that the dog can be left alone, which may be as short as 5 minutes. While the owner is determining that, he or she should also keep a behavioral diary in which the dog's actions, as well as the owner's in relation to the dog, are all recorded. The behavioral diary may contain instances of abnormal behavior by the dog or counterproductive treatment of the dog by the owner. When the longest period of time that the dog can be left alone has been determined, the behavior modification procedure can begin.

The dog is given a new toy and perhaps a new rug to lie on. The toy is only given to the dog when the owner leaves. The owner should handle the toy so that he has the owner's scent. The radio or, if possible, the television set is left on while the dog is alone. All these things—the toy, bed, and sounds—are to increase the amount of stimulation provided by the dog's restricted environment. They also serve as a cue to the dog that things are going to be different—in particular, that his behavior is going to be different in the new environment. On the first and on all subsequent days of training, the owners make all the usual preparations for leaving for the day. Coats are put on, keys collected, briefcases and handbags picked up, and doors locked. The owners then leave. They drive off in the car, if that is their usual means of departure. They return in 5 minutes, and if the dog has not been destructive, they praise him mildly, put away their coats, and so on. The next day the procedure is repeated, except that the dog is left alone for 10 minutes. Each day the length of time the dog is left alone is increased. The increase should be gradual. Many owners will be so elated by 10 minutes of good behavior in their absence that they leave the dog for 6 hours and return to find a torn-up cushion or worse. The veterinarian would do well to give the client a schedule to follow to avoid misunderstanding or mistakes (see Appendix 1).

The scheduled absences should be 1 minute, 2 minutes, 5 minutes, 10 minutes, 20 minutes, 30 minutes, and so on. It is obvious that the behavior modification method takes a long time. Unless the owners are willing to devote their month's vacation to the procedure, it will be exceedingly difficult for them to engage in their usual occupations and still modify the dog's behavior. There are two solutions that have been used by owners who can use the behavior modification schedule only on weekends. One solution is to leave the dog in the car during the day if the weather is cool enough. In the summer, temperatures in a parked car can be lethally high. Another solution is to leave the dog somewhere outside the home or at least outside the room or rooms where he is left in the owners' absence. The dog can be taken to friends who stay home or have other dogs in order to keep the problem dog company. If there is no other solution, the dog can be kept in a cage when the owners are away from home.

It is important that the dog get enough exercise to help attenuate his destructive tendencies, and exercise is particularly important if the dog must be closely confined during most of the day. The dog should also be obedience trained. Regular exercise periods and obedience training should be instituted, preferably in the morning and evening. These exercise periods serve two functions: the obvious one of ensuring that the muscles of the dog are active and also that of providing a time when the dog will regularly receive attention from the owner. One daily session of reviewing his training will provide another opportunity for the dog to be rewarded for good behavior.

The owner should be warned that the dog may be destructive during the behavior modification training period. In that event, the owner should not punish the dog. The dog should not be rewarded with any attention at all, but should be ignored. Any mess that the dog has made should be cleaned up while the dog is not present. The next day, the period of absence should be shortened and the training resumed.

The theory behind behavior modification of destructive behavior is that the dog learns that (1) the environment has changed (is more stimulating with toys and music), (2) nondestructive, not destructive, behavior will be rewarded, and (3) he will receive attention from the owners without having to resort to destructiveness. The behavior modification method is theoretically the best, but it is only 50% successful in our experience. Few owners have the patience to continue the training. A modification of the method in which many short absences are used, rather than absences of gradually increasing length, may be effective.

Denning.   It is sometimes possible to reduce destructiveness in the owners' absence by a method called "denning," which depends on gradual reduction of confinement. The theory is that the dog will become used to the crate and treat it like a den in which he can retreat when he is anxious. Alternatively, the contrast between confinement and the relative freedom of the house may make the dog more content in the house. This technique can only be used on dogs that do not injure themselves or soil in a crate. An airline crate is much more effective than a wire cage because it is darker and because the dog cannot catch his feet in the wire. If an open wire cage must be used, cover it with a blanket or towels so that light can enter only from the front, but the dog may pull the covering into the cage and chew it. Put food and water and an attractive toy in the cage. Put the dog in the cage after a play session with the toy or for his dinner the first few times. Try to associate the cage with pleasant things. The entire denning process takes 6 weeks. The first few days are the most important because that is when the dog may bark; therefore, the process should begin on a weekend or at some time when the owners can devote all of their time to it. Put the dog in the cage. If he barks or howls, punish by throwing a light chain (the collar) against the cage or by throwing water on the dog. Such methods will not hurt the dog but will startle him. If the owner lets the dog out when he barks, he will learn to bark many times in hopes of the reward of freedom or attention. The dog should live in the cage for the first 2 weeks and should only be taken out to eliminate (three or four walks on a leash per day) and for obedience training (at least 20 minutes per day, preferably in several short sessions). During the second 2 weeks, the dog should be allowed out of the cage when the owners are at home and awake. Leave the cage door open when the dog is outside the cage so that he can return when he wishes. Continue leash walks and obedience sessions and begin desensitization. During the third 2-week period, the cage should be open and the dog free to go in and out of it as long as the owners are home, whether they are awake or asleep. Continue desensitization training during this period. After 6 weeks, the dog should no longer be confined in the cage at all, but the cage should remain with the door open for him to use.

Operant Conditioning. Another solution to destructive behavior is to provide the animal with a task that will command his attention and give him the opportunity to use his paws and/or muzzle. The solution is to operantly condition the dog to obtain his food by pressing a panel with his muzzle or a lever with his paws. The dog can be easily trained, as many laboratory dogs have been, to press a panel a number of times for a few pieces of dog chow. He will have to work several hours a day to obtain his nutritional requirements. The apparatus must be constructed to fit the dog, and it must be made of rugged parts that a dog could not destroy. Operant conditioning has been used with considerable success to treat and prevent stereotyped behavior in zoo animals,[1270] but we have not been very successful, perhaps because most destructive dogs will not eat when alone.

Dog Doors. As noted previously, many dogs show a particular kind of destructive behavior, barrier frustration. When possible these dogs should be provided with a "dog door" so that they can go in and out of a fenced yard at will. They should not run free, but may be content to be able to choose whether they are inside or outside.

Drug Treatment. The most effective and practical treatment is amitriptyline (2–3 mg/kg every 12 or 24 hours). This can be combined with denning or desensitization. Other antidepressants or antianxiety drugs may be used but are more expensive.

Self-Mutilation

Self-mutilation can vary from excessive grooming to the commonly encountered lick granuloma or even to life-threatening self-inflicted wounds. Many dogs lick their paws until they are hyperemic and the fur is discolored. Whenever dogs bite or scratch at their coats, a physical basis, such as impacted anal sacs, flea-bite allergy, or dermatitis, should be considered.

Occasionally, however, there is a behavioral basis for such actions. A German shepherd developed a granulomatous lesion on his foreleg. The dog had begun to lick the affected paw when the little boy with whom the dog normally spent much time was in the hospital for a few days. The abnormal behavior ceased, and the lesion healed with application of topical corticosteroids and the return of the child. The behavior and the lesion recurred when the child started school. Again, corticosteroid medication and adjustment of the dog to the child's new schedule were followed by healing of the lesion.

Self-mutilation may be considered a form of obsessive compulsive disease. Self-mutilation, both acral lick granulomas[1180] and tail chewing[1097,1394] have responded to clomipramine in gradually increasing doses from 1 to 3 mg/kg per day. Dodman et al.[386] have used the opiate blocker naltrexone to treat lick granulomas.

Shock collars have been used to punish dogs when they lick a granulomatous lesion.[422] First, all secondary infections must be eliminated. The dog must wear an Elizabethan collar until the lesion has healed and continue to wear it whenever the owner cannot be observing the dog. The shock collar should be worn before punishment is instigated so that the dog will not connect the presence of the collar with the pain of a shock. The Elizabethan collar is removed, and whenever the dog licks his leg, the owner should shock him. This method might also be used for the more acute and more common problem of the dog chewing out his stitches or eating his cast. The postsurgical type of self-mutilation is not a strongly entrenched behavior pattern and should be more easily suppressed than long-standing licking problems. Extreme cases of self-mutilation during which the dog barks or screams are usually accompanied by abnormal electroencephalograms, indicating that antiepileptic drugs may attenuate the problem.

Tail Chasing

Tail chasing can be a form of play or a sign of gastrointestinal parasites in puppies, but it is a behavior problem in adult dogs. It is most common in bull terriers. The cause is unknown. Trauma to the tail caused by self-mutilation makes it difficult to determine whether a lesion in the area preceded the behavior. Restraint seems to exacerbate the condition. This has been noted in Scottish terriers.[1391] In mild cases, eliminating cage confinement and counterconditioning the dog to obey a command has been successful. In cases that do not respond to behavioral therapy, naloxone has been used successfully in one case[256] and barbiturates in several others.[380,381]

Phobias

Dogs have an interesting variety of individual phobias, though potentially phobic stimuli seem to be limited. Dogs do not develop phobias to trees, for example. Many dogs are afraid of thunder; others of gunfire; some of street noises or of riding in a car. Occasionally, the phobia can be traced to a fear and/or pain-producing experience. For example, as a young dog, a cocker spaniel loved to ride in cars, but after the car in which he was riding was involved in a serious accident that injured the dog slightly, the dog was extremely frightened of riding in cars and had to be heavily tranquilized when automobile transportation was necessary.

Storm-Related Phobias

Fear of the loud noise of thunder is understandable, and most dogs that are afraid of thunder simply slink under a bed. Some dogs will become frantic, however, and try to escape from the pen or house and, if they succeed, run long distances. Most dogs with storm-related phobias appear to be afraid of the noise of thunder, but in one case, that of an English setter, the animal appeared to be responding to changes in barometric pressure. He would scratch, chew on his kennel, and generally make escape attempts before any storm, rain or snow. He would calm down if the owner was present. He was a good hunting dog and was not afraid of gunfire. His early history was unknown, but he may have been abused.

Treatment. The classic treatment for thunder phobias is progressive desensitization. The dog is made to lie quietly near the owner, and a very good quality tape recording of thunder is played to the dog. The sound source should be as high above the dog as possible. It is important that the sound will elicit the behavior. Play the sound at high volume a day before desensitization is to begin. If the dog does not react, another stimulus sound will be necessary. At first, the volume is kept very low. If the dog shows no signs of apprehension, he is rewarded with food and the volume is raised slightly. The daily desensitization sessions should be short, only 10 minutes. Gradually, the volume is raised to that occurring during a nearby thunderstorm. Each day, the initial volume is just below the loudest volume tolerated on the previous day. Fear of guns can be treated in a similar manner. Place a starter's pistol inside a nest of cardboard boxes. Reward the dog for not flinching when the gun is fired. Gradually remove one box at a time until the dog is calm even when the gun is fired without muffling.[617] Progressive desensitization can be used for nonsound-related fears as well. Fear of cars can be treated by first getting the dog to relax by gentle praise near, and then in, the car. When the dog can sit for a few minutes in a quiet car without trembling, the engine is started, but the car is not moved. When the dog accepts the noise of the engine, the car is moved a few yards. Gradually, the rides are lengthened. By breaking the fearful event into as many subevents as possible and patiently exposing the dog to each in turn under relaxed and rewarding circumstances, phobias may be overcome. Owners must be advised not to try to calm the dog because the soothing may reward the dog for acting in a wild or frightened manner. The benzodiazepine drugs (diazepam 0.11–0.45 mg/kg orally) are most useful for treating dogs with noise phobias.[1442] They can be used on the Fourth of July and during the hunting season or whenever thunderstorms are predicted. Clorazepate is a sustained-release preparation that can be given on days on which storms are predicted.

A combination of phenobarbital and the beta-blocker, propranolol, has also been successful in treatment of thunderphobia. The tricyclic antidepressant amitriptyline can be administered daily throughout the storm season with or without clorazepate when a storm is predicted.

Running and Car Chasing

The best way to prevent running away and car chasing is to keep dogs on a leash, under voice control, or in a sturdy pen at all times. Giving a dog freedom to roam is giving him freedom to die under the wheels of a car. Once a dog has learned to chase cars or to roam, he can often find ways to escape confinement, so behavioral means of attenuating those activities, as well as restraint should be used.

Backyard Problems

Escaping

The most common backyard problem is the dog that escapes. A high enough fence can hold a small dog, but a large dog can sometimes jump or climb nearly any fence, no matter how high and expensive.

One suggestion[270] is to erect a small barrier 0.9–1.8 m (1–2 yd) or so in front of the fence so that the dog cannot gather momentum for his jump or to dig a ditch to serve the same purpose. Many dogs respond well to dog doors that enable them to enter and leave the house as they please. Occasionally, the destructive dog, as well as the one that digs in the yard, will do neither if he is not barricaded in one location or the other. The best way to deal with a roaming dog or one that escapes from a fenced yard is to keep the dog indoors and walk him on a leash. Roaming can be prevented by boundary training. The dog is taught that he cannot step across a visible or invisible barrier or he will be punished verbally. The dog should be on a leash while he is boundary trained so that he cannot keep running. The most common response of an owner is to punish a dog for running away or for failing to come after he has returned. The dog learns that if he comes when called, he will be punished; consequently, it learns not to come.

Dogs usually have one of two motives for roaming: sex or food. Castration reduces the first motivation,[674] and ad libitum feeding will reduce the second. In the latter case, the risk of obesity is less than the risk that the dog will be hit by a car.

A method that has been suggested for a roaming dog that does not respond to these simple remedies is as follows: Fast the dog for 24 hours and then feed him a very palatable food a spoonful at a time every 15 minutes for 2 days and then on a variable schedule of 15 to 45 minutes during the next 2 days and at increasing intervals on the following days.[617] This method relies on the difficulty of extinguishing behavior rewarded on a variable interval schedule (see Chapter 7, Learning).

A more difficult problem is the dog that is kept indoors and walked on a leash but sneaks out the door. Teaching the dog to sit and stay before exiting and that the door will be slammed if he precedes the owner may help.

Digging

Digging holes in the yard is another closely related backyard behavior problem. Dogs that dig are frequently those that are put in the yard as a punishment for bad indoor behavior. If the indoor behavior can be improved, isolation punishment will not be necessary. Dogs also dig holes to keep cool or to catch rodents. Providing a cool shelter should reduce digging for the first reason, and eliminating the rodents should reduce digging for the second. Putting chicken wire where the dog digs may also deter him.[404]

In cases of backyard digging, Campbell[270] recommends the same routine as that recommended for destructiveness because he feels that the etiology of the two behaviors is the same. Koehler[822] recommends filling the hole with water and submerging the dog's head in it. This only teaches the dog to dig elsewhere and to fear water. Some dogs dig up flower beds in imitation of their owners. The solution for these dogs is simple: do not let them watch people working in the garden.

A few dogs dig imaginary holes in the house or outdoors. For example, a well-trained Doberman pinscher began to dig holes. He looked as if he were digging mice out of their holes. While engaged in this behavior he would barely react to his name, would perform none of his obedience routines, and would become increasingly rigid in posture and unresponsive as stimulation (calling, patting, and so forth) increased. If the dog was put into a stay position by the owner when the dog was not digging, he would not hold the stay for more than a few minutes before digging began, whereas, formerly, the dog would maintain a stay for an hour. That is probably a seizure disorder or an obsessive compulsive problem.

For more-serious digging problems, dogs should be treated as they are treated for destructiveness in the house. In some cases, provision of a dog door that allows the dog to enter and leave the house at will is sometimes successful.[1030]

Jumping Up on People

Jumping up on people is a very common minor behavior problem. It is minor as long as the dog is small and only soils trousers or tears stockings when he jumps up. The problem becomes major when the dog is large and knocks down children, or even adults. Typical solutions are to place a knee in the oncoming dog's chest. This may be effective if everyone on whom the dog jumps can apply this treatment. Otherwise, the dog will only learn not to jump on robust adults and will continue to jump on children and old people. Stepping on the dog's hind feet is also recommended, but that maneuver requires agility and may result in fractured canine toes. Holding the dog's paws and wheelbarrowing him backward on his hind feet and throwing him away from you while saying, "Off," is useful because the "Off" command can be given by the owner or the intended victim as the dog approaches. Campbell[270] recommends squatting down when the dog is approaching to jump. The rationale is that the dog will not jump up if the human is at his level. Some dogs still jump up on a squatting person and can do more damage to the head and neck than if the person were standing.

The best therapy is counterconditioning, teaching the dog to do something incompatible with jumping such as sitting. This is simply obedience training, using a command that the dog consistently obeys in order to rid the dog of the habit of jumping. Turning one's back on the dog, i.e., withdrawing attention when he jumps, is helpful in many cases.

Fly Catching

Some dogs develop a behavior that resembles snapping at flies when there are no flies present. The snapping at air occurs several times a minute. It probably represents a form of psychomotor epilepsy, but needs more thorough study.

Vomiting

Carnivores vomit frequently. This is an advantage to a scavenger animal that may ingest spoiled food. Vomiting food is part of normal canine maternal behavior also. Owners may complain that the dog vomits "for spite" when they are away or to attract attention. Vomiting usually does attract attention, so the dog is rewarded for his behavior with the owner's attention. In one case, the dog vomited into the owner's shoe. This behavior is probably akin to urination on beds, a response to the owner's scent, in this case a maternal response.

Psychosomatic Problems

Fox[492] and Tanzer and Lyons[1380] present a number of cases of dogs that exhibit "sympathy" lameness or seem to enjoy receiving medication.

CATS

Destructiveness

In general, cats develop behavior problems much less frequently than dogs. Feline destructive behavior appears to fall into three categories: (1) clawing (discussed in Chapter 2, Aggression and Social Structure); (2) wool sucking (a vice of Siamese cats in particular), which is sometimes transferred to synthetics to the great detriment of sweaters and upholstery (wool sucking may or may not be related to early weaning); and (3) plant eating. Wool sucking and plant eating have already been discussed in Chapter 8 (Ingestive Behavior: Food and Water Intake).

Withdrawal

A change in the cat's environment is not always followed by inappropriate urination. Another response can be withdrawal from the owner. In one case, the cat was taken from a suburban setting, where he was free to go out of doors, to a high-rise apartment. It had been an affectionate cat, but ignored the owner for weeks after the move. The owner then boarded the cat for 2 weeks; on his return to the apartment, he was again affectionate. The stress of boarding probably was responsible for the cat's adaptation to the less stressful apartment.

HORSES

These large, powerful animals, which are trained to work closely with humans, may develop a wide range of irritating or even dangerous behaviors. These abnormal behaviors are commonly referred to as vices, but this term implies that the horse is making a moral decision. The term *problem* is used here. See the section on cribbing in Chapter 8 (Ingestive Behavior: Food and Water Intake) for a discussion of the factors that are associated with stable problems. More roughage, straw bedding, more contact with other horses, and more time out of the stall prevent or attenuate these problems.

Behavior Problems in the Stable

Locomotion Problems

Stall Walking and Weaving.   This behavior may often occur as part of a herd-rejoining behavior, in which case, more visual contact with other horses or a stall companion will help. In other cases, it is claustrophobia, in that the confinement itself, with or without other horses, is a problem. Finally, stall walking may occur as a stereotypy, a repetitive functionless behavior. This form of stall walking is usually slower than the others but is more difficult to interrupt. Horses that constantly circle their stalls may lose condition or fail to obtain it because they expend more energy walking than they ingest. Their performance is usually affected as well. Restraining the horse by tying it often converts a stall-walking horse into a weaver. Weaving is a behavior in which the horse stands in one spot but shifts his weight and his head from side to side, lifting each hoof in turn and walking in place. The cause of the behavior appears to be confinement. The treatment is to maintain the animal on pasture with a run-out shed for shelter. Other treatments are stall toys and more work. The size of the stall does not seem to influence the stall-walking behavior because a horse given access to an entire barn still circled in one corner. Stress appears to aggravate the problem; horses circle more frequently the evening before a hunt or show. Bagshaw et al.[99] did not find any calming effect of tryptophan.

Weaving may be similar in etiology to stall walking. It is a stereotypy that appears in many zoo animals as well. Confinement and frustration are causes, and horses in pain may also weave. The form of self-stimulation probably offers the horse some comfort, as rocking does to humans. Tail rubbing is also probably a form of comforting self-stimulation, but medical causes must be eliminated. There is some evidence that stall walking and weaving may be inherited and possibly related to endogenous opiate production.[384,1429]

Pawing.   Pawing has been described by Ödberg[1101] as a response to frustration, a displacement activity that originated from the activity of uncovering food buried under snow. Horses have been noted to paw in a variety of situations: when restrained from moving, when eating grain, when expecting feed, at a recumbent foal that does not stand, and in order to reach another horse. The most extreme forms of pawing occur in horses that have barrier frustration, that is they are seeking to escape from their stalls. A typical case is that of a standardbred that had spent most of his life on pasture with a run-out shed for shelter. When confined in a box stall for training, he dug a hole measuring 1.5 m (4 ft) in depth in his stall. Some horses habituate to stall confinement eventually; others may do extensive damage to themselves and the stalls as they try to escape by jumping over the walls. Pawing is a less dangerous activity, but it damages dirt or clay floors. The second cause of pawing is an attempt to reach another horse. Horses are herd animals, and stallions, in particular, try to reach one another to play if they are young or to fight if they are mature. If dirt floors are replaced with concrete, the horses will stop pawing, but their motivation has not changed. The stallions may rear up to reach one another over stall walls that do not reach the ceiling or lean out the front of the stalls to make contact if that is possible. A concrete floor can be hazardous when slippery and is more stressful to the horses' limbs and feet because it is a rigid surface.

Pawing in anticipation of food is similar to kicking in anticipation of food and is discussed under Stall Kicking. Pawing at recumbent foals may serve to stimulate the foal but can injure a foal, especially if the foal is unable to rise and the pawing continues for some time. The reason horses paw when eating grain is unknown. One hypothesis is that they may be responding to the highly palatable food that can be prehended quickly, two unnatural, and therefore frustrating, qualities.

Stall Kicking.   Aggressive kicking is discussed in Chapter 2 (Aggression and Social Behavior). Only kicking directed against stall walls is considered here. Most horses kick stall walls with their hooves; a few knock their hocks against the wall. Both activities produce unwanted concussion on the horses' bones and joints. Kicking can damage the walls as well. A small hole made by kicking is often enlarged by wood chewing. Stall kicking, like weaving, may also be a form of self-stimulation. The horse kicks to hear the sound his hoof makes as it strikes the wood. Sometimes stabling a horse on a wooden floor that makes a similar sound as the horse walks on it will eliminate the kicking.

The most common form of pawing and kicking is that that the

owner has operantly conditioned. The horse will tend to paw or kick at feeding time because he is frustrated to see or smell food (or the giver of food) but not be able to eat. This is reflected physiologically in the increase in heart rate observed in horses that see feed. The horse is, of course, fed, so the animal's behavior has been positively reinforced. He has learned that kicking or pawing are followed by food. The horse paws, and food appears. He will begin to paw or kick earlier and earlier. In effect, he is increasing the fixed ratio of number of responses (paws or kicks) for every reward. The longer the horse kicks before feeding, the longer it will take to extinguish the behavior. In order to extinguish the behavior, the owner should feed the horse only when he does not kick. The process will go much more quickly if many small meals a day are given. The horse must at first only refrain from kicking for 2 seconds before he is given a half cup of feed. Gradually, the criterion is raised so that there must be no kicking for 5 then 10 then 30 seconds before food will be given. Only when the horse will refrain from kicking for a short time for several feedings or trials should a longer time be demanded. The training will go faster if the horse is taught a countercommand such as "stand" for a food reward at the same time.

Trailering Problems

Loading

Many horses exhibit undesirable behavior in relationship to ground transport. The most common undesirable response is failure to load. Both innate behaviors and learning contribute to loading problems. The properties of a trailer that release the horse's innate fears are (1) the dark interior of the trailer; (2) the hollow sound of hooves on the ramp, an indication of poor or insecure footing; and (3) the instability of the ramp and vehicle. In addition, horses are generally neophobic, that is, they are afraid of new or strange things.

Experience or learning can also play a part in loading problems. A horse that already has had unpleasant experiences with loading or riding in a trailer will also be understandably difficult to load. Hitting the horse may cause him to leap forward into a trailer, but the animal will associate pain with trailers and may be even more difficult to load on the next occasion. Learning to dislike loading is even more likely to occur if the horse injured itself while resisting loading. For example, the horse may rear back just as he reaches the entrance to the trailer and strike his head on the roof.

Far more common are less dramatic negative experiences. The horse may have been thrown against the side of the trailer many times. He may have lost his balance or he may have struck his head. He may even have experienced motion sickness, which would be difficult to diagnose in an animal that does not vomit, such as the horse. The horse will have learned to associate one or more of these unpleasant experiences with trailers.

The first approach to solving a loading problem is to train the horse to move in response to a touch on his body. This procedure, if correctly done, takes only a few minutes. The handler, preferably a person who has not tried to load the horse, holds the horse with a lead rope and halter. A frightening but nonpainful stimulus, such as a lunge whip with a cloth tied to the end, is used to tap the horse. The horse should move away from the stimulus. The horse can be encouraged to walk forward, to stop, to move the hindquarters toward the handler, and to move the hindquarters away from the handler. After practicing these movements, the horse is walked to the trailer and urged to approach it, but not allowed to enter. After a few repetitions, the horse is tapped on the hindquarters to encourage him to enter the trailer. The first entrance may only be partial, only the forefeet. Later, the horse can be allowed to enter the trailer with all four feet many times before the trip begins and to back off quietly. This method has been most successful, but it requires patience.

One could desensitize and countercondition the horse by rewarding it for loading itself. Desensitization takes time—days and weeks— and therefore is not a solution for the acute problem of how to load a horse, which is covered here. Owners of horses with trailer problems should begin the desensitization program a month or more in advance of the proposed trailer trip. The horse and the trailer should be placed in a paddock. The trailer should be secured so that it does not tip, and the wheels should be blocked. The horse's food, all of it, should be put at the bottom of the trailer ramp. Each day the feed, hay and grain, should be placed a little farther up on the ramp. Next, the feed is placed on the floor of the trailer. Day by day, the feed should be placed farther and farther into the trailer, forcing the horse to load himself in order to eat. The horse should obtain no other feed except that in the trailer during the desensitization process. Very few horses will starve to death rather than enter a trailer. Once the horse is loading himself into the trailer, he should be led on to the trailer for all his meals. Some horses will learn to load themselves but will still object to being led onto a trailer. The loading problem may not be completely solved even if the horse loads easily in his home paddock. The same horse may be reluctant to enter the same trailer when it is parked in a strange place. A few days of confining the horse in a paddock at another farm will help the horse generalize the lesson that getting on a trailer is safe and desirable. Another method to encourage a horse to enter a trailer is to put it in an unfamiliar place in such a way that entering the trailer is the only means of escape.

The rear-facing trailers sometimes offer a solution for horses that will not enter conventional trailers. The ramp of rear-facing trailers can be made into a horizontal platform. The horse can step onto the platform and can be backed into the trailer. Riding in a rear-facing trailer is different and less stressful than riding in conventional trailers, so the horse will not associate riding in it with his previous experiences in conventional trailers and may result in easy loading. He should not refuse to back into the trailer the second time it is attempted if he does not dislike or fear the sensations of riding in the trailer.

There are some solutions to an acute trailer problem in which the horse must be loaded that do not cause the horse to associate being loaded with pain. If a horse must be forced onto a trailer, pushing is to be recommended over hitting. Long ropes tied to each side of the trailer held by two people can be crossed behind the horse and used to apply pressure to its rump. The ropes must be soft to avoid rope burns. The ends attached to the trailer should discourage the horse from leaping to either side of the ramp. The ropes should be held so that they can be dropped if the horse becomes entangled. Sometimes, two people can join hands behind the horse and use their arms to push against the horse's rump. This will encourage the horse that is only mildly reluctant to enter the trailer.

Horses are herd animals. Advantage can be taken of equine gregariousness to solve immediate trailer problems. A horse that is reluctant to enter a trailer may be willing to follow another. Naturally, this method—utilizing social facilitation of behavior—is unlikely to be successful if the horse is already very excited. The same characteristic of horses—social facilitation of behavior—can be used to prevent trailer problems. A foal should be loaded beside his mother several times in his first few months because he will follow his dam onto the trailer as he follows her elsewhere. Of course, this method should not be used if the mare herself does not load. Bad behavior as well as good behavior can be learned by observation. If the foal's mother is reluctant to load, the foal can still be trained to lead relatively easily after weaning is complete. With the foal correctly haltered and on a lead rope, gently encourage the foal to put one hoof on the ramp for a handful of grain. Gradually, encourage him to go farther and farther up the ramp. Talk softly and encouragingly and immediately reward any progress with a little rub or grain. Full loading may take several sessions. Once the foal has gotten on the trailer the first time, feed him his daily grain ration in the trailer several times before traveling with it. Never use hitting, pushing, or pulling during these training sessions. Repeat the process in as many different vehicles and in as many different sites as possible to help the foal generalize. Before the foal goes on his first long trailer ride, he should take several short trips one-half to one mile, ending with a grazing session. Gradually, take the foal on longer trips. The etiology of some horses' trailer problems may be a first trailer experience that consisted of 4 hours of fear and exhaustion while trying to balance on a strange surface. Several short trips help the horse to habituate to the trailer and the sensation of a moving vehicle while the animal is rested. The time involved can save much more time later; a foal properly introduced to trailering should load easily as an adult even when trailering episodes are years apart.

Sedatives such as xylazine can be used in the acute situation. The sedative will be far more effective if the animal has not become excited or frightened before administration. It might also be possible to desensitize a horse to the loading procedure by repeatedly sedating him and loading him several times a day for a week or so, but in some experimental situations other species cannot remember in the undrugged state what they learned in the drugged state; the phenomenon is called *state dependent learning.* Sedatives have several disadvantages: a sedated horse cannot perform properly, and in some shows, cannot perform legally. Furthermore, a sedated horse is more at risk of losing its balance in a moving trailer. A way to avoid state dependent learning is to decrease the dose of the sedative gradually over several sessions.

Moving Trailer Problems—Scrambling

The next class of behavior problems involves horses that enter trailers without hesitating but misbehave when the trailer moves. The horse can be badly injured and/or the trailer can be damaged. There are several causes of struggling. Most are related to the horse's inability to keep his balance in the trailer. Smith et al.[1331] have found that when given a choice of direction of travel in a stock trailer, horses are not uniform in their choice. Some horses stand facing backward, but more stand at an angle and they change their position during travel. Kusunose and Torikai[850] have shown that thoroughbreds also orient away from the direction of travel. The forward movement of the trailer is not a familiar force for a horse to resist, nor one that the horse evolved to counter. For this reason Cregier[325] has advocated trailers designed to transport horses facing towards the rear. The horse can brace against the forces of the moving trailer because it is the same action the horse would use to stop himself.

There have only been a few studies of the behavior and physiology of transported horses. They are more likely to strike their heads when facing forward in a standard two-horse trailer than when facing backward. Several investigators have found that heart rate increases during trailering and cortisol levels rise.[298,1451] Horses stop eating and adopt a base-wide posture with their limbs abducted, indicating that they are trying to maintain their balance.

If the horse is struggling because it is losing its balance or is afraid of doing so, simple changes in the trailer may be all that are necessary. Removing the center partition of a two-horse trailer will allow the horse to plant his feet more widely. A layer of sand topped with wood chips on the floor of the trailer will prevent the horse from slipping. Alternative solutions are to transport the horse in a large horse van or a stock trailer. More horses travel well in stock trailers than in standard two-horse trailers. If a two-horse trailer must be used, there should be no bulkhead or compartment on which the horse can strike his head. A padded bar at the horse's chest level will restrain the animal but allow him to move his head without striking it.

A horse may be reacting to erratic movements of the trailer. This can be ascertained if the horse scrambles only when a certain person is driving. A rarer cause of misbehavior is electric shock. If the horse only struggles when the brakes are applied, the wiring should be examined.

Horses may struggle in trailers because they are losing their balance as has already been mentioned, but some horses may be reacting in anticipation of the journey's end. Horses participating in such high-speed events as barrel racing or games are the most likely to scramble on the way to a competition. These horses could be desensitized by taking them on many trailer rides that end not with competing, but with grazing or a leisurely trail ride. Care must be taken to treat the horse in the same manner before a "therapeutic" trailer trip as before a ride to a show. The same tack and the same grooming routine should be used in both cases so the horse cannot discriminate between trips that end in shows and those that end at less exciting destinations. There are horses that do not scramble in trailers but do sweat or paw. These horses may be experiencing motion sickness. Although most horses travel better with another horse, a few are aggressive toward the other horse.

Horses that load and ride well but do not stand quietly in a stationary trailer are still another problem. They will move, kick, and struggle when the trailer stops at a tollgate or a traffic light. Small food rewards for standing quietly during practice trailering sessions can be used to treat the problem. The reason why horses fret in stationary trailers is unknown but may be the same as those for horses that struggle in the moving trailer. Better footing, driving ability, or a different type of trailer may be necessary to eliminate the problem.

A final problem is the horse that will not leave the trailer. Although horses can sometimes be backed where they will not walk, a horse that reaches back with a hind hoof and encounters nothing but air will be afraid to back off. Trailers with walk-through construction enabling the horse to enter from the back and exit from the front are helpful. To solve the immediate problem, the horse should be allowed to turn his head so that he can see where he is backing. It may be necessary to remove the center partition of a two-horse trailer so that the horse has room to do so. Horses are more reluctant to back off step-up trailers, those without a ramp. A loading dock of dirt can be made to give the animal a solid place to put his hind feet.

In summary, the best approach to the treatment of stable problems and trailer problems is to remove the cause of the problem, that is to change the horse's motivation. This approach is more apt to be successful than punishment or physical restraint of the horse. Careful, early handling and optimal housing can prevent the emergence of these problems.

Behavior Problems under Saddle

Head Shyness

The horse shown in Figure 9.1 illustrates head shyness, a common equine behavior problem. Head shyness is usually, but not always, secondary to mismanagement of the horse when he was first handled. Occasionally, progressive desensitization may be used by a gentle and patient owner to overcome the vice, but while the horse is becoming accustomed to handling of his ears and poll, he will still be difficult to bridle, and attempting to do so will undo the desensitization. Horses will often tolerate handling of their head and ears best when they are hot, sweaty, and, apparently, itchy. Putting on the bridle with one check piece unfastened may be tolerated. Rubbing the bit with molasses rewards the horse for accepting the bit.

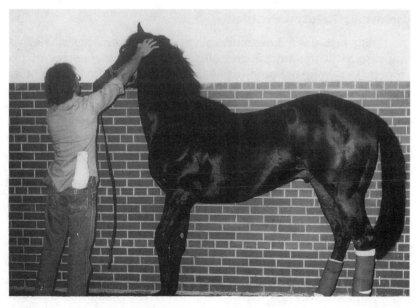

FIG. 9.1. A horse that kicks stall walls, thus aggravating hock injury. The horse is also head shy.

Bucking, Shying, and Grazing

Vices under saddle can vary from bucking and running away to grazing. The former habits are quite dangerous, and treatments as bizarre as injections of succinylcholine have been used to cure horses of bucking, but nothing discourages the child who is learning to ride more than the pony that puts its head down to graze at every opportunity. Muzzling the pony may discourage it from attempts to eat. The habit that is most apt to dislodge the rider is the simple fear response of shying. The good rider knows what objects are most likely to elicit a shying response in a particular horse, is prepared, and tries to distract the animal.

Phobias

Some horses have fears that might be classified as phobias. A horse that was badly stung by bees became uncontrollable whenever he heard any insect buzz. The horse could be desensitized by playing bee noises to him, in the same manner as dogs are desensitized to thunder. This technique may also be used to accustom horses to band music, applause, and the other frightening aspects of parades and horse shows. The horse can be urged toward a frightening object and rewarded with both removal of the negative reinforcement (kicking) and a positive reinforcement, a pat.

Behavior-Related Problems

Many horses will move much more quickly toward their stable than away from it. These animals are called "barn rats." Other horses will trot smartly in a group of horses but must be pummeled into moving off by themselves. Such behavior is hardly surprising, as horses are herd animals and group, or allelomimetic, behavior is part of their evolutionary heritage. The horse racing back to the barn is rejoining its herd, if other horses are there, or is simply returning to its home range. Consideration of feral horse behavior indicates that many of the vices that appear to be spiteful, lazy, or stupid, are in reality perfectly normal adaptive behaviors for an animal that must find food (graze while bridled), avoid predators (shy), remain with its herd, or rejoin it if separated (barn rat).

Physiologically Based Problems

Other common habits of horses may be explained on the basis of their physiology. Horses refuse jumps because their vision is such that they cannot see the jump when they are close to it. Some horses are very hard to canter on the left lead but take the right easily. This is because some horses (70%) show definite handedness[578] even if they have been carefully schooled to canter both clockwise and counterclockwise.

The myriad other vices and training problems of horses are beyond the scope of this book. Behavioral history forms for horses are given in Appendix 2C, 2D, and 2E.

CATTLE AND OTHER FARM ANIMALS

The behavior problems of farm animals have been reviewed by Fraser[504] and by Kiley-Worthington.[799]

Cattle

Problems Related to Changes in Management

Dairy cattle present few behavior problems in general. This is probably due to selection on the part of the dairy farmer for tractable animals as well as high producers. The biggest problems that arise with dairy cattle concern changes in their management. When cattle that have formerly been milked and fed simultaneously in a stanchion barn are placed in free stalls and are not fed concentrates when they are milked, they may enter the milking parlor only reluctantly. Milking time can be prolonged for hours. Apparently, the milking procedure and the reward of relieving pressure on the udder is not enough to induce cattle to be milked. One solution is to feed grain during milking. Even if the cow requires more concentrates than she can consume while being milked, a portion of her grain can be fed at milking and the rest afterwards. Other management-related behavior problems have involved electrically operated squeeze gates that push the cattle closest to the gate when the cattle ahead balk. Time can be lost and injuries sustained if mechanical devices are not built with the behavior of the normal cow in mind.

Cattle that are always milked or handled from one side may become frightened or aggressive when handled from the other side. A veterinarian or stockperson should note the position of the milking machine outlet in stanchioned cattle and approach the cow from the same side. Milk production may be affected by milking from the "wrong" (the unfamiliar) side. Such obvious stresses as isolation and being chased by a dog can lower milk production,[1473] but the handler attitude can also affect production either positively or negatively. If the handler is satisfied with the job, the cows will give more milk and will approach the milking area more quickly.[1287]

Kicking

Beef cattle cause more problems for the average veterinarian because they have neither been selected for tractability nor handled often.

A beef heifer brought from range or pasture and placed in a box stall is as dangerous as an ill-mannered horse. Such animals should be treated with caution because even when heavily tranquilized they can and will kick. Cattle do not always "cowkick." They can place a well-aimed blow backward as well as forward. Such cattle should be fed only when they stand quietly. The food is the reward for nonkicking behavior.

Problems Related to Environment

Cattle not used to stanchions, such as those living in loose-housing facilities, may find it difficult to lie down or arise when first stanchioned. Hospitalized cattle may be particularly affected because the illness for which they were hospitalized will be compounded by the stress of lack of rest.

Calves and adult bulls show a variety of oral "vices" such as bar licking and tongue rolling. Perhaps the most compelling evidence for the protective effect of stereotypies is the finding of Wiepkema et al.[1481] that veal calves that indulged in tongue rolling had a lower incidence of abomasal ulcers than calves that had no vices.

Pigs

Few pigs are kept as individuals, so most of the their behavior problems concern group interactions. Aggression has been discussed in Chapter 2 (Aggression and Social Structure) and sexual behavior in Chapter 4 (Sexual Behavior).

Other problems seen are rubbing the snout on the floor or on the flank of another pig. Bar biting by confined sows is believed to be a consequence of restraint or of food deprivation and may release endogenous opiates because treatment of the sow with the opiate blocker naloxone will halt the behavior.[330]

Sheep and Goats

The only miscellaneous problem of sheep that has not been discussed elsewhere in this book is wool chewing. This behavior is of unknown etiology.

Goats frequently become nuisances, especially if they are kept as pets and inadequately restrained. The behavior of the goat is not abnormal. He is an animal that browses normally, but he may browse on the ornamental flower beds or the crops. Goats are herd animals, and a solitary goat may seek human company if he has been kept as a pet since he was a kid. Normal caprine feeding, investigatory, and allelomimetic behavior may seem very abnormal to the naive goat owner.

APPENDIX 1

Separation Anxiety

Separation anxiety in dogs can be manifested many different ways. Dogs are social animals that, by nature, need the company of their "pack." This is what attracts us to dogs as pets. Yet we expect them to accept long hours of solitude every day while we work or go to school; it is not surprising that some dogs exhibit anxiety when left alone. The causes of problem behavior may or may not be clear. Dogs that were previously well behaved when alone may become anxious as a result of a change in the owner's schedule. After being boarded, many dogs will exhibit anxiety when left alone. Still others may grow increasingly anxious for no obvious reason.

Separation anxiety is expressed in several ways. Some dogs may be destructive, especially at points of exit such as doorways or windows, or they may chew and destroy the owners' personal articles such as clothing or bedding. Other dogs may bark all day or urinate and defecate in the house.

In treatment of separation anxiety, as with many behavior problems, it is important to treat the dog as an individual with his own particular anxiety-inducing stimuli, as well as his own limits. There is no one program of behavior modification that will suit all problems. The general protocol, however, is similar for all dogs. Treatment usually consists of a combination of gradual desensitization to the owner's departures with temporary antianxiety medication. Some dogs do well with only medication, whereas others do not require it. Medication is usually given for a few months while graduated departures are practiced; then the dog is gradually weaned from the drug.

During training, if at all possible, the dog should not be left alone. The purpose of graduated departures is to allow your dog to feel safe when left alone. Because you always return fairly soon (immediately at first), leaving the dog to the point of anxiety may set things back. If it is not possible to take the dog to work, leave the dog with a neighbor or relative or use "day-care" kennels, and so forth. The dog may be left alone but without routine departure cues (see below).

When you are home, begin to discourage any clinging behavior the dog tends to show. If the dog follows you everywhere, practice putting the dog in a down stay in a comfortable corner or in his or her bed. If the dog insists on lying under your legs while you read, have her lie 3 feet away instead. Praise, but be a little less exuberant than usual. If your dog's obedience skills are poor, one of the most important things you can do is teach the basic commands (good for quality time and for general control). We will provide information about training if necessary. Although obedience is very important, it is not a cure for separation anxiety.

For 15 minutes before leaving the house and 15 minutes after coming home, you should practically ignore your dog. You can say hello, briefly pet and then walk away. This is an important point, because many of us unintentionally encourage anxiety by reassuring the dog constantly before we leave and then saying hello with an edge of hysteria, as if we have not seen the dog for months. Generally, downplay both greetings and departures.

During desensitization, think of all the cues you usually give your dog when you leave the house. Dogs can tell the difference between weekdays and weekends just by the clothes you are wearing, or the difference between departure for 8 hours of work versus a short trip to the grocery store, so they are very sensitive to cues. Make a list: do you shut lights, pack a lunch, jingle keys, put on your coat and hat, turn on the answering machine, shut the TV, brush your teeth, pick up the cat's food, go out the garage door instead of the front door, and so forth?

Once you have your departure cues listed, begin to use them without actually leaving the house. For example, begin to ignore the dog (put the dog into a down stay if necessary but remember that later, when you are really leaving, you have to release the stay well before you leave the house), put on your coat, shut lights, pick up keys, then put down keys, turn on lights, take off the coat. Do not greet the dog enthusiastically. You can praise the dog but be nonchalant about it. During practice in the beginning, it may be very helpful to put the dog into a stay so that when you are done with the exercise, you can praise (quietly) for staying.

It is helpful to offer a special object or toy that the dog gets *only* when you are practicing departures (and also later when you are actually leaving). Do not use this toy if you actually leave the dog for a long period before he is ready, or it will be associated with anxiety. A sterilized marrow bone (from a pet shop) filled with cream cheese, processed cheese, or peanut butter is often a good distractor. The radio or TV may be left on, but only if you have not already been doing this (if you have, try changing it, that is, leave the TV on instead of the radio). Again, do not leave the TV on if you have to leave the dog alone before he is ready. During the initial exercises (key jingling, and so on), you do not have to give the dog the bone, but start giving it when you are practicing even momentary departures.

Punishment is not effective with separation anxiety. No punishment is effective after the fact; it must be administered within a half second of the act. If you come home to find something destroyed, it is too late to act on it. But with anxious dogs, punishment does not solve the problem even if given in a timely fashion because we have not changed the source of the anxiety.

The following is an example of a graduated departures format. It is important to remember that each dog progresses at his or her own pace, so use this only as a guide. To assess the dog's progress, watch his or her behavior right before leaving and after returning home. For some departures, one family member may try to stay behind and observe, although many dogs will not exhibit separation anxiety if not entirely alone. Both predeparture and greeting behaviors should be used when weighing the dog's reaction; *only* when the dog does not appear anxious or exhibit exaggerated greeting should you proceed to the next step. Exaggerated or prolonged greeting behavior usually will last for more than several minutes. Always wait until your dog has calmed down before repeating a step or proceeding.

An Example of Graduated Departures Format for Use in Treatment of Separation Anxiety

Day 1–3

Separation cues with no departure. Repeat any steps in any order, for example, pick up keys, put on coat, put down keys, take off coat. Later, put on coat, shut off lights, walk to door, walk back, take off coat, turn on lights, and so on. Do this randomly and as frequently as possible. During predeparture practice sessions, the dog can be placed in a down stay and released and praised (briefly) when done with each step.

Day 4–7

Practice the cues and walking to door, rattling the doorknob, and returning. Then rattle the doorknob and open and shut the door. Combine all cues, including keys, coat, lights, radio, and bone; then rattle the doorknob and open and shut the door.

Day 7–14

Begin departures. If your dog is particularly anxious, even with the previous exercises, you should start with very brief exits, for example, 10 seconds and return. If your dog tends to tolerate brief departures, you can start with 1–5 minutes. If there is no evidence of anxiety at 5 minutes, start there. Do exercises in clusters of 10–20 minutes, if possible. Examples follow:

a. 10 s, 20 s, 10 s, do not leave (open and shut door), 20 s, 30 s, 5 s, do not leave, 30 s, 10 s, 45 s, do not leave, 5 s, 20 s, 1 min
b. Do not leave, 30 s, 10 s, 1 min, 10 s, 30 s, 1 min, do not leave, 30 s, 1.5 min, 10 s, 2 min, do not leave, 20 s
c. 30 s, 1 min, 5 s, 2 min, do not leave, 30 s, 2 min, 2.5 min, 10 s, 1 min, 3 min, 1 min, 30 s, 3 min, 20 s
d. 1 min, 10 s, 2 min, do not leave, 1 min, 3 min, 1 min, 5 min, 1 min, 30 s, 3 min, 30 s
e. 1 min, 3 min, 1 min, 10 s, 5 min, 1 min, 1 min, do not leave, 3 min
f. 2 min, 5 min, 1 min, 7 min, 2 min, 3 min, do not leave, 5 min
g. 1 min, 3 min, 30 s, 5 min, 2 min, 10 min

Day 15–30

At this point, if your dog has responded well, continue increasing increments as suggested here. Even if your dog is doing well, *do not increase duration too rapidly*. Because most anxiety-related behaviors occur within 30 minutes of the owner's departure, dogs that can be left for 30 minutes usually will tolerate larger increments of departure. When your dog can be alone for 90 minutes, he or she usually will tolerate several hours; but even then, you should not depart abruptly for 8 hours at once. Again, proceed at your dog's own pace. For example, you may have to repeat (a) several times before proceeding to (b). Begin to introduce car sounds, for example, when you are out for 5 minutes or more (not every time), open and shut the car door. Then begin to start the engine, but do not go anywhere; alternate this activity with just slamming the door or with doing neither. When you are up to 15-minute departures (with no problems), drive the car around the block, and so forth. If your dog reacts badly to the car leaving, go back several steps and do not add car stimuli until later. Then reintroduce car stimuli more slowly.

a. 2 min, 5 min, 2 min, 7 min, 5 min, 10 min
b. 5 min, 3 min, 10 min, 5 min, 10 min
c. 5 min, 7 min, 2 min, 10 min, 1 min, 5 min
d. 5 min, 10 min, 5 min, 10 min
e. 1 min, 10 min, 2 min, 13 min
f. 2 min, 5 min, 15 min, 5 min, 10 min
g. 1 min, 10 min, 12 min, 15 min, 5 min
h. 5 min, 13 min, 1 min, 17 min, 2 min
i. 7 min, 1 min, 15 min, 3 min, 20 min
j. 10 min, 5 min, 17 min, 3 min, 20 min, 1 min
k. 5 min, 15 min, 20 min, 2 min, 2 min
l. 10 min, 5 min, 20 min, 5 min
m. 5 min, 20 min, 1 min, 25 min
n. 10 min, 1 min, 30 min, 2 min
o. 5 min, 30 min, 1 min, 1 min

At this point, you can start using larger increments, for example (in minutes): 30, 45, 20, 45, 30, 50, 10, 60, 45, 15, 5, 45, 10, 60, 20, and 75. There is no set schedule for this desensitization process. You may have to repeat one step many times or make up your own randomized schedule around a given time frame.

Do not hesitate to contact an animal behavior clinic if you have questions. Not all dogs are responsive to graduated departures, but with some work you can make a difference.

APPENDIX 2

Animal Ownership, History, and Behavior Records

ANIMAL BEHAVIOR CLINIC
College of Veterinary Medicine
Cornell University
Ithaca, NY 14853-6401
Fax (607) 253-3846

Canine Behavioral History

General Information

Date: _____   Clinic # _____

Recorder: _____

Client's name _____   Name of pet:_____

Address: _____   Breed: _____

_____   Date of Birth: _____

Home phone: _____   Sex: _____ neutered/spayed _____

Work/Day phone:_____

Who is your regular veterinarian?

Dr. _____

Clinic Name: _____

Address: _____

_____

Phone: _____

1. Main problem is: a) Aggression
                        b) Destructive
                        c) Barking
                        d) Phobia
                        e) House soiling

2. Secondary problem(s) is (are):   a) Aggression
                                      b) Destructive
                                      c) Barking
                                      d) Phobia
                                      e) House soiling

3. How frequently does the problem occur:

a) Number of times daily: _____

b) Number of times weekly:_____

c) Number of times monthly: _____

Chronology of the Behavior Problem

4. Age of onset?
    a) <6 months
    b) 6 months –1 year
    c) 1–2 years
    d) >2 years

Describe several examples in detail including the time, date, place, victim (if aggression), people present and owner's response:

1. Most recent incident: (Date:)

2. Second to last incident: (Date:)

3. Third to last incident: (Date:)

Other significant incidents:

What have you done so far to try to correct the problem?
    a) Punished dog verbally
    b) Punished dog by striking
    c) Turned dog on his back
    d) Treated with drugs
    e) Consulted a behaviorist

Home Environment

Indicate the number of each type of person in the household:
- a) _____adults
- b) _____children <5 years
- c) _____children 5–10 years
- d) _____children 10–20 years
- e) _____elderly >70 years

Indicate other dogs in the household (provide name, age, breed, and sex):

Indicate other animal species (provide name, species, and age):

What is your dog's relationship to the other animals?
- a) Friendly
- b) Hostile
- c) Fearful

What type of area do you live in?   a) City/Town   b) Suburbs   c) Rural

What type of dwelling do you live in?   a) House   b) Apartment

Dog's Background

Reasons for obtaining this dog?
- a) Companion
- b) Protection
- c) Work
- d) Other

Where did you get this dog? a) SPCA   b) Breeder – referral   c) Pet store
d) Friend   e) Stray

How many other owners has this dog had?   a) 0   b) 1   c) >1

At what age was your pet neutered/spayed?
   a) <6 months
   b) 6 months – 2 years
   c) >2 years

Diet and Feeding

What do you feed your dog?
   a) Dry food
   b) Canned food
   c) Semimoist
   d) Dry and canned

Daily feeding schedule: a) 1 meal   b) 2 meals   c) 3 meals   d) free choice

Who feeds the dog?
   Location _____

What is your dog's favorite treat? _____

When does your dog receive treats?   a) Schedule   b) As a reward

Daily Schedule – Typical 24-hr Day

Please describe a typical 24-hour day in your dog's life:

How does the dog behave with <u>familiar</u> visitors?
   a) No response
   b) Wags tail
   c) Jumps up
   d) Barks
   e) Barks and lunges

How does the dog behave with <u>unfamiliar</u> visitors (children or adults)?
   a) No response
   b) Wags tail
   c) Jumps up
   d) Barks
   e) Barks and lunges

Please describe your dog's daily exercise, including the amount of time of each:
   1. Leash walks   _____
   2. Supervised <u>walks</u> off leash _____
   3. Unsupervised free roaming   _____
   4. Loose in yard   _____
   5. Playing outdoors   _____
   6. Playing indoors   _____
   7. Other (describe:)   _____

Total <u>active time</u> (circle one):   0 min   15 min   30 min   1 hr   >1 hr

How many hours per day does your dog spend outdoors (outside house)?
   a) 0   b) <1   c) 1–11   d) 12   e) >12

How do you play with your dog?
   a) Pet
   b) Throw toys
   c) Wrestle
   d Tug o' war

What toys does the dog have?
   a) None
   b) Throw toys
   c) Chew toys
   d) Other

Does your dog ever eliminate (urinate or defecate) in the house?
                        ___Yes   ____No
   If yes, does your dog: a) Urinate   b) Defecate   c) Both

Where does your dog sleep at night:
   a) Outside
   b) In crate
   c) Free in house
   d) In bed

Where is your dog when alone in the house?

Where is your dog when you have guests?

How does your dog behave while you are leaving the house?
   a) No response
   b) Tail and ears down
   c) Whines
   d) Barks

How does your dog behave when you return?
   a) No response
   b) Wags tail
   c) Barks
   d) Jumps up
   e) Very excited for 25 minutes

Obedience Training

What basic obedience training has your dog had?

   a) None   b) Trained at home   c) Started obedience classes but didn't
                                          finish
   d) Graduated obedience class   e) Private trainer

How old was the dog when obedience training started?

Who in the family is the primary trainer?

What <u>percent</u> of the time does your dog obey the following commands, for <u>each member</u> of the family:

Family Member	a) Sit	b) Down	c) Stay	d) Heel	e) Heel (don't pull)
	100%	100%	100%	100%	100%
	50%	50%	50%	50%	50%
	0%	0%	0%	0%	0%

Does your dog jump up on you or others without permission?
                                                     ____Yes ____No

Does your dog paw at you or at others?   ____Yes ____No

Does your dog lick you?   ____Yes ____No

Does your dog mount people?   ____Yes ____No

If yes, whom does he or she mount?

Does your dog mount other animals or objects? ____Yes   ____No
Please describe:

Does your dog ever bark at you? ＿＿＿No ＿＿＿Yes. When? Please describe:

Does your dog bark at other times? Please describe:

What is your dog's activity level in general: a) Low   b) Average   c) High
d) Excessive

Medical History

Is your dog on any medication now, for this or other problems?

Has your dog been on medication in the past?

Date of most recent rabies vaccination: ＿＿＿＿＿＿＿＿＿＿＿＿＿(1 year, 3 year)

AGGRESSION SCREEN
Animal Behavior Clinic
Cornell University

GR –growl
SL –snarl/bare teeth
SB –snap/bite
NR –no reaction
NA –not applicable

Owner: _____
Pet: _____
Date _____

	GR	SL	SB	NR	NA
1. Pet dog					
2. Hug dog					
3. Kiss dog					
4. Lift dog					
5. Call off furniture					
6. Push/pull off furniture					
7. Approach on furniture					
8. Disturb while resting/sleeping					
9. Approach while eating					
10. Touch while eating					
11. Take dog food away					
12. Take human food away					
13. Take water dish away					
14. Take rawhide					
15. Take biscuit/cookie					
16. Take real bone					
17. Take toy/object					
18. Approach when dog has any object/toy/bone					
19. Verbally punish					
20. Physically punish					
21. Visual threat					
22. Speak to dog (normal tone)					
23. Stare at dog					
24. Bend over dog					
25. Push on shoulders or back					

	GR	SL	SB	NR	NA
26. Approach dog near spouse					
27. Enter room					
28. Leave room					
29. Reach toward dog					
30. Leash restraint					
31. Collar restraint					
32. Scruff restraint					
33. Put leash on/take off					
34. Put collar on/take off					
35. Bathe dog					
36. Towel dog					
37. Groom/brush dog					
38. Dog at groomer's					
39. Trim nails					
40. Leash/collar correction					
41. Response to "sit"					
42. Response to "down"					
43. Dog at veterinary clinic					
44. Unfamiliar adult enters house or yard					
45. Unfamiliar child enters house or yard					
46. Familiar adult enters house or yard					
47. Familiar child enters house or yard					
48. Response to toddlers/babies					
49. Dog in car at tollbooths, gas stations					
50. Unfamiliar adult approaches owner, dog on leash					
51. Unfamiliar child approaches owner, dog on leash					
52. Dog in house, sees people outside					
53. Response to other dogs, while on leash					
54. Response to other dogs, while not on leash					

Where are you on a scale of 1 to 5 as follows:

1. I am here only out of curiosity—problem is not serious.
2. I would like to change the problem, but it is not serious.
3. The problem is **serious** and I would like to change it, but if it remains unchanged that's all right.
4. The problem is **very serious** and I would like to change it, but if it remains unchanged I will keep my dog.
5. The problem is **very serious** and I would like to change it; if it remains unchanged, I will have my dog euthanized or give him/her up.

FOR AGGRESSION (TOWARD PEOPLE) (Skip this section if aggression is not the problem):

Aggression is directed at (circle all that apply):
  a) Owner
  b) Other people in family
  c) Other people outside the family
  d) Other dogs
  e) Other animals

Can you predict when your dig will be aggressive? _____Yes   _____No

Please answer yes or no to these characteristics of your dog's aggressive behavior:

  _____Attacks are sudden and surprising
  _____Episodes appear unprovoked
  _____The dog is abruptly docile after an episode
  _____The dog appears "sorry" afterward
  _____The dog appears disoriented afterward
  _____Episodes are associated with a "glazed" or "absent" expression
  _____I can usually tell what will set off my dog
  _____The aggressive behavior is new and uncharacteristic

Has your dog bitten and broken skin? _____Yes _____No

Number of bites that broke skin:_____

Total number of bites (that <u>did or did not</u> break skin):_____

Total number of episodes of aggression (growling, snapping, biting):
_____

Describe typical episode (e.g., does dog growl, lunge or bite, and in what circumstance?):

If the dog is in the above situation 10 times, in how many of those times is aggression seen (e.g., all=100%, just one=10%)?

What parts of the body has the dog bitten?
   a) Feet or legs
   b) Hands or arms
   c) Face
   d) Buttocks or back

Did your dog bite as a puppy? _____Yes _____No
If yes, please describe, including age:

How old was your dog the first time he/she growled at a person?

What was the circumstance?

How old was your dog the first time he/she snapped or bit at a person?

What was the circumstance?

ANIMAL BEHAVIOR CLINIC
College of Veterinary Medicine
Cornell University
Ithaca, NY 14853-6401
Fax (607) 253-3846

Cat Owner Questionnaire

General Information

Client's Name: _____

Address: _____

_____

Home phone: (   ) _____

Work/Day phone: _____

Name of pet:_____

1. Breed: a)___DSH   b)___Persian   c)___ Abyssinian
          d)___Siamese   e)___Japanese bobtail   f)___ Other

2. Age: a)___ <1 yr   b)___2–5 yrs   c)___5–10 yrs
         d)___ >10 yrs

3. Sex: a)___Intact male   b)___Intact female
         c)___Neutered male   d)___Spayed female

4. Who referred you to us? _____

5. Who is your regular veterinarian?
          Dr._____
          Clinic Name: _____
          Address: _____
          _____
          Phone: (   ) _____

Behavior Problem

6. Main behavior problem (mark those that apply):
     a)____House soiling   d)____ Aggression toward people
     b)____Fighting   e)____ Scratching furniture
     c)____Meowing   f)____ Eating wool/nonfood
     g)____Other (describe)_____

7. How old was the cat when you first noticed the problem? _____

8. How frequently does the problem occur? _____

9. Describe several examples in detail (give time, location, and people/animals present):
Most recent incident and date:

Second to last incident and date:

Third to last incident and date:

Other significant incidents:

10. What have you done to correct the problem?

11. How do you discipline your cat for this problem?
    a)____ Yell   b)____ Rub nose in mess   c)____Hit
    d)____ Isolate   e)____ Spray bottle   f)____Do nothing

Elimination Behavior

12. Does your cat use a litter pan?   ____Yes   ____No
13. Does cat ever eliminate in the house but outside the litter pan?
                                  ____Yes   ____No
14. If yes, does your cat use it to   a)____urinate, b)____defecate,
    or (c)____both
15. Does your cat spray?   ____Yes   ____No
16. How many litter pans do you have? ____
17. Locations of litter pans?

18. What kind of litter pans are they (indicate kind and number of each)?
    a) Commercial litter pan _____
    b) Commercial pan with removable "lip" _____
    c) Covered box, "cave"-type front door _____
    d) Covered box, "Booda"-type (cat crawls into hole) _____
    e) Dishpan_____
    f) Cardboard box _____
    g) Other _____
19. Do you use a liner?   ____Yes   ____No

20. What type of litter is used?
        a)____ Clay
        b)____ Scented clay
        c)____ Clumping
        d)____ Scented clumping
        e)____ Other
21. How often is the litter scooped?
22. How often is the litter replaced?
23. Does the cat cover urine in the box?   ____Yes   ____No
24. Does the cat cover feces in the box?   ____Yes   ____No

Diet and Feeding

25. What do you feed your cat?
        a)____ Dry food only   c)____Canned food only
        b)____ Dry and canned food

26. How often is your cat fed?
        a)____Free choice   b)____1 meal/day   c)____2 meals/day
        d)____3 meals/day
27. Where does your cat drink?
        a)____Water bowl   b)____Faucet   c)____Toilet   d)____Other
28. Who feeds the cat?
29. What is your cat's favorite treat?

Home Environment

30. Indicate the number of each type of person in the household:
        a)____Adults
        b)____Children <5 years
        c)____Children 5–10 years
        d)____Children 10–20 years
        e)____Elderly >70 years
31. Indicate the number of other animals in the household:
        a)____Cats
        b)____Dogs
        c)____Other
32. Give name and age of other cats:

33. Your cat's relationship to other cats in the household is best described as:
        a)____Friendly   b)____Hostile   c)____Neutral
        d)____Fighter   e)____Hider

Daily Schedule

34. How do you play with your cat? a)____Pet   b)____Throw toys
        c)____Play fight d)____Pull toys
35. Type of toys? a)____Ball   b)____Stuffed toys   c)____Hanging toys
36. Does your cat go outside?   ____Yes   ____No

Social Behavior

37. How does your cat respond to cats seen out a window or in the yard?
    a)____ Neutral
    b)____ Hisses
    c)____ Caterwauls
    d)____ Attacks

38. When does your cat meow?   a)____To be fed   b)____For attention
39. How would you describe your cat's personality (mark all that apply)?
    ____Shy   ____Fearful   ____Loving   ____Playful
    ____Bold   ____Pushy   ____Friendly

40. How does your cat react to visitors?
    a)____ Jumps in lap
    b)____ Bunts or rubs
    c)____ Stays in room but does not approach
    d)____ Hides
    e)____ Other

Sexual Behavior

41. At what age was your cat spayed/neutered?
42. Does your cat mount other cats?   ____Yes   ____No
43. Other animals?   ____Yes   ____No
44. People?   ____Yes   ____No
45. Objects?   ____Yes   ____No

Medical History

46. Has your cat had any illnesses?

47. Is your cat currently on medication? If so, what kind?

48. Has your cat been on medication in the past? If so, what kind?

Date_____

ANIMAL BEHAVIOR CLINIC

**College of Veterinary Medicine
Cornell University
Ithaca, New York 14853-6401
(607) 253-3846 Fax**

BEHAVIORAL HISTORY FORM FOR HORSES

Owner's	Veterinarian's	Horse's
Name _____	Name _____	Name _____
Address _____	Address _____	Breed _____
_____	_____	Sex, Age _____
Telephone _____	Telephone _____	Color _____

A. Main behavior problem:

   1. Chief complaint?

   2. When did problem begin?

   3. When does horse misbehave? How often and under what circumstances?

   4. Has there been a change in the frequency of appearance of the problem?

   5. What has been done so far to correct the problem?

B. Horse's environment:

   1. Type of housing (stall, pasture, run-out shed)?

   2. Diet?

   3. Exercise (hrs per wk ridden, hrs per wk in paddock, type of bit used, martingale)?

4. Other horses in environment and relations between horses (friendly, aggressive, neutral)?

5. Other animals in environment?

C. Early history:

1. Why was horse obtained? Is it still used for this purpose?

2. Source of animal?

3. Age at weaning?

4. Age when obtained by present owner?

5. Were there previous owners?

6. Do related horses have similar problems?

D. Education:

1. Age at halter breaking?

2. Method of training to saddle or harness, age when training began?

3. Other types of training methods (Circle all that apply):

   Driving   Jumping   Dressage   Games   Trail riding   Cutting

E. Other behavior problems:

1. Shying, how often and at what? Any other phobias?

2. Head shy?

3. Resentful of grooming?

4. Aggression toward humans or animals (dogs, cows, etc.)?

5. Aggression toward other horses (threatens, strikes, bites, kicks, chases)?

6. Misbehavior under saddle (circle appropriate behavior):

    Moves while rider mounts   Backs in harness   Bucks   Rears

    Wants to lead or will only follow other horses   Runs away

    Slow to leave and quick to return to barn   Hard to keep on right or left

    Other

7. Barn vices (circle appropriate one):

    Cribs   Chews wood   Paws   Kicks stall

8. Sexual behavior:

    Excessive   Inadequate   Abnormal?

9. Maternal behavior:
    Excessive   Inadequate   Abnormal?

F. Physical history:

    1. Present medical problems?

    2. Past medical problems?

    3. Drug history?

    4. Results of diagnostic tests?

G. Stallion behavior questions:

    1. Was he raised alone or in the company of other foals?

    2. Was he ever punished for acting like a stallion (studdish)?

    3. Have stallion rings or a brush to prevent masturbation been used in the past or present?

    4. How many mares has he bred? How many have you attempted to have bred?

5. Has he ever been allowed to breed an unrestrained mare?

6. Has he watched other stallions breed?

7. How many mares have been used to "tease" him?

8. Does he engage in the following (circle all that apply):
    Flehmen (horse laugh)?   Lick the mare's legs or hindquarters?
    Attempt to mount?   Mount?   Intromit?   Ejaculate?

9. Has he ever been used as a "teaser"?

10. Does he attempt to bite or kick the mare before or after breeding?

CLINIC # _____
DATE _____

ANIMAL BEHAVIOR CLINIC
College of Veterinary Medicine
Cornell University
Ithaca, New York 14853-6401
Fax (607) 253-3846

HISTORY FORM FOR FOAL REJECTING MARES

Owner's	Veterinarian's	Horse's
Name _____	Name _____	Name _____
Address _____	Address _____	Breed _____
_____	_____	Sex, Age _____
Telephone _____	Telephone _____	Color _____

1. Is this the mare's first foal?
   If not, how long has it been since she last foaled?
   How many foals has she had?
   How has she behaved toward previous foals (has she rejected a foal before)?

2. Did her dam show normal maternal behavior?
   Toward the mare?
   Toward other foals?

3. Date and hour of foal's birth?

4. Was the birth observed?

5. Was the veterinarian present?

6. Was the birth assisted (that is, was the foal pulled)?

7. Were there any complications during parturition?

8. Where did foaling occur?
   If outside, were other horses present?

9. If in a stall, was the stall rebedded after parturition?
   If yes, how long after the foal's birth?

10. Were the placenta and fetal membranes (afterbirth) removed?

11. How soon after the foal's birth did anyone enter the stall and for how long?

12. Was the foal's umbilicus treated with iodine?
    If so, when?

13. Was an enema administered?
    If so, when?

14. How long before the mare stood after parturition?

15. How long until the foal stood after parturition?

16. How long until the foal first nursed?

17. Was the nursing assisted?

18. Were there other horses in the barn?
    If so, were they visible to the mare?

19. How many people visited the mare and foal during the first 24 hours?

20. How many people entered the stall and how often was the stall entered during parturition?
    During the first 24 hours?

21. How did the mare react toward humans during parturition?
    During the first 24 hours?

22. How does the mare react to dogs?
    To other foals?

23. Describe in detail the mare's behavior toward the foal.

24. What treatment has been attempted (twitching, tranquilizers)?

DATE _____

ANIMAL BEHAVIOR CLINIC
College of Veterinary Medicine
Cornell University
Ithaca, New York 14853-6401
(607) 253-3846 Fax

BEHAVIORAL HISTORY FORM FOR HORSES
WITH TRAILER PROBLEMS

Owner's   Veterinarian's   Horse's
Name _____   Name _____   Name _____

Address _____   Address _____   Breed _____

_____   _____   Sex, Age _____

Telephone _____   Telephone _____   Color _____

---

1. What is main problem?

2. Why is horse trailered?

    a. To show
    b. To trail ride
    c. To race
    d. To hunt
    e. Other

3. When did problem begin?

4. Did horse previously trailer well?

5. When does problem occur?

    a. Loading?
    b. Traveling?
       1) On curves
       2) When braking
       3) As soon as trailer moves
       4) After traveling ____minutes or ____miles
    c. Unloading?
    d. Standing in stationary trailer?
    e. Other?

6. Does the problem occur:

   a. On way to event?
   b. On way home?
   c. Both?
   d. Other?

7. Does change of driver or trailer have any effect?

8. Type of trailer:

   Brand _____ Model _____ Year _____ Height, Width _____

   a. One horse
   b. Two horse
      1) Can partition be removed?
      2) Does partition reach floor?
      3) Can horse turn neck to face out?
   c. Gooseneck
   d. Van
   e. Livestock
   f. Other

9. Type of entrance – exit:

   a. Ramp
   b. Walk through
   c. Step down
   d. Other

10. Type of flooring?

11. Does horse face:

    a. Direction of travel?
    b. Opposite to direction of travel?
    c. Sideways?
    d. Other?

12. Does trailer have doors:

    a. In front?
    b. One or both sides?

13. Is trailer lighted inside?

14. Are there side or front windows?

15. Has the horse been injured while traveling or in loading or unloading?

16. Has there been an accident while horse was in trailer?

17. History of problem. Describe last three instances in detail.

18. What treatments have been used?

   a. Tranquilizer
   b. Cattle prod
   c. All feeding in trailer—horse must enter to eat for 1 week
   d. Other

19. Did sire, dam, and siblings:

   a. Travel well?
   b. Have similar problems?
   c. Unknown?

APPENDIX 3

Canine and Ovine Brains

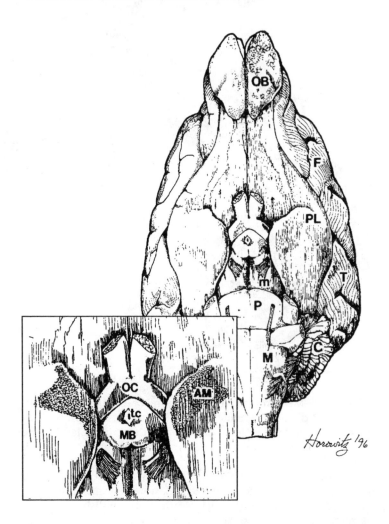

APPENDIX 3A. Ventral view of the canine brain with pituitary gland removed at the junction of infundibulum and tuber cinereum. *Inset* is an enlargement of the hypothalamic region. AM, amygdaloid body (its position superimposed on the piriform lobe); C, cerebellum; F, frontal lobe of the cerebral hemisphere; M, medulla oblongata; m, midbrain; MB, mammillary body; OB, olfactory bulb; OC, optic chiasm; P, pons; PL, piriform lobe; T, temporal lobe of the cerebral hemisphere; tc, tuber cinereum. (From A. Horowitz, unpublished, reprinted with permission.)

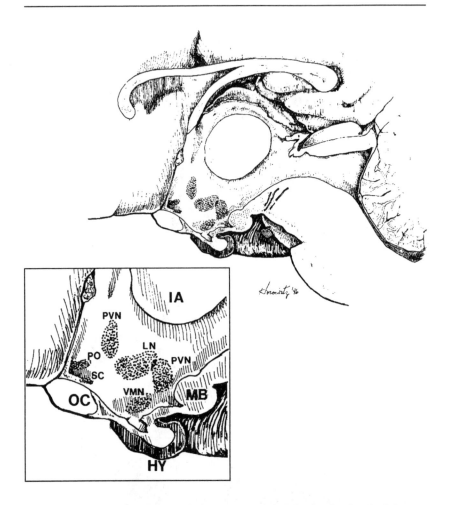

APPENDIX 3B.   Median section of the canine brain at the level of the third ventricle and mesencephalic aqueduct, with the hypophysis intact. The tuber cinereum is unlabeled. The *inset* is of the hypothalamic region and shows the relative positions of specific hypothalamic nuclei. HY, hypophysis (pituitary); IA, interthalamic adhesion; LN, lateral nucleus; MB mammillary body; OC, optic chiasm; PO, medial preoptic nucleus; PVN, rostral and caudal periventricular nuclei; SC, suprachiasmatic nucleus; VMN, ventromedial nucleus. (From A. Horowitz, unpublished, reprinted with permission.)

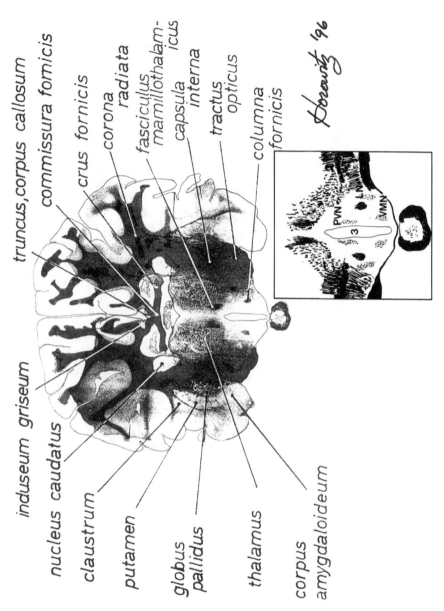

APPENDIX 3C. Transverse section of the ovine brain at the level of the hypothalamus. *Inset* shows the relative position of three of the hypothalamic nuclei shown in median section in Appendix *3B*. The designation of the nuclei is the same as in *3B*; the lower part of the third ventricle is designated "3." Note the positions of lateral and ventromedial nuclei with respect to the column of the fornix (*columna fornicis*). (From A. Horowitz, unpublished, reprinted with permission.)

GLOSSARY

achondroplasia—shortened limbs due to genetic defect in conversion of cartilage to bone

ACTH—(adrenocorticotropic hormone) secreted by the anterior pituitary and acts primarily on the adrenal center to stimulate production of corticosteroids, e.g., cortisol

ad libitum—free choice

adrenocortical—pertaining to the cortex of the adrenal gland

aliphatic—pertaining to a fatty series of hydrocarbons with an open or straight chain of carbons, as opposed to alicyclic acids, which have ring structures

allogroom—groom another animal

analgesia—insensitivity to pain without loss of consciousness

androstenol—a metabolite of testosterone found in the saliva, preputial fluid, and tissues of boars

anestrus—a period of sexual inactivity between two estrous cycles

anosmia—lack of sense of smell

anoxia—lack of oxygen

antigonadotropic—a chemical that inhibits gonadotropin release or action

apnea—not breathing

artificial insemination—(AI) process by which semen obtained from a donor animal is deposited within the vagina or cervix of the recipient female

atonia—lack of muscle tone

balanoposthitis—inflammation of the glans penis and prepuce

barrier frustration—the behavior of dogs, usually in the owner's absence, in which the dog claws or bites at exits, such as doors and windows, and may escape

behavior modification—changing an animal's behavior by application of learning principles such as habituation and counterconditioning

bigeminal—twin; occurring in twos

break point—the highest number of responses that an animal will make for a particular reward. The break point would be higher for a preferred food reward than for a nonpreferred one.

buccostomy—the surgical creation of a permanent buccal (oral) fistula

bulbectomy—cutting the connection between the olfactory bulb and the rest of the brain (see Appendix 3)

buller steer—a steer who attracts other steers to mount him

bunting
—in cats, an affiliative marking behavior of rubbing the side of the head on objects, cats, or people
—in cattle, an agonistic behavior in which one animal strikes the other with its head

catalepsy—a condition characterized by waxy rigidity of the limbs, which may be placed in various positions that are maintained for a time

central pattern generator—a neural circuit that controls motor movements and which is independent of higher centers; walking is an example

cerebral ventricles (see Appendix 3)—cavities within the cerebrum that are filled with cerebrospinal fluid

cerebrospinal fluid—fluid that bathes the brain and spinal cord

cervical—pertaining to the neck or to the neck of any organ or structure, for example, cervical vertebrae or cervix of the uterus

cholinergic—acting like acetylcholine, the neurotransmitter released by the parasympathetic nerves and at the nerve-muscle junction

choroid—a vascular membrane containing large pigmented cells that lies between the retina and the sclera or outer coat of the eye

circadian rhythm—cycles of approximately 24 hours that are endogenously generated by an organism

circannual cycle—seasonal, as in the breeding season of horses

circatrigentian rhythms—cycle of approximately 30 days

cochlear—pertaining to the cochlea, a spirally wound tube similar in form to a snail shell, which forms part of the inner ear

concaveation—acceptance of an offspring after a period of forced association

conditioned response—a response that becomes associated with a previously unrelated stimulus through repeated presentation of the stimulus simultaneously with a stimulus that normally elicits the response

conditioned stimulus—a stimulus that will elicit a response after repeated simultaneous presentations with a stimulus that normally elicits the response

conditioned taste aversion—avoidance or dislike of the taste of a food previously associated with illness, especially nausea

conspecifics—animals of the same species

Coolidge effect—stimulation of male libido by a novel female

coprophagia—eating feces

copulatory lock or tie—the condition that occurs in dogs after intromission. The male cannot withdraw the penis because the vagina has constricted around it.

corpus luteum—the structure remaining in the ovary after rupture of a follicle or ovulation. It produces progesterone.

corticosteroid—steroid hormones produced by the adrenal cortex

corticosterone—a steroid hormone produced by the adrenal cortex

cortisol—a steroid hormone produced by the adrenal cortex

courting grunts—the *chant de coer* of the boar; may stimulate the immobility response of a sow

diabetes insipidus—a lack of antidiuretic hormone or vasopressin, causing elimination of a large volume of dilute urine

diabetes mellitus—a lack of insulin-causing hyperglycemia (high blood sugar) and glucosuria (a large volume of urine containing glucose)

diazepam (Valium)—a benzodiazepine, hypnotic sedative

diurnal—during the daylight hours

dopamine—a neurotransmitter found in the brain

-ectomy—removal

electroencephalogram (EEG)—measurements of the electrical activity of the brain measured through needles inserted in the scalp

emetic—a substance that stimulates vomiting

endogenous—produced or synthesized within the animal, as opposed to exogenous, introduced from or produced outside

endometrial epithelium—the lining of the uterus

epimeletic—caregiving behavior

estradiol—a steroid hormone produced by the ovary

estrogen—a steroid hormone produced by the ovary

et-epimeletic—care-soliciting behavior

false heat—behavioral signs of estrus or heat without physiological signs

flehmen—lip curling; turning up of the upper lip by ungulates in response to an odor; usually accompanied by dorsiflexion of the neck

flunixin meglumine—a nonsteroidal anti-inflammatory drug used to treat pain in horses

fluoxetine (Prozac)—a specific serotonin reuptake blocker used to treat obsessive compulsive behaviors and aggression

follicle-stimulating hormone (FSH)—a gonadotropin secreted by the anterior pituitary that stimulates the release of estrogen from the ovary in the female and sperm production in the male

foundered—a horse afflicted with laminitis, a condition in which the third phalanx separates from the hoof wall

freemartin cow—female cow born twin to a male

fundamental frequency—lowest frequency at which a sound can be perceived

gastritis—inflammation of the stomach

glycogen—the chief carbohydrate storage material in animals

gustatory—referring to the sense of taste

gyrus (plural, gyri)—the raised protuberances on the brain (see Appendix 3)

habituation—a lack of responsiveness of the central nervous system to a stimulus of long duration

hackles—hair that can be raised on the neck and back

high frequency—above the range of human hearing; about 30 Khz

hypermetropic—farsighted

hyperthermia—increased body temperature; fever

hypophysis—pituitary

hypothermia—condition of lowered body temperature

impotence—lack of ability to copulate due to failure to initiate an erection or to maintain an erection

imprinting—a species-specific rapid type of learning during a critical period of early life in which social attachment and identification are established

infradian—cycles of rhythms less frequent than every 24 hours

inguinal—situated in the groin, the junction between the hind leg and the abdomen

inhibin—a steroid hormone produced by Sertoli cells of the testes that inhibits production of follicle-stimulating hormone intromission—insertion of the penis into the vagina during copulation

ischiocavernosus—a muscle that originates on the ischium of the pelvis and inserts on the corpus cavernosus of the penis

ketamine—a dissociative anesthetic

leptin—ob receptor protein; the signal from fat cells to the brain that inhibits food intake and stimulates heat production

leptospirosis—a bacteria disease caused by *Leptospira*

lissencephaly—an abnormality of the brain whereby there are no gyri or sulci, i.e., the brain is smooth and, therefore, has a smaller surface area

luteinizing hormone (LH)—a hormone produced by the anterior pituitary that stimulates progesterone production and formation of the corpus luteum in the female and testosterone production by the Leydig cells of the testes of the male

macrosmatic—having a better sense of smell than humans

Magnus reflex—turning the head to one side results in flexion of the limb on that side and extension of the opposite limb

masseter muscle—major muscle closing the jaw

masturbation—self-stimulation of the penis; stallions strike the erect penis against their bellies; bulls thrust with the pelvis

metritis—inflammation of the uterus

microfilaria—the prelarval stage of Filarioidea—small filaria found in blood

micturition—urination

muscle guarding—contraction of the muscles, usually of the abdomen to protect the animal from painful manipulations

myectomy—removal of the muscle

   ischiocavernosus myectomy—removal of the ischiocavernosus muscle that originates on the ischium of the pelvis and inserts on the corpus cavernosus of the penis

myelin—substance of the cell membrane of Schwann's cells that coils to form the myelin sheath around nerves outside the brain

myoepithelial—pertaining to tissue composed of contractile epithelial cells

myopic—nearsighted

myotomy—the dissection of muscular tissue

narcolepsy—a sleep disorder characterized by uncontrolled loss of muscle tone (cataplexy)

navicular disease—lameness caused by pathological changes in the navicular bone, a boat-shaped sesamoid bone located posterior to the second and third phalanges in the hoof

nepetalactone—the chief constituents of the aromatic volatile oil from the leaves and tops of *Nepeta cataries* (catnip)

neurectomy—the excision of part of a nerve

nictitating membrane—the third eyelid attached medially

nyctohermerol—both daily and nightly

nymphomania—exaggerated sexual desire in the female

olfactory bulbectomy—removal of the olfactory bulb

omohyoideus—a muscle in the neck of the horse originating on the fascia of the shoulder and attaching to the hyoid bone

onychectomy—removal of the nail; declawing

operant conditioning—learning to perform a behavior, operate on the environment, in order to receive a reinforcement or reward

orchiectomy—excision of one or both testes

otic—pertaining to the ear

otitis—inflammation of the internal or external meatus (ear canal) but not affecting the eardrum or inner ear

-otomy—cutting

ovariectomy—excision of one or both ovaries

oxytocin—hormone produced in the hypothalamus and stored in the posterior pituitary; causes uterine contractions and milk letdown

parabiotic—surgically joined animals

parasympathetic—division of autonomic nervous system

pathognomonic—distinctly characteristic of a particular disease

periventricular—around a ventricle

phenylbutazone—analgesic, antipyretic, anti-inflammatory drug

pheromone—a chemical substance secreted by one individual that has a behavioral and/or physiological affect on another animal

pherphenazine—long-acting (weeks' duration) phenothiazine tranquilizer

piloerection—hair standing on end

pinnae—the large projecting cartilaginous part of the external ear

pituitary or hypophysis—the gland located at the base of the brain that produces a variety of hormones. The anterior pituitary or adenohypophysis produces and releases adrenocorticotropic hormone, thyrotropic hormone, follicle-stimulating hormone, luteinizing hormone, and prolactin; the posterior pituitary or neurohypophysis stores and releases oxytocin and vasopressin.

polydipsia (Greek, "many drinks")—increased water intake

polypnea—breathing quickly

posterior pituitary—see *pituitary*

postprandial—after eating

preputial—pertaining to the prepuce, the fold of skin covering the penis

primiparous—giving birth to the first offspring

proceptive behavior—attraction of the female to the male

progesterone—a steroid hormone produced by the corpus luteum of the ovary

progestins—female hormones or synthetic ones having the action of progesterone, a steroid female sex hormone

progressive desensitization—exposure of an animal to gradual increase in the strength of a stimuli, beginning with a stimulus below the animal's threshold of response. The animal is rewarded for not responding. If done properly, the animal will not respond even when the stimulus is as strong as that which initially elicited fear or aggression.

prolactin—hormone of anterior pituitary that stimulates and sustains lactation

pruritus—itching

pseudopregnancy—false pregnancy

psychic heat—see *false heat*

pyometra—an accumulation of pus within the uterus

queen—a breeding female cat

ram effect—the accelerated onset of estrus in ewes by the addition of a ram or
his odor

ramped retina—a retina that is slanted rather than straight

ratio of response to reward

fixed—a given number of responses is always followed by a reward

variable—the reward is given after a variable number of responses, for example, after three, then after seven, and then after four

reinforcement—a reward, the giving of which causes the frequency of the response to a particular stimulus to increase

positive reinforcement—a reward contingent on the animal's making a response; positive reinforcements can be food

negative reinforcement—something aversive that will cease when the animal makes a response or performs a behavior. Electric shock or ear
pinch are common negative reinforcements. Not to be confused with
punishment, an aversive event that occurs contingent on the animal's
making a response

reticular activating system—an area in the midbrain and hindbrain that is involved in arousal from sleep

rooting—digging with the snout

scotopic—pertaining to night vision

self-mutilation—injuring of the animal by its own biting, scratching, or rubbing behavior

shaping—rewarding first a general approximation of the desired behavior, then
only rewarding behaviors that are closer to the desired behavior, and finally
rewarding only the desired behavior. The horse is rewarded for approaching
a panel, then for standing with its head close to the panel, and finally only
when he presses the panel.

silent heats—physiological heat or estrus without behavioral signs of heat or estrus

sinus arrhythmia—the increase in heart rate with inspiration

somatotropin—growth hormone

sonographic—pertaining to a sonogram; a sonogram is the plotting of the frequency of a sound versus time

split estrus—short periods of sexual receptivity interspersed with periods of
nonreceptivity

spraying—a feline behavior of urinating onto vertical objects. The cat twitches
his tail, which is held vertically while spraying.

stallion rings—bands placed around the penis, which will constrict the erect
but not the nonerect penis, that are used to inhibit masturbation or sexual
arousal in stallions

stereoscopic—pertaining to vision in which objects are seen in three dimen-
sions

stereotypy (plural, stereotypies)—a repetitious, apparently functionless, behavior such as cribbing in horses or tongue rolling in cattle

sternohyoideus—a muscle that originates on the sternum and shoulder and inserts on the hyoid bone

sternothyrohyoideus—a muscle that originates on the sternum and inserts on
the thyroid cartilage of the sternum

GLOSSARY

sulcus (plural, sulci)—the depressions in the brain (see Appendix 3)
suprachiasmatic nucleus—an area of the brain in the hypothalamus above the optic chiasm believed to be involved in circadian rhythms

tachycardia—slow heartbeat
tapetum—the iridescent pigment found in the choroidea of the eye of many mammals, including cats and dogs but not humans, which shines in the dark when a light reflects from it
terpenoid—pertaining to cyclic tertiary alcohol found in essential oils of plants
testosterone—principal steroid androgenic hormone produced by the Leydig cells in response to stimulation of luteinizing hormone
tetracycline—a group of antibiotics isolated from *Streptomyces* spp.
theophylline—a chemical stimulant related to caffeine
tractomy—cutting the tract
  olfactory tractomy—cutting the connection between the olfactory bulb and the rest of the brain

ultradian rhythm—cycles of less than 24 hours
unconditioned stimulus—a stimulus that naturally elicits a response. For example, dogs salivate when they see meat.
urolithiasis—formation of urinary calculi; kidney or bladder stones

vasopressin—stimulates contraction of blood vessels and retention of urine
vestibular—pertaining to the eighth cranial nerve, which carries sensory information from the inner ear and functions to maintain balance
vomeronasal organ—a paired tubular sensory organ lying between the hard palate and the nasal cavity on the midline

winking—rhythmic eversion of the clitoris of the mare, a sign of heat or estrus

zeitgeber (German, "time giver")—an environmental agent or event (the occurrence of light or dark being the most important one) that provides the stimulus that sets biological clocks

REFERENCES

1. Abraham, S., R. Baker, D. A. Denton, F. Kraintz, L. Kraintz, and L. Purser. 1973. Components in the regulation of salt balance: Salt appetite studied by operant behaviour. *Aust. J. Exp. Biol. Med. Sci.* 51:65–81.

2. Abraham, S. F., R. M. Baker, E. H. Blaine, D. A. Denton, and M. J. McKinley. 1975. Water drinking induced in sheep by angiotensin—a physiological or pharmacological effect? *J. Comp. Physiol. Psychol.* 88:503–518.

3. Adamec, R. E. 1976. The interaction of hunger and preying in the domestic cat (Felis catus): An adaptive hierarchy? *Behav. Biol.* 18:263–272.

4. Adamec, R. E. 1990. Individual differences in temporal lobe sensory processing of threatening stimuli in the cat. *Physiol. Behav.* 49:455–464.

5. Adamec, R. E. 1991. Anxious personality in the cat: Its ontogeny and physiology. In *Psychopathology and the Brain*. B. J. Carroll and J. E. Barrett, eds. New York, NY: Raven Press, Ltd, pp. 153–168.

6. Adamec, R. E., C. Stark-Adamec, and K. E. Livingston. 1983. The expression of an early developmentally emergent defensive bias in the adult domestic cat (Felis catus) in non-predatory situations. *Appl. Anim. Ethol.* 10:89–108.

7. Adams, D. K. 1929. Experimental studies of adaptive behavior in cats. *Comp. Psychol. Monogr.* 6:1–168.

8. Adams, G. J., and K. G. Johnson. 1993. Sleep-wake cycles and other nighttime behaviours of the domestic dog Canis familiaris. *Appl. Anim. Behav. Sci.* 36:233–248.

9. Adams, G. J., and K. G. Johnson. 1994. Behavioural responses to barking and other auditory stimuli during night-time sleeping and waking in the domestic dog (Canis familiaris). *Appl. Anim. Behav. Sci.* 39:151–162.

10. Adams, G. J., and K. G. Johnson. 1994. Sleep, work, and the effects of shift work in drug detector dogs Canis familiaris. *Appl. Anim. Behav. Sci.* 41:115–126.

11. Adams, G. J., and K. G. Johnson. 1995. Guard dogs: Sleep, work and the behavioural responses to people and other stimuli. *Appl. Anim. Behav. Sci.* 46:103–115.

12. Adams, T. 1963. Hypothalamic temperature in the cat during feeding and sleep. *Science* 139:609–610.

13. Adler, L. L., and H. E. Adler. 1977. Ontogeny of observational learning in the dog (Canis familiaris). *Dev. Psychobiol.* 10:267–271.

14. Agrawal, H. C., M. W. Fox, and W. A. Himwich. 1967. Neurochemical and behavioral effects of isolation-rearing in the dog. *Life Sci.* 6:71–78.

15. Alavi, F. K., J. P. McCann, A. Mauromoustakis, and S. Sangiah. 1993. Feeding behavior and its responsiveness to naloxone differ in lean and obese sheep. *Physiol. Behav.* 53:317–323.

16. Albright, J. L. 1969. Social environment and growth. In *Animal Growth And Nutrition*. E. S. E. Hafez and I. A. Dyer, eds. Philadelphia, PA: Lea & Febiger, pp. 106–120.

17. Albright, J. L., W. P. Gordon, W. C. Black, J. P. Dietrich, W. W. Snyder, and C. E. Meadows. 1966. Behavioral responses of cows to auditory training. *J. Dairy Sci.* 49:104–106.

18. Aldinger, S. M., V. C. Speer, V. W. Hays, and D. V. Catron. 1959. Effect of saccharin on consumption of starter rations by baby pigs. *J. Anim. Sci.* 18:1350–1355.

19. Aldis, O. 1975. *Play Fighting*. New York, NY: Academic Press.

20. Aldrich, C. G., M. T. Rhodes, J. L. Miner, M. S. Kerley, and P. Y. Paterson. 1993. The effects of endophyte-infected tall fescue consumption and use of a dopamine antagonist on intake, digestibility, body temperature, and blood constituents in sheep. *J. Anim. Sci.* 71:158–163.

21. Alexander, G. 1977. Role of auditory and visual cues in mutual recognition between ewes and lambs in Merino sheep. *Appl. Anim. Ethol.* 3:65–81.

22. Alexander, G., and L. R. Bradley. 1985. Fostering in sheep IV. Use of restraint. *Appl. Anim. Behav. Sci.* 14:355–364.

23. Alexander, G., and E. E. Shillito Walser. 1978. Visual discrimination between ewes by lambs. *Appl. Anim. Ethol.* 4:81–85.

24. Alexander, G., and E. E. Shillito. 1977. Importance of visual cues from various body regions in maternal recognition of the young in Merino sheep (Ovis aries). *Appl. Anim. Ethol.* 3:137–143.

25. Alexander, G., and E. E. Shillito. 1977. The importance of odour, appearance and voice in maternal recognition of the young in Merino sheep (Ovis aries). *Appl. Anim. Ethol.* 3:127–135.

26. Alexander, G., and E. E. Shillito. 1978. Maternal responses in merino ewes to artificially coloured lambs. *Appl. Anim. Ethol.* 4:141–152.

27. Alexander, G., J. P. Signoret, and E. S. E. Hafez. 1974. Sexual and maternal behavior. In *Reproduction In Farm Animals*. E. S. E. Hafez, ed. Philadelphia, PA: Lea & Febiger, pp. 222–254.

28. Alexander, G., and D. Stevens. 1981. Recognition of washed lambs by merino ewes. *Appl. Anim. Ethol.* 7:77–86.

29. Alexander, G., and D. Stevens. 1985. Fostering in sheep. III. Facilitation by the use of odorants. *Appl. Anim. Behav. Sci.* 14:345–354.

30. Alexander, G., D. Stevens, and L. R. Bradley. 1983. Washing lambs and confinement as aids to fostering. *Appl. Anim. Ethol.* 10:251–261.

31. Alexander, G., D. Stevens, and L. R. Bradley. 1985. Fostering in sheep. I. Facilitation by use of textile lamb coats. *Appl. Anim. Behav. Sci.* 14:315–334.

32. Alexander, G., D. Stevens, and L. R. Bradley. 1988. Maternal behaviour in ewes following caesarian section. *Appl. Anim. Behav. Sci.* 19:273–277.

33. Alexander, G., D. Stevens, R. Kilgour, H. de Langen, B. E. Mottershead, and J. J. Lynch. 1983. Separation of ewes from twin lambs: Incidence in several breeds. *Appl. Anim. Ethol.* 10:301–317.

34. Alexander, G., and D. Williams. 1964. Maternal facilitation of sucking drive in newborn lambs. *Science* 146:665–666.

35. Alexander, G., and D. Williams. 1966. Teat-seeking activity in lambs during the first hours of life. *Anim. Behav.* 14:166–176.

36. Algers, B., P. Jensen, and L. Steinwall. 1990. Behaviour and weight changes at weaning and regrouping of pigs in relation to teat quality. *Appl. Anim. Behav. Sci.* 26:143–155.

37. Algers, B., S. Rojanasthien, and K. Uvnäs-Moberg. 1990. The relationship between teat stimulation, oxytocin release and grunting rate in the sow during nursing. *Appl. Anim. Behav. Sci.* 26:267–276.

38. Allan, C. J., P. J. Holst, and G. N. Hinch. 1991. Behaviour of parturient Australian bush goats. I. Doe behaviour and kid vigour. *Appl. Anim. Behav. Sci.* 32:55–64.

39. Allen, B. D., J. F. Cummings, and A. de Lahunta. 1974. The effects of prefrontal lobotomy on aggressive behavior in dogs. *Cornell Vet.* 64:201–216.

40. Allin, J. T., and E. M. Banks. 1972. Functional aspects of ultrasound production by infant albino rats (Rattus norvegicus). *Anim. Behav.* 20:175–185.

41. Allison, T., and D. V. Cicchetti. 1976. Sleep in mammals: Ecological and constitutional correlates. *Science* 194:732–734.

42. Altmann, M. 1941. Interrelations of the sex cycle and the behavior of the sow. *J. Comp. Psychol.* 31:481–498.

43. Ames, D. R., and L. A. Arehart. 1972. Physiological response of lambs to auditory stimuli. *J. Anim. Sci.* 34:994–998.

44. Anand, B. K., and J. R. Brobeck. 1951. Hypothalamic control of food intake in rats and cats. *Yale J. Biol. Med.* 24:123–140.

45. Anand, B. K., S. Dua, and K. Shoenberg. 1955. Hypothalamic control of food intake in cats and monkeys. *J. Physiol.* 127:143–152.

46. Anderson, D. M., C. V. Hulet, S. K. Hamadeh, J. N. Smith, and L. W. Murray. 1990. Diet selection of bonded and non-bonded free-ranging sheep and cattle. *Appl. Anim. Behav. Sci.* 26:231–242.

47. Anderson, D. M., C. V. Hulet, J. N. Smith, W. L. Shupe, and L. W. Murray. 1987. Heifer disposition and bonding of lambs to heifers. *Appl. Anim. Behav. Sci.* 19:27–30.

48. Anderson, D. M., C. V. Hulet, J. N. Smith, W. L. Shupe, and L. W. Murray. 1992. An attempt to bond weaned 3-month-old beef heifers to yearling ewes. *Appl. Anim. Behav. Sci.* 34:181–188.

49. Anderson, D. M., and N. S. Urquhart. 1986. Using digital pedometers to monitor travel of cows grazing arid rangeland. *Appl. Anim. Behav. Sci.* 16:11–23.

50. Anderson, O. D., and R. Parmenter. 1941. A long-term study of the experimental neurosis in the sheep and dog. *Psychosom. Med. Monogr.* 2(3–4):1–150.

51. Anderson, R. S. 1974. Obesity in the dog and cat. In *The Veterinary Annual 1973.* C. S. G. Grunsell and F. W. G. Hill, eds. Bristol, UK: John Wright and Sons, pp. 182–186.

52. Andersson, B., O. Augustinsson, E. Bademo, J. Junkergard, C. Kvart, G. Nyman, and M. Wiberg. 1987. Systemic and centrally mediated angiotensin II effects in the horse. *Acta Physiol. Scand.* 129:143–149.

53. Andersson, B., L. Eriksson, and R. Oltner. 1970. Further evidence for angiotensin-sodium interaction in central control of fluid balance. *Life Sci.* 9(1):1091–1096.

54. Andersson, B., and S. M. McCann. 1955. A further study of polydipsia evoked by hypothalamic stimulation in the goat. *Acta Physiol. Scand.* 33:333–346.

55. Angel, C., O. D. Murphree, and D. C. De Lucia. 1974. The effects of chlordiazepoxide, amphetamine and cocaine on bar-press behavior in normal and genetically nervous dogs. *Res. Nerv. Syst.* 35:220–223.

56. Anonymous., 1975. Behaviour of boars. TDC article. *Vet. Rec.* 96:221.

57. Apple, J. K., and J. V. Craig. 1992. The influence of pen size on toy preference of growing pigs. *Appl. Anim. Behav. Sci.* 35:149–155.

58. Appleby, M. C., E. A. Pajor, and D. Fraser. 1991. Effects of management options on creep feeding by piglets. *Anim. Prod.* 53:361–366.

59. Appleby, M. C., E. A. Pajor, and D. Fraser. 1992. Individual variation in feeding and growth of piglets: Effects of increased access to creep food. *Anim. Prod.* 55:147–152.

60. Appleby, M. C., and G. M. Wood-Gush. 1988. Effect of earth as an additional stimulus on the behaviour of confined piglets. *Behav. Proc.* 17:83–91.

61. Araba, B. D., and S. L. Crowell-Davis. 1994. Dominance relationships and aggression of foals (Equus caballus). *Appl. Anim. Behav. Sci.* 41:1–25.

62. Arave, C. W., and J. L. Albright. 1976. Social rank and physiological traits of dairy cows as influenced by changing group membership. *J. Dairy Sci.* 59:974–981.

63. Arave, C. W., R. C. Lamb, M. J. Arambel, D. Purcell, and J. L. Walters. 1992. Behavior and maze learning ability of dairy calves as influenced by housing, sex and sire. *Appl. Anim. Behav. Sci.* 33:149–163.

64. Archer, M. 1973. The species preferences of grazing horses. *J. Br. Grassland Soc.* 28:123–128.

65. Archibald, J. 1974. *Canine Surgery*. 2nd ed. Santa Barbara, CA: American Veterinary Publications.

66. Arendt, J., A. M. Symons, C. A. Laud, and S. J. Pryde. 1983. Melatonin can induce early onset of the breeding season in ewes. *J. Endocrinol.* 97:395–400.

67. Arey, D. S., and M. F. Franklin. 1995. Effects of straw and unfamiliarity on fighting between newly mixed growing pigs. *Appl. Anim. Behav. Sci.* 45:23–30.

68. Arey, D. S., A. M. Petchey, and V. R. Fowler. 1991. The preparturient behaviour of sows in enriched pens and the effect of pre-formed nests. *Appl. Anim. Behav. Sci.* 31:61–68.

69. Arey, D. S., A. M. Petchey, and V. R. Fowler. 1992. The peri-parturient behaviour of sows housed in pairs. *Appl. Anim. Behav. Sci.* 34:49–59.

70. Arnold, G. W. 1966. The special senses in grazing animals. I. Sight and dietary habits in sheep. *Aust. J. Agric. Res.* 17:521–529.

71. Arnold, G. W. 1966. The special senses in grazing animals. II. Smell, taste, and touch and dietary habits in sheep. *Aust. J. Agric. Res.* 17:531–542.

72. Arnold, G. W. 1984. Comparison of the time budgets and circadian patterns of maintenance activities in sheep, cattle and horses grouped together. *Appl. Anim. Behav. Sci.* 13:19–30.

73. Arnold, G. W., C. A. P. Boundy, P. D. Morgan, and G. Bartle. 1975. The roles of sight and hearing in the lamb in the location and discrimination between ewes. *Appl. Anim. Ethol.* 1:167–176.

74. Arnold, G. W., and M. L. Dudzinski. 1978. *Ethology of Free-Ranging Domestic Animals*. Amsterdam, The Netherlands: Elsevier Scientific Publishing Co.

75. Arnold, G. W., and A. Grassia. 1982. Ethogram of agonistic behaviour for thoroughbred horses. *Appl. Anim. Ethol.* 8:5–25.

76. Arnold, G. W., and R. A. Maller. 1974. Some aspects of competition between sheep for supplementary feed. *Anim. Prod.* 19:309–319.

77. Arnold, G. W., and P. D. Morgan. 1975. Behaviour of the ewe and lamb at lambing and its relationship to lamb mortality. *Appl. Anim. Ethol.* 2:25–46.

78. Arnold, G. W., and P. J. Pahl. 1974. Some aspects of social behaviour in domestic sheep. *Anim. Behav.* 22:592–600.

79. Arnold, G. W., S. R. Wallace, and R. A. Maller. 1979. Some factors involved in natural weaning processes in sheep. *Appl. Anim. Ethol.* 5:43–50.

80. Arnold, G. W., S. R. Wallace, and W. A. Rea. 1981. Associations between individuals and home-range behaviour in natural flocks of three breeds of domestic sheep. *Appl. Anim. Ethol.* 7:239–257.

81. Arnould, C., V. Piketty, and F. Lévy. 1991. Behaviour of ewes at parturition toward amniotic fluids from sheep, cows and goats. *Appl. Anim. Behav. Sci.* 32:191–196.

82. Aronson, L. R., and M. L. Cooper. 1966. Seasonal variation in mating behavior in cats after desensitization of glans penis. *Science* 152:226–230.

83. Aronson, L. R., and M. L. Cooper. 1974. Olfactory deprivation and mating behavior in sexually experienced male cats. *Behav. Biol.* 11:459–480.

84. Aronson, L. R., and M. L. Cooper. 1977. Central versus peripheral genital desensitization and mating behavior in male cats: Tonic and phasic effects. *Ann. N. Y. Acad. Sci.* 290:299–313.

85. Asa, C. S., D. A. Goldfoot, and O. J. Ginther. 1979. Sociosexual behavior and the ovulatory cycle of ponies (Equus caballus) observed in harem groups. *Horm. Behav.* 13:46–65.

86. Asa, C. S., D. A. Goldfoot, and O. J. Ginther. 1983. Assessment of the sexual behavior of pregnant mares. *Horm. Behav.* 17:405–413.

87. Aschoff, J. 1965. *Circadian Clocks. Proceedings of the Feldafing Summer School*. Amsterdam, The Netherlands: North Holland Publishing Co.

88. Ashmead, D. H., R. K. Clifton, and E. P. Reese. 1986. Development of auditory localization in dogs: Single source and precedence effect sounds. *Dev. Psychobiol.* 19:91–103.

88a. Askew, H. R. 1996. *Treatment of Behavior Problems in Dogs and Cats. A Guide for the Small Animal Veterinarian.* Oxford, U.K.: Blackwell Science.

89. Atkeson, F. W., A. O. Shaw, and H. W. Cave. 1942. Grazing habits of dairy cattle. *J. Dairy Sci.* 25:779–784.

90. Auffray, P. 1969. Effets des lesions des noyaux ventro-medians hypothalamiques sur la prise d'aliment chez le porc. *Ann. Biol. Anim. Biochim. Biophys.* 9:513–526.

91. Auffray, P., and J.-C. Marcilloux. 1983. An analysis of feeding patterns in the adult pig. *Reprod. Nutr. Develop.* 22:517–524.

92. Back, D. G., B. W. Pickett, J. L. Voss, and G. E. Seidel. 1974. Observations on the sexual behavior of nonlactating mares. *J. Am. Vet. Med. Assoc.* 165:717–720.

93. Bacon, W. E. 1973. Aversive conditioning in neonatal kittens. *J. Comp. Physiol. Psychol.* 83:306–313.

94. Bacon, W. E., and W. C. Stanley. 1963. Effect of deprivation level in puppies on performance maintained by a passive person reinforcer. *J. Comp. Physiol. Psychol.* 56:783–785.

95. Badamana, M. S., J. D. Sutton, J. D. Oldham, and A. Mowlem. 1990. The effect of amount of protein in the concentrates on hay intake and rate of passage, diet digestibility and milk production in British Saanen goats. *Anim. Prod.* 51:333–342.

96. Bado, A., M. J. M. Lewin, and M. Dubrasquet. 1989. Effects of bombesin on food intake and gastric acid secretion in cats. *Am. J. Physiol.* 256:R181–R186.

97. Bado, A., M. Rodriguez, M. J. M. Lewin, J. Martinez, and M. Dubrasquet. 1988. Cholecystokinin suppresses food intake in cats: Structure-activity characterization. *Pharm. Biochem. Behav.* 31:297–303.

98. Baer, K. L., G. D. Potter, T. H. Friend, and B. V. Beaver. 1983. Observation effects on learning in horses. *Appl. Anim. Ethol.* 11:123–129.

99. Bagshaw, C. S., S. L. Ralston, and H. Fisher. 1994. Behavioral and physiological effect of orally administered tryptophan on horses subjected to acute isolation stress. *Appl. Anim. Behav. Sci.* 40:1–12.

100. Baile, C. A., and J. M. Forbes. 1974. Control of feed intake and regulation of energy galance in ruminants. *Physiol. Rev.* 54:160–214.

101. Baile, C. A., C. L. McLaughlin, F. C. Buonomo, R. J. Lauterio, L. Marson, and M. A. Della Fera. 1987. Opioid peptides and the control of feeding in sheep. *Fed. Proc.* 46:173–177.

102. Baile, C. A., C. L. McLaughlin, and M. A. Della Fera. 1986. Role of cholecystokinin and opioid peptides in control of food intake. *Physiol. Rev.* 66:171–234.

103. Baile, C. A., and W. H. Pfander. 1966. A possible chemo sensitive regulatory mechanism of ovine feed intake. *Am. J. Physiol.* 210:1243–1248.

104. Baile, C. A., C. W. Simpson, L. F. Krabill, and F. H. Martin. 1972. Adrenergic agonists and antagonists and feeding in sheep and cattle. *Life Sci.* 11(I):661–668.

105. Bailey, C. J., and L. W. Porter. 1955. Relevant cues in drive discrimination in cats. *J. Comp. Physiol. Psychol.* 48:180–182.

105a. Bailey, P. J., A. H. Bishop, and C. T. Boord. 1974. Grazing behaviour of steers. *Proc. Aust. Soc. Anim. Prod.* 10:303–306.

106. Baker, A. E. M., and B. H. Crawford. 1986. Observational learning in horses. *Appl. Anim. Behav. Sci.* 15:7–13.

107. Baker, A. E. M., and G. E. Seidel. 1985. Why do cows mount other cows? *Appl. Anim. Behav. Sci.* 13:237–241.

108. Baker, G. J., and J. Kear-Colwell. 1974. Aerophagia (windsucking) and aversion therapy in the horse. *Proc. Am. Assoc. Eq. Pract.* 20:127–130.

109. Balch, C. C. 1955. Sleep in ruminants. *Nature* 175:940–941.

110. Baldwin, B. A. 1969. The study of behaviour in pigs. *Br. Vet. J.* 125:281–288.

111. Baldwin, B. A. 1977. Ability of goats and calves to distinguish between conspecific urine samples using olfaction. *Appl. Anim. Ethol.* 3:145–150.

112. Baldwin, B. A. 1979. Operant studies on shape discrimination in goats. *Physiol. Behav.* 23:455–459.

113. Baldwin, B. A. 1981. Shape discrimination in sheep and calves. *Anim. Behav.* 29:830–834.

114. Baldwin, B. A., D. J. Conner, and G. B. Meese. 1974. Sensory reinforcement in the pig. *J. Physiol.* 242:27P.

115. Baldwin, B. A., and T. R. Cooper. 1979. The effects of olfactory bulbectomy on feeding behaviour in pigs. *Appl. Anim. Ethol.* 5:153–159.

116. Baldwin, B. A., and C. de la Riva. 1995. Effects of the 5-HT1A agonist 8-OH-DPAT on operant feeding in pigs. *Physiol. Behav.* 58:611–613.

117. Baldwin, B. A., I. S. Ebenezer, and C. de la Riva. 1990. Effects of intracerebroventricular injection of muscimol or GABA on operant feeding in pigs. *Physiol. Behav.* 48:417–421.

118. Baldwin, B. A., I. S. Ebenezer, and C. de la Riva. 1990. Effects of intracerebroventricular injection of dynorphin, leumorphin and a neo-endorphin on operant feeding in pigs. *Physiol. Behav.* 48:821–824.

119. Baldwin, B. A., W. L. Grovum, C. A. Baile, and J. R. Brobeck. 1975. Feeding following intraventricular injection of Ca++, Mg++ or pentobarbital in pigs. *Pharm. Biochem. Behav.* 3:915–918.

120. Baldwin, B. A., and D. L. Ingram. 1968. Factors influencing behavioral thermoregulation in the pig. *Physiol. Behav.* 3:409–415.

121. Baldwin, B. A., C. L. McLaughlin, and C. A. Baile. 1977. The effect of ablation of the olfactory bulbs on feeding behaviour in sheep. *Appl. Anim. Ethol.* 3:151–161.

122. Baldwin, B. A., and G. B. Meese. 1977. The ability of sheep to distinguish between conspecifics by means of olfaction. *Physiol. Behav.* 18:803–808.

123. Baldwin, B. A., and G. B. Meese. 1977. Sensory reinforcement and illumination preference in the domesticated pig. *Anim. Behav.* 25:497–507.

124. Baldwin, B. A., and G. B. Meese. 1979. Social behaviour in pigs studied by means of operant conditioning. *Anim. Behav.* 27:947–957.

125. Baldwin, B. A., and R. F. Parrott. 1979. Studies on intracranial electrical self-stimulation in pigs in relation to ingestive and exploratory behaviour. *Physiol. Behav.* 22:723–730.

126. Baldwin, B. A., and R. F. Parrott. 1985. Effects of intracerebroventricular injection of naloxone on operant feeding and drinking in pigs. *Pharm. Biochem. Behav.* 22:37–40.

127. Baldwin, B. A., and E. E. Shillito. 1974. The effects of ablation of the olfactory bulbs on parturition and maternal behaviour in Soay sheep. *Anim. Behav.* 22:220–223.

128. Baldwin, B. A., and D. B. Stephens. 1970. Operant conditioning procedures for producing emotional responses in pigs. *J. Physiol.* 210:127P–128P.

129. Baldwin, B. A., and D. B. Stephens. 1973. The effects of conditioned behaviour and environmental factors on plasma corticosteroid levels in pigs. *Physiol. Behav.* 10:267–274.

130. Baldwin, B. A., and S. N. Thornton. 1986. Operant drinking in pigs following intracerebralventricular injections of hypertonic solutions and angiotensin II. *Physiol. Behav.* 36:325–328.

131. Baldwin, B. A., and J. O. Yates. 1977. The effects of hypothalamic temperature variation and intracarotid cooling on behavioural thermoregulation in sheep. *J. Physiol.* 265:705–720.

132. Bane, A. 1954. Studies on monozygous cattle twins. XV. Sexual functions of bulls in relation to heredity, rearing intensity and somatic conditions. *Acta Agric. Scand.* 4:95–208.

133. Banks, E. M. 1964. Some aspects of sexual behavior in domestic sheep, Ovis aries. *Behaviour* 23:249–279.

134. Barber, J. A., and S. L. Crowell-Davis. 1994. Maternal behavior of Belgian (Equus caballus) mares. *Appl. Anim. Behav. Sci.* 41:161–189.

135. Bareham, J. R. 1975. The effect of lack of vision on suckling behaviour of lambs. *Appl. Anim. Ethol.* 1:245–250.

136. Bareham, J. R. 1976. The behaviour of lambs on the first day after birth. *Br. Vet. J.* 132:152–162.

137. Barnes, R. H., A. U. Moore, and W. G. Pond. 1970. Behavioral abnormalities in young adult pigs caused by malnutrition in early life. *J. Nutr.* 100:149–155.

138. Barnett, J. H., P. H. Hemsworth, C. G. Winfield, and C. Hansen. 1986. Effects of social environment on welfare status and sexual behaviour of female pigs. I. Effects of group size. *Appl. Anim. Behav. Sci.* 16:249–257.

139. Barnett, J. L., G. M. Cronin, T. H. McCallum, and E. A. Newman. 1993. Effects of 'chemical intervention' techniques on aggression and injuries when grouping unfamiliar adult pigs. *Appl. Anim. Behav. Sci.* 36:135–148.

140. Barnett, J. L., G. M. Cronin, T. H. McCallum, and E. A. Newman. 1993. Effects of pen size/shape and design on aggression when grouping unfamiliar adult pigs. *Appl. Anim. Behav. Sci.* 36:111–122.

141. Barnett, J. L., G. M. Cronin, T. H. McCallum, and E. A. Newman. 1994. Effects of food and time of day on aggression when grouping unfamiliar adult pigs. *Appl. Anim. Behav. Sci.* 39:339–347.

142. Baron, A., C. N. Stewart, and J. M. Warren. 1957. Patterns of social interaction in cats (Felis domestica). *Behaviour* 11:56–66.

143. Barrett, P., and P. Bateson. 1978. The development of play in cats. *Behaviour* 66:106–120.

144. Bartoshuk, L. M., M. A. Harned, and L. H. Parks. 1971. Taste of water in the cat: Effects on sucrose preference. *Science* 171:699–701.

145. Bateson, P. 1978. Sexual imprinting and optimal outbreeding. *Nature* 273:659–660.

146. Bateson, P. 1979. How do sensitive periods arise and what are they for? *Anim. Behav.* 27:470–486.

147. Bateson, P., M. Mendl, and J. Feaver. 1990. Play in the domestic cat is enhanced by rationing of the mother during lactation. *Anim. Behav.* 40:514–525.

148. Baumgardt, B. R., and A. D. Peterson. 1970. Hyperphagia in sheep induced by infusion of the ventriculo cisternal system with a depressant [abstract]. *Fed. Proc.* 29:760.

149. Beach, F. A. 1968. Coital behavior in dogs. III. Effects of early isolation on mating in males. *Behaviour* 30:218–238.

150. Beach, F. A. 1970. Coital behavior in dogs. VI. Long-term effects of castration upon mating in the male. *J. Comp. Physiol. Psychol. Monogr.* 70(3):1–32.

151. Beach, F. A. 1970. Coital behaviour in dogs. VIII. Social affinity, dominance and sexual preference in the bitch. *Behaviour* 36:131–148.

152. Beach, F. A. 1974. Effects of gonadal hormones on urinary behavior in dogs. *Physiol. Behav.* 12:1005–1013.

153. Beach, F. A., M. G. Buehler, and I. F. Dunbar. 1983. Development of attraction to estrous females in male dogs. *Physiol. Behav.* 31:293–297.

154. Beach, F. A., I. F. Dunbar, and M. G. Buehler. 1982. Sexual characteristics of female dogs during successive phases of the ovarian cycle. *Horm. Behav.* 16:414–442.

155. Beach, F. A., and R. W. Gilmore. 1949. Response of male dogs to urine from females in heat. *J. Mammal.* 30:391–392.

156. Beach, F. A., A. I. Johnson, J. J. Anisko, and I. F. Dunbar. 1977. Hormonal control of sexual attraction in pseudohermaphroditic female dogs. *J. Comp. Physiol. Psychol.* 91:711–715.

157. Beach, F. A., and R. E. Kuehn. 1970. Coital behavior in dogs. X. Effects of androgenic stimulation during development on feminine mating responses in females and males. *Horm. Behav.* 1:347–367.

158. Beach, F. A., R. E. Kuehn, R. H. Sprague, and J. J. Anisko. 1972. Coital behavior in dogs. XI. Effects of androgenic stimulation during development on masculine mating responses in females. *Horm. Behav.* 3:143–168.

159. Beach, F. A., and B. J. Le Boeuf. 1967. Coital behaviour in dogs. I. Preferential mating in the bitch. *Anim. Behav.* 15:546–558.

160. Beach, F. A., and A. Merari. 1968. Coital behavior in dogs, IV. Effects of progesterone in the bitch. *Proc. Natl. Acad. Sci.* 61:442–446.

161. Beamer, W., G. Bermant, and M. T. Clegg. 1969. Copulatory behaviour of the ram, Ovis aries. II. Factors affecting copulatory satiation. *Anim. Behav.* 17:706–711.

162. Beattie, V. E., N. Walker, and I. A. Sneddon. 1995. Effect of rearing environment and change of environment on the behaviour of gilts. *Appl. Anim. Behav. Sci.* 46:57–65.

163. Beauchemin, K. A., S. Zelin, D. Genner, and J. G. Buchanan-Smith. 1989. An automatic system for quantification of eating and ruminating activities of dairy cattle housed in stalls. *J. Dairy Sci.* 72:2746–2759.

164. Beaudet, R., A. Chalifoux, and A. Dallaire. 1994. Predictive value of activity level and behavioral evaluation on future dominance in puppies. *Appl. Anim. Behav. Sci.* 40:273–284.

165. Beaver, B. 1980. *Veterinary Aspects of Feline Behavior*. St. Louis, MO: C.V. Mosby.

166. Beaver, B. V. 1983. Clinical classification of canine aggression. *Appl. Anim. Ethol.* 10:35–43.

167. Beaver, B. V., M. Fischer, and C. E. Atkinson. 1992. Determination of favorite components of garbage by dogs. *Appl. Anim. Behav. Sci.* 34:129–136.

168. Beck, A. M. 1973. *The Ecology of Stray Dogs. A Study of Free-Ranging Urban Animals*. Baltimore, MD: York Press.

169. Becker, B. A., J. J. Ford, R. K. Christenson, R. C. Manak, G. L. Hahn, and J. A. DeShazor. 1985. Cortisol response of gilts in tether stalls. *J. Anim. Sci.* 60:264–270.

170. Becker, R. F., J. E. King, and J. E. Markee. 1962. Studies on olfactory discrimination in dogs: II. Discriminatory behavior in a free environment. *J. Comp. Physiol. Psychol.* 55:773–780.

171. Beckett, S. D., R. S. Hudson, D. R. Walker, and R. C. Purohit. 1978. Effect of local anesthesia of the penis and dorsal penile neurectomy on the mating ability of bulls. *J. Am. Vet. Med. Assoc.* 173:838–839.

172. Beilharz, R. G., D. F. Butcher, and A. E. Freeman. 1966. Social dominance and milk production in Holsteins. *J. Dairy Sci.* 49:887–892.

173. Beilharz, R. G., and D. R. Cox. 1967. Social dominance in swine. *Anim. Behav.* 15:117–122.

174. Beilharz, R. G., and P. J. Mylrea. 1963. Social position and behaviour of dairy heifers in yards. *Anim. Behav.* 11:522–528.

175. Beilharz, R. G., and P. J. Mylrea. 1963. Social position and movement orders of dairy heifers. *Anim. Behav.* 11:529–533.

176. Beilharz, R. G., and K. Zeeb. 1982. Social dominance in dairy cattle. *Appl. Anim. Ethol.* 8:79–97.

177. Bekoff, M. 1974. Social play and play-soliciting by infant canids. *Am. Zool.* 14:323–340.

178. Bekoff, M. 1977. Social communication in canids: Evidence for the evolu-

tion of a stereotyped mammalian display. *Science* 197:1097–1099.

179. Bekoff, M., H. L. Hill, and J. B. Mitton. 1975. Behavioural taxonomy in canids by discriminant function analyses. *Science* 190:1223–1225.

180. Belkin, M., U. Yinon, L. Rose, and I. Reisert. 1977. Effect of visual environment on refractive error of cats. *Doc. Ophthalmol.* 42:433–437.

181. Bell, F. R. 1959. Preference thresholds for taste discrimination in goats. *J. Agric. Sci.* 52:125–128.

182. Bell, F. R. 1960. The electroencephalogram of goats during somnolence and rumination. *Anim. Behav.* 8:39–42.

183. Bell, F. R. 1963. Alkaline taste in goats assessed by the preference test technique. *J. Comp. Physiol. Psychol.* 56:174–178.

184. Bell, F. R., and H. L. Williams. 1959. Threshold values for taste in monozygotic twin calves. *Nature* 183:345–346.

185. Bellinger, L. L., G. J. Trietley, and L. L. Bernardis. 1976. Failure of portal glucose and adrenaline infusions or liver denervation to affect food intake in dogs. *Physiol. Behav.* 16:299–304.

186. Bellinger, L. L., and F. E. Williams. 1990. The effect of portal infusions of epinephrine on ingestion, plasma glucose and insulin in dogs. *Physiol. Behav.* 48:479–483.

187. Bennett, M., K. A. Houpt, and H. N. Erb. 1988. Effects of declawing on feline behavior. *Companion Anim. Pract.* 2:7–12.

188. Berg, I. A. 1944. Development of behavior: The micturition pattern in the dog. *J. Exp. Psychol.* 34:343–368.

189. Berger, J. 1977. Organizational systems and dominance in feral horses in the Grand Canyon. *Behav. Ecol. Sociobiol.* 2:131–146.

190. Berger, J. 1986. *Wild Horses of the Great Basin.* Chicago, IL: The University of Chicago Press.

191. Berger, J., and C. Cunningham. 1987. Influence of familiarity on frequency of inbreeding in wild horses. *Evolution* 41:229–231.

192. Berggren-Thomas, B., and W. D. Hohenboken. 1986. The effects of sirebreed, forage availability and weather on the grazing behavior of crossbred ewes. *Appl. Anim. Behav. Sci.* 15:217–228.

193. Berkson, G. 1968. Maturation defects in kittens. *Am. J. Ment. Defic.* 72:757–777.

194. Berman, M., and I. Dunbar. 1983. The social behavior of free-ranging urban dogs. *Appl. Anim. Ethol.* 10:5–17.

195. Bermant, G., M. T. Clegg, and W. Beamer. 1969. Copulatory behaviour of the ram, Ovis aries. I. A normative study. *Anim. Behav.* 17:700–705.

196. Bielanski, W., and S. Wierzbowski. 1961. Depletion test in stallions. *Proc. IVth Int. Cong. Anim. Reprod.*, pp. 279–282.

197. Bigelow, J. A., and T. R. Houpt. 1988. Feeding and drinking patterns in young pigs. *Physiol. Behav.* 43:99–109.

198. Billing, A. E., and M. A. Vince. 1987. Teat-seeking behaviour in newborn lambs. I. Evidence for the influence of maternal skin temperature. *Appl. Anim. Behav. Sci.* 18:301–313.

199. Billing, A. E., and M. A. Vince. 1987. Teat-seeking behaviour in newborn lambs. II. Evidence for the influence of the dam's surface textures and degree of surface yield. *Appl. Anim. Behav. Sci.* 18:315–325.

200. Bines, J. A., S. Suzuki, and C. C. Balch. 1969. The quantitative significance of long-term regulation of food intake in the cow. *Br. J. Nutr.* 23:695–704.

201. Biquand, S., and V. Biquand-Guyot. 1992. The influence of peers, lineage and environment on food selection of the criollo goat (Capra hircus). *Appl. Anim. Behav. Sci.* 34:231–245.

202. Bitterman, M. E. 1965. Phyletic differences in learning. *Am. Psychol.* 20:396–410.

203. Björk, A., N. G. Olsson, E. Christensson, K. Martinsson, and O. Olsson. 1988. Effects of amperozide on biting behavior and performance in restricted-fed pigs following regrouping. *J. Anim. Sci.* 66:669–675.

204. Blair, R., and J. FitzSimons. 1970. A note on the voluntary feed intake and growth of pigs given diets containing an extremely bitter compound. *Anim. Prod.* 12:529–530.

205. Blakemore, C., and R. C. Van Sluyters. 1975. Innate and environmental factors in the development of the kitten's visual cortex. *J. Physiol.* 248:663–716.

206. Bland, K. P., and B. M. Jubilan. 1987. Correlation of flehmen by male sheep with female behaviour and oestrus. *Anim. Behav.* 35:735–738.

207. Blaxter, K. L. 1944. Food preferences and habits in dairy cows. *Br. Soc. Anim. Prod., Second Meeting,* pp. 85–94.

208. Blaxter, K. L., V. R. Fowler, and J. C. Gill. 1982. A study of the growth of sheep to maturity. *J. Agric. Sci.* 98:405–420.

209. Bleicher, N. 1962. Behavior of the bitch during parturition. *J. Am. Vet. Med. Assoc.* 140:1076–1082.

210. Blissitt, M. J., K. P. Bland, and D. F. Cottrell. 1990. Olfactory and vomeronasal chemoreception and the discrimination of oestrous and non-oestrous ewe urine odours by the ram. *Appl. Anim. Behav. Sci.* 27:325–335.

211. Blockey, M. A. B. 1981. Modification of a serving capacity test for beef bulls. *Appl. Anim. Ethol.* 7:321–336.

212. Blockey, M. A. B. 1981. Further studies on the serving capacity test for beef bulls. *Appl. Anim. Ethol.* 7:337–350.

213. Blockey, M. A. B. 1981. Development of a serving capacity test for beef bulls. *Appl. Anim. Ethol.* 7:307–319.

214. Blom, A. K., K. Halse, and K. Hove. 1976. Growth hormone, insulin and sugar in the blood plasma of bulls. Interrelated diurnal variations. *Acta Endocrinol.* 82:758–766.

215. Bøe, K. 1991. The process of weaning in pigs: When the sow decides. *Appl. Anim. Behav. Sci.* 30:47–59.

216. Bøe, K. 1993. Maternal behaviour of lactating sows in a loose-housing system. *Appl. Anim. Behav. Sci.* 35:327–338.

217. Bøe, K. 1994. Variation in maternal behaviour and production of sows in integrated loose housing systems in Norway. *Appl. Anim. Behav. Sci.* 41:53–62.

218. Bøe, K., and P. Jensen. 1995. Individual differences in suckling and solid food intake by piglets. *Appl. Anim. Behav. Sci.* 42:183–192.

219. Boissy, A., and M.-F. Bouissou. 1995. Assessment of individual differences in behavioural reactions of heifers exposed to various fear-eliciting situations. *Appl. Anim. Behav. Sci.* 46:17–31.

220. Boivin, X., P. Le Neindre, J. M. Chupin, J. P. Garel, and G. Trillat. 1992. Influence of breed and early management on ease of handling and open-field behaviour of cattle. *Appl. Anim. Behav. Sci.* 32:313–323.

221. Boivin, X., P. LeNeindre, and J. M. Chupin. 1992. Establishment of cattle-human relationships. *Appl. Anim. Behav. Sci.* 32:325–335.

222. Booth, W. D., and B. A. Baldwin. 1980. Lack of effect on sexual behaviour or the development of testicular function after removal of olfactory bulbs in prepubertal boars. *J. Reprod. Fert.* 58:173–182.

223. Borchelt, P. L. 1983. Aggressive behavior of dogs kept as companion animals: Classification and influence of sex, reproductive status and breed. *Appl. Anim. Ethol.* 10:45–61.

224. Borchelt, P. L. 1991. Cat elimination behavior problems. *Vet. Clin. N. Am. :Sm. Anim. Prac.* 21:257–264.

225. Borchelt, P. L., R. Lockwood, A. M. Beck, and V. L. Voith. 1983. Attacks by packs of dogs involving predation on human beings. *Publ. Hlth. Rep.* 98:57–66.

226. Borchelt, P. L., and V. L. Voith. 1981. Elimination behavior problems in cats. *Comp. Contin. Ed.* 3:730–737.

227. Borchelt, P. L., and V. L. Voith. 1985. Punishment. *Comp. Contin. Ed.* 9:780–791.

228. Borchelt, P. L., and V. L. Voith. 1985. Aggressive behavior in dogs and cats. *Comp. Contin. Ed.* 11:949–957.

229. Bordi, A., G. DeRosa, F. Napolitano, M. Litterio, V. Marino, and R. Rubino. 1994. Postpartum development of the mother-young relationship in goats. *Appl. Anim. Behav. Sci.* 42:145–152.

230. Borg, K. E., K. L. Esbenshade, and B. H. Johnson. 1993. Effects of the peri-pubertal rearing environment on endocrine and behavioural responses to oestrous female exposure in the mature bull. *Appl. Anim. Behav. Sci.* 35:245–253.

231. Bottoms, G. D., O. F. Roesel, F. D. Rausch, and E. L. Akins. 1972. Circadian variation in plasma cortisol and corticosterone in pigs and mares. *Am. J. Vet. Res.* 33:785–790.

232. Bouissou, M.-F. 1965. Observations sur la hierarchie sociale chez les bovins domestiques. *Ann. Biol. Anim. Biochim. Biophys.* 5:327–339.

233. Bouissou, M.-F. 1971. Effet de l'absence d'informations optiques et de con-tact physique sur la manifestation des relations hierarchiques chez les bovins domes-tiques. *Ann. Biol. Anim. Biochim. Biophys.* 11:191–198.

234. Bouissou, M.-F. 1972. Influence of body weight and presence of horns on social rank in domestic cattle. *Anim. Behav.* 20:474–477.

235. Bouissou, M.-F. 1978. Effects of injections of testosterone proprionate on dominance relationships in a group of cows. *Horm. Behav.* 11:388–400.

236. Bouissou, M.-F., and V. Gaudioso. 1982. Effect of early androgen treatment on subsequent social behavior in heifers. *Horm. Behav.* 16:132–146.

237. Bowersox, S. S., T. L. Baker, and W. C. Dement. 1984. Sleep-wakefulness pat-terns in the aged cat. *Electroenceph. clin. Neurophysiol.* 58:240–252.

238. Boy, V., and P. Duncan. 1979. Time budgets of Camargue horses I. Develop-mental changes in the time budgets of foals. *Behaviour* 71:187–202.

239. Boyd, L. E. 1980. *The Natality, Foal Survivorship, and Mare-Foal Behavior of Feral Horses in Wyoming's Red Desert.* Laramie, WY: M.S. Thesis, University of Wyoming.

239a. Boyd, L. E. 1988. Time budgets of adult Przewalski horses: Effects of sex, re-productive status and enclosure. *Appl. Anim. Behav. Sci.* 21:19–39.

240. Boyd, L. E., D. A. Carbonaro, and K. A. Houpt. 1988. The 24-hour time bud-get of Przewalski horses. *Appl. Anim. Behav. Sci.* 21:5–17.

241. Brakel, W. J., and R. A. Leis. 1976. Impact of social disorganization on be-havior, milk yield, and body weight of dairy cows. *J. Dairy Sci.* 59:716–721.

242. Braude, R., M. J. Newport, and J. W. G. Porter. 1971. Artificial rearing of pigs. 3. The effect of heat treatment on the nutritive value of spray-dried whole-milk pow-der for the baby pig. *Br. J. Nutr.* 25:113–125.

243. Bray, A. R., and M. Wodzicka-Tomaszewska. 1974. Perinatal behaviour and progesterone and corticosteroid levels in sheep. *Proc. Aust. Soc. Anim. Prod.* 10:318–321.

244. Breland, K., and M. Breland. 1966. *Animal Behavior.* New York, NY: Macmil-lan Co.

245. Bressers, H. P. M., J. H. A. Te Brake, B. Engel, and J. P. T. M. Noordhuizen.

1993. Feeding order of sows at an individual electronic feed station in a dynamic group-housing system. *Appl. Anim. Behav. Sci.* 36:123–134.

246. Brindley, E. L., D. J. Bullock, and F. Maisels. 1989. Effects of rain and fly harassment on feeding behaviour of free-ranging feral goats. *Appl. Anim. Behav. Sci.* 24:31–41.

247. Brisbin, I. L., Jr., and S. N. Austad. 1991. Testing the individual odour theory of canine olfaction. *Appl. Anim. Behav. Sci.* 42:63–69.

248. Bristol, F. 1982. Breeding behaviour of a stallion at pasture with 20 mares in synchronized oestrus. *J. Reprod. Fert. Suppl.* 32:71–77.

249. Brobeck, J. 1955. Neural regulation of food intake. *Ann. N. Y. Acad. Sci.* 63:44–55.

250. Bronson, F. H., and W. K. Whitten. 1968. Oestrus-accelerating pheromone of mice: Assay, androgen-dependency, and presence in bladder urine. *J. Reprod. Fert.* 15:131–134.

251. Bronson, R. T. 1979. Brain weight-body weight scaling in breeds of dogs and cats. *Brain Behav. Evol.* 16:227–236.

252. Broom, D. M., and J. D. Leaver. 1978. Effects of group-rearing or partial isolation on later social behaviour of calves. *Anim. Behav.* 26:1255–1263.

253. Brouns, F., and S. A. Edwards. 1994. Social rank and feeding behaviour of group-housed sows fed competitively or ad libitum. *Appl. Anim. Behav. Sci.* 39:225–235.

254. Brouns, F., S. A. Edwards, and P. R. English. 1994. Effect of dietary fibre and feeding system on activity and oral behaviour of group housed gilts. *Appl. Anim. Behav. Sci.* 39:215–223.

255. Brown, R. F., K. A. Houpt, and H. F. Schryver. 1976. Stimulation of food intake in horses by diazepam and promazine. *Pharm. Biochem. Behav.* 5:495–497.

256. Brown, S. A., S. Crowell-Davis, T. Malcolm, and P. Edwards. 1987. Naloxone-responsive compulsive tail chasing in a dog. *J. Am. Vet. Med. Assoc.* 190:884–886.

257. Brownlee, A. 1954. Play in domestic cattle in Britain: An analysis of its nature. *Br. Vet. J.* 110:48–68.

258. Bryant, M. J. 1975. A note on the effect of rearing experience upon the development of sexual behaviour in ram lambs. *Anim. Prod.* 21:97–99.

259. Bryant, M. J., and R. Ewbank. 1972. Some effects of stocking rate and group size upon agonistic behaviour in groups of growing pigs. *Br. Vet. J.* 128:64–70.

260. Bryant, M. J., and T. Tompkins. 1973. Sexual behaviour of sheep. *Vet. Rec.* 93:253.

261. Buchenauer, V. D., and B. Fritsch. 1980. Zum farbsehvermogen von hausziegen (Capra bircus L.). *Z. Tierpsychol.* 53:225–230.

262. Buddenberg, B. J., C. J. Brown, Z. B. Johnson, and R. S. Honea. 1986. Maternal behavior of beef cows at parturition. *J. Anim. Sci.* 62:42–47.

263. Burger, J. F. 1952. Sex physiology of pigs. *Onderstepoort J. Vet. Res. Suppl.* 2:1–218.

264. Burritt, E. A., and F. D. Provenza. 1991. Ability of lambs to learn with a delay between food ingestion and consequences given meals containing novel and familiar foods. *Appl. Anim. Behav. Sci.* 32:179–189.

265. Burritt, E. A., and F. D. Provenza. 1992. Lambs form preferences for nonnutritive flavors paired with glucose. *J. Anim. Sci.* 70:1133–1136.

266. Busnel, R.-G. 1963. *Acoustic Behaviour of Animals.* Amsterdam, The Netherlands: Elsevier Publishing Co.

267. Cairns, R. B., and D. L. Johnson. 1965. The development of interspecies social attachments. *Psychonom. Sci.* 2:337–338.

268. Campbell, R. G. 1976. A note on the use of a feed flavour to stimulate the feed intake of weaner pigs. *Anim. Prod.* 23:417–419.

269. Campbell, S. S., and I. Tobler. 1984. Animal sleep: A review of sleep duration across phylogeny. *Neurosci. Behav. Rev.* 8:269–300.

270. Campbell, W. E. 1975. *Behavior Problems in Dogs.* Santa Barbara, CA: American Veterinary Publications.

271. Campbell, W. E. 1989. *Better Behavior in Dogs and Cats.* Loveland, CO: Alpine Publications, Inc.

272. Campitelli, S., C. Carenzi, and M. Verga. 1982. Factors which influence parturition in the mare and development of the foal. *Appl. Anim. Ethol.* 9:7–14.

273. Canali, E., M. Verga, M. Montagna, and A. Baldi. 1986. Social interactions and induced behavioural reactions in milk-fed female calves. *Appl. Anim. Behav. Sci.* 16:207–215.

274. Candland, D. K., and D. Milne. 1966. Species differences in approach-behavior as a function of developmental environment. *Anim. Behav.* 14:539–545.

275. Carlstead, K. 1986. Predictability of feeding: Its effect on agonistic behaviour and growth in grower pigs. *Appl. Anim. Behav. Sci.* 16:25–38.

276. Caro, T. M. 1980. Predatory behaviour and social play in kittens. *Behaviour* 76:1–24.

277. Caro, T. M. 1980. Predatory behaviour in domestic cat mothers. *Behaviour* 74:128–148.

278. Caro, T. M. 1980. Effects of the mother, object play and adult experience on predation in cats. *Behav. Neural Biol.* 29:29–51.

279. Carson, K., and D. G. M. Wood-Gush. 1983. The nursing behavior of thoroughbred foals. *Eq. vet. J.* 15:257–262.

280. Carson, K., and D. G. M. Wood-Gush. 1983. Equine behaviour: I. A review of the literature on social and dam foal behavior. *Appl. Anim. Ethol.* 10:165–178.

281. Carter, R. R., and W. L. Grovum. 1990. Factors affecting the voluntary intake of food by sheep. 5. The inhibitory effect of hypertonicity in the rumen. *Br. J. Nutr.* 64:285–299.

282. Carver, D. S., and H. N. Waterhouse. 1962. The variation in the water consumption of cats. *Proc. Anim. Care Panel* 12:267–270.

283. Cassini, M. H., and G. Hermitte. 1992. Patterns of environmental use by cattle and consumption of supplemental food blocks. *Appl. Anim. Behav. Sci.* 32:297–312.

283a. Castle, M. E. and R. J. Halley. 1953. The grazing behaviour of dairy cattle at the National Institute for Research in Dairying. *Br. J. Anim. Behav.* 1:139–143.

284. Castonguay, T. W. 1981. Dietary dilution and intake in the cat. *Physiol. Behav.* 27:547–549.

285. Castrén, H., B. Algers, A.-M. de Passillé, J. Rushen, and K. Uvnäs-Moberg. 1993. Preparturient variation in progesterone, prolactin, oxytocin and somatostatin in relation to nest building in sows. *Appl. Anim. Behav. Sci.* 38:91–102.

286. Cerny, V. A. 1977. Failure of dihydrotestosterone to elicit sexual behaviour in the female cat. *J. Endocrinol.* 75:173–174.

287. Cervantes, M., R. Ruelas, and C. Beyer. 1983. Serotonergic infleunces on EEG synchronization induced by milk drinking in the cat. *Pharm. Biochem. Behav.* 18:851–855.

288. Chakraborty, P. K., W. B. Panko, and W. S. Fletcher. 1980. Serum hormone concentrations and their relationships to sexual behavior at the first and second estrous cycles of the Labrador bitch. *Biol. Reprod.* 22:227–232.

288a. Chamberlain, B., F. R. Ervin, R. O. Pihl, et al. 1987. The effect of raising or lowering tryptophan levels on aggression in Vervet monkeys. *Pharm. Biochem. Behav.* 28:503–510.

289. Champion, R. A., S. M. Rutter, P. D. Penning, and A. J. Rook. 1994. Temporal variation in grazing behaviour of sheep and the reliability of sampling periods. *Appl. Anim. Behav. Sci.* 42:99–108.

290. Chase, L. E., P. J. Wangsness, and B. R. Baumgardt. 1976. Feeding behavior of steers fed a complete mixed ration. *J. Dairy Sci.* 59:123–128.

291. Chemineau, P. 1986. Sexual behavior and gonadal activity during the year in the tropical Creole meat goat I. Female estrous behavior and ovarian activity. *Reprod. Nutr. Develop.* 26:441–452.

292. Chepko, B. D. 1971. A preliminary study of the effects of play deprivation on young goats. *Z. Tierpsychol.* 28:517–526.

293. Chesler, P. 1969. Maternal influence in learning by observation in kittens. *Science* 166:901–903.

294. Christensen, H. R., G. W. Seifert, and T. B. Post. 1982. The relationship between a serving capacity test and fertility of beef bulls. *Aust. Vet. J.* 58:241–244.

295. Christie, D. W., and E. T. Bell. 1972. Studies on canine reproductive behaviour during the normal oestrous cycle. *Anim. Behav.* 20:621–631.

296. Church, S. C., J. A. Allen, and J. W. S. Bradshaw. 1994. Anti-apostatic food selection by the domestic cat. *Anim. Behav.* 48:747–749.

297. Cizek, L. J., R. E. Semple, K. C. Huang, and M. I. Gregersen. 1951. Effect of extracellular electrolyte depletion on water intake in dogs. *Am. J. Physiol.* 164:415–422.

298. Clark, D. K., T. H. Friend, and G. Dellmeier. 1993. The effect of orientation during trailer transport on heart rate, cortisol and balance in horses. *Appl. Anim. Behav. Sci.* 38:179–189.

299. Clark, G. I., and W. N. Boyer. 1993. The effects of dog obedience training and behavioural counselling upon the human-canine relationship. *Appl. Anim. Behav. Sci.* 37:147–159.

300. Clark, J. R., R. W. Bell, L. F. Tribble, and A. M. Lennon. 1985. Effects of composition and density of the group on the performance, behaviour and age at puberty in swine. *Appl. Anim. Behav. Sci.* 14:127–135.

301. Clarke, I. J., and R. J. Scaramuzzi. 1978. Sexual behaviour and LH secretion in spayed androgenized ewes after a single injection of testosterone or oestradiol-17ß. *J. Reprod. Fert.* 52:313–320.

302. Clegg, M. T., W. Beamer, and G. Bermant. 1969. Copulatory behaviour of the ram, Ovis aries. III. Effects of pre-and postpubertal castration and androgen replacement therapy. *Anim. Behav.* 17:712–717.

303. Clutton-Brock, T. H., P. J. Greenwood, and R. P. Powell. 1976. Ranks and relationships in Highland ponies and Highland cows. *Z. Tierpsychol.* 41:202–216.

304. Cohen-Tannoudji, J., J. Einhorn, and J. P. Signoret. 1994. Ram sexual pheromone: First approach of chemical identification. *Physiol. Behav.* 56:955–961.

305. Cohen-Tannoudji, J., A. locatelli, and J. P. Signoret. 1986. Non-pheromonal stimulation by the male of LH release in the anoestrous ewe. *Physiol. Behav.* 36:921–924.

306. Cohn, R. 1956. A contribution to the study of color vision in cat. *J. Neurophysiol.* 19:416–423.

307. Coile, D. C., C. H. Pollitz, and J. C. Smith. 1989. Behavioral determination of critical flicker fusion in dogs. *Physiol. Behav.* 45:1087–1092.

308. Cole, D. D., and J. N. Shafer. 1966. A study of social dominance in cats. *Behaviour* 27:39–53.

309. Cole, D. J. A., J. E. Duckworth, and W. Holmes. 1967. Factors affecting voluntary feed intake in pigs. I. The effect of digestible energy content of the diet on the intake of castrated male pigs housed in holding pens and in metabolism crates. *Anim. Prod.* 9:141–148.

310. Collard, R. R. 1967. Fear of strangers and play behavior in kittens with varied social experience. *Child. Dev.* 38:877–891.

311. Collias, N. E. 1956. The analysis of socialization in sheep and goats. *Ecology* 37:228–239.

312. Collis, K. A. 1976. An investigation of factors related to the dominance order of a herd of dairy cows of similar age and breed. *Appl. Anim. Ethol.* 2:167–173.

313. Collis, K. A., S. J. Kay, A. J. Grant, and A. J. Quick. 1979. The effect on social organization and milk production of minor group alterations in dairy cattle. *Appl. Anim. Ethol.* 5:103–111.

314. Concannon, P. W., W. Hansel, and W. J. Visek. 1975. The ovarian cycle of the bitch: Plasma estrogen, LH and progesterone. *Biol. Reprod.* 13:112–121.

315. Cooling, M. J., and M. D. Day. 1975. Drinking behaviour in the cat induced by renin, angiotensin I, II and isoprenaline. *J. Physiol.* 244:325–336.

316. Cooper, R. A., S. Evans, and J. A. Kirk. 1991. Effects of water additives on water consumption, urine output and urine mineral levels in Angora goats. *Anim. Prod.* 52:609.

317. Coppinger, R., J. Lorenz, J. Glendinning, and P. Pinardi. 1983. Attentiveness of guarding dogs for reducing predation on domestic sheep. *J. Range Manag.* 36:275–279.

318. Corbett, J. L. 1953. Grazing behaviour in New Zealand. *Br. J. Anim. Behav.* 1:67–71.

319. Coren, S. 1994. *The Intelligence of Dogs: Canine Consciousness and Capabilities.* New York, NY: The Free Press.

320. Corson, S. A., E. OL. Corson, V. Kirilcuk, J. Kirilcuk, W. Knopp, and L. E. Arnold. 1972. Differential effects of amphetamines on several types of hyperkinetic and normal dogs and on learning disability [abstract]. *Psychopharmacology* 26 (Suppl): 55.

321. Cory, V. L. 1927. Activities of livestock on the range. *Tex. Agric. Exp. Sta. Bull.* 367:5–47.

322. Cowlishaw, S. J., and F. E. Alder. 1960. The grazing preferences of cattle and sheep. *J. Agric. Sci. (Camb.)* 54:257–265.

323. Crawford, M. A., M. D. Kittleson, and G. D. Fink. 1984. Hypernatremia and adipsia in a dog. *J. Am. Vet. Med. Assoc.* 184:818–821.

324. Creel, S. R., and J. L. Albright. 1988. The effects of neonatal social isolation on the behavior and endocrine function of Holstein calves. *Appl. Anim. Behav. Sci.* 21:293–306.

325. Cregier, S. W. 1982. Reducing equine hauling stress: A review. *J. Eq. Vet. Sci.* 2:187–198.

326. Cresswell, E. 1959. A cattle rangemeter. *Anim. Behav.* 7:244.

327. Cresswell, E. 1960. Ranging behaviour studies with Romney Marsh and Cheviot sheep in New Zealand. *Anim. Behav.* 8:32–28.

328. Cronin, G. M., B. N. Schirmer, T. H. McCallum, J. A. Smith, and K. L. Butler. 1993. The effects of providing sawdust to pre-parturient sows in farrowing crates on sow behaviour, the duration of parturition and the occurrence of intra-partum stillborn piglets. *Appl. Anim. Behav. Sci.* 36:301–315.

329. Cronin, G. M., and J. A. Smith. 1992. Effects of accommodation type and straw bedding around parturition and during lactation on the behaviour of primiparous sows and survival and growth of piglets to weaning. *Appl. Anim. Behav. Sci.* 33:191–208.

330. Cronin, G. M., P. R. Wiepkema, and J. M. van Ree. 1986. Endorphins implicated in stereotypies of tethered sows. *Experientia* 42:198–199.

331. Crowell-Davis, S. L. 1985. Nursing behaviour and maternal aggression among Welsh ponies (Equus caballus). *Appl. Anim. Behav. Sci.* 14:11–25.

332. Crowell-Davis, S. L. 1986. Spatial relations between mares and foals of the Welsh pony (Equus caballus). *Anim. Behav.* 34:1007–1015.

333. Crowell-Davis, S. L. 1994. Daytime rest behavior of the Welsh pony (Equus caballus) mare and foal. *Appl. Anim. Behav. Sci.* 40:197–210.

334. Crowell-Davis, S. L., K. Barry, J. M. Ballam, and D. P. Laflamme. 1995. The effect of caloric restriction on the behavior of pen-housed dogs: Transition from unrestricted to restricted diet. *Appl. Anim. Behav. Sci.* 43:27–41.

335. Crowell-Davis, S. L., and A. B. Caudle. 1989. Coprophagy by foals: Recognition of maternal feces. *Appl. Anim. Behav. Sci.* 24:267–272.

336. Crowell-Davis, S. L., and K. A. Houpt. 1985. Coprophagy by foals: Effect of age and possible function. *Eq. vet. J.* 17:17–19.

337. Crowell-Davis, S. L., and K. A. Houpt. 1985. The ontogeny of flehmen in horses. *Anim. Behav.* 33:739–774.

338. Crowell-Davis, S. L., K. A. Houpt, and J. S. Burnham. 1985. Snapping by foals of Equus caballus. *Z. Tierpsychol.* 69:42–54.

339. Crowell-Davis, S. L., K. A. Houpt, and C. M. Carini. 1986. Mutual grooming and nearest-neighbor relationships among foals of Equus caballus. *Appl. Anim. Behav. Sci.* 15:113–123.

340. Crowell-Davis, S. L., K. A. Houpt, and J. Carnevale. 1985. Feeding and drinking behavior of mares and foals with free access to pasture and water. *J. Anim. Sci.* 60:883–889.

341. Crowell-Davis, S. L., K. A. Houpt, and L. Kane. 1987. Play development in Welsh pony (Equus caballus) foal. *Appl. Anim. Behav. Sci.* 18:119–131.

341a. Culley, M. J. 1938. Grazing habits of range cattle. *J. Forestry* 36:715–717.

342. Cummings, B. J., E. Head, W. Ruehl, N. W. Milgram, and C. W. Cotman. 1996. The canine as an animal model of human aging and dementia. *Neurobiol. Aging* 17:259–269.

343. Curtis, Q. F. 1937. Experimental neurosis in the pig [abstract]. *Psychol. Bull.* 34:723.

344. Czarkowska, J. 1983. Changes of some postural reflxes during the first postnatal weeks in the dog. *Acta Neurobiol.* 43:27–35.

345. Dabrowska, B., W. Harmata, Z. Lenkiewicz, Z. Schiffer, and R. J. Wojtusiak. 1981. Colour perception in cows. *Behav. Proc.* 6:1–10.

346. Dallaire, A., and Y. Ruckebusch. 1974. Sleep and wakefulness in the housed pony under different dietary conditions. *Can. J. Comp. Med.* 38:65–71.

347. Dallaire, A., and Y. Ruckebusch. 1974. Sleep patterns in the pony with observations on partial perceptual deprivation. *Physiol. Behav.* 12:789–796.

348. Dalton, D. C., M. E. Pearson, and M. Sheard. 1967. The behaviour of dairy bulls kept in groups. *Anim. Prod.* 9:1–5.

349. Daniels, T. J. 1983. The social organization of free-ranging urban dogs II. Estrus groups and the mating system. *Appl. Anim. Ethol.* 10:365–373.

350. Daniels, T. J. 1983. The social organization of free-ranging urban dogs I. Non–estrous social behavior. *Appl. Anim. Ethol.* 10:341–363.

351. Dantzer, R. 1976. Effect of diazepam on performance of pigs in a progressive ratio schedule. *Physiol. Behav.* 17:161–163.

352. Dantzer, R. 1977. Effects of diazepam on conditioned suppression in pigs. *J. Pharmacol.* 8:405–414.

353. Dantzer, R., and B. A. Baldwin. 1974. Effects of chlordiazepoxide on heart rate and behavioural suppression in pigs subjected to operant conditioning procedures. *Psychopharmacology* 37:169–177.

354. Dantzer, R., and B. A. Baldwin. 1974. Changes in heart rate during suppression of operant responding in pigs. *Physiol. Behav.* 12:385–391.

355. Dantzer, R., P. Mormede, and B. Favre. 1976. Fear-dependent variations in continuous avoidance behaviour of pigs. II. Effects of diazepam on acquisition and performance of Pavlovian fear conditioning and plasma corticosteroid levels. *Psychopharmacology* 49:75–78.

356. Dards, J. L. 1983. The behaviour of dockyard cats: Interactions of adult males. *Appl. Anim. Behav. Sci.* 10:133–153.

357. Darke, P. G. G. 1978. Obesity in small animals. *Vet. Rec.* 102:545–546.

358. Davies, D. A. R., P. M. Lerman, and M. M. Crosse. 1974. Food preferences after weaning of artificially reared lambs. *J. Agric. Sci. (Camb.)* 82:469–471.

359. Davis, C. N., L. E. Davis, and T. E. Powers. 1975. Comparative body compositions of the dog and goat. *Am. J. Vet. Res.* 36:309–311.

360. Davis, J. L., and R. A. Jensen. 1976. The development of passive and active avoidance learning in the cat. *Dev. Psychobiol.* 9:175–179.

361. Davis, K. L., J. C. Gurski, and J. P. Scott. 1977. Interaction of separation distress with fear in infant dogs. *Dev. Psychobiol.* 10:203–212.

362. Dawson, W. M., and R. L. Revens. 1946. Varying susceptibility in pigs to alarm. *J. Comp. Physiol. Psychol.* 39:297–305.

363. Day, J. E. L., I. Kyriazakis, and A. B. Lawrence. 1995. The effect of food deprivation on the expression of foraging and exploratory behaviour in the growing pig. *Appl. Anim. Behav. Sci.* 42:193–206.

364. De Boer, J. N. 1977. The age of olfactory cues functioning in chemocommunication among male domestic cats. *Behav. Proc.* 2:209–225.

365. de Passille, A. M. B., J. H. M. Metz, P. Mekking, and P. R. Wiepkema. 1992. Does drinking milk stimulate sucking in young calves? *Appl. Anim. Behav. Sci.* 34:23–36.

366. De Vuyst, A., G. Thines, L. Henriet, and M. Soffie. 1964. Influence des stimulations auditives sur le comportement sexuel du taureau. *Experientia* 20:648–650.

367. deLahunta, A. 1977. *Veterinary Neuroanatomy and Clinical Neurology.* Philadelphia, PA: W.B. Saunders Company.

368. Delius, K., M. Günderoth-Palmowski, I. Krause, and W. Engelmann. 1984. Effects of lithium salts on the behaviour and the circadian system of Mesocricetus auratus W. *J. interdiscipl. Cycle Res.* 15:289–299.

369. Delude, L. A. 1986. Activity patterns and behavior of sled dogs. *Appl. Anim. Behav. Sci.* 15:161–168.

370. Denton, D. 1982. *The Hunger For Salt.* Berlin, Germany: Springer-Verlag.

371. Deswysen, A. G., W. C. Ellis, and K. R. Pond. 1987. Interrelationships among voluntary intake, eating and ruminating behavior and ruminal motility of heifers fed corn silage. *J. Anim. Sci.* 64:835–841.

372. Devilat, J., W. G. Pond, and P. D. Miller. 1970. Dietary amino acid balance in growing-finishing pigs: Effect on diet preference and performance. *J. Anim. Sci.* 30:536–543.

373. Diakow, C. 1971. Effects of genital desensitization on mating behavior and ovulation in the female cat. *Physiol. Behav.* 7:47–54.

374. Dickson, D. P., G. R. Barr, L. P. Johnson, and D. A. Wieckert. 1970. Social dominance and temperament of Holstein cows. *J. Dairy Sci.* 53:904–907.

375. Dickson, D. P., G. R. Barr, and D. A. Weickert. 1967. Social relationship of dairy cows in a feed lot. *Behaviour* 29:195–203.

376. Dinger, J. E., and E. E. Noiles. 1986. Effects of controlled exercise on libido in 2 year old stallions. *J. Anim. Sci.* 62:1220–1223.

377. Dinius, D. A., and C. A. Baile. 1977. Beef cattle response to a feed intake stimulant given alone and in combination with a propionate enhancer and an anabolic agent. *J. Anim. Sci.* 45:147–153.

378. Distel, R. A., E. A. Laca, T. C. Griggs, and M. W. Demment. 1995. Patch selection by cattle: Maximization of intake rate in horizontally heterogeneous pastures. *Appl. Anim. Behav. Sci.* 45:11–21.

379. Distel, R. A., J. J. Villalba, and H. E. Laborde. 1994. Effects of early experience on voluntary intake of low-quality roughage in sheep. *J. Anim. Sci.* 72:1191–1195.

380. Dodman, N. H., R. Bronson, and J. Gliatto. 1993. Tail chasing in a Bull Terrier. *J. Am. Vet. Med. Assoc.* 202:758–760.

381. Dodman, N. H., K. E. Knowles, L. Shuster, A. A. Moon-Fanelli, A. S. Tidwell, and C. L. Keen. 1996. Behavioral changes associated with suspected complex partial seizures in Bull Terriers. *J. Am. Vet. Med. Assoc.* 208:688–691.

382. Dodman, N. H., K. A. Miczek, K. Knowles, J. G. Thalhammer, and L. Shuster. 1992. Phenobarbital-responsive episodic dyscontrol (rage) in dogs. *J. Am. Vet. Med. Assoc.* 201:1580–1583.

383. Dodman, N. H., J. A. Normile, L. Shuster, and W. Rand. 1994. Equine self-mutilation syndrome (57 cases). *J. Am. Vet. Med. Assoc.* 204:1219–1223.

383a. Dodman, N.H., I. Reisner, L. Shuster, W. Rand, U. A. Luescher, I. Robinson, and K.A. Houpt. 1996. Effect of dietary protein content on behavior in dogs. *J. Am. Vet. Med. Assoc.* 208:376–379.

384. Dodman, N. H., L. Schuster, M. H. Court, and R. Dixon. 1987. An investigation into the use of narcotic antagonists in the treatment of a sterotypic behaviour pattern (crib-biting) in the horse. *Am. J. Vet. Res.* 48:311–319.

385. Dodman, N. H., L. Shuster, M. H. Court, and J. Patel. 1988. Use of a narcotic antagonist (nalmefene) to suppress self-mutilative behavior in a stallion. *J. Am. Vet. Med. Assoc.* 192:1585–1587.

386. Dodman, N. H., L. Shuster, S. D. White, M. H. Court, D. Parker, and R. Dixon. 1988. Use of narcotic antagonists to modify stereotypic self-licking, self-chewing, and scratching behavior in dogs. *J. Am. Vet. Med. Assoc.* 193:815–819.

387. Donaldson, L. E., and J. W. James. 1963. A connection between pregnancy and crush order in cows. *Anim. Behav.* 11:286.

388. Donovan, C. A. 1967. Some clinical observations on sexual attraction and deterrence in dogs and cattle. *Vet. Med. Sm. Anim. Clin.* 62:1047–1051.

389. Donovan, G. A., L. Badinga, R. J. Collier, C. J. Wilcox, and R. K. Braun. 1986. Factors influencing passive transfer in dairy calves. *J. Dairy Sci.* 69:754–759.

390. Doran, C. W. 1943. Activities and grazing habits of sheep on summer ranges. *J. Forestry* 41:253–258.

391. Dorries, K. M., E. Adkins-Regan, and B. P. Halpern. 1991. Sex difference in olfactory sensitivity to the boar chemosignal, androstenone, in the domestic pig. *Anim. Behav.* 42:403–411.

392. Dorries, K. M., E. Adkins-Regan, and B. P. Halpern. 1995. Olfactory sensitivity to the pheromone, androstenone, is sexually dimorphic in the pig. *Physiol. Behav.* 57:255–259.

393. Doty, R. L., and I. Dunbar. 1974. Attraction of beagles to conspecific urine, vaginal and anal sac secretion odors. *Physiol. Behav.* 12:825–833.

394. Doty, R. W., L. T. Rutledge,Jr., and R. M. Larsen. 1956. Conditioned reflexes established to electrical stimulation of cat cerebral cortex. *J. Neurophysiol.* 19:401–415.

395. Dougherty, C. T., F. W. Knapp, P. B. Burrus, D. C. Willis, and N. W. Bradley. 1993. Face flies (Musca autumnalis de Geer) and the behavior of grazing beef cattle. *Appl. Anim. Behav. Sci.* 35:313–326.

396. Dougherty, D. M., and P. Lewis. 1993. Generalization of a tactile stimulus in horses. *J. Exp. Anal. Behav.* 59:521–528.

397. Dourmad, J. Y. 1993. Standing and feeding behaviour of the lactating sow: Effect of feeding level during pregnancy. *Appl. Anim. Behav. Sci.* 37:311–319.

398. Dove, H., R. G. Beilharz, and J. L. Black. 1974. Dominance patterns and positional behaviour of sheep in yards. *Anim. Prod.* 19:157–168.

399. Ducker, M. J., P. T. Kendall, R. G. Hemingway, and T. H. McClelland. 1981. An evaluation of feedblocks as a means of providing supplementary nutrients to ewes grazing upland/hill pastures. *Anim. Prod.* 33:51–57.

400. Duckworth, J. E., and D. W. Shirlaw. 1958. A study of factors affecting feed intake and the eating behaviour of cattle. *Anim. Behav.* 6:147–154.

401. Dufty, J. H. 1971. Determination of the onset of parturition in Hereford cattle. *Aust. Vet. J.* 47:77–82.

402. Dufty, J. H. 1972. Clinical studies on bovine parturition. Maternal causes of dystocia and stillbirth in an experimental herd of Hereford cattle. *Aust. Vet. J.* 48:1–6.

403. Dufty, J. H. 1973. Clinical studies on bovine parturition—foetal aspects. *Aust. Vet. J.* 49:177–182.

404. Dunbar, I., and G. Bohnenkamp. 1985. *Behavior Booklets*. Oakland, CA: James and Kenneth.

405. Dunbar, I., and M. Buehler. 1980. A masking effect of urine from male dogs. *Appl. Anim. Ethol.* 6:297–301.

406. Dunbar, I. F. 1978. Olfactory preferences in dogs: The response of male and female beagles to conspecific urine. *Biol. Behav.* 3:273–286.

407. Dunbar, I. F., and M. Carmichael. 1981. The response of male dogs to urine from other males. *Behav. Neural Biol.* 31:465–470.

408. Dunbar, R. I. M., D. Buckland, and D. Miller. 1990. Mating strategies of male feral goats: A problem in optimal foraging. *Anim. Behav.* 40:653–667.

409. Duncan, P. 1980. Time-budgets of Camargue horses II. Time-budgets of adult horses and weaned sub-adults. *Behaviour* 72:26–49.

410. Duncan, P. 1985. Time-budgets of Camargue horses III. Environmental influences. *Behaviour* 92:188–208.

411. Duncan, P., and P. Cowtan. 1980. An unusual choice of habitat helps Camargue horses to avoid blood-sucking horse-flies. *Biol. Behav.* 5:55–60.

412. Duncan, P., P. H. Harvey, and S. M. Wells. 1984. On lactation and associated behaviour in a natural herd of horses. *Anim. Behav.* 32:255–263.

413. Dunne, H. W., and A. D. Leman. 1975. *Diseases of Swine*. 4th ed. Ames, IA: Iowa State University Press.

414. Duquette, P. F., and L. A. Muir. 1979. Monitoring the effects of selected compounds on feeding behaviour of sheep. *J. Anim. Sci.* 49:1120–1124.

415. Durrer, J. L., and J. P. Hannon. 1962. Seasonal variations in caloric intake of dogs living in an arctic environment. *Am. J. Physiol.* 202:375–378.

416. Dworkin, S. 1939. Conditioning neuroses in dog and cat. *Psychosom. Med.* 1:388–396.

417. Dyck, G. W., E. E. Swierstra, R. M. McKay, and K. Mount. 1987. Effect of location of the teat suckled, breed and parity on piglet growth. *Can. J. Anim. Sci.* 67:929–939.

418. Dykeman, R. A., O. D. Murphree, and J. E. Peters. 1969. Like begets like: Behavioral test, classical autonomic and motor conditioning and operant conditioning in two strains of pointer dogs. *Ann. N. Y. Acad. Sci.* 159:976–1007.

419. Eccles, R. 1982. Autonomic innervation of the vomeronasal organ of the cat. *Physiol. Behav.* 28:1011–1015.

420. Echeverri, A. C., H. W. Gonyou, and A. W. Ghent. 1992. Preparturient behavior of confined ewes: Time budgets, frequencies, spatial distribution and sequential analysis. *Appl. Anim. Behav. Sci.* 34:329–344.

421. Eckstein, P., and S. Zuckerman. 1956. The oestrous cycle in the mammalia. In *Marshall's Physiology of Reproduction*. 3rd ed. A. S. Parkes, ed. London, UK: Longmans, Green and Co., pp. 226–396.

422. Eckstein, R. A., and B. L. Hart. 1996. Treatment of canine acral lick dermatitis by behavior modification using electronic stimulation. *J. Am. Anim. Hosp. Assoc.* 32:225–230.

423. Edwards, S. A. 1982. Factors affecting time to first suckling in dairy calves. *Anim. Prod.* 34:339–346.

424. Edwards, S. A., and D. M. Broom. 1982. Behavioural interactions of dairy cows with their newborn calves and the effects of parity. *Anim. Behav.* 30:525–535.

425. Ehret, C. F., V. R. Potter, and K. W. Dobra. 1975. Chronotypic action of theophylline and of phenobarbital as circadian Zeit-gebers in the rat. *Science* 188:1212–1215.

426. Eldridge, F., and Y. Suzuki. 1976. A mare mule—dam or foster mother? *J. Hered.* 67:353–360.

427. Ellendorff, F., N. Parvizi, D. K. Pomerantz, A. Hartjen, A. Konig, D. Smidt, and F. Elsaesser. 1975. Plasma luteinizing hormone and testosterone in the adult male pig: 24 hour fluctuations and the effect of copulation. *J. Endocrinol.* 67:403–410.

428. Elliot, J. M., H. W. Symonds, and B. Pike. 1985. Effect on feed intake of infusing sodium propionate or sodium acetate into a mesenteric vein of cattle. *J. Dairy Sci.* 68:1165–1170.

429. Elliot, O., and J. A. King. 1960. Effect of early food deprivation upon later consummatory behavior in puppies. *Psychol. Rep.* 6:391–400.

430. Elliot, O., and J. P. Scott. 1961. The development of emotional distress reactions to separation, in puppies. *J. Genet. Psychol.* 99:3–22.

431. Elliott, J. A., M. H. Stetson, and M. Menaker. 1972. Regulation of testis function in golden hamsters: A circadian clock measures photoperiodic time. *Science* 178:771–773.

432. Ely, F., and W. E. Petersen. 1941. Factors involved in the ejection of milk. *J. Dairy Sci.* 24:211–223.

433. England, G. J. 1954. Observations on the grazing behaviour of different breeds of sheep at Pantyrhuad Farm, Carmarthenshire. *Br. J. Anim. Behav.* 2:56–60.

434. Entsu, S., H. Dohi, and A. Yamada. 1992. Visual acuity of cattle determined by the method of discrimination learning. *Appl. Anim. Behav. Sci.* 34:1–10.

435. Erlinger, L. L., D. R. Tolleson, and C. J. Brown. 1990. Comparison of bite size, biting rate and grazing time of beef heifers from herds distinguished by mature size and rate of maturity. *J. Anim. Sci.* 68:3578–3587.

436. Escós, J., C. L. Alados, and J. Boza. 1993. Leadership in a domestic goat herd. *Appl. Anim. Behav. Sci.* 38:41–47.

437. Esselmont, R. J., R. G. Glencross, M. J. Bryant, and G. S. Pope. 1980. A quantitative study of pre-ovulatory behaviour in cattle (British Friesian heifers). *Appl. Anim. Ethol.* 6:1–17.

438. Estep, D. Q., S. L. Crowell-Davis, S.-A. Earl-Costello, and S. A. Beatey. 1993. Changes in the social behaviour of drafthorse (Equus caballus) mares coincide with foaling. *Appl. Anim. Behav. Sci.* 35:199–213.

439. Evans, J. W., C. M. Winget, C. De Roshia, and D. C. Holley. 1976. Ovulation and equine body temperature and heart rate circadian rhythms. *J. interdiscipl. Cycle Res.* 7:25–37.

440. Everitt, G. C., and D. S. M. Phillips. 1971. Calf rearing by multiple suckling and the effects on lactation performance of the cow. *Proc. N. Z. Soc. Anim. Prod.* 31:22–40.

441. Ewbank, R. 1963. Predicting the time of parturition in the normal cow: A study of the precalving drop in body temperature in relation to the external signs of imminent calving. *Vet. Rec.* 75:367–370.

442. Ewbank, R. 1964. Observations on the suckling habits of twin lambs. *Anim. Behav.* 12:34–37.

443. Ewbank, R. 1967. Nursing and suckling behaviour amongst Clun Forest ewes and lambs. *Anim. Behav.* 15:251–258.

444. Ewbank, R. 1967. Behavior of twin cattle. *J. Dairy Sci.* 50:1510–1512.

445. Ewbank, R. 1973. Abnormal behaviour and pig nutrition. An unseccessful attempt to induce tail biting by feeding a high energy, low fibre vegetable protein ration. *Br. Vet. J.* 129:366–369.

446. Ewbank, R., and M. J. Bryant. 1972. Aggressive behaviour amongst groups of domesticated pigs kept at various stocking rates. *Anim. Behav.* 20:21–28.

447. Ewbank, R., and A. C. Mason. 1967. A note on the sucking behaviour of twin lambs reared as singles. *Anim. Prod.* 9:417–420.

448. Ewbank, R., and G. B. Meese. 1971. Aggressive behaviour in groups of domesticated pigs on removal and return of individuals. *Anim. Prod.* 13:685–693.

449. Ewbank, R., G. B. Meese, and J. E. Cox. 1974. Individual recognition and the dominance hierarchy in the domesticated pig. The role of sight. *Anim. Behav.* 22:473–480.

450. Ewer, R. F. 1959. Suckling behaviour in kittens. *Behaviour* 15:146–162.

451. Ewer, R. F. 1973. *The Carnivores*. Ithaca, NY: Cornell University Press.

452. Ezeh, P. I., L. J. Myers, L. A. Hanrahan, R. J. Kemppainen, and K. A. Cummins. 1992. Effects of steroids on the olfactory function of the dog. *Physiol. Behav.* 51:1183–1187.

453. Fagen, R. M., and T. K. George. 1977. Play behavior and exercise in young ponies (Equus caballus L.). *Behav. Ecol. Sociobiol.* 2:267–269.

454. Faichney, G. J. 1992. Consumption of solid feed by lambs during their transition from pre-ruminant to full ruminant function. *Appl. Anim. Behav. Sci.* 34:85–91.

455. Farner, D. S. 1961. Comparative physiology: Photoperiodicity. *Ann. Rev. Physiol.* 23:71–96.

456. Farningham, D. A. H., J. G. Mercer, and C. B. Lawrence. 1993. Satiety signals in sheep: Involvement of CCK, propionate, and vagal CCK binding sites. *Physiol. Behav.* 54:437–442.

457. Fay, P. K., V. T. McElligott, and K. M. Havstad. 1989. Containment of free-ranging goats using pulsed-radio-wave-activated shock collars. *Appl. Anim. Behav. Sci.* 23:165–171.

458. Feaver, J., M. Mendl, and P. Bateson. 1986. A method for rating the individual distinctiveness of domestic cats. *Anim. Behav.* 34:1016–1025.

459. Feddes, J. J. R., B. A. Young, and J. A. DeShazer. 1989. Influence of temperature and light on feeding behaviour in pigs. *Appl. Anim. Behav. Sci.* 23:215–222.

460. Feh, C. 1990. Long-term paternity data in relation to different aspects of rank for Camargue stallions, Equus caballus. *Anim. Behav.* 40:995–996.

461. Feh, C., and J. de Mazières. 1993. Grooming at a preferred site reduces heart rate in horses. *Anim. Behav.* 46:1191–1194.

462. Feist, J. D., and D. R. McCullough. 1975. Reproduction in feral horses. *J. Reprod. Fert. Suppl.* 23:13–18.

463. Feist, J. D., and D. R. McCullough. 1976. Behavior patterns and communication in feral horses. *Z. Tierpsychol.* 41:337–371.

464. Feldman, H. N. 1993. Maternal care and differences in the use of nests in the domestic cat. *Anim. Behav.* 45:13–23.

465. Feldman, H. N. 1994. Domestic cats and passive submission. *Anim. Behav.* 47:457–459.

466. Feldmann, B. M. 1974. The problem of urban dogs. *Science* 185:903.

467. Feldmann, B. M., and T. H. Carding. 1973. Free-roaming urban pets. *Health Serv. Rep.* 88:956–962.

468. Fernandez, X., M.-C. Meunier-Salaün, and P. Mormede. 1994. Agonistic behavior, plasma stress hormones, and metabolites in response to dyadic encounters in domestic pigs: Interrelationships and effect of dominance status. *Physiol. Behav.* 56:841–847.

469. Ferreira, A., A. Carrau, E. Rodas, E. Rubianes, and A. Benech. 1992. Diazepam facilitates acceptance of alien lambs by postparturient ewes. *Physiol. Behav.* 41:1117–1121.

470. Ferrell, F. 1984. Preference for sugars and nonnutritive sweetners in young beagles. *Neurosci. Biobehav. Rev.* 8:199–203.

471. Finger, K. H., and H. Brummer. 1969. Beobachtungen uber das Saugverhalten mutterlos aufgezogener Kalber. *Duet. Tierarztl. Wochenschr.* 76:665–667.

472. Firth, E. C. 1980. Bilateral ventral accessory neurectomy in windsucking horses. *Vet. Rec.* 106:30–32.

473. Fisher, R. B., and M. L. G. Gardner. 1976. A diurnal rhythm in the absorption of glucose and water by isolated rat small intestine. *J. Physiol.* 254:821–825.

474. Fiske, J. C., and G. D. Potter. 1979. Discrimination reversal learning in yearling horses. *J. Anim. Sci.* 49:583–588.

475. Fitzgerald, J. A., A. Perkins, and K. Hemenway. 1993. Relationship of sex and number of siblings in utero with sexual behavior of mature rams. *Appl. Anim. Behav. Sci.* 38:283–290.

476. Fitzsimons, J. T., and E. Szczepanska-Sadowska. 1974. Drinking and antidiuresis elicited by isoprenaline in the dog. *J. Physiol.* 239:251–267.

477. Fletcher, I. C., and D. R. Lindsay. 1968. Sensory involvement in the mating behaviour of domestic sheep. *Anim. Behav.* 16:410–414.

478. Folman, Y., and R. Volcani. 1966. Copulatory behaviour of the prepubertally castrated bull. *Anim. Behav.* 14:572–573.

479. Fonberg, E. 1976. The relation between alimentary and emotional amygdalar regulation. In *Hunger: Basic Mechanisms And Clinical Implications.* D. Novin, W. Wyrwicka, and G. A. Bray, eds. New York, NY: Raven Press, pp. 61–75.

480. Fontenot, J. P., and R. E. Blaser. 1965. Symposium on factors influencing the voluntary intake of herbage by ruminants: Selection and intake by grazing animals. *J. Anim. Sci.* 24:1202–1208.

481. Foot, J. Z., and A. J. F. Russel. 1978. Pattern of intake of three roughage diets by nonpregnant, nonlactating Scottish blackface ewes over a long period and the effects of previous nutritional history on current intake. *Anim. Prod.* 26:203–215.

482. Forbes, J. M. 1986. *The Voluntary Food Intake of Farm Animals.* London, UK: Butterworths.

483. Forbes, J. M. 1995. *Voluntary Food Intake and Diet Selection in Farm Animals.* Wallingford, UK: CAB International.

484. Ford, J. J., and R. K. Christenson. 1981. Glucocorticoid inhibition of estrus in ovariectomized pigs: Relationship to progesterone action. *Horm. Behav.* 15:427–435.

485. Ford, J. J., and H. S. Teague. 1978. Effect of floor space restriction on age at puberty in gilts and on performance of barrows and gilts. *J. Anim. Sci.* 47:828–832.

486. Forkman, B., I. L. Furuhaug, and P. Jensen. 1995. Personality, coping patterns, and aggression in piglets. *Appl. Anim. Behav. Sci.* 45:31–42.

487. Forssell, G. 1926. The new surgical treatment against crib-biting. *Vet. J.* 82:538–548.

488. Foss, I., and G. Flottorp. 1974. A comparative study of the development of hearing and vision in various species commonly used in experiments. *Acta Oto-Laryngol.* 77:202–214.

489. Foster, J. A., M. Morrison, S. J. Dean, M. Hill, and H. Frenk. 1981. Naloxone suppresses food/water consumption in the deprived cat. *Pharm. Biochem. Behav.* 14:419–421.

490. Foutz, A. S., M. M. Mitler, and W. C. Dement. 1980. Narcolepsy. *Vet. Clin. N. Am. :Sm. Anim. Prac.* 10:65–80.

491. Fowler, D. G., and L. D. Jenkins. 1976. The effects of dominance and infertility of rams on reproductive performance. *Appl. Anim. Ethol.* 2:327–337.

492. Fox, M. W. 1968. *Abnormal Behavior in Animals.* Philadelphia, PA: W.B. Saunders.

493. Fox, M. W. 1970. Reflex development and behavioral organization. In *Developmental Neurobiology.* W. A. Himwich, ed. Springfield, IL: Charles C. Thomas, pp. 553–580.

494. Fox, M. W. 1971. *Integrative Development of Brain and Behavior in the Dog.* Chicago, IL: University of Chicago Press.

495. Fox, M. W. 1972. *Understanding Your Dog.* New York, NY: Coward, McCann and Geoghegan.

496. Fox, M. W. 1975. The behaviour of cats. In *The Behaviour Of Domestic Animals.* 3rd ed. E. S. E. Hafez, ed. Baltimore, MD: Williams & Wilkins, pp. 410–436.

497. Fox, M. W., and M. Bekoff. 1975. The behaviour of dogs. In *The Behaviour Of Domestic Animals*. 3rd ed. E. S. E. Hafez, ed. Baltimore, MD: Williams & Wilkins, pp. 370–409.

498. Fox, M. W., and J. W. Spencer. 1967. Development of the delayed response in the dog. *Anim. Behav.* 15:162–168.

499. Fox, M. W., and G. Stanton. 1967. A developmental study of sleep and wakefulness in the dog. *J. Sm. Anim. Prac.* 8:605–611.

500. Fox, M. W., and D. Stelzner. 1966. Approach/withdrawal variables in the development of social behaviour in the dog. *Anim. Behav.* 14:362–366.

501. Fox, M. W., and D. Stelzner. 1966. Behavioural effects of differential early experience in the dog. *Anim. Behav.* 14:273–281.

502. Francis-Smith, K., and D. G. M. Wood-Gush. 1977. Coprophagia as seen in thoroughbred foals. *Eq. vet. J.* 9:155–157.

503. Franke Stevens, E. 1990. Instability of harems of feral horses in relation to season and presence of subordinate stallions. *Behaviour* 112:149–161.

504. Fraser, A. F. 1963. Behavior disorders in domestic animals. *Cornell Vet.* 53:213–223.

505. Fraser, A. F. 1968. *Reproductive Behaviour in Ungulates*. New York, NY: Academic Press.

506. Fraser, A. F. 1974. *Farm Animal Behaviour*. Baltimore, MD: Williams & Wilkins.

507. Fraser, D. 1973. The nursing and suckling behaviour of pigs. I. The importance of stimulation of the anterior teats. *Br. Vet. J.* 129:324–336.

508. Fraser, D. 1974. The behaviour of growing pigs during experimental social encounters. *J. Agric. Sci. (Camb.)* 82:147–163.

509. Fraser, D. 1974. The vocalizations and other behaviour of growing pigs in an "open field" test. *Appl. Anim. Ethol.* 1:3–16.

510. Fraser, D. 1975. The nursing and suckling behaviour of pigs. IV. The effect of interrupting the sucking stimulus. *Br. Vet. J.* 131:549–559.

511. Fraser, D. 1975. The nursing and suckling behaviour of pigs. III. Behaviour when milk ejection is elicited by manual stimulation of the udder. *Br. Vet. J.* 131:416–426.

512. Fraser, D. 1975. The effect of straw on the behaviour of sows in tether stalls. *Anim. Prod.* 21:59–68.

513. Fraser, D. 1977. Some behavioural aspects of milk ejection failure by sows. *Br. Vet. J.* 133:126–133.

514. Fraser, D. 1987. Attraction to blood as a factor in tail-biting by pigs. *Appl. Anim. Behav. Sci.* 17:61–68.

515. Fraser, D. 1987. Mineral-deficient diets and the pig's attraction to blood: Implications for tail-biting. *Can. J. Anim. Sci.* 67:909–918.

516. Fraser, D., and J. Rushen. 1993. A colostrum feeder for newborn lambs. *Appl. Anim. Behav. Sci.* 35:267–276.

517. Freedman, D. G. 1958. Constitutional and environmental interactions in rearing of four breeds of dogs. *Science* 127:585–586.

518. Freedman, D. G., J. A. King, and E. Elliot. 1961. Critical period in the social development of dogs. *Science* 133:1016–1017.

519. Freeman, N. C. G., and J. S. Rosenblatt. 1978. The interrelationship between thermal and olfactory stimulation in the development of home orientation in newborn kittens. *Dev. Psychobiol.* 11:437–457.

520. Freeman, N. C. G., and J. S. Rosenblatt. 1978. Specificity of litter odors in the control of home orientation among kittens. *Dev. Psychobiol.* 11:459–468.

521. Fretz, P. B. 1977. Behavioral virilization in a brood mare. *Appl. Anim. Ethol.* 3:277–280.

522. Friend, D. W. 1973. Self-selection of feeds and water by unbred gilts. *J. Anim. Sci.* 37:1137–1141.

523. Friend, T. H., L. O'Connor, D. Knabe, and G. Dellmeier. 1989. Preliminary trials of a sound-activated device to reduce crushing of piglets by sows. *Appl. Anim. Behav. Sci.* 24:23–29.

524. Friend, T. H., and C. E. Polan. 1974. Social rank, feeding behavior, and free stall utilization by dairy cattle. *J. Dairy Sci.* 57:1214–1220.

525. Fuller, C. A., F. M. Sulzman, and M. C. Moore-Ede. 1978. Thermoregulation is impaired in an environment without circadian time cues. *Science* 199:794–796.

526. Fuller, J. L. 1956. Photoperiodic control of estrus in the basenji. *J. Hered.* 47:179–180.

527. Fuller, J. L. 1967. Experiential deprivation and later behavior. *Science* 158:1645–1652.

528. Fuller, J. L., C. A. Easler, and E. M. Banks. 1950. Formation of conditioned avoidance responses in young puppies. *Am. J. Physiol.* 160:462–466.

529. Funston, R. N., D. D. Kress, K. M. Havstad, and D. E. Doombos. 1991. Grazing behavior of rangeland beef cattle differing in biological type. *J. Anim. Sci.* 69:1435–1442.

530. Furuse, M., S. I. Yang, Y. H. Choi, N. Kawamura, A. Takahashi, and J. Okomura. 1991. A note on plasma cholecystokinin concentration in dairy cows. *Anim. Prod.* 53:123–125.

531. Gadbury, J. C. 1975. Some preliminary field observations on the order of entry of cows into herringbone parlours. *Appl. Anim. Ethol.* 1:275–281.

532. Gaebelein, C. J., R. A. Galosy, L. Botticelli, J. L. Howard, and P. A. Obrist. 1977. Blood pressure and cardiac changes during signalled and unsignalled avoidance in dogs. *Physiol. Behav.* 19:69–74.

533. Ganjam, V. K., and R. M. Kenney. 1975. Androgens and oestrogens in normal and cryptorchid stallion. *J. Reprod. Fert. Suppl.* 23:67–73.

534. Garbarg, M., C. Julien, and J.-C. Schwartz. 1974. Circadian rhythm of histamine in the pineal gland. *Life Sci.* 14:539–543.

535. Garcia, J., W. G. Hankins, and K. W. Rusiniak. 1974. Behavioral regulation of the milieu interne in man and rat. *Science* 185:824–831.

536. Garcia, M. C., S. M. McDonnell, R. M. Kenney, and H. G. Osborne. 1986. Bull sexual behavior tests: Stimulus cow affects performance. *Appl. Anim. Behav. Sci.* 16:1–10.

537. Gardner, L. P. 1937. The responses of cows in a discrimination problem. *J. Comp. Psychol.* 23:35–57.

538. Gardner, L. P. 1937. The responses of cows to the same signal in different positions. *J. Comp. Psychol.* 23:333–350.

539. Gardner, L. P. 1937. The responses of horses in a discrimina tion problem. *J. Comp. Psychol.* 23:13–34.

540. Gardner, L. P. 1937. Responses of horses to the same signal in different positions. *J. Comp. Psychol.* 23:305–332.

541. Gardner, L. P. 1945. Responses of sheep in a discrimination problem with variations of the position of the signal. *J. Comp. Physiol. Psychol.* 38:343–351.

542. Garner, F. H. 1963. The palatability of herbage plants. *J. Br. Grassland Soc.* 18:79–89.

543. Gary, L. A., G. W. Sherritt, and E. B. Hale. 1970. Behavior of Charolais cattle on pasture. *J. Anim. Sci.* 30:203–206.

544. Geary, T. W., and J. J. Reeves. 1992. Relative importance of vision and olfaction for detection of estrus by bulls. *J. Anim. Sci.* 70:2726–2731.

545. Geist, V. 1971. *Mountain Sheep. A Study in Behavior and Evolution.* Chicago, IL: University of Chicago Press.

546. George, J. M., and I. A. Barger. 1974. Observations of bovine parturition. *Proc. Aust. Soc. Anim. Prod.* 10:314–317.

547. Ghosh, B., D. K. Choudhuri, and B. Pal. 1984. Some aspects of the sexual behaviour of stray dogs, Canis familiaris. *Appl. Anim. Behav. Sci.* 13:113–127.

548. Gibbs, J., R. C. Young, and G. P. Smith. 1973. Cholecystokinin decreases food intake in rats. *J. Comp. Physiol. Psychol.* 84:488–495.

549. Giebel, H.-D. 1958. Visuelles Lernvermogen bei Einhufern. *Zool. Jahrb.* 67:487–520.

550. Gilbert, B. J.,Jr. and C. W. Arave. 1986. Ability of cattle to distinguish among different wavelengths of light. *J. Dairy Sci.* 69:825–832.

551. Gill, J., K. Skwarlo, and A. Flisinska-Bojanowska. 1974. Diurnal and seasonal changes in carbohydrate metabolism in the blood of Thoroughbred horses. *J. interdiscipl. Cycle Res.* 5:355–361.

552. Gill, J. C., and W. Thomson. 1956. Observations on the behaviour of suckling pigs. *Br. J. Anim. Behav.* 4:46–51.

553. Gillham, S. B., N. H. Dodman, L. Shuster, R. Kream, and W. Rand. 1994. The effect of diet on cribbing behavior and plasma ß-endorphin in horses. *Appl. Anim. Behav. Sci.* 41:147–153.

554. Gleitman, H. 1974. Getting animals to understand the experimen ter's instructions. *Anim. Learn. Behav.* 2:1–5.

555. Glencross, R. G., R. J. Esslemont, M. J. Bryant, and G. S. Pope. 1981. Relationships between the incidence of pre-ovulatory behaviour and the concentrations of oestradiol-17ß and progesterone in bovine plasma. *Appl. Anim. Ethol.* 7:141–148.

556. Goatcher, W. D., and D. C. Church. 1970. Taste responses in ruminants. II. Reactions of sheep to acids, quinine, urea and sodium hydroxide. *J. Anim. Sci.* 30:784–790.

557. Goatcher, W. D., and D. C. Church. 1970. Taste responses in ruminants. III. Reactions of pygmy goats, normal goats, sheep and cattle to sucrose and sodium chloride. *J. Anim. Sci.* 31:364–372.

558. Goatcher, W. D., and D. C. Church. 1970. Taste responses in ruminants. IV. Reactions of pygmy goats, normal goats, sheep and cattle to acetic acid and quinine hydrochloride. *J. Anim. Sci.* 31:373–382.

559. Goddard, M. E., and R. G. Beilharz. 1982. Genetics of traits which determine the suitability of dogs as guide-dogs for the blind. *Appl. Anim. Ethol.* 9:299–315.

560. Gonyou, H. W., R. P. Chapple, and G. R. Frank. 1992. Productivity, time budgets and social aspects of eating in pigs penned in groups of five or individually. *Appl. Anim. Behav. Sci.* 34:291–301.

561. Gonyou, H. W., and J. M. Stookey. 1985. Behavior of parturient ewes in group-lambing pens with and without cubicles. *Appl. Anim. Behav. Sci.* 14:163–171.

562. Gonyou, H. W., and W. R. Stricklin. 1984. Diurnal behavior of feedlot bulls during winter and spring in northern latitudes. *J. Anim. Sci.* 58:1075–1083.

563. Goodwin, M., K. M. Gooding, and F. Regnier. 1979. Sex pheromone in the dog. *Science* 203:559–561.

564. Gould, S. J. 1978. Women's brains. *Natur. Hist.* 87(8):44–50.

565. Grace, J., and M. Russek. 1969. The influence of previous experience on the taste behavior of dogs toward sucrose and saccharin. *Physiol. Behav.* 4:553–558.

566. Grandage, J. 1972. The erect dog penis: A paradox of flexible rigidity. *Vet. Rec.* 91:141–147.

567. Grandin, T. 1993. Behavioral agitation during handling of cattle is persistent over time. *Appl. Anim. Behav. Sci.* 36:1–9.

568. Grandin, T., M. J. Deesing, J. J. Struthers, and A. M. Swinker. 1995. Cattle with hair whorl patterns above the eyes are more behaviorally agitated during restraint. *Appl. Anim. Behav. Sci.* 46:117–123.

569. Graves, H. B. 1984. Behavior and ecology of wild and feral swine (Sus scrofa). *J. Anim. Sci.* 58:482–492.

570. Green, J. S., R. A. Woodruff, and T. T. Tueller. 1984. Livestock-guarding dogs for predator control: Costs, benefits and practicality. *Wildl. Soc. Bull.* 12:44–50.

571. Greene, W. A., L. Mogil, and R. H. Foote. 1978. Behavioral characteristics of freemartins administered estradiol, estrone, testosterone, and dihydrotestosterone. *Horm. Behav.* 10:71–84.

572. Greet, T. R. C. 1982. Windsucking treated by myectomy and neurectomy. *Eq. vet. J.* 14:299–301.

573. Gregory, P. C., M. McFadyen, and D. V. Rayner. 1989. Relation between gastric emptying and short-term regulation of food intake in the pig. *Physiol. Behav.* 45:677–683.

574. Griffith, M. K., and J. E. Minton. 1992. Effect of light intensity on circadian profiles of melatonin, prolactin, ACTH, and cortisol in pigs. *J. Anim. Sci.* 70:492–498.

575. Grossman, M. I., G. M. Cummins, and A. C. Ivy. 1947. The effect of insulin on food intake after vagotomy and sympathectomy. *Am. J. Physiol.* 149:100–102.

576. Grubb, P. 1974. Social organization of Soay sheep and the behaviour of ewes and lambs. In *Island Survivors: The Ecology Of The Soay Sheep Of St. Kilda.* P. A. Jewell, C. Milner, and J. M. Boyd, eds. London, UK: The Athlone Press of the University of London, pp. 131–159.

577. Grubb, P. 1974. The rut and behaviour of Soay rams. In *Island Survivors: The Ecology Of The Soay Sheep Of St. Kilda.* P. A. Jewell, C. Milner, and J. M. Boyd, eds. London, UK: The Athlone Press of the University of London, pp. 195–223.

578. Grzimek, B. 1949. Rangordnungsversuche mit Pferden. *Z. Tierpsychol.* 6:455–464.

579. Grzimek, B. 1952. Versuche uber das Farbsehen von Pflanzenessern. I. Das farbige Sehen (und die Sehscharfe) von Pferden. *Z. Tierpsychol.* 9:23–39.

580. Gubernick, D. J. 1980. Maternal "imprinting" or maternal "labelling" in goats? *Anim. Behav.* 28:124–129.

581. Gubernick, D. J., K. C. Jones, and P. H. Klopfer. 1979. Maternal imprinting in goats. *Anim. Behav.* 27:314–315.

582. Gurr, M. I., J. Kirtland, M. Phillip, and M. P. Robinson. 1977. The consequences of early overnutrition for fat cell size and number: The pig as an experimental model for human obesity. *Int. J. Obesity* 1:151–170.

583. Gustavson, C. R., J. Garcia, W. G. Hankins, and K. W. Rusiniak. 1974. Coyote predation control by aversive conditioning. *Science* 184:581–583.

584. Haag, E. L., R. Rudman, and K. A. Houpt. 1980. Avoidance, maze learning and social dominance in ponies. *J. Anim. Sci.* 50:329–335.

585. Hafez, E. S. E. 1952. Studies on the breeding season and reproduction of the ewe. Part V. Mating behaviour and pregnancy diagnosis. *J. Agric. Sci.* 42:255–265.

586. Hafez, E. S. E. 1974. *Reproduction in Farm Animals.* Philadelphia, PA: Lea & Febiger.

587. Hafez, E. S. E. 1975. *The Behaviour of Domestic Animals.* Baltimore, MD: Williams & Wilkins.

588. Hafez, E. S. E., and M.-F. Bouissou. 1975. The behaviour of cattle. In *The Behaviour Of Domestic Animals.* 3rd ed. E. S. E. Hafez, ed. Baltimore, MD: Williams & Wilkins, pp. 203–245.

589. Hafez, E. S. E., and J. A. Lineweaver. 1968. Suckling behaviour in natural and artificially fed neonate calves. *Z. Tierpsychol.* 25:187–198.

590. Hafez, E. S. E., M. W. Schein and R. Ewbank. 1969. The behaviour of cattle. In *The Behaviour Of Domestic Animals.* 2nd ed. E. S. E. Hafez, ed. Baltimore, MD: Williams & Wilkins, pp. 235–295.

591. Hafez, E. S. E., and J. P. Signoret. 1969. The behaviour of swine. In *The Behaviour Of Domestic Animals.* 2nd ed. E. S. E. Hafez, ed. Baltimore, MD: Williams & Wilkins, pp. 349–390.

592. Hale, E. B. 1966. Visual stimuli and reproductive behavior in bulls. *J. Anim. Sci. (Suppl.)* 25:36–44.

593. Hale, L. A., and S. E. Huggins. 1980. The electroencephalogram of the normal 'grade' pony in sleep and wakefulness. *Comp. Biochem. Physiol.* 66A:251–257.

594. Hall, S. J. G. 1986. Chillingham cattle: Dominance and affinities and access to supplementary food. *Ethology* 71:201–215.

595. Hall, S. J. G. 1989. Chillingham cattle: Social and maintenance behaviour in an ungulate that breeds all year round. *Anim. Behav.* 38:215–225.

596. Hamilton, G. V. 1911. A study of trial and error reactions in mammals. *J. Anim. Behav.* 1:33–66.

597. Hamm, D. 1977. A new surgical procedure to control crib-biting. *Proc. 23rd Annu. Meet. Am Assoc. Eq. Pract.*, pp. 301–302.

598. Hammell, D. L., D. D. Kratzer, and W. J. Bramble. 1975. Avoidance and maze learning in pigs. *J. Anim. Sci.* 40:573–579.

599. Hancock, J. 1950. Grazing habits of dairy cows in New Zealand. *Emp. J. Exp. Agric.* 18:249–263.

599a. Hancock, J. 1954. Studies of grazing behaviour in relation to grassland management. I. Variations in grazing habits of dairy cattle. *J. Agric. Sci.* 44:420–433.

600. Hansen, K. E., and S. E. Curtis. 1980. Prepartal activity of sows in stall or pen. *J. Anim. Sci.* 51:456–460.

600a. Hardison, W. A., H. L. Fisher, G. G. Graf, and N. R. Thompson. 1956. Some observations on the behavior of grazing lactating cows. *J. Dairy Sci.* 39:1735–1741.

600b. Harker, K. W., J. I. Taylor, and D. H. L. Rollinson. 1954. Studies on the habits of Zebu cattle. I. Preliminary observations on grazing habits. *J. Agric. Sci.* 44:193–198.

601. Harker, K. W., J. I. Taylor, and D. H. L. Rollinson. 1956. Studies on the habits of Zebu cattle. V. Night paddocking and its effect on the animal. *J. Agric. Sci.* 47:44–49.

602. Harlow, H. F., M. K. Harlow and E. W. Hansen. 1963. The maternal affectional system of rhesus monkeys. In *Maternal Behavior In Mammals*. H. L. Rheingold, ed. New York, NY: John Wiley & Sons, pp. 254–281.

603. Harlow, H. F., and P. Settlage. 1939. The effect of curariza tion of the fore part of the body upon the retention of conditioned responses in cats. *J. Comp. Psychol.* 27:45–48.

604. Harrington, F. H. 1986. Timber wolf howling playback studies: Discrimination of pup from adult howls. *Anim. Behav.* 34:1575–1577.

605. Harrington, F. H., and L. D. Mech. 1978. Howling at two Minnesota wolf pack summer homesites. *Can. J. Zool.* 56:2024–2028.

606. Harris, D., P. J. Imperato, and B. Oken. 1974. Dog bites—an unrecognized epidemic. *Bull. N. Y. Acad. Med.* 50:981–1000.

607. Harrison, J., and J. Buchwald. 1983. Eyeblink conditioning deficits in the old cat. *Neurobiol. Aging* 4:45–51.

608. Hart, B. L. 1968. Role of prior experience in the effects of castration on sexual behavior of male dogs. *J. Comp. Physiol. Psychol.* 66:719–725.

609. Hart, B. L. 1970. Mating behavior in the female dog and the effects of estrogen on sexual reflexes. *Horm. Behav.* 1:93–104.

610. Hart, B. L. 1974. Environmental and hormonal influences on urine marking behavior in the adult male dog. *Behav. Biol.* 11:167–176.

611. Hart, B. L. 1978. *Feline Behavior. A Practitioner Monograph*. Santa Barbara, CA: Veterinary Practice Publishing Co.

612. Hart, B. L. 1980. Objectionable urine spraying and urine marking in cats: Evaluation of progestin treatment in gonadectomized males and females. *J. Am. Vet. Med. Assoc.* 177:529–533.

613. Hart, B. L. 1981. Olfactory tractotomy for control of objectionable urine spraying and urine marking in cats. *J. Am. Vet. Med. Assoc.* 179:231–234.

614. Hart, B. L., and R. E. Barrett. 1973. Effects of castration on fighting, roaming,

and urine spraying in adult male cats. *J. Am. Vet. Med. Assoc.* 163:290–292.

615. Hart, B. L., and L. Cooper. 1984. Factors relating to urine spraying and fighting in prepubetally gonadectomized male and female cats. *J. Am. Vet. Med. Assoc.* 184:1255–1258.

616. Hart, B. L., R. A. Eckstein, K. L. Powell, and N. H. Dodman. 1993. Effectiveness of buspirone on urine spraying and inappropriate urination in cats. *J. Am. Vet. Med. Assoc.* 203:254–258.

617. Hart, B. L., and L. A. Hart. 1985. *Canine and Feline Behavioral Therapy.* Philadelphia, PA: Lea & Febiger.

618. Hart, B. L., and L. A. Hart. 1988. *The Perfect Puppy: How To Choose Your Dog By Its Behavior.* New York, NY: W.H. Freeman and Company.

619. Hart, B. L., and C. M. Haugen. 1971. Scent marking and sexual behavior maintained in anosmic male dogs. *Commun. Behav. Biol.* 6:131–135.

620. Hart, B. L., and T. O. A. C. Jones. 1975. Effects of castration on sexual behavior of tropical male goats. *Horm. Behav.* 6:247–258.

621. Hart, B. L., and M. G. Leedy. 1985. Analysis of the catnip reaction: Mediation by olfactory system, not vomeronasal organ. *Behav. Neural Biol.* 44:38–46.

622. Hart, B. L., and V. L. Voith. 1978. Changes in urine spraying, feeding and sleep behavior of cats following medial preoptic-anterior hypothalamic lesions. *Brain Res.* 145:406–409.

623. Hartman, D. A., and W. G. Pond. 1960. Design and use of a milking machine for sows. *J. Anim. Sci.* 19:780–785.

624. Haskins, R. 1977. Effect of kitten vocalizations on maternal behavior. *J. Comp. Physiol. Psychol.* 91:830–838.

625. Haskins, R. 1979. A casual analysis of kitten vocalization: An observational and experimental study. *Anim. Behav.* 27:726–736.

626. Hatch, R. C. 1972. Effect of drugs on catnip (Nepeta cataria) induced pleasure behavior in cats. *Am. J. Vet. Res.* 33:143–155.

627. Haugse, C. N., W. E. Dinusson, D. O. Erickson, J. N. Johnson, and M. L. Buchanan. 1965. A day in the life of a pig. *N. Dak. Farm Res.* 23(12):18–23.

628. Hawkes, J., M. Hedges, P. Daniluk, H. F. Hintz, and H. F. Schryver. 1985. Feed preferences of ponies. *Eq. vet. J.* 17:20–22.

629. Hawking, F. 1971. Circadian rhythms in monkeys, dogs and other animals. *J. interdiscipl. Cycle Res.* 2:153–156.

630. Hawking, F. 1971. Circadian rhythms of parasites. *J. interdiscipl. Cycle Res.* 2:157–160.

631. Hayes, K. E. N., and O. J. Ginther. 1989. Relationships between estrous behavior in pregnant mares and the presence of a female conceptus. *J. Eq. Vet. Sci.* 9:316–318.

632. Hayman, R. H. 1964. Exercise of mating preference by a Merino ram. *Nature* 203:160–162.

633. Heady, H. F. 1964. Palatability of herbage and animal preference. *J. Range Manag.* 17:76–82.

634. Hebel, R. 1976. Distribution of retinal ganglion cells in five mammalian species (pig, sheep, ox, horse, dog). *Anat. Embryol.* 150:45–51.

635. Hedlund, L., M. M. Lischko, M. D. Rollag, and G. D. Niswender. 1977. Melatonin: Daily cycle in plasma and cerebrospinal fluid of calves. *Science* 195:686–687.

636. Hegsted, D. M., S. N. Gershoff, and E. Lentini. 1956. The development of palatability tests for cats. *Am. J. Vet. Res.* 17:733–737.

637. Hein, A., and R. Held. 1967. Dissociation of the visual placing response into elicited and guided components. *Science* 158:390–392.

637a. Hein, M. A. 1935. Grazing time of beef steers on permanent pastures. *J. Am. Soc. Agron.* 27:675–679.

638. Heird, J. C., A. M. Lennon, and R. W. Bell. 1981. Effects of early experience on the learning ability of yearling horses. *J. Anim. Sci.* 53:1204–1209.

639. Heird, J. C., D. D. Whitaker, R. W. Bell, C. B. Ramsey, and C. E. Lokey. 1986. The effects of handling at different ages on the subsequent learning ability of 2-year-old horses. *Appl. Anim. Behav. Sci.* 15:15–25.

640. Heitzman, R. J. 1978. The use of hormones to regulate the utilization of nutrients in farm animals: Current farm practices. *Proc. Nutr. Soc.* 37:289–293.

641. Hellekant, G., C. H. A. F. Segerstad, and T. W. Roberts. 1994. Sweet taste in the calf: III. Behavioral responses to sweeteners. *Physiol. Behav.* 56:555–562.

642. Hemsworth, P. H., and J. L. Barnett. 1992. The effects of early contact with humans on the subsequent level of fear of humans in pigs. *Appl. Anim. Behav. Sci.* 35:83–90.

643. Hemsworth, P. H., J. L. Barnett, and C. Hansen. 1987. The influence of inconsistent handling by humans on the behaviour, growth and corticosteroids of young pigs. *Appl. Anim. Behav. Sci.* 17:245–252.

644. Hemsworth, P. H., J. L. Barnett, C. Hansen, and C. G. Winfield. 1986. Effects of social environment on welfare status and sexual behaviour of female pigs. II. Effects of space allowance. *Appl. Anim. Behav. Sci.* 16:259–267.

645. Hemsworth, P. H., R. G. Beilharz, and D. B. Galloway. 1977. Influence of social conditions during rearing on the sexual behaviour of the domestic boar. *Anim. Prod.* 24:245–251.

646. Hemsworth, P. H., G. M. Cronin, C. Hansen, and C. G. Winfield. 1984. The effects of two oestrus detection procedures and intense boar stimulation near the time of oestrus on mating efficiency of the female pig. *Appl. Anim. Behav. Sci.* 12:339–347.

647. Hemsworth, P. H., J. K. Findlay, and R. G. Beilharz. 1978. The importance of physical contact with other pigs during rearing on the sexual behaviour of the male domestic pig. *Anim. Prod.* 27:201–207.

648. Hemsworth, P. H., C. Hansen, and C. G. Winfield. 1989. The influence of mating conditions on the sexual behaviour of male and female pigs. *Appl. Anim. Behav. Sci.* 23:207–214.

649. Hemsworth, P. H., E. O. Price, and A. J. Tilbrook. 1992. Influence of the sexual motivation of the boar on the sexual partner preferences of oestrous gilts. *Appl. Anim. Behav. Sci.* 33:209–215.

650. Hemsworth, P. H., C. G. Winfield, J. L. Barnett, B. Schirmer, and C. Hansen. 1986. A comparison of the effects of two estrus detection procedures and two housing systems on the oestrus dectection rate of female pigs. *Appl. Anim. Behav. Sci.* 16:345–351.

651. Hemsworth, P. H., C. G. Winfield, R. G. Beilharz, and D. B. Galloway. 1977. Influence of social conditions post-puberty on the sexual behaviour of the domestic male pig. *Anim. Prod.* 25:305–309.

652. Hemsworth, P. H., C. G. Winfield, and P. D. Mullaney. 1976. A study of the development of the teat order in piglets. *Appl. Anim. Ethol.* 2:225–233.

653. Hendricks, J. C., A. R. Morrison, G. L. Farnbach, S. A. Steinberg, and G. Mann. 1981. A disorder of rapid eye movement sleep in a cat. *J. Am. Vet. Med. Assoc.* 178:55–57.

654. Hendriks, W. H., M. F. Tarttelin, and P. J. Moughan. 1995. Twenty-four hour feline excretion patterns in entire and castrated cats. *Physiol. Behav.* 58:467–469.

655. Henry, Y., B. Sève, Y. Colléaux, P. Ganier, C. Saligaut, and P. Jégo. 1992. Interactive effects of dietary levels of tryptophan and protein on voluntary feed intake and growth performance in pigs, in relation to plasma free amino acids and hypothalamic serotonin. *J. Anim. Sci.* 70:1873–1887.

656. Hepper, P. G. 1986. Sibling recognition in the domestic dog. *Anim. Behav.* 34:288–289.

657. Herbel, C. H., and A. B. Nelson. 1966. Species preference of Hereford and Santa Gertrudis cattle on a southern New Mexico range. *J. Range Manag.* 19:177–181.

658. Herd, R. M. 1988. A technique for cross-mothering beef calves which does

not affect growth. *Appl. Anim. Behav. Sci.* 19:239–244.

659. Herring, S. W., and R. P. Scapino. 1973. Physiology of feeding in miniature pigs. *J. Morphol.* 141:427–460.

660. Hersher, L., J. B. Richmond and A. U. Moore. 1963. Maternal behavior in sheep and goats. In *Maternal Behavior In Mammals*. H. L. Rheingold, ed. New York, NY: John Wiley & Sons, pp. 203–232.

661. Hessing, M. J. C., A. M. Hagels, J. A. M. van Beek, P. R. Wiepkema, G. P. Schouten, and R. Krukow. 1993. Individual behavioural characteristics in pigs. *Appl. Anim. Behav. Sci.* 37:285–295.

662. Hetts, S., J. Derrell Clark, J. P. Calpin, C. E. Arnold, and J. M. Mateo. 1992. Influence of housing conditions on beagle behaviour. *Appl. Anim. Behav. Sci.* 34:137–155.

663. Hetzer, H. O., and W. R. Harvey. 1967. Selection for high and low fatness in swine. *J. Anim. Sci.* 26:1244–1251.

664. Hill, J. O., E. J. Pavlik, G. L. Smith,III, G. M. Burghardt, and P. B. Coulson. 1976. Species-characteristic responses to catnip by undomesticated felids. *J. Chem. Ecol.* 2:239–253.

665. Hinch, G. N., J. J. Lynch, and C. J. Thwaites. 1982. Patterns and frequency of social interactions in young grazing bulls and steers. *Appl. Anim. Ethol.* 9:15–30.

666. Hirsch, E., C. Dubose, and H. L. Jacobs. 1978. Dietary control of food intake in cats. *Physiol. Behav.* 20:287–295.

667. Hite, M., H. M. Hanson, N. R. Bohidar, P. A. Conti, and P. A. Mattis. 1977. Effect of cage size on patterns of activity and health of beagle dogs. *Lab. Anim. Sci.* 27:60–64.

668. Hoagland, T. A., and M. A. Diekman. 1982. Influence of supplemental lighting during increasing daylength on libido and reproductive hormones in prepubertal boars. *J. Anim. Sci.* 55:1483–1489.

669. Hoffman, M. P., and H. L. Self. 1973. Behavioral traits of feedlot steers in Iowa. *J. Anim. Sci.* 37:1438–1445.

670. Hoffman, R. 1985. On the development of social behavior in immature males of a feral horse population (Equus przewalski f. caballus). *Zeitschrift Saugetierkunde* 50:302–314.

671. Hoffman, R. M., D. S. Kronfeld, J. L. Holland, and K. M. Greiwe-Crandell. 1995. Preweaning diet and stall weaning method influences on stress response in foals. *J. Anim. Sci.* 73:2922–2930.

671a. Holder, J. M. 1960. Observations on the grazing behaviour of lactating dairy cattle in a subtropical environment. *J. Agric. Sci.* 55:261–267.

672. Holmes, J. H., and L. J. Cizek. 1951. Observations on sodium chloride depletion in the dog. *Am. J. Physiol.* 164:407–414.

673. Holmes, L. N., G. K. Song, and E. O. Price. 1987. Head partitions facilitate feeding by subordinate horses in the presence of dominant pen-mates. *Appl. Anim. Behav. Sci.* 19:179–182.

674. Hopkins, S. G., T. A. Schubert, and B. L. Hart. 1976. Castration of adult male dogs: Effects on roaming, aggression, urine marking, and mounting. *J. Am. Vet. Med. Assoc.* 168:1108–1110.

675. Hoppenbrouwers, T., and M. B. Sterman. 1975. Development of sleep state patterns in the kitten. *Exp. Neurol.* 49:822–838.

676. Horrell, I., and J. Hodgson. 1992. The bases of sow-piglet identification. 2. Cues used by piglets to identify their dam and home pen. *Appl. Anim. Behav. Sci.* 33:329–343.

677. Horrell, I., and J. Hodgson. 1992. The bases of sow-piglet identification 1. The identification by sows of their own piglets and the presence of intruders. *Appl. Anim. Behav. Sci.* 33:319–327.

678. Hosoi, E., L. R. Rittenhouse, D. M. Swift, and R. W. Richards. 1995. Foraging strategies of cattle in a Y-maze: Influence of food availability. *Appl. Anim. Behav. Sci.* 43:189–196.

679. Houpt, K. A. 1977. Horse behavior: Its relevancy to the equine practitioner. *J. Eq. Med. Surg.* 1:87–94.

680. Houpt, K. A. 1977. The physiology of hunger and palatability in animals. In *Proceedings of the Cornell Conference for Feed Manufacturers.* Ithaca, NY: Cornell University, pp. 113–119.

681. Houpt, K. A. 1983. Disruption of the human-companion animal bond: Aggressive behavior in dogs. In *New Perspectives on Our Lives with Companion Animals.* A. H. Katcher and A. M. Beck, eds, pp. 197–204.

682. Houpt, K. A., B. Coren, H. F. Hintz, and J. E. Hilderbrant. 1979. Effect of sex and reproductive status on sucrose preference, food intake, and body weight of dogs. *J. Am. Vet. Med. Assoc.* 174:1083–1085.

683. Houpt, K. A., and H. Hintz. 1978. Palatability and canine food preferences. *Canine Pract.* 5:29–35.

684. Houpt, K. A., and H. F. Hintz. 1982. Some effects of maternal deprivation on maintenance behavior, spatial relationships and responses to environmental novelty in foals. *Appl. Anim. Ethol.* 9:221–230.

685. Houpt, K. A., H. F. Hintz, and P. Shepherd. 1978. The role of olfaction in canine food preferences. *Chem. Senses Flavor* 3:281–290.

686. Houpt, K. A., and T. R. Houpt. 1976. Comparative aspects of the ontogeny of taste. *Chem. Senses Flavor* 2:219–228.

687. Houpt, K. A., and T. R. Houpt. 1988. Social and illumination preferences of mares. *J. Anim. Sci.* 66:2159–2164.

688. Houpt, K. A., T. R. Houpt, and W. G. Pond. 1977. Food intake controls in the suckling pig: Glucoprivation and gastro-intestinal factors. *Am. J. Physiol.* 232:E510–E514.

689. Houpt, K. A., T. R. Houpt, and W. G. Pond. 1979. The pig as a model for the study of obesity and of control of food intake: A review. *Yale J. Biol. Med.* 52:307–329.

690. Houpt, K. A., and R. R. Keiper. 1982. The position of the stallion in the equine dominance hierarchy of feral and domestic ponies. *J. Anim. Sci.* 54:945–950.

691. Houpt, K. A., K. Law, and V. Martinisi. 1978. Dominance hierarchies in domestic horses. *Appl. Anim. Ethol.* 4:273–283.

692. Houpt, K. A., N. Northrup, T. Wheatley, and T. R. Houpt. 1991. Thirst and salt appetite in horses treated with furosemide. *J. Appl. Physiol.* 71(6):2380–2386.

693. Houpt, K. A., M. S. Parson, and H. F. Hintz. 1982. Learning ability of orphan foals, of normal foals and of their mothers. *J. Anim. Sci.* 55:1027–1032.

694. Houpt, K. A., W. Rivera, and L. Glickstein. 1989. The flehmen response of bulls and cows. *Theriogenology* 32:343–350.

695. Houpt, K. A., and G. Wollney. 1989. Frequency of masturbation and time budgets of dairy bulls used for semen production. *Appl. Anim. Behav. Sci.* 24:217–225.

696. Houpt, K. A., and T. R. Wolski. 1979. Equine maternal behavior and aberrations. *Equine Pract.* 1:7–20.

697. Houpt, K. A., and T. R. Wolski. 1980. Stability of equine hierarchies and prevention of dominance related aggression. *Eq. vet. J.* 12:18–24.

698. Houpt, K. A., D. M. Zahorik, and J. A. Swartzman-Andert. 1990. Taste aversion learning in horses. *J. Anim. Sci.* 68:2340–2344.

699. Houpt, T. R. 1974. Stimulation of food intake in ruminants by 2-deoxy-D-glucose and insulin. *Am. J. Physiol.* 227:161–167.

700. Houpt, T. R. 1984. Controls of feeding in pigs. *J. Anim. Sci.* 59:1345–1353.

701. Houpt, T. R., S. M. Anika, and K. A. Houpt. 1979. Preabsorptive intestinal satiety controls of food intake in pigs. *Am. J. Physiol.* 236:R328–R337.

702. Houpt, T. R., B. A. Baldwin, and K. A. Houpt. 1983. Effects of duodenal osmotic loads on spontaneous meals in pigs. *Physiol. Behav.* 30:787–795.

703. Houpt, T. R., and H. H. Hance. 1969. Effect of 2-deoxy-D glucose on food intake by the goat, rabbit and dog [abstract]. *Fed. Proc.* 28:648.

704. Houpt, T. R., K. A. Houpt, and A. A. Swan. 1983. Duodenal osmoconcentration and food intake in pigs after ingestion of hypertonic nutrients. *Am. J. Physiol.* 245:R181–F189.

705. Houpt, T. R., L. C. Weixler, and D. W. Troy. 1986. Water drinking induced by gastric secretagogues in pigs. *Am. J. Physiol.* 251:R157–R164.

706. Hubrecht, R. C. 1993. A comparison of social and environmental enrichment methods for laboratory housed dogs. *Appl. Anim. Behav. Sci.* 37:345–361.

707. Hubrecht, R. C., J. A. Serpell, and T. B. Poole. 1992. Correlates of pen size and housing conditions on the behaviour of kennelled dogs. *Appl. Anim. Behav. Sci.* 34:365–383.

708. Hudson, S. J. 1977. Multiple fostering of calves onto nurse cows at birth. *Appl. Anim. Ethol.* 3:57–63.

709. Hudson, S. J., and M. M. Mullord. 1977. Investigations of maternal bonding in dairy cattle. *Appl. Anim. Ethol.* 3:271–276.

710. Hughes, G. P., and D. Reid. 1951. Studies on the behaviour of cattle and sheep in relation to the utilization of grass. *J. Agric. Sci.* 41:350–366.

711. Hughes, P. E., P. H. Hemsworth, and C. Hansen. 1985. The effects of supplementary olfactory and auditory stimuli on the stimulus value and mating success of the young boar. *Appl. Anim. Behav. Sci.* 14:245–252.

712. Hulet, C. V. 1966. Behavioral, social and psychological factors affecting mating time and breeding efficiency in sheep. *J. Anim. Sci. (Suppl.)* 25:5–20.

713. Hulet, C. V., G. Alexander and E. S. E. Hafez. 1975. The behaviour of sheep. In *The Behaviour of Domestic Animals*. E. S. E. Hafez, ed. Baltimore, MD: Williams & Wilkins.

714. Hulet, C. V., D. M. Anderson, J. N. Smith, W. L. Shupe, C. A. J. Taylor, and L. W. Murray. 1989. Bonding of goats to sheep and cattle for protection from predators. *Appl. Anim. Behav. Sci.* 22:261–267.

715. Hulet, C. V., R. L. Blackwell, and S. K. Ercanbrack. 1964. Observations on sexually inhibited rams. *J. Anim. Sci.* 23:1095–1097.

716. Hulet, C. V., R. L. Blackwell, S. K. Ercanbrack, D. A. Price, and L. O. Wilson. 1962. Mating behavior of the ewe. *J. Anim. Sci.* 21:870–874.

717. Hunter, R. F., and G. E. Davies. 1963. The effect of method of rearing on the social behaviour of Scottish Blackface hoggets. *Anim. Prod.* 5:183–194.

718. Hunter, W. S. 1917. The delayed reaction in a child. *Psychol. Rev.* 24:74–87.

718a. Hunthausen, W. 1997. Effects of aggressive behavior on canine welfare. *J.A.V.M.A.* 210:1134-1136.

719. Hunthausen, W., and G. Landsberg. 1993. *Providing Behavior Services In Veterinary Practices*. Denver, CO: American Animal Hospital Association.

720. Hurnik, J. F., G. J. King, and H. A. Robertson. 1975. Estrous and related behaviour in postpartum Holstein cows. *Appl. Anim. Ethol.* 2:55–68.

721. Hutson, G. D. 1980. The effect of previous experience on sheep movement through yards. *Appl. Anim. Ethol.* 6:233–240.

722. Hutson, G. D. 1985. The influence of barley food rewards on sheep movement through a handling system. *Appl. Anim. Behav. Sci.* 14:263–273.

723. Hutson, G. D., M. F. Argent, L. G. Dickenson, and B. G. Luxford. 1992. Influence of parity and time since parturition on responsiveness of sows to a piglet distress call. *Appl. Anim. Behav. Sci.* 34:303–313.

724. Hutson, G. D., and M. J. Haskell. 1990. The behaviour of farrowing sows

with free and operant access to an earth floor. *Appl. Anim. Behav. Sci.* 26:363–372.

725. Hutson, G. D., E. O. Price, and L. G. Dickenson. 1993. The effect of playback volume and duration on the response of sows to piglet distress calls. *Appl. Anim. Behav. Sci.* 37:31–37.

726. Hutson, G. D., J. L. Wilkinson, and B. G. Luxford. 1991. The response of lactating sows to tactile, visual and auditory stimuli associated with a model piglet. *Appl. Anim. Behav. Sci.* 32:129–137.

727. Igel, G. J., and A. D. Calvin. 1960. The development of affectional responses in infant dogs. *J. Comp. Physiol. Psychol.* 53:302–305.

728. Illius, A. W., N. B. Haynes, and G. E. Lamming. 1976. Effects of ewe proximity on peripheral plasma testosterone levels and behaviour in the ram. *J. Reprod. Fert.* 48:25–32.

729. Illmann, G., and J. Madlafousek. 1995. Occurrence and characteristics of unsuccessful nursings in minipigs during the first week of life. *Appl. Anim. Behav. Sci.* 44:9–18.

730. Illmann, G., and M. Spinka. 1993. Maternal behaviour of dairy heifers and sucking of their newborn calves in group housing. *Appl. Anim. Behav. Sci.* 36:91–98.

731. Ingram, D. L., and M. J. Dauncy. 1985. Circadian rhythms in the pig. *Comp. Biochem. Physiol.* 82A:1–5.

732. Ingram, D. L., M. J. Dauncy, and K. F. Legge. 1985. Synchronization of motor activity in young pigs to a non-circadian rhythm without affecting food intake and growth. *Comp. Biochem. Physiol.* 80A:363–368.

733. Ingram, D. L., and K. F. Legge. 1974. Effects of environmental temperature on food intake in growing pigs. *Comp. Biochem. Physiol.* 48(A):573–581.

734. Ingram, D. L., and D. B. Stephens. 1979. The relative importance of thermal, osmotic and hypovolaemic factors in the control of drinking in the pig. *J. Physiol.* 293:501–512.

735. Inselman-Temkin, B. R., and J. P. Flynn. 1973. Sex-dependent effects of gonadal and gonadotropic hormones on centrally—elicited attack in cats. *Brain Res.* 60:393–410.

736. Irwin, M. R., D. R. Melendy, M. S. Amoss, and D. P. Hutcheson. 1979. Roles of predisposing factors and gonadal hormones in the buller syndrome of feedlot steers. *J. Am. Vet. Med. Assoc.* 174:367–370.

737. Izard, M. K., and J. G. Vandenbergh. 1982. Priming pheromones from oestrous cows increase synchronization of oestrus in dairy heifers after PGF-2a injection. *J. Reprod. Fert.* 66:189–196.

738. Jackson, B., and A. Reed. 1969. Catnip and the alteration of consciousness. *J. Am. Med. Assoc.* 207:1349–1350.

739. Jackson, H. M., and D. W. Robinson. 1971. Evidence for hypothalamic a and ß adrenergic receptors involved in the control of food intake of the pig. *Br. Vet. J.* 127:li–liii.

740. Jackson, S. A., R. A. Rich, and S. L. Ralston. 1984. Feeding behavior and feed efficiency in groups of horses as a function of feeding frequency and use of alfalfa hay cubes. *J. Anim. Sci.* 59 (Suppl. 1):152–153.

741. Jalowiec, J. E., J. Panksepp, H. Shabshelowitz, A. J. Zolovick, W. Stern, and P. J. Morgane. 1973. Suppression of feeding in cats following 2-deoxy-D-glucose. *Physiol. Behav.* 10:805–807.

742. James, W. T. 1951. Social organization among dogs of different temperaments, terriers and beagles, reared together. *J. Comp. Physiol. Psychol.* 44:71–77.

743. James, W. T., and T. F. Gilbert. 1955. The effect of social facilitation on food intake of puppies fed separately and together for the first 90 days of life. *Br. J. Anim. Behav.* 3:131–133.

744. Janowitz, H. D., and M. I. Grossman. 1949. Some factors affecting the food intake of normal dogs and dogs with esophagostomy and gastric fistula. *Am. J. Physiol.* 159:143–148.

745. Janowitz, H. D., and M. I. Grossman. 1949. Effect of variations in nutritive density on intake of food of dogs and rats. *Am. J. Physiol.* 158:184–193.

746. Jarvis, A. M., and M. S. Cockram. 1995. Some factors affecting resting behaviour of sheep in slaughterhouse lairages after transport from farms. *Anim. Welfare* 4:53–60.

747. Jensen, P. 1980. An ethogram of social interacion patterns in group-housed dry sows. *Appl. Anim. Ethol.* 6:341–350.

748. Jensen, P. 1984. Effects of confinement on social interaction patterns in dry sows. *Appl. Anim. Behav. Sci.* 12:93–101.

749. Jensen, P. 1986. Observations on the maternal behavior of free-ranging domestic pigs. *Appl. Anim. Behav. Sci.* 16:131–142.

750. Jensen, P. 1993. Nest building in domestic sows: The role of external stimuli. *Anim. Behav.* 45:351–358.

751. Jensen, P., K. Floren, and B. Hobroh. 1987. Peri-parturient changes in behaviour in free-ranging domestic pigs. *Appl. Anim. Behav. Sci.* 17:69–76.

752. Jensen, P., G. Stangel, and B. Algers. 1991. Nursing and suckling behaviour of semi-naturally kept pigs during the first 10 days postpartum. *Appl. Anim. Behav. Sci.* 31:195–209.

753. Jensen, P., K. Vestergaard, and B. Algers. 1993. Nestbuilding in free-ranging domestic sows. *Appl. Anim. Behav. Sci.* 38:245–255.

754. Jeppesen, L. E. 1982. Teat-order in groups of piglets reared on an artificial sow. I. Formation of teat-order and influence of milk yield on teat preference. *Appl. Anim. Ethol.* 8:335–345.

755. Jeppesen, L. E. 1982. Teat-order in groups of piglets reared on an artificial sow. II. Maintenance of teat-order with some evidence for the use of odour cues. *Appl. Anim. Ethol.* 8:347–355.

756. Jewell, P. A., S. J. G. Hall, and M. M. Rosenberg. 1986. Multiple mating and siring success during natural oestrus in the ewe. *J. Reprod. Fert.* 77:81–89.

757. John, E. R., P. Chesler, F. Bartlett, and I. Victor. 1968. Observation learning in cats. *Science* 159:1589–1491.

758. Johnson, B. F., and C. Chura. 1974. Diurnal variation in the effect of tolbutamide. *Am. J. Med. Sci.* 268:93–96.

759. Johnsson, A., W. Engelmann, B. Pflug, and W. Klemke. 1980. Influence of lithium ions on human circadian rhythms. *Z. Naturforsch.* 35:503–507.

760. Johnstone-Wallace, D. B., and K. Kennedy. 1944. Grazing management practices and their relationship to the behaviour and grazing habits of cattle. *J. Agric. Sci.* 34:190–197.

761. Jones, C. G., K. D. Maddever, D. L. Court, and M. Phillips. 1966. The time taken by cows to eat concentrates. *Anim. Prod.* 8:489–497.

762. Jordan, W. A., E. E. Lister, J. M. Wauthy, and J. C. Comeau. 1973. Voluntary roughage intake by nonpregnant and pregnant or lactating beef cows. *Can. J. Anim. Sci.* 53:733–738.

763. Juarbe-Díaz, S. V., and K. A. Houpt. 1996. Comparison of two antibarking collars for treatment of nuisance barking. *J. Am. Anim. Hosp. Assoc.* 32:231–235.

764. Kalmus, H. 1955. The discrimination by the nose of the dog of individual human odours and in particular of the odours of twins. *Br. J. Anim. Behav.* 3:25–31.

765. Kanarek, R. B. 1975. Availability and caloric density of the diet as determinants of meal patterns in cats. *Physiol. Behav.* 15:611–618.

766. Kanno, Y. 1977. Experimental studies on body temperature rhythm in dogs. I. Application of cosinor method to body temperature rhythm in dogs. *Jpn. J. Vet. Sci.* 39:69–76.

767. Karas, G. G., R. I. Willham, and D. F. Cox. 1962. Avoidance learning in swine. *Psychol. Rep.* 11:51–54.

768. Kare, M. R., W. C. Pond, and J. Campbell. 1965. Observations on the taste reactions in pigs. *Anim. Behav.* 13:265–269.

769. Karlander, S., J. Mansson, and G. Tufvesson. 1965. Buccostomy as a method of treatment for aerophagia (windsucking) in the horse. *Nordisk Vet.-Med.* 17:455–458.

770. Karn, H. W., and H. R. Malamud. 1939. The behavior of dogs on the double alternation problem in the temporal maze. *J. Comp. Psychol.* 27:461–466.

771. Karn, J. F., and D. C. Clanton. 1974. Electronically controlled individual cattle feeding [abstract]. *J. Anim. Sci.* 39:136.

772. Karsh, E. B., and D. C. Turner. 1988. The human-cat relationship. In *The Domestic Cat. The Biology Of Its Behaviour.* D. C. Turner and P. Bateson, eds. Cambridge, UK: Cambridge University Press, pp. 159–177

773. Kaseda, Y., and A. M. Khalil. 1996. Harem size and reproductive success of stallions in Misaki feral horses. *Appl. Anim. Behav. Sci.* 47:163–174.

774. Katz, L. S., E. O. Price, S. J. R. Wallach, and J. J. Zenchak. 1988. Sexual performance of rams reared with or without females after weaning. *J. Anim. Sci.* 66:1166–1173.

775. Katz, R. J., and E. Thomas. 1976. Effects of para-chlorophenylalanine upon brain stimulated affective attack in the cat. *Pharm. Biochem. Behav.* 5:392–94.

776. Keiper, R. R. 1976. Social organization of feral ponies. *Proc. Penn. Acad. Sci.* 50:69–70.

776a. Keiper, R. R. 1981. Time bugets of EIA positive ponies. *Equine Pract.* 3:6–10.

777. Keiper, R. R. 1985. *The Assateague Ponies.* Centreville, MD: Tidewater Press.

778. Keiper, R. R., and J. Berger. 1982. Refuge-seeking and pest avoidance by feral horses in desert and island environments. *Appl. Anim. Ethol.* 9:111–120.

779. Keiper, R. R., and K. A. Houpt. 1984. Reproduction in feral horses: An eight year study. *Amer. J. Vet. Res.* 45:991–995.

779a. Keiper, R. R. and M. A. Keenan. 1980. Nocturnal activity patterns of feral horses. *J. Mammal.* 61:116–118.

780. Keiper, R. R., and H. Receveur. 1992. Social interactions of free-ranging Przewalski horses in semi-reserves in the Netherlands. *Appl. Anim. Behav. Sci.* 33:303–318.

781. Keiper, R. R., and H. H. Sambraus. 1986. The stability of equine dominance hierarchies and the effects of kinship, proximity and foaling status on hierarchy rank. *Appl. Anim. Behav. Sci.* 16:121–130.

782. Kendrick, K. M., K. Atkins, M. R. Hinton, K. B. Broad, C. Fabre-Nys, and B. Keverne. 1995. Facial and vocal discrimination in sheep. *Anim. Behav.* 49:1665–1676.

783. Kendrick, K. M., and B. A. Baldwin. 1986. Characterization of neuronal responses in the zona incerta of the subthalamic region of the sheep during the ingestion of food and liquid. *Neurosci. Lett.* 63:237–242.

784. Kendrick, K. M., and B. A. Baldwin. 1986. The activity of neurones in the lateral hypothalamus and zona incerta of the sheep responding to the sight or approach of food is modified by learning and satiety and reflects food preferences. *Brain Res.* 375:320–328.

785. Kendrick, K. M., and E. B. Keverne. 1991. Importance of progesterone and estrogen priming for theinduction of maternal behavior by vaginocervical stimulation in sheep: Effects of maternal experience. *Physiol. Behav.* 49:745–750.

786. Kendrick, K. M., E. B. Keverne, and B. A. Baldwin. 1987. Intracerebroventricular oxytocin stimulates maternal behaviour in sheep. *Neuroendocrinology* 46:56–61.

787. Kendrick, K. M., E. B. Keverne, B. A. Baldwin, and D. F. Sharman. 1986. Cerebrospinal fluid levels of acetylcholinesterase, monoamines and oxytocin during

labour, parturition, vaginal stimulation, lamb separartion and suckling in sheep. *Neuroendocrinology* 44:149–156.

788. Kennedy, J. M., and B. A. Baldwin. 1972. Taste preferences in pigs for nutritive and non-nutritive sweet solutions. *Anim. Behav.* 20:706–718.

789. Kenny, F. J., and P. V. Tarrant. 1987. The behaviour of young Friesian bulls during social re-grouping at an abattoir. Influence of an overhead electrified wire grid. *Appl. Anim. Behav. Sci.* 18:233–246.

790. Kent, J. P. 1984. A note on multiple fostering of calves on to nurse cows at a few days post-partum. *Appl. Anim. Behav. Sci.* 12:183–186.

791. Kerruish, B. M. 1955. The effect of sexual stimulation prior to service on the behaviour and conception rate of bulls. *Br. J. Anim. Behav.* 3:125–130.

792. Keverne, E. B., F. Levy, P. Poindron, and D. R. Lindsay. 1982. Vaginal stimulation: An important determinant of maternal bonding in sheep. *Science* 219:81–83.

793. Key, C., and R. M. MacIver. 1977. Factors affecting sexual preferences in sheep [abstract]. *Appl. Anim. Ethol.* 3:291.

794. Khalaf, F. 1969. Hyperphagia and aphagia in swine with induced hypothalamic lesions. *Res. Vet. Sci.* 10:514–517.

795. Kiddy, C. A., D. S. Mitchell, D. J. Bolt, and H. W. Hawk. 1978. Detection of estrus-related odors in cows by trained dogs. *Biol. Reprod.* 19:389–395.

796. Kiley, M. 1972. The vocalizations of ungulates, their causation and function. *Z. Tierpsychol.* 31:171–222.

797. Kiley, M. 1976. Fostering and adoption in beef cattle. *Br. Cattle Breeders Club Dig.* 31:42–55.

798. Kiley-Worthington, M. 1976. The tail movements of ungulates, canids and felids with particular reference to their causation and function as displays. *Behaviour* 56:69–115.

799. Kiley-Worthington, M. 1977. *Behavioural Problems of Farm Animals*. Stocksfield, UK: Oriel Press.

800. Kiley-Worthington, M., and P. Savage. 1978. Learning in dairy cattle using a device for economical management of behaviour. *Appl. Anim. Ethol.* 4:119–124.

801. Kilgour, R. 1972. Some observations on the suckling activity of calves on nurse cows. *Proc. N. Z. Soc. Anim. Prod.* 32:132–136.

802. Kilgour, R. 1981. Use of the Hebb-Williams closed-field test to study the learning ability of Jersey cows. *Anim. Behav.* 29:850–860.

803. Kilgour, R., and D. N. Campin. 1973. The behaviour of entire bulls of different ages at pasture. *Proc. N. Z. Soc. Anim. Prod.* 33:125–138.

804. Kilgour, R., and T. H. Scott. 1959. Leadership in a herd of dairy cows. *Proc. N. Z. Soc. Anim. Prod.* 19:36–43.

805. Kilgour, R., B. H. Skarsholt, J. F. Smith, K. J. Bremner, and M. C. L. Morrison. 1977. Observations on the behaviour and factors influencing the sexually-active group in cattle. *Proc. N. Z. Soc. Anim. Prod.* 37:128–135.

806. Kilgour, R., and C. G. Winfield. 1977. Pen-mating of pedigree sheep. *N. Z. J. Agric.* 134:25–27.

807. Kilgour, R., C. G. Winfield, K. J. Bremner, M. M. Mullord, H. de Langen, and S. J. Hudson. 1976. Behaviour of early-weaned calves in indoor individual cubicles and group pens. *N. Z. Vet. J.* 23:119–123.

808. Kim, F. B., R. E. Jackson, G. D. Gordon, and M. S. Cockram. 1994. Resting behaviour of sheep in a slaughterhouse lairage. *Appl. Anim. Behav. Sci.* 40:45–54.

809. King, J. E., R. F. Becker, and J. E. Markee. 1964. Studies on olfactory discrimination in dogs: (3) Ability to detect human odour trace. *Anim. Behav.* 12:311–315.

810. Kirkpatrick, J. F., R. Vail, S. Devous, S. Schwend, C. B. Baker, and L. Wiesner. 1976. Diurnal variation of plasma testosterone in wild stallions. *Biol. Reprod.* 15:98–101.

811. Kirkwood, R. N., J. M. Forbes, and P. E. Hughes. 1981. Influence of boar con-

tact on attainment of puberty in gilts after removal of the olfactory bulbs. *J. Reprod. Fert.* 61:193–196.

812. Kitchell, R. L. 1972. Dogs know what they like. *Friskies Res. Dig.* 8(3):1–4.

813. Klemm, W. R., C. J. Sherry, L. M. Schake, and R. F. Sis. 1983. Homosexual behavior in feedlot steers: An aggression hypothesis. *Appl. Anim. Ethol.* 11:187–195.

814. Kling, A., and D. Coustan. 1964. Electrical stimulation of the amygdala and hypothalamus in the kitten. *Exp. Neurol.* 10:81–89.

815. Kling, A., J. K. Kovach and T. J. Tucker. 1969. The behaviour of cats. In *The Behaviour of Domestic Animals*. 2nd ed. E. S. E. Hafez, ed. Baltimore: Williams and Wilkins, pp. 482–512.

816. Klingel, H. 1974. A comparison of the social behaviour of the Equidae. In *The Behaviour of Ungulates and Its Relation to Management*. V. Geist and F. Walther, eds. Morges, Switzerland: International Union for Conservation of Nature and Natural Resources.

817. Klopfer, F. D. 1961. Early experience and discrimination learning in swine [abstract]. *Am. Zool.* 1:366.

818. Klopfer, F. D. 1966. Visual learning in swine. In *Swine in Biomedical Research*. L. K. Bustad, R. O. McClellan, and M. P. Burns, eds. Richland, WA: Bettelle Memorial Institute Pacific Northwest Laboratory.

819. Klopfer, P. H., and J. Gamble. 1966. Maternal "imprinting" in goats: The role of chemical senses. *Z. Tierpsychol.* 23:588–592.

820. Knecht, C. D., J. E. Oliver, R. Redding, R. Selcer, and G. Johnson. 1973. Narcolepsy in a dog and a cat. *J. Am. Vet. Med. Assoc.* 162:1052–1053.

821. Knight, T. W., and P. R. Lynch. 1980. Source of ram pheromones that stimulate ovulation in the ewe. *Anim. Reprod. Sci.* 3:133–136.

822. Koehler, W. 1962. *The Koehler Method of Dog Training*. New York, NY: Howell Book House.

823. Koepke, J. E., and K. H. Pribram. 1971. Effect of milk on the maintenance of sucking behavior in kittens from birth to six months. *J. Comp. Physiol. Psychol.* 75:363–377.

824. Kolb, B., and A. J. Nonneman. 1975. The development of social responsiveness in kittens. *Anim. Behav.* 23:368–374.

825. Kondo, S., and J. F. Hurnik. 1990. Stabilization of social hierarchy in dairy cows. *Appl. Anim. Behav. Sci.* 27:287–297.

826. Kondo, S., N. Kawakami, H. Kohama, and S. Nishino. 1983. Changes in activity spatial pattern and social behavior in calves after grouping. *Appl. Anim. Ethol.* 11:217–228.

827. Konrad, K. W., and M. Bagshaw. 1970. Effect of novel stimuli on cats reared in a restricted environment. *J. Comp. Physiol. Psychol.* 70:157–164.

828. Korda, P., and J. Brewinska. 1977. Effect of stimuli emitted by sucklings on tactile contact of the bitches with sucklings and on number of licking acts. *Acta Neurobiol. Exp.* 37:99–115.

829. Kovach, J. K., and A. Kling. 1967. Mechanisms of neonate sucking behaviour in the kitten. *Anim. Behav.* 15:91–101.

830. Kovalcik, K., and M. Kovalcik. 1986. Learning ability and memory testing in cattle of different ages. *Appl. Anim. Behav. Sci.* 15:27–29.

831. Krabill, L. F., P. J. Wangsness, and C. A. Baile. 1978. Effects of elfazepam on digestibility and feeding behavior in sheep. *J. Anim. Sci.* 46:1356–1359.

832. Kratzer, D. D. 1969. Effects of age on avoidance learning in pigs. *J. Anim. Sci.* 28:175–179.

833. Kratzer, D. D., W. M. Netherland, R. E. Pulse, and J. P. Baker. 1977. Maze learning in quarter horses. *J. Anim. Sci.* 45:896–902.

834. Krebs, J. R., and N. B. Davies. 1978. *Behavioural Ecology. An Evolutionary Approach*. Sunderland, MA: Sinauer Associates.

835. Kristal, M. B., A. C. Thompson, S. B. Heller, and B. R. Komisaruk. 1986. Placenta ingestion enhances analgesia produced by vaginal/cervical stimulation in rats. *Physiol. Behav.* 36:1017–1020.

836. Kristula, M. A., and S. M. McDonnell. 1994. Drinking water temperature affects consumption of water during cold weather in ponies. *Appl. Anim. Behav. Sci.* 41:155–160.

837. Krohn, C. C., and L. Munksgaard. 1993. Behaviour of dairy cows kept in extensive (loose housing/pasture) or intensive (tie stall) environments. II. Lying and lying-down behaviour. *Appl. Anim. Behav. Sci.* 37:1–16.

838. Krohn, C. C., L. Munksgaard, and B. Jonasen. 1992. Behaviour of dairy cows kept in extensive (loose housing/pasture) or intensive (tie stall) environments. I. Experimental procedure, facilities, time budgets—diurnal and seasonal conditions. *Appl. Anim. Behav. Sci.* 34:37–47.

839. Kronberg, S. L., R. B. Muntifering, and E. L. Ayers. 1993. Feed aversion learning in cattle with delayed negative consequences. *J. Anim. Sci.* 71:1767–1770.

839a. Kropp, J. R., J. W. Holloway, D. F. Stephens, L. Knori, R. D. Morrison, and R. Totusek. 1973. Range behavior of Hereford, Hereford x Holstein and Holstein nonlactating heifers. *J. Anim. Sci.* 36:797–802.

840. Krzak, W. E., H. W. Gonyou, and L. M. Lawrence. 1991. Wood chewing by stabled horses: Diurnal pattern and effects of exercise. *J. Anim. Sci.* 69:1053–1058.

841. Kuhn, G., and W. Hardegg. 1988. Effects of indoor and outdoor maintenance of dogs upon food intake, body weight, and different blood parameters. *Z. Versuchstierkd.* 31:205–214.

842. Kuipers, M., and T. S. Whatson. 1979. Sleep in piglets: An observational study. *Appl. Anim. Ethol.* 5:145–151.

843. Kuo, Z. Y. 1930. The genesis of the cat's responses to the rat. *J. Comp. Psychol.* 11:1–35.

844. Kurz, J. C., and R. L. Marchinton. 1972. Radiotelemetry studies of feral hogs in South Carolina. *J. Wildl. Manag.* 36:1240–1248.

845. Kusunose, R. 1992. Diurnal pattern of cribbing in stabled horses. *Jpn. J. Equine Sci.* 3:173–176.

846. Kusunose, R., H. Hatakeyama, F. Ichikawa, K. Kubo, A. Kiguchi, Y. Asai, and K. Ito. 1986. Behavioral studies on yearling horses in field environments 2. Effects of the group size on the behavior of horses. *Bull. Equine Res. Inst.* 23:1–6.

847. Kusunose, R., H. Hatakeyama, F. Ichikawa, H. Oki, Y. Asai, and K. Ito. 1987. Behavioral studies on yearling horses in field environments 3. Effects of the pasture shape on the behavior of horses. *Bull. Equine Res. Inst.* 24:1–5.

848. Kusunose, R., H. Hatakeyama, K. Kubo, A. Kiguchi, Y. Asai, Y. Fujii, and K. Ito. 1985. Behavioral studies on yearling horses in field environments 1. Effects of the field size on the behavior of horses. *Bull. Equine Res. Inst.* 22:1–7.

849. Kusunose, R., and H. Sawazaki. 1984. The behavioral development of thoroughbred foals and the relationship between dams and foals. *Jap. J. Zootech. Sci.* 55:263–271.

850. Kusunose, R., and K. Torikai. 1996. Behavior of untethered horses during vehicle transport. *J. Eq. Sci.* 7:21–26

851. Laca, E. A., E. D. Ungar, and M. W. Demment. 1994. Mechanisms of handling time and intake rate of a large mammalian grazer. *Appl. Anim. Behav. Sci.* 39:3–19.

852. Lammers, G. J., and A. De Lange. 1986. Pre- and post-farrowing behaviour in primiparous domesticated pigs. *Appl. Anim. Behav. Sci.* 15:31–43.

852a. Lampkin, G. H., J. Quarterman, and M. Kidner. 1958. Observations on the grazing habits of grade and Zebu steers in a high altitude temperate climate. *J. Agric. Sci.* 50:211–218.

852b. Larsen, H. J. 1963. Feeding habits of grazing and green feeding cows [abstract]. *J. Anim. Sci.* 22:1134.

853. Launchbaugh, K. L., and F. D. Provenza. 1994. The effect of flavor concentration and toxin dose on the formation and generalization of flavor aversions in lambs. *J. Anim. Sci.* 72:10–13.

854. Laundre, J. 1977. The daytime behavior of domestic cats in a free-roaming population. *Anim. Behav.* 25:990–998.

855. Laut, J. E., K. A. Houpt, H. F. Hintz, and T. R. Houpt. 1985. The effects of caloric dilution on meal patterns and food intake of ponies. *Physiol. Behav.* 35:549–554.

856. Lawrence, A. B. 1990. Mother-daughter and peer relationships of Scottish hill sheep. *Anim. Behav.* 39:481–486.

857. Lawrence, A. B., J. C. Petherick, K. McLean, C. L. Gilbert, C. Chapman, and J. A. Russell. 1992. Naloxone prevents interruption of parturition and increases plasma oxytocin following environmental disturbance in parturient sows. *Physiol. Behav.* 52:917–923.

858. Lawson, D. C., S. S. Schiffman, and T. N. Pappas. 1993. Short-term oral sensory deprivation: Possible cause of binge eating in sham-feeding dogs. *Physiol. Behav.* 53:1231–1234.

859. Lazo, A. 1994. Social segregation and the maintenance of social stability in a feral cattle population. *Anim. Behav.* 48:1133–1141.

860. Le Boeuf, B. J. 1967. Interindividual associations in dogs. *Behaviour* 29:268–295.

861. Le Boeuf, B. J. 1970. Copulatory and aggressive behavior in the prepuberally castrated dog. *Horm. Behav.* 1:127–136.

862. Le Neindre, P., G. Trillat, J. Sapa, F. Ménissier, J. N. Bonnet, and J. M. Chupin. 1995. Individual differences in docility in Limousin cattle. *J. Anim. Sci.* 73:2249–2253.

863. Lees, J. L., and M. Weatherhead. 1970. A note on mating preferences of Clun Forest ewes. *Anim. Prod.* 12:173–175.

864. Lehner, P. N., C. McCluggage, D. R. Mitchell, and D. H. Neil. 1983. Selected parameters of the Fort Collins, Colorado, dog population, 1979–1980. *Appl. Anim. Ethol.* 10:19–25.

865. Lenhardt, M. L. 1977. Vocal contour cues in maternal recognition of goat kids. *Appl. Anim. Ethol.* 3:211–219.

866. Levine, A. S., C. E. Sievert, J. E. Morley, B. A. Gosnell, and S. E. Silvis. 1984. Peptidergic regulation of feeding in the dog (Canis familiaris). *Peptides* 5:675–679.

867. Levy, F., R. Gervais, U. Kindermann, M. Litterio, P. Poindron, and R. Porter. 1991. Effects of early post-partum separation on maintenance of maternal responsiveness and selectivity in parturient ewes. *Appl. Anim. Behav. Sci.* 31:101–110.

868. Levy, F., A. Locatelli, V. Piketty, Y. Tillet, and P. Poindron. 1995. Involvement of the main but not the accessory olfactory system in maternal behavior of primiparous and multiparous ewes. *Physiol. Behav.* 57:97–104.

869. Levy, F., and P. Poindron. 1987. The importance of amniotic fluids for the establishment of maternal behaviour in experienced and inexperienced ewes. *Anim. Behav.* 35:1188–1192.

870. Levy, F., P. Poindron, and P. LeNeindre. 1983. Attraction and repulsion by amniotic fluids and their olfactory control in the ewe around parturition. *Physiol. Behav.* 31:687–692.

871. Lewis, N. J., and J. F. Hurnick. 1985. The development of nursing behaviour in swine. *Appl. Anim. Behav. Sci.* 14:225–232.

871a. Lewis, R. C. and J. D. Johnson. 1954. Observations of dairy cow activities in loose-housing. *J. Dairy Sci.* 37:269–275.

872. Leyhausen, P. 1973. Addictive behavior in free ranging animals. In *Bayer Symposium IV, Psychic Dependence*. L. Goldberg and F. Hoffmeister, eds. Berlin, Germany: Springer-Verlag.

873. Leyhausen, P. 1975. *Verhaltensstudien an Katzen*. Berlin, Germany: Paul Parey.

874. Leyhausen, P. 1979. *Cat Behavior*. New York, NY: Garland STPM Press.

875. Liberg, O. 1983. Courtship behaviour and sexual selection in the domestic cat. *Appl. Anim. Ethol.* 10:117–132.

876. Lickliter, R. E. 1984. Mother-infant spatial relationships in domestic goats. *Appl. Anim. Behav. Sci.* 13:93–100.

877. Lickliter, R. E. 1984. Hiding behavior in domestic goat kids. *Appl. Anim. Behav. Sci.* 12:245–251.

878. Lickliter, R. E. 1985. Behavior associated with parturition in the domesticated goat. *Appl. Anim. Behav. Sci.* 13:335–345.

879. Lickliter, R. E. 1987. Activity patterns and companion preferences of domestic goat kids. *Appl. Anim. Behav. Sci.* 19:137–145.

880. Lickliter, R. E., and J. R. Heron. 1984. Recognition of mother by newborn goats. *Appl. Anim. Behav. Sci.* 12:187–192.

881. Liddell, H. S. 1926. A laboratory for the study of conditioned motor reflexes. *Am. J. Psychol.* 37:418–419.

882. Liddell, H. S. 1926. The effect of thyroidectomy on some unconditioned responses of the sheep and goat. *Am. J. Physiol.* 75:579–590.

883. Liddell, H. S. 1954. Conditioning and emotions. *Sci. Am.* 190(1):48–57.

884. Liddell, H. S., and O. D. Anderson. 1931. A comparative study of the conditioned motor reflex in the rabbit, sheep, goat, and pig [abstract]. *Am. J. Physiol.* 97:539–540.

885. Liddell, H. S., W. T. James, and O. D. Anderson. 1934. The comparative physiology of the conditioned motor reflex based on experiments with the pig, dog, sheep, goat and rabbit. *Comp. Psychol. Monogr.* 11:1–89.

886. Lidfors, L., and P. Jensen. 1988. Behaviour of free-ranging beef cows and calves. *Appl. Anim. Behav. Sci.* 20:237–247.

887. Lidfors, L. M. 1993. Cross-sucking in group-housed dairy calves before and after weaning off milk. *Appl. Anim. Behav. Sci.* 38:15–24.

888. Lien, J., and F. D. Klopfer. 1978. Some relations between stereotyped suckling in piglets and exploratory behaviour and discrimination reversal learning in adult swine. *Appl. Anim. Ethol.* 4:223–233.

889. Lindahl, I. L. 1964. Time of parturition in ewes. *Anim. Behav.* 12:231–234.

890. Lindsay, D. R. 1965. The importance of olfactory stimuli in the mating behaviour of the ram. *Anim. Behav.* 13:75–78.

891. Lindsay, D. R. 1966. Modification of behavioural oestrus in the ewe by social and hormonal factors. *Anim. Behav.* 14:73–83.

892. Lindsay, D. R. 1966. Mating behaviour of ewes and its effect on mating efficiency. *Anim. Behav.* 14:419–424.

893. Lindsay, D. R., and I. C. Fletcher. 1968. Sensory involvement in the recognition of lambs by their dams. *Anim. Behav.* 16:415–417.

894. Lindsay, D. R., and I. C. Fletcher. 1972. Ram-seeking activity associated with oestrous behaviour in ewes. *Anim. Behav.* 20:452–456.

895. Lindsay, D. R., and T. J. Robinson. 1961. Studies on the efficiency of mating in the sheep. II. The effect of freedom of rams, padddock size, and age of ewes. *J. Agric. Sci.* 57:141–145.

896. Lindsay, D. R., and T. J. Robinson. 1961. Studies on the efficiency of mating in the sheep. I. The effect of paddock size and number of rams. *J. Agric. Sci.* 57:137–140.

897. Line, S. W., B. L. Hart, and L. Sanders. 1985. Effect of prepubertal versus postpubertal castration on sexual and aggressive behavior in male horses. *J. Am. Vet. Med. Assoc.* 186:249–251.

898. Liptrap, R. M., and J. I. Raeside. 1978. A relationship between plasma concentrations of testosterone and cortico-steroids during sexual and aggressive behaviour in the boar. *J. Endocrinol.* 76:75–85.

899. Littlejohn, A., and R. Munro. 1972. Equine recumbency. *Vet. Rec.* 90:83–85.

900. Lockwood, R. 1987. Pit bull terriers. *Arthrozoos* 1:193–194.

900a. Lofgreen, G. P., J. H. Meyer, and J. L. Hull. 1957. Behavior patterns of sheep and cattle being fed pasture or soilage. *J. Anim. Sci.* 16:773–780.

901. Lohse, C. L. 1974. Preferences of dogs for various meats. *J. Am. Anim. Hosp. Assoc.* 10:187–192.

902. Lorenz, K. 1952. *King Solomon's Ring. New Light on Animal Ways.* New York, NY: Thomas Y. Crowell Co.

903. Lorenz, K. Z. 1957. Companionship in bird life. In *Instinctive Behavior. The Development of a Modern Concept.* C. H. Schiller and K. S. Lashley, eds. New York, NY: International Universities Press, pp. 83–128.

904. Lou, Z., and J. F. Hurnik. 1994. An ellipsoid farrowing crate: Its ergonomical design and effects on pig productivity. *J. Anim. Sci.* 72:2610–2616.

905. Lowman, B. G., M. S. Hankey, N. A. Scott, and D. W. Deas. 1981. Influence of time of feeding on time of parturition in beef cows. *Vet. Rec.* 109:557–559.

906. Luescher, U. A. 1993. Hyperkinesis in dogs: Six case reports. *Can. Vet. J.* 34:368–370.

907. Lunstra, D. D., G. W. Boyd, and L. R. Corah. 1989. Effects of natural mating stimuli on serum luteinizing hormone, testosterone and estradiol-17ß in yearling beef bulls. *J. Anim. Sci.* 67:3277–3288.

908. Lustgarten, C., G. D. Bottoms, and J. R. Shaskas. 1973. Experimental adrenalectomy of pigs. *Am. J. Vet. Res.* 34:279–282.

909. Lynch, J. J., and G. Alexander. 1976. The effect of gramineous windbreaks on behaviour and lamb mortality among shorn and unshorn Merino sheep during lambing. *Appl. Anim. Ethol.* 2:305–325.

910. Lynch, J. J., G. N. Hinch, and D. B. Adams. 1992. *The Behaviour of Sheep. Biological Principles And Implications for Production.* Oxon, UK: CAB International.

911. Lynch, J. J., and J. F. McCarthy. 1967. The effect of petting on a classically conditioned emotional response. *Behav. Res. Ther.* 5:55–62.

912. Lyons, D. M., E. O. Price, and G. P. Moberg. 1988. Social modulation of pituitary-adrenal responsiveness and individual differences in behavior of young domestic goats. *Physiol. Behav.* 43:451–458.

913. Macaulay, A. S., G. L. Hahn, D. H. Clark, and D. V. Sisson. 1995. Comparison of calf housing types and tympanic temperature rhythms in Holstein calves. *J. Dairy Sci.* 78:856–862.

914. MacDonald, D. 1981. The behaviour and ecology of farm cats. In *The Ecology and Control of Feral Cats.* Potters Bar, UK: Universities Federation for Animal Welfare, pp. 23–29.

915. Macfarlane, J. S. 1974. The effect of two post-weaning management systems on the social and sexual behaviour of Zebu bulls. *Appl. Anim. Ethol.* 1:31–34.

916. Mackenzie, S. A., E. A. B. Oltenacu, and K. A. Houpt. 1986. Canine behavioral genetics—A review. *Appl. Anim. Behav. Sci.* 15:365–393.

917. Mackenzie, S. A., E. A. B. Oltenacu, and E. Leighton. 1985. Heritable estimate for temperament scores in German shepherd dogs and its genetic correlation with hip dysplasia. *Behav. Genet.* 15:475–482.

918. Maddison, S., and B. A. Baldwin. 1983. Diencephalic neuronal activity during acquisition and ingestion of food in sheep. *Brain Res.* 278:195–206.

919. Mader, D. R., and E. O. Price. 1980. Discrimination learning in horses: Effects of breed, age and social dominance. *J. Anim. Sci.* 50:962–965.

920. Mader, D. R., and E. O. Price. 1984. The effects of sexual stimulation on the sexual performance of hereford bulls. *J. Anim. Sci.* 59:294–300.

921. Maier, N. R. F., and T. C. Schneirla. 1964. *Principles of Animal Psychology.* New York, NY: Dover Publications.

922. Maier, S. F., and M. E. P. Seligman. 1976. Learned helpless ness: Theory and evidence. *J. Exp. Psychol. (General)* 105:3–46.

923. Mal, M. E., and C. A. McCall. 1996. The influence of handling during different ages on a halter training test in foals. *Appl. Anim. Behav. Sci.* 50:115–120.

924. Mal, M. E., C. A. McCall, K. A. Cummins, and M. C. Newland. 1994. Influence of preweaning handling methods on post-weaning learning ability and manageability of foals. *Appl. Anim. Behav. Sci.* 40:187–195.

925. Malbert, C. H., and Y. Ruckebusch. 1989. Hyperphagia induced by pylorectomy in sheep. *Physiol. Behav.* 45:495–499.

926. Malm, K. 1995. Regurgitation in relation to weaning in the domestic dog: A questionnaire study. *Appl. Anim. Behav. Sci.* 43:111–122.

927. Malm, K., and P. Jensen. 1993. Regurgitation as a weaning strategy—a selective review on an old subject in a new light. *Appl. Anim. Behav. Sci.* 36:47–64.

928. Malpass, J. P., and B. J. Weigler. 1994. A simple and effective environmental enrichment device for ponies in long-term indoor confinement. *Contemporary Topics* 33:74–76.

929. Manson, F. J., and M. C. Appleby. 1990. Spacing of dairy cows at a food trough. *Appl. Anim. Behav. Sci.* 26:69–81.

930. Marcella, K. L. 1983. A note on canine aggression towards veterinarians. *Appl. Anim. Ethol.* 10:155–157.

931. Marcuse, F. L., and A. U. Moore. 1944. Tantrum behavior in the pig. *J. Comp. Psychol.* 37:235–241.

932. Marcuse, F. L., and A. U. Moore. 1946. Motor criteria of discrimination. *J. Comp. Psychol.* 39:25–27.

933. Marder, A. R. 1991. Psychotropic drugs and behavioral therapy. *Vet. Clin. N. Am. :Sm. Anim. Prac.*

934. Marinier, S. L., and A. J. Alexander. 1992. Use of field observations to measure individual grazing ability in horses. *Appl. Anim. Behav. Sci.* 33:1–10.

935. Marinier, S. L., A. J. Alexander, and G. H. Waring. 1988. Flehmen behaviour in the domestic horse: Discrimination of conspecific odours. *Appl. Anim. Behav. Sci.* 19:227–237.

936. Marten, G. C., and J. E. Donker. 1964. Selective grazing induced by animal excreta. I. Evidence of occurrence and superficial remedy. *J. Dairy Sci.* 47:773–776.

937. Martin, F. H., J. R. Seoane, and C. A. Baile. 1973. Feeding in satiated sheep elicited by intraventricular injections of CSF from fasted sheep. *Life Sci.* 13:177–184.

938. Martin, J. E., and S. A. Edwards. 1994. Feeding behaviour of outdoor sows: The effects of diet quantity and type. *Appl. Anim. Behav. Sci.* 41:63–74.

939. Martin, J. T. 1975. Movement of feral pigs in North Canterbury, New Zealand. *J. Mammal.* 56:914–915.

940. Martin, P. 1984. The time and energy costs of play behaviour in the cat. *Z. Tierpsychol.* 64:298–312.

941. Martin, P. 1986. An experimental study of weaning in the domestic cat. *Behaviour* 99:221–249.

942. Martin, P., and P. Bateson. 1985. The ontogeny of locomotor play behaviour in the domestic cat. *Anim. Behav.* 33:502–510.

943. Martin, R. J., J. L. Gobble, T. H. Hartsock, H. B. Graves, and J. H. Ziegler. 1973. Characterization of an obese syndrome in the pig. *Proc. Soc. Exp. Biol. Med.* 143:198–203.

944. Martins, T. 1949. Disgorging of food to the puppies by the lactating dog. *Physiol. Zool.* 22:169–172.

945. Mason, E. 1970. Obesity in pet dogs. *Vet. Rec.* 86:612–616.

946. Mateo, J. M., D. Q. Estep, and J. S. McCann. 1991. Effects of differential handling on the behaviour of domestic ewes (Ovis aries). *Appl. Anim. Behav. Sci.* 32:45–54.

947. Mattner, P. E., A. W. H. Braden, and K. E. Turnbull. 1967. Studies in flock mating of sheep. 1. Mating behaviour. *Aust. J. Exp. Agric. Anim. Husb.* 7:103–109.

948. Mayes, E., and P. Duncan. 1986. Temporal patterns of feeding in free-ranging horses. *Behaviour* 96:105–129.

949. McBane, S. 1987. *Behaviour Problems of Horses.* North Pomfret, VT: David and Charles.

950. McBride, G. 1963. The "teat order" and communication in young pigs. *Anim. Behav.* 11:53–56.

951. McBride, G., J. W. James, and N. Hodgens. 1964. Social behaviour of domestic animals. IV. Growing pigs. *Anim. Prod.* 6:129–139.

952. McBride, G., J. W. James, and G. S. F. Wyeth. 1965. Social behaviour of domestic animals. VII. Variation in weaning weight in pigs. *Anim. Prod.* 7:67–74.

953. McCall, C. A. 1989. *Behaviour Problems of Horses.* North Pomfret, VT: David and Charles. 304. pp.

954. McCall, C. A. 1991. Utilizing taped stallion vocalizations as a practical aid in estrus detection in mares. *Appl. Anim. Behav. Sci.* 28:305–310.

955. McCall, C. A., G. D. Potter, T. H. Friend, and R. S. Ingram. 1981. Learning abilities in yearling horses using the Hebb-Williams closed field maze. *J. Anim. Sci.* 53:928–933.

956. McCall, C. A., G. D. Potter, and J. L. Kreider. 1985. Locomotor, vocal and other behavioural responses to varying methods of weaning foals. *Appl. Anim. Behav. Sci.* 14:27–35.

957. McCall, C. A., M. A. Salters, and S. M. Simpson. 1993. Relationship between number of conditioning trials per training session and avoidance learning in horses. *Appl. Anim. Behav. Sci.* 36:291–299.

958. McClure, S. R., M. K. Chaffin, and B. V. Beaver. 1992. Nonpharmacologic management of stereotypic self-mutilative behavior in a stallion. *J. Am. Vet. Med. Assoc.* 200:1975–1977.

959. McConnell, P. B. 1990. Acoustic structure and receiver response in domestic dogs, Canis familiaris. *Anim. Behav.* 39:897–904.

960. McConnell, P. B., and J. R. Baylis. 1985. Interspecific communication in cooperative herding: Acoustic and visual signals from human shepherds and herding dogs. *Z. Tierpsychol.* 67:302–328.

961. McCune, S. 1995. The impact of paternity and early socialisation on the development of cats' behaviour to people and novel objects. *Appl. Anim. Behav. Sci.* 45:109–124.

962. McDonald, C. L., R. G. Beilharz, and J. C. McCutchan. 1981. Training cattle to control by electric fences. *Appl. Anim. Ethol.* 7:113–121.

963. McDonnell, S. 1986. Reproductive behavior of the stallion. *Vet. Clin. N. Am. :Eq. Prac.* 2535–556.

964. McDonnell, S. M. 1992. Sexual behavior dysfunction in stallions. In *Current Therapy In Equine Medicine.* 3rd ed. N. E. Robinson, ed. Philadelphia, PA: W.B. Saunders Company, pp. 668–671

965. McDonnell, S. M., N. K. Diehl, M. C. Garcia, and R. M. Kenney. 1989. Gonadotropin releasing hormone (GnRH) affects precopulatory behavior in testosterone-treated geldings. *Physiol. Behav.* 45:145–149.

966. McDonnell, S. M., and J. C. S. Haviland. 1995. Agonistic ethogram of the equid bachelor band. *Appl. Anim. Behav. Sci.* 43:147–188.

967. McDonnell, S. M., R. M. Kenney, P. E. Meckley, and M. C. Garcia. 1985. Con-

ditioned suppression of sexual behavior in stallions and reversal with diazepam. *Physiol. Behav.* 34:951–956.

968. McDonnell, S. M., R. M. Kenney, P. E. Meckley, and M. C. Garcia. 1986. Novel environment suppression of stallion sexual behavior and effects of diazepam. *Physiol. Behav.* 37:503–505.

969. McDonnell, S. M., and S. C. Murray. 1995. Bachelor and harem stallion behavior and endocrinology. *Biol. Reprod. Mono.* 1:577–590.

970. McDougall, K. D., and W. McDougal. 1931. Insight and foresight in various animals—monkey, racoon, rat, and wasp. *J. Comp. Psychol.* 11:237–273.

971. McEachron, D. L., D. F. Kripke, R. Hawkins, E. Haus, D. Pavlinac, and L. Deftos. 1982. Lithium delays biochemical circadian rhythms in rats. *Neuropsychobiology* 8:12–29.

972. McGeer, E. G., and P. L. McGeer. 1966. Circadian rhythm in pineal tyrosine hydroxylase. *Science* 153:73–74.

973. McGinty, D. J., M. Stevenson, T. Hoppenbrouwers, R. M. Harper, M. B. Sterman, and J. Hodgman. 1977. Polygraphic studies of kitten development: Sleep state patterns. *Dev. Psychobiol.* 10:455–469.

974. McGlone, J. J. 1985. A quantitative ethogram of aggressive and submissive behaviors in recently regrouped pigs. *J. Anim. Sci.* 61:559–565.

975. McGlone, J. J. 1985. Olfactory cues and pig agonistic behavior: Evidence for a submissive pheromone. *Physiol. Behav.* 34:195–198.

976. McGlone, J. J. 1986. Influence of resources on pig aggression and dominance. *Behav. Proc.* 12:135–144.

977. McGlone, J. J., and S. E. Curtis. 1985. Behavior and performance of weanling pigs in pens equipped with hide areas. *J. Anim. Sci.* 60:20–24.

978. McGlone, J. J., K. W. Kelley, and C. T. Gaskins. 1980. Lithium and porcine aggression. *J. Anim. Sci.* 51:447–455.

979. McGlone, J. J., and J. L. Morrow. 1987. Individual differences among mature boars in T-maze preference for estrous or non-estrous sows. *Appl. Anim. Behav. Sci.* 17:77–82.

980. McGlone, J. J., R. I. Nicholson, J. M. Hellman, and D. N. Herzog. 1993. The development of pain in young pigs associated with castration and attempts to prevent castration-induced behavioral changes. *J. Anim. Sci.* 71:1441–1446.

981. McGreevy, P. D., P. J. Cripps, N. P. French, L. E. Green, and C. J. Nicol. 1995. Management factors associated with stereotypic and redirected behaviour in the Thoroughbred horse. *Eq. vet. J.* 27:86–91.

982. McGreevy, P. D., N. P. French, and C. J. Nichol. 1995. The prevalence of abnormal behaviours in dressage, eventing and endurance horses in relation to stabling. *Vet. Rec.* 137:36–37.

983. McGreevy, P. D., J. D. Richardson, C. J. Nichol, and J. G. Lane. 1995. Radiographic and endoscopic study of horses performing an oral based stereotypy. *Eqine vet. J.* 27:92–95.

984. McGuire, R. A., W. M. Rand, and R. J. Wurtman. 1973. Entrainment of the body temperature rhythm in rats: Effect of color and intensity of environmental light. *Science* 181:956–957.

985. McKinley, P. E. 1982. *Cluster Analysis of the Domestic Cat's Vocal Repertoire.* College Park, MD: Ph.D. Thesis, University of Maryland.

986. McLaughlin, C. L., C. A. Baile, L. L. Buckholtz, and S. K. Freeman. 1983. Preferred flavors and performance of weanling pigs. *J. Anim. Sci.* 56:1287–1293.

987. McLaughlin, C. L., L. F. Krabill, G. C. Scott, and C. A. Baile. 1976. Chemical stimulants of feeding animals [abstract]. *Fed. Proc.* 35:579.

988. McPhee, C. P., G. McBride, and J. W. James. 1964. Social behaviour of domestic animals. III. Steers in small yards. *Anim. Prod.* 6:9–15.

989. Mech, L. D. 1975. Hunting behavior in two similar species of social canids. In *The Wild Canids. Their Systematics, Behavioral Ecology and Evolution.* M. W. Fox, ed. New York, NY: Van Nostrand Reinhold Co., pp. 363–368.

990. Meese, G. B., and B. A. Baldwin. 1975. The effects of ablation of the olfactory bulbs on aggressive behaviour in pigs. *Appl. Anim. Ethol.* 1:251–262.

991. Meese, G. B., and B. A. Baldwin. 1975. Effects of olfactory bulb ablation on maternal behaviour in sows. *Appl. Anim. Ethol.* 1:379–386.

992. Meese, G. B., D. J. Conner, and B. A. Baldwin. 1975. Ability of the pig to distinguish between conspecific urine samples using olfaction. *Physiol. Behav.* 15:121–125.

993. Meese, G. B., and R. Ewbank. 1973. The establishment and nature of the dominance hierarchy in the domesticated pig. *Anim. Behav.* 21:326–334.

994. Meese, G. B., and R. Ewbank. 1973. Exploratory behaviour and leadership in the domesticaged pig. *Br. Vet. J.* 129:251–262.

995. Meier, G. W. 1961. Infantile handling and development in Siamese kittens. *J. Comp. Physiol. Psychol.* 54:284–286.

996. Meier, G. W., and J. L. Stuart. 1959. Effects of handling on the physical and behavioral development of Siamese kittens. *Psychol. Rep.* 5:497–501.

997. Meikle, D. B., L. C. Drickamer, S. H. Vessey, T. L. Rosenthal, and K. S. Fitzgerald. 1993. Maternal dominance rank and secondary sex ratio in domestic swine. *Anim. Behav.* 46:79–85.

998. Melese-d'Hospital, P. 1996. Eliminating urine odors in the home. In *Readings In Companion Animal Behavior.* V. L. Voith and P. L. Borchelt, eds. Trenton, NJ: Veterinary Learning Systems, pp. 191–197

999. Melrose, D. R., H. C. B. Reed, and R. L. S. Patterson. 1971. Androgen steroids associated with boar odour as an aid to the detection of oestrus in pig artificial insemination. *Br. Vet. J.* 127:497–502.

1000. Melzack, R. 1962. Effects of early perceptual restriction on simple visual discrimination. *Science* 137:978–979.

1001. Melzack, R., and T. H. Scott. 1957. The effects of early experience on the response to pain. *J. Comp. Physiol. Psychol.* 50:155–161.

1002. Mendl, M., and R. Harcourt. 1988. Individuality in the domestic cat. In *The Domestic Cat: The Biology Of Its Behaviour.* D. C. Turner and P. Bateson, eds. Cambridge, UK: Cambridge University Press, pp. 41–54.

1003. Mendl, M., A. J. Zanella, and D. M. Broom. 1992. Physiological and reproductive correlates of behavioural strategies in female domestic pigs. *Anim. Behav.* 44:1107–1121.

1004. Mendl, M., A. J. Zanella, D. M. Broom, and C. T. Whittemore. 1995. Maternal social status and birth sex ratio in domestic pigs: An analysis of mechanisms. *Anim. Behav.* 50:1361–1370.

1005. Merrick, A. W., and D. W. Scharp. 1971. Electroencephalography of resting behavior in cattle, with observations on the question of sleep. *Am. J. Vet. Res.* 32:1893–1897.

1006. Mersmann, H. J., M. D. MacNeil, S. C. Seideman, and W. G. Pond. 1987. Compensatory growth in finishing pigs after feed restriction. *J. Anim. Sci.* 64:752–764.

1007. Mertens, D. R. 1987. Predicting intake and digestibility using mathematical models of ruminal function. *J. Anim. Sci.* 64:1548–1558.

1008. Metz, J. H. M. 1985. The reaction of cows to a short-term deprivation of lying. *Appl. Anim. Behav. Sci.* 13:301-307.

1009. Metz, J. H. M., and H. W. Gonyou. 1990. Effect of age and housing conditions on the behavioural and haemolytic reaction of piglets to weaning. *Appl. Anim. Behav. Sci.* 27:299-309.

1010. Metz, J. H. M., and P. Mekking. 1984. Crowding phenomena in dairy cows as related to available idling space in a cubicle housing system. *Appl. Anim. Behav. Sci.* 12:63-78.

1011. Michael, R. P. 1973. The effects of hormones on sexual behavior in female cat and rhesus monkey. In *Handbook of Physiology, Section 7, Edocrinology. Volume II, Female Reproductive System. Part 1*. R. O. Greep and E. B. Astwood, eds. Washington, DC: American Physiological Society, pp. 187-221.

1012. Michell, A. R. 1992. Sodium preference in sheep excreting sodium predominantly in urine or faeces. *Physiol. Behav.* 52:285-286.

1013. Michell, A. R., and P. Moss. 1988. Salt appetite during pregnancy in sheep. *Physiol. Behav.* 42:491-493.

1014. Miles, R. C. 1958. Learning in kittens with manipulatory, exploratory, and food incentives. *J. Comp. Physiol. Psychol.* 51:39-42.

1015. Milgram, N. W., E. Head, E. Weiner, and E. Thomas. 1994. Cognitive functions and aging in the dog: Acquisition of nonspatial visual tasks. *Behav. Neurosci.* 108:57-68.

1016. Miller, E. R., S. Vathana, F. F. Green, J. R. Black, D. R. Romsos, and D. E. Ullrey. 1974. Dietary caloric density and caloric intake in the pig [abstract]. *J. Anim. Sci.* 39:980.)

1017. Miller, R. M. 1991. *Imprint Training Of The Newborn Foal*. Colorado Springs, CO: The Western Horseman, Inc.

1018. Miller, R. R. 1981. Male aggresion, dominance, and breeding behavior in Red Desert feral horses. *Z. Tierpsychol.* 57:340-351.

1019. Miller, R. R., and R. H. Denniston, II. 1979. Interband dominance in feral horses. *Z. Tierpsychol.* 51:41-47.

1020. Mirza, S. N., and F. D. Provenza. 1992. Effects of age and conditions of exposure on maternally mediated food selection by lambs. *Appl. Anim. Behav. Sci.* 33:35-42.

1021. Mirza, S. N., and F. D. Provenza. 1994. Socially induced food avoidance in lambs: Direct or indirect maternal influence? *J. Anim. Sci.* 72:899-902.

1022. Mistlberger, R. E., T. A. Houpt, and M. C. Moore-Ede. 1990. Food-anticipatory rhythms under 24-hour schedules of limited access to single macronutrients. *J. Biol. Rhythms* 5:35-46.

1023. Mitler, M. M., O. Soave, and W. C. Dement. 1976. Narcolepsy in seven dogs. *J. Am. Vet. Med. Assoc.* 168:1036-1038.

1024. Moelk, M. 1944. Vocalizing in the house-cat: A phonetic and functional study. *Am. J. Psychol.* 57:184-205.

1025. Mohan Raj, A. B., W. J. McCaughey, W. McLauchlan, D. J. Kilpatrick, and S. . McGaughey. 1991. Behavioural response to mixing of entire bulls, vasectomised bulls and steers. *Appl. Anim. Behav. Sci.* 31:157-168.

1026. Mohr, E., and H. Krzywanek. 1990. Variations in core-temperature rhythms in unrestrained sheep. *Physiol. Behav.* 48:467-473.

1027. Mohr, R. G., and H. Krzywanek. 1995. Endogenous oscillator and regulatory mechanisms of body temperature in sheep. *Physiol. Behav.* 57:339-347.

1028. Molliver, M. E. 1963. Operant control of vocal behavior in the cat. *J. Exp. Anal. Behav.* 6:197-202.

1029. Moltz, H. 1960. Imprinting: Empirical basis and theoretical significance. *Psychol. Bull.* 57:291-314.

1030. Monks of New Skete, 1978. *How To Be Your Dog's Best Friend*. Boston, MA: Little Brown.

1031. Monteiro, L. S. 1972. The control of appetite in lactating cows. *Anim. Prod.* 14:263-281.

1032. Montgomery, G. G. 1957. Some aspects of the sociality of the domestic horse. *Trans. Kansas Acad. Sci.* 60:419-424.

1033. Moore, A. S., H. W. Gonyou, and A. W. Ghent. 1993. Integration of newly introduced and resident sows following grouping. *Appl. Anim. Behav. Sci.* 38:257-267.

1034. Moore, A. S., H. W. Gonyou, J. M. Stookey, and D. G. McLaren. 1994. Effect of group composition and pen size on behaviour, productivity and immune response of growing pigs. *Appl. Anim. Behav. Sci.* 40:13-30.

1035. Moore, A. U., and F. L. Marcuse. 1945. Salivary, cardiac and motor indices of conditioning in two sows. *J. Comp. Psychol.* 38:1-16.

1036. Moore, C. L., W. G. Whittlestone, M. Mullord, P. N. Priest, R. Kilgour, and J. L. Albright. 1975. Behavior responses of dairy cows trained to activate a feeding device. *J. Dairy Sci.* 58:1531-1535.

1037. Moore, R. M., R. B. Zehmer, J. I. Moulthrop, and R. L. Parker. 1977. Surveillance of animal-bite cases in the United States, 1971-1972. *Arch. Environ. Health* 32:267-270.

1037a. Moorefield, J. G. and H. H. Hopkins. 1951. Grazing habits of cattle in a mixed-prairie pasture. *J. Range Manag.* 4:151-157.

1038. Morag, M. 1967. Influence of diet on the behaviour pattern of sheep. *Nature* 213:110.

1039. Morgan, M., and K. A. Houpt. 1989. Feline behavior problems: The influence of declawing. *Anthrozoos* 3:50-53.

1040. Morgan, P. D., and G. W. Arnold. 1974. Behavioural relationships between Merino ewes and lambs during the four weeks after birth. *Anim. Prod.* 19:169-176.

1041. Morgan, P. D., C. A. P. Boundy, G. W. Arnold, and D. R. Lindsay. 1975. The roles played by the senses of the ewe in the location and recognition of lambs. *Appl. Anim. Ethol.* 1:139-150.

1042. Mormede, P., and R. Dantzer. 1977. Effects of dexamethasone on fear conditioning in pigs. *Behav. Biol.* 21:225-235.

1043. Mormede, P., and R. Dantzer. 1977. Experimental studies on avoidance behaviour in pigs. *Appl. Anim. Ethol.* 3:173-185.

1044. Morrison, A. R. 1983. A window on the sleeping brain. *Sci. Am.* 248:86-94.

1045. Morrison, S. R., H. F. Hintz, and R. L. Givens. 1968. A note on effect of exercise on behaviour and performance of confined swine. *Anim. Prod.* 10:341-344.

1046. Morrow, D. A. 1976. Fat cow syndrome. *J. Dairy Sci.* 59:1625-1629.

1047. Morrow, D. A., D. Hillman, A. W. Dade, and H. Kitchen. 1979. Clinical investigation of a dairy herd with the fat cow syndrome. *J. Am. Vet. Med. Assoc.* 174:161-167.

1048. Morrow-Tesch, J., and J. J. McGlone. 1990. Sources of maternal odors and the development of odor preferences in baby pigs. *J. Anim. Sci.* 68:3563-3571.

1049. Morrow-Tesch, J., and J. J. McGlone. 1990. Sensory systems and nipple attachment behavior in neonatal pigs. *Physiol. Behav.* 47:1-4.

1050. Morrow-Tesch, J. L., J. J. McGlone, and J. L. Salak-Johnson. 1994. Heat and social stress effects on pig immune measures. *J. Anim. Sci.* 72:2599-2609.

1051. Moulton, D. G., E. H. Ashton, and J. T. Eayrs. 1960. Studies in olfactory acuity. 4. Relative detectability of n-aliphatic acids by the dog. *Anim. Behav.* 8:117-128.

1052. Mount, L. E. 1979. *Adaptation to Thermal Environment: Man and His Productive Animals.* Baltimore, MD: University Park Press.

1053. Mount, N. C., and M. F. Seabrook. 1993. A study of aggression when group housed sows are mixed. *Appl. Anim. Behav. Sci.* 36:377-383.

1054. Mugford, R. A. 1977. External influences on the feeding of carnivores. In *The Chemical Senses and Nutrition.* M. R. Kare and O. Maller, eds. New York, NY: Academic Press, pp. 25-50.

1055. Munkenbeck, N. 1983. *A Test For Color Vision And a Spectral Sensitivity Curve in the Sheep (Ovis aries).* Ithaca, NY: M.S. Thesis, Cornell University.

1056. Munro, J. 1956. Observations on the suckling behaviour of young lambs. *Br. J. Anim. Behav.* 4:34–36.

1057. Murphey, R. M., F. A. M. Duarte, W. Coelho Novaes, and M. C. Torres Penedo. 1981. Age group differences in bovine investigatory behavior. *Dev. Psychobiol.* 14:117–125.

1058. Murphey, R. M., C. R. Ruiz-Miranda, and F. A. de Moura Duarte. 1990. Maternal recognition in Gyr (Bos indicus) calves. *Appl. Anim. Behav. Sci.* 27:183–191.

1059. Murphree, O. D., J. E. Peters, and R. A. Dykman. 1969. Behavioral comparisons of nervous, stable, and crossbred pointers at ages 2, 3, 6, 9, and 12 months. *Cond. Reflex.* 4:20–23.

1060. Myers, G. C. 1916. The importance of primacy in the learning of a pig. *J. Anim. Behav.* 6:64–69.

1061. Myers, R. D., and D. C. Mesker. 1960. Operant responding in a horse under several schedules of reinforcement. *J. Exp. Anal. Behav.* 3:161–164.

1062. Mylrea, P. J., and R. G. Beilharz. 1964. The manifestation and detection of oestrous in heifers. *Anim. Behav.* 12:25–30.

1063. Nagamachi, Y. 1972. Effect of satiety center damage on food intake, blood glucose and gastric secretion in dogs. *Am. J. Dig. Dis. New Series* 17:139–148.

1064. Napolitano, F., V. Marino, G. De Rosa, R. Capparelli, and A. Bordi. 1995. Influence of artificial rearing on behavioral and immune response of lambs. *Appl. Anim. Behav. Sci.* 45:245–253.

1065. Natoli, E. 1985. Spacing pattern in a colony of urban stray cats (Felis catus L.) in the historic centre of Rome. *Appl. Anim. Behav. Sci.* 14:289–304.

1066. Natoli, E. 1990. Mating strategies in cats: A comparison of the role and importance of infanticide in domestic cats, Felis catus L. and lions, Panthera leo L. *Anim. Behav.* 40:183–186.

1067. Natoli, E., and E. De Vito. 1991. Agonistic behaviour, dominance rank and copulatory success in a large multi-male feral cat, Felis catus L, colony in central Rome. *Anim. Behav.* 42:227–241.

1068. Neathery, M. W. 1971. Acceptance of orphan lambs by tranquilized ewes (Ovis aries). *Anim. Behav.* 19:75–79.

1069. Neff, W. D., and I. T. Diamond. 1958. The neural basis of auditory discrimination. In *Biological and Biochemical Bases of Behavior.* H. F. Harlow and C. N. Woolsey, eds. Madison, WI: University of Wisconsin Press, pp. 101–126.

1070. Neitz, J., T. Geist, and G. H. Jacobs. 1989. Color vision in the dog. *Visual Neurosci.* 3:119–125.

1071. Netto, W. J., and D. J. U. Planta. 1996. Behavioral testing for aggression in the domestic dog. (Unpublished)

1072. Newberry, R. C., and D. G. M. Wood-Gush. 1986. Social relationships of piglets in a seminatural environment. *Anim. Behav.* 34:1311–1318.

1073. Newman, J. A., P. D. Penning, A. J. Parson, A. Harvey, and R. J. Orr. 1994. Fasting affects intake behaviour and diet preference of grazing sheep. *Anim. Behav.* 47:185–193.

1074. Nikitopoulou, G., and J. L. Crammer. 1976. Change in diurnal temperature rhythm in manic-depressive illness. *Br. Med. J.* 1:1311–1314.

1075. Noble, M., and C. K. Adams. 1963. Conditioning in pigs as a function of the interval between CS and US. *J. Comp. Physiol. Psychol.* 56:215–219.

1076. Noda, K., and K. Chikamori. 1976. Effect of ammonia via prepyriform cortex on regulation of food intake in the rat. *Am. J. Physiol.* 231:1263–1266.

1077. Nolte, D. L., and F. D. Provenza. 1992. Food preferences in lambs after exposure to flavors in solid foods. *Appl. Anim. Behav. Sci.* 32:337–347.

1078. Nonneman, A. J., and J. M. Warren. 1977. Two-cue learning by brain-damaged cats. *Physiol. Psychol.* 5:397–402.

1079. Nowak, R. 1991. Senses involved in discrimination of merino ewes at close contact and from a distance by their newborn lambs. *Anim. Behav.* 42:357–366.

1080. Noyes, L. 1976. A behavioural comparison of gnotobiotic with normal neonate pigs, indicating stress in the former. *Appl. Anim. Ethol.* 2:113–121.

1081. O'Brien, P. H. 1984. Leavers and stayers: Maternal postpartum strategies in feral goats. *Appl. Anim. Behav. Sci.* 12:233–243.

1082. O'Brien, P. H. 1984. Feral goat home range: Influence of social class and environmental variables. *Appl. Anim. Behav. Sci.* 12:373–385.

1083. O'Brien, P. H. 1988. Feral goat social organization: A review and comparative analysis. *Appl. Anim. Behav. Sci.* 21:209–221.

1084. O'Connell, J. M., P. S. Giller, and W. J. Meaney. 1993. Weanling training and cubicle usage as heifers. *Appl. Anim. Behav. Sci.* 37:185–195.

1085. O'Connor, C. E., A. B. Lawrence, and D. G. M. Wood-Gush. 1992. Influence of litter size and parity on maternal behaviour at parturition in Scottish blackface sheep. *Appl. Anim. Behav. Sci.* 33:345–355.

1085a. O'Donnell, T. G. and G. A. Walton. 1969. Some observations on the behaviour and hill-pasture utilization of Irish cattle. *J. Br. Grassland Soc.* 24:128–133.

1086. O'Farrell, V., and E. Peachey. 1990. Behavioural effects of overiohysterectomy on bitches. *J. Sm. Anim. Prac.* 31:595–598.

1087. Oberosler, R., C. Carenzi, and M. Verga. 1982. Dominance hierarchies of cows on Alpine pastures as related to phenotype. *Appl. Anim. Ethol.* 8:67–77.

1088. Occupational Safety and Health Administration, 1972. Federal Regulation 37(202), Part II. In *Occupational Safety and Health Standards*. Washington, DC: Department of Labor, pp. 22,102–22,356.

1089. Odde, K. G., G. H. Kiracofe, and R. R. Schalles. 1985. Suckling behavior in range beef calves. *J. Anim. Sci.* 61:307–309.

1090. Offord, K. P., L. D. Satter, and D. A. Wieckert. 1969. Study of behavioral conditioning and feed intake in dairy heifers [abstract]. *J. Dairy Sci.* 52:918.

1091. Olm, D. D., and K. A. Houpt. 1988. Feline house-soiling problems. *Appl. Anim. Behav. Sci.* 20:335–345.

1092. Orgeur, P. 1991. Identification of sexual receptivity in ewes by young sexually inexperienced rams. *Appl. Anim. Behav. Sci.* 31:83–90.

1093. Orgeur, P. 1995. Sexual play behavior in lambs androgenized in utero. *Physiol. Behav.* 47:185–187.

1094. Orgeur, P., P. Mimouni, and J. P. Signoret. 1990. The influence of rearing conditions on the social relationships of young male goats (Capra hircus). *Appl. Anim. Behav. Sci.* 27:105–113.

1095. Ortega-Reyes, L., and F. D. Provenza. 1993. Amount of experience and age affect the development of foraging skills of goats browsing blackbrush (Coleogyne ramosissima). *Appl. Anim. Behav. Sci.* 36:169–183.

1096. Over, R., J. Cohen-Tannoudji, M. Dehnhard, R. Claus, and J. P. Signoret. 1990. Effect of pheromones from male goats on LH-secretion in anoestrous ewes. *Physiol. Behav.* 48:665–668.

1097. Overall, K. L. 1992. Recognition, diagnosis, and management of obsessive-compulsive disorders. *Canine Pract.* 17:39–43.

1097a. Overall, K. L. 1997. *Clinical Behavioral Medicine for Small Animals*. St. Louis: Mosby.

1098. Owen, J. B., and W. J. Ridgman. 1967. The effect of dietary energy content on the voluntary intake of pigs. *Anim. Prod.* 9:107–113.

1099. Owen, R. R., F. J. McKeating, and D. W. Jagger. 1980. Neurectomy in wind-sucking horses. *Vet. Rec.* 106:134–135.

1100. Owens, J. L., T. N. Edey, B. M. Bindon, and L. R. Piper. 1984. Parturient behaviour and calf survival in a herd selected for twinning. *Appl. Anim. Behav. Sci.* 13:321–333.

1101. Ödberg, F. O. 1973. An interpretation of pawing by the horse (Equus caballus Linnaeus), displacement activity and original functions. *Saugetierkund. Mitteil.* 21:1–12.

1102. Ödberg, F. O., and K. Francis-Smith. 1977. Studies on the formation of un-grazed eliminative areas in fields used by horses. *Appl. Anim. Ethol.* 3:27–34.

1103. Packwood, J., and B. Gordon. 1975. Stereopsis in normal domestic cat, Siamese cat, and cat raised with alternating monocular occlusion. *J. Neurophysiol.* 38:1485–1499.

1104. Pajor, E. A., D. Fraser, and D. L. Kramer. 1991. Consumption of solid food by suckling pigs: Individual variation and relation to weight gain. *Appl. Anim. Behav. Sci.* 32:139–155.

1105. Palazzolo, D. L., and S. K. Quadri. 1987. The effects of aging on the circa-dian rhythm of serum cortisol in the dog. *Exp. Gerontol.* 22:379–387.

1106. Palen, G. F., and G. V. Goddard. 1966. Catnip and oestrous behaviour in the cat. *Anim. Behav.* 14:372–377.

1107. Panaman, R. 1981. Behavior and ecology of free-ranging female farm cats (Felis catus L.). *Z. Tierpsychol.* 56:59–73.

1108. Pappas, T. N., R. L. Melendez, K. M. Strah, and H. T. Debas. 1985. Chole-cystokinin is not a peripheral satiety signal in the dog. *Am. J. Physiol.* G733–G738.

1109. Parrott, R. F. 1993. Peripheral and central effects of CCK receptor agonists on operant feeding in pigs. *Physiol. Behav.* 53:367–372.

1110. Parrott, R. F., and B. A. Baldwin. 1984. Olfactory stimuli and intermale ag-gression in androgen-treated castrated sheep. *Aggress. Behav.* 10:115–122.

1111. Parrott, R. F., and W. D. Booth. 1984. Behavioural and morphological ef-fects of 5 a-dihydrotestosterone and oestradiol-17ß in the prepubertally castrated boar. *J. Reprod. Fert.* 71:453–461.

1112. Parsons, S. D., and G. L. Hunter. 1967. Effect of the ram on duration of oestrus in the ewe. *J. Reprod. Fert.* 14:61–70.

1113. Pavlov, I. P. 1927. *Conditioned Reflexes. An Investigation of the Physiological Ac-tivity of the Cerebral Cortex.* London, UK: Oxford University Press.

1114. Pedersen, L. J., T. Rojkittikhun, S. Einarsson, and L.-E. Edqvist. 1993. Post-weaning grouped sows: Effects of aggression on hormonal patterns and oestrous be-haviour. *Appl. Anim. Behav. Sci.* 38:25–39.

1115. Pekas, J. C. 1983. A method for direct gastric feeding and the effect of vol-untary ingestion in young swine. *Appetite* 4:23–30.

1116. Pekas, J. C. 1985. Animal growth during liberation from appetite suppres-sion. *Growth* 49:19–27.

1117. Pekas, J. C., and W. E. Trout. 1993. Cholecystokinin octapeptide immu-nization: Effect on growth of barrows and gilts. *J. Anim. Sci.* 71:2499–2505.

1118. Penning, P. D., A. J. Parsons, J. A. Newman, R. J. Orr, and A. Harvey. 1993. The effects of group size on grazing time in sheep. *Appl. Anim. Behav. Sci.* 37:101–109.

1119. Penning, P. D., A. J. Parsons, R. J. Orr, A. Harvey, and R. A. Champion. 1995. Intake and behaviour responses by sheep, in different physiological states, when grazing monocultures of grass or white clover. *Appl. Anim. Behav. Sci.* 45:63–78.

1120. Penning, P. D., A. J. Rook, and R. J. Orr. 1991. Patterns of ingestive behviour of sheep continuously stocked on monocultures of ryegrass or white clover. *Appl. Anim. Behav. Sci.* 31:237–250.

1121. Penny, R. H. C., F. W. G. Hill, J. E. Field, and J. T. Plush. 1972. Tailbiting in pigs: A possible sex incidence. *Vet. Rec.* 91:482–483.

1122. Pepelko, W. E., and M. T. Clegg. 1965. Influence of season of the year upon patterns of sexual behavior in male sheep. *J. Anim. Sci.* 24:633–637.

1123. Pepelko, W. E., and M. T. Clegg. 1965. Studies of mating behaviour and some factors influencing the sexual response in the male sheep Ovis aries. *Anim. Be-hav.* 13:249–258.

1124. Perez, O., N. Jimenez de Perez, P. Poindron, P. Le Neindre, and J. P. Ravault. 1985. Influence of management conditions after calving on mother-young relationships and PRL response to mammaray stimulation in the cow. *Reprod. Nutr. Develop.* 25:605–618.

1125. Perkins, A., and J. A. Fitzgerald. 1994. The behavioral component of the ram effect: The influence of ram sexual behavior on the induction of estrus in anovulatory ewes. *J. Anim. Sci.* 72:51–55.

1126. Persson, N. 1962. Self-stimulation in the goat. *Acta Physiol. Scand.* 55:276–285.

1127. Petchey, A. M., and J. Abdulkader. 1991. Intake and behaviour of cattle at different food barriers. *Anim. Prod.* 52:576–577.

1128. Petersen, H. V., K. Vestergaard, and P. Jensen. 1989. Integration of piglets into social groups of free-ranging domestic pigs. *Appl. Anim. Behav. Sci.* 23:223–236.

1129. Petersen, V. 1994. The development of feeding and investigatory behaviour in free-ranging domestic pigs during their first 18 weeks of life. *Appl. Anim. Behav. Sci.* 42:87–98.

1130. Petersen, V., H. B. Simonsen, and L. G. Lawson. 1995. The effect of environmental stimulation on the development of behaviour in pigs. *Appl. Anim. Behav. Sci.* 45:215–224.

1131. Petit, M. 1972. Emploi du temps des troupeaux de vaches-meres et de leurs veaux sur les paturages d'altitude de l'Aubrac. *Ann. Zootech.* 21:5–27.

1132. Pettijohn, T. F., T. W. Wong, P. D. Ebert, and J. P. Scott. 1977. Alleviation of separation distress in 3 breeds of young dogs. *Dev. Psychobiol.* 10:373–381.

1133. Pettyjohn, J. D., J. P. Everett,Jr., and R. D. Mochrie. 1963. Responses of dairy calves to milk replacer fed at various concentrations. *J. Dairy Sci.* 46:710–714.

1134. Pfaffenberger, C. J., and J. P. Scott. 1959. The relationship between delayed socialization and trainability in guide dogs. *J. Genet. Psychol.* 95:145–155.

1135. Phillips, C. J. C., and S. A. Schofield. 1990. The effect of environment and stage of the oestrous cycle on the behaviour of dairy cows. *Appl. Anim. Behav. Sci.* 27:21–31.

1136. Phillips, C. J. C., and L. Weiguo. 1991. Brightness discrimination abilities of calves relative to those of humans. *Appl. Anim. Behav. Sci.* 31:25–33.

1137. Phillips, P. A., D. Fraser, and B. K. Thompson. 1991. Preference by sows for a partially enclosed farrowing crate. *Appl. Anim. Behav. Sci.* 32:35–43.

1138. Pick, D. F., G. Lovell, S. Brown, and D. Dail. 1994. Equine color perception revisited. *Appl. Anim. Behav. Sci.* 42:61–65.

1139. Pickerel, T. M., Crowell-Davis.S.L., A. B. Caudle, and D. Q. Estep. 1993. Sexual preference of mares (Equus caballus) for individual stallions. *Appl. Anim. Behav. Sci.* 38:1–13.

1140. Pickett, B. W., L. C. Faulkner, and J. L. Voss. 1975. Effect of season on some characteristics of stallion semen. *J. Reprod. Fert. Suppl.* 23:25–28.

1141. Pickett, B. W., J. L. Voss, and E. L. Squires. 1977. Impotence and abnormal sexual behavior in the stallion. *Theriogenology* 8:329–347.

1142. Podberscek, A. L., J. K. Blackshaw, and A. W. Beattie. 1991. The behaviour of laboratory colony cats and their reactions to a familiar and unfamiliar person. *Appl. Anim. Behav. Sci.* 31:119–130.

1143. Pollard, J. C. 1992. Effects of litter size on the vocal behaviour of ewes. *Appl. Anim. Behav. Sci.* 34:75–84.

1144. Pollard, J. S., M. D. Baldock, and R. F. Lewis. 1971. Learning rate and use of visual information in five animal species. *Aust. J. Psychol.* 23:29–34.

1145. Pond, W. G., and J. H. Maner. 1974. *Swine Production in Temperate and Tropical Environments.* San Francisco, CA: W.H. Freeman and Co.

1146. Porter, R. H., R. Nowak, and P. Orgeur. 1995. Influence of a conspecific age-mate on distress bleating by lambs. *Appl. Anim. Behav. Sci.* 45:239–244.

1147. Prescott, C. W. 1973. Reproduction patterns in the domestic cat. *Aust. Vet. J.* 49:126–129.

1148. Presicce, G. A., C. C. Brockett, T. Cheng, R. H. Foote, G. F. Rivard, and W. R. Klemm. 1993. Behavioral responses of bulls kept under artificial breeding conditions to compounds presented for olfaction, taste or with topical nasal application. *Appl. Anim. Behav. Sci.* 37:273–284.

1149. Price, E. O., J. K. Blackshaw, A. Blackshaw, R. Borgwardt, M. R. Dally, and R. H. Bondurant. 1994. Sexual responses of rams to ovariectomized and intact estrous ewes. *Appl. Anim. Behav. Sci.* 42:67–71.

1150. Price, E. O., R. Borgwardt, and M. R. Dally. 1992. Measures of libido and their relation to serving capacity in the ram. *J. Anim. Sci.* 70:3376–3380.

1151. Price, E. O., R. Borgwardt, and M. R. Dally. 1993. Effect of ewe restraint on the libido and serving capacity of rams. *Appl. Anim. Behav. Sci.* 35:339–345.

1152. Price, E. O., G. D. Hutson, M. I. Price, and R. Borgwardt. 1994. Fostering in swine as affected by age of offspring. *J. Anim. Sci.* 72:1697–1701.

1153. Price, E. O., L. S. Katz, G. P. Moberg, and S. J. R. Wallach. 1986. Inability to predict sexual and aggressive behavior by plasma concentration of testosterone and luteinizing hormone in Hereford bulls. *J. Anim. Sci.* 62:613–617.

1154. Price, E. O., L. S. Katz, S. J. R. Wallach, and J. J. Zenchak. 1988. The relationship of male-male mounting to the sexual preferences of young rams. *Appl. Anim. Behav. Sci.* 21:347–355.

1155. Price, E. O., C. L. Martinez, and B. L. Coe. 1984. The effects of twinning on mother-offspring behavior in range beef cattle. *Appl. Anim. Behav. Sci.* 13:309–320.

1156. Price, E. O., and V. M. Smith. 1984. The relationhip of male-male mounting to mate choice and sexual preference in male dairy goats. *Appl. Anim. Behav. Sci.* 13:71–82.

1157. Price, E. O., V. M. Smith, and L. S. Katz. 1984. Sexual stimulation of male dairy goats. *Appl. Anim. Behav. Sci.* 13:83–92.

1158. Price, E. O., J. Thos, and G. B. Anderson. 1981. Maternal responses of confined beef cattle to single versus twin calves. *J. Anim. Sci.* 53:934–939.

1159. Price, E. O., and S. J. R. Wallach. 1990. Rearing bulls with females fails to enhance sexual performance. *Appl. Anim. Behav. Sci.* 26:339–347.

1160. Price, E. O., and S. J. R. Wallach. 1991. Development of sexual and aggressive behaviors in Hereford bulls. *J. Anim. Sci.* 69:1019–1027.

1161. Price, E. O., and S. J. R. Wallach. 1991. Effects of group size and the male-to-female ratio on the sexual performance and aggressive behavior of bulls in serving capacity tests. *J. Anim. Sci.* 69:1034–1040.

1162. Price, E. O., S. J. R. Wallach, and G. V. Silver. 1990. The effects of long-term individual vs group housing on the sexual behavior of beef bulls. *Appl. Anim. Behav. Sci.* 27:277–285.

1163. Prince, J. H. 1977. The eye and vision. In *Duke's Physiology of Domestic Animals.* 9th ed. M. J. Swenson, ed. Ithaca, NY: Cornell University Press, pp. 696–712.

1164. Provenza, F. D., and J. C. Malechek. 1986. A comparison of food selection and foraging behavior in juvenile and adult goats. *Appl. Anim. Behav. Sci.* 16:49–61.

1165. Provenza, F. D., L. Ortega-Reyes, C. B. Scott, J. J. Lynch, and E. A. Burritt. 1994. Antiemetic drugs attenuate food aversions in sheep. *J. Anim. Sci.* 72:1989–1994.

1166. Purcell, D., and C. W. Arave. 1991. Isolation vs group rearing in monozygous twin heifer calves. *Appl. Anim. Behav. Sci.* 31:147–156.

1167. Purcell, D., C. W. Arave, and J. L. Walters. 1988. Relationship of three measures of behavior to milk production. *Appl. Anim. Behav. Sci.* 21:307–313.

1168. Putnam, P. A., and R. E. Davis. 1963. Ration effects on drylot steer feeding patterns. *J. Anim. Sci.* 22:437–443.

1169. Ralphs, M. H., and C. D. Cheney. 1993. Influence of cattle age, lithium chloride dose level, and food type in the retention of food aversions. *J. Anim. Sci.* 71:373–379.

1170. Ralston, S. L. 1984. Controls of feeding in horses. *J. Anim. Sci.* 59:1354–1361.

1171. Ralston, S. L., and C. A. Baile. 1983. Effects of intragastric loads of xylose, sodium chloride and corn oil on feeding behavior of ponies. *J. Anim. Sci.* 56:302–308.

1172. Ralston, S. L., D. E. Freeman, and C. A. Baile. 1983. Volatile fatty acids and the role of the large intestine in the control of feed intake in ponies. *J. Anim. Sci.* 57:815–825.

1172a. Ralston, S. L., G. Van den Broek, and C. A. Baile. 1979. Feed intake patterns and associated blood glucose, free fatty acid and insulin changes in ponies. *J. Anim. Sci.* 49:838–845.

1173. Ramirez, A., A. Quiles, M. Hevia, and F. Sotillo. 1995. Behavior of the Murciano-Granadina goat in the hour before parturition. *Appl. Anim. Behav. Sci.* 44:29–35.

1174. Ramsay, D. J., B. J. Rolls, and R. J. Wood. 1975. The relationship between elevated water intake and oedema associated with congestive cardiac failure in the dog. *J. Physiol.* 244:303–312.

1175. Ramsay, D. J., B. J. Rolls, and R. J. Wood. 1977. Thirst following water deprivation in dogs. *Am. J. Physiol.* 232:R93–R100.

1176. Randall, G. C. B. 1972. Observations on parturition in the sow. I. Factors associated with the delivery of the piglets and their subsequent behaviour. *Vet. Rec.* 90:178–182.

1177. Randall, R. P., W. A. Schurg, and D. C. Church. 1978. Response of horses to sweet, salty, sour and bitter solutions. *J. Anim. Sci.* 47:51–55.

1178. Randall, W., and V. Lakso. 1968. Body weight and food intake rhythms and their relationship to the behavior of cats with brain stem lesions. *Psychonom. Sci.* 11:33–34.

1179. Randall, W., R. Swenson, V. Parsons, J. Elbin, and M. Trulson. 1975. The influence of seasonal changes in light on hormones in normal cats and in cats with lesions of the superior colliculi and pretectum. *J. Interdiscipl. Cycle Res.* 6:253–266.

1180. Rapoport, J. L., D. H. Ryland, and M. Kriete. 1992. Drug treatment of canine acral lick. An animal model of obsessive-compulsive disorder. *Arch. Gen. Psych.* 49:517–521.

1181. Rashotte, M. E., J. C. Smith, T. Austin, T. W. Castonguay, and L. Jonsson. 1984. Twenty-four-hour free feeding patterns of dogs eating dry food. *Neurosci. Biobehav. Rev.* 8:205–210.

1182. Rasmussen, O. G., E. M. Banks, T. H. Berry, and D. E. Becker. 1962. Social dominance in gilts. *J. Anim. Sci.* 21:520–522.

1183. Ray, D. E., and C. B. Roubicek. 1971. Behavior of feedlot cattle during two seasons. *J. Anim. Sci.* 33:72–76.

1184. Rayner, D. V., and S. Miller. 1993. Voluntary intake and gastric emptying in pigs: Effects of fat and a CCK inhibitor. *Physiol. Behav.* 54:917–922.

1185. Redding, R. W. 1975. Prefrontal lobotomy. In *Current Techniques in Amall Animal Surgery*. M. J. Bojrab, ed. Philadelphia, PA: Lea & Febiger, pp. 3–9.

1186. Reed, H. C. B., D. R. Melrose, and R. L. S. Patterson. 1974. Androgen steroids as an aid to the detection of oestrus in pig artificial insemination. *Br. Vet. J.* 130:61–67.

1187. Reinhardt, V., F. M. Mutiso, and A. Reinhardt. 1978. Social behaviour and social relationships between female and male prepubertal bovine calves (Bos indicus). *Appl. Anim. Ethol.* 4:43–54.

1188. Reinhardt, V., F. M. Mutiso, and A. Reinhardt. 1978. Resting habits of Zebu cattle in a nocturnal enclosure. *Appl. Anim. Ethol.* 4:261–271.

1189. Reisner, I. 1991. The pathophysiologic basis of behavior problems. *Vet. Clin. N. Am. :Sm. Anim. Prac.* 21:207–224.

1190. Reisner, I., K. A. Houpt, H. N. Erb, and F. W. Quimby. 1994. Friendliness to humans and defensive aggression in cats: The influence of handling and paternity. *Physiol. Behav.* 55:1119–1124.

1191. Reisner, I. R., H. N. Erb, and K. A. Houpt. 1994. Risk factors for behavior-related euthanasia among dominant-aggressive dogs: 110 cases (1989–1992). *J. Am. Vet. Med. Assoc.* 205:855–863.

1192. Reisner, I. R., J. J. Mann, M. Stanley, Y.-y. Huang, and K. A. Houpt. 1996. Comparison of cerebropsinal fluid monoamine metabolite levels in dominant-aggressive and non-aggressive dogs. *Brain Res.* 714:57–64.

1193. Remmers, J. E., and H. Gautier. 1972. Neural and mechanical mechanisms of feline purring. *Respir. Physiol.* 16:351–361.

1194. Rensch, B. 1956. Increase of learning capability with increase of brain-size. *Am. Natur.* 90:81–95.

1195. Reppert, S. M., H. G. Artman, S. Swaminathan, and D. A. Fisher. 1981. Vasopressin exhibits a rhythmic daily pattern in cerebrospinal fluid, but not in blood. *Science* 213:1256–1257.

1196. Rheingold, H. L. 1963. Maternal behavior in the dog. In *Maternal Behavior in Mammals*. H. L. Rheingold, ed. New York, NY: John Wiley & Sons, pp. 169–202.

1197. Rheingold, H. L., and C. O. Eckerman. 1971. Familiar social and nonsocial stimuli and the kitten's response to a strange environment. *Dev. Psychobiol.* 4:71–89.

1198. Riches, J. H., and R. H. Watson. 1954. The influence of the introduction of rams on the incidence of oestrus in Merino ewes. *Aust. J. Agric. Res.* 5:141–147.

1199. Richman, L. M., D. E. Johnson, and R. F. Angell. 1994. Evaluation of a positive conditioning technique for influencing big sagebrush (Artemisia tridentata subspp. wyomingensis) consumption by goats. *Appl. Anim. Behav. Sci.* 40:229–240.

1200. Ringo, J., M. L. Wolbarsht, H. G. Wagner, R. Crocker, and F. Amthor. 1977. Trichromatic vision in the cat. *Science* 198:753–755.

1201. Riol, J. A., J. M. Sanchez, V. G. Eguren, and V. R. Gaudioso. 1989. Colour perception in fighting cattle. *Appl. Anim. Behav. Sci.* 23:199–206.

1202. Robert, S., J. J. Matte, C. Farmer, C. L. Girard, and G. P. Martineau. 1993. High-fibre diets for sows: Effects of stereotypies and adjunctive drinking. *Appl. Anim. Behav. Sci.* 37:297–309.

1203. Roberts, S. J. 1971. *Veterinary Obstetrics and Genital Diseases (Theriogenology)*. 2nd ed. Ithaca, NY: Stephen J. Roberts.

1204. Roberts, W. W. 1958. Rapid escape learning without avoidance learning motivated by hypothalamic stimulation in cats. *J. Comp. Physiol. Psychol.* 51:391–399.

1205. Roberts, W. W. 1958. Both rewarding and punishing effects from stimulation of psoterior hypothalamus of cat with same electrode at same intensity. *J. Comp. Physiol. Psychol.* 51:400–407.

1206. Robinson, D. W. 1975. Food intake regulation in pigs. IV. The influence of dietary threonine imbalance on food intake, dietary choice and plasma acid patterns. *Br. Vet. J.* 131:595–600.

1207. Rogers, V. P., G. T. Hartke and R. L. Kitchell. 1967. Behavioral technique to analyze a dog's ability to discriminate flavors in commercial food products. In *Olfaction and Tast II*. Y. Hayashi, ed. Oxford, UK: Pergamon Press, pp. 353–359.

1208. Rohde Parfet, K. A., and H. W. Gonyou. 1991. Attraction of newborn piglets to auditory, visual, olfactory and tactile stimuli. *J. Anim. Sci.* 69:125–133.

1209. Rohde, K. A., and H. W. Gonyou. 1987. Strategies of teat-seeking behavior in neonatal pigs. *Appl. Anim. Behav. Sci.* 19:57–72.

1210. Romeyer, A., P. Poindron, and P. Orgeur. 1994. Olfaction mediates the establishment of selective bonding in goats. *Physiol. Behav.* 56:693–700.

1211. Romeyer, A., P. Poindron, R. H. Porter, F. Levy, and P. Orgeur. 1994. Establishment of maternal bonding and its mediation by vaginocervical stimulation in goats. *Physiol. Behav.* 55:395–400.

1212. Romeyer, A., R. H. Porter, F. Levy, R. Nowak, P. Orgeur, and P. Poindron. 1993. Maternal labelling is not necessary for the establishment of discrimination between kids by recently parturient goats. *Anim. Behav.* 46:705–712.

1213. Romsos, D. R., and D. Ferguson. 1983. Regulation of protein intake in adult dogs. *J. Am. Vet. Med. Assoc.* 182:41–43.

1214. Rook, A. J., and P. D. Penning. 1991. Synchronisation of eating, ruminating and idling activity by grazing sheep. *Appl. Anim. Behav. Sci.* 32:157–166.

1215. Root, M. V., S. D. Johnston, and P. N. Olson. 1996. Effect of prepuberal and postpuberal gonadectomy on heat production measured by indirect calorimetry in male and female domestic cats. *Am. J. Vet. Res.* 57:371–374.

1216. Rose, G. H., and J. P. Collins. 1975. Light-dark discrimination and reversal learning in early postnatal kittens. *Dev. Psychobiol.* 8:511–518.

1217. Rose, J. E. 1968. Discussion following paper, Cortical Representation by E.F. Evans. In *Hearing Mechanisms in Vertebrates. Ciba Foundation Symposium*. A. V. S. De Reuck and J. Knight, eds. London, UK: Churchill Ltd., pp. 287–295.

1218. Rosenblatt, J. S. 1965. The basis of synchrony in the behavioral interaction between the mother and her offspring in the laboratory rat. In *Determinants of Infant Behaviour III*. B. M. Foss, ed. New York, NY: John Wiley & Sons, pp. 3–45.

1219. Rosenblatt, J. S. 1965. Effects of experience on sexual behavior in male cats. In *Sex and Behavior*. F. A. Beach, ed. New York, NY: John Wiley & Sons., pp. 416–439.

1220. Rosenblatt, J. S. 1971. Suckling and home orientation in the kitten. A comparative developmental study. In *The Biopsychology of Development*. E. Tobach, L. R. Aronson, and E. Shaw, eds. New York, NY: Academic Press, pp. 345–410.

1221. Rosenblatt, J. S., and L. R. Aronson. 1958. The decline of sexual behavior in male cats after castration with special reference to the role of prior sexual experience. *Behaviour* 12:285–338.

1222. Rosenblatt, J. S., and L. R. Aronson. 1958. The influence of experience on the behavioural effects of androgen in prepuberally castrated male cats. *Anim. Behav.* 6:171–182.

1223. Rosenblatt, J. S., and T. C. Schneirla. 1962. The behaviour of cats. In *The Behaviour of Domestic Animals*. E. S. E. Hafez, ed. Baltimore: Williams and Wilkins, pp. 453–488.

1224. Ross, S. 1951. Sucking behavior in neonate dogs. *J. Abnorm. Soc. Psychol.* 46:142–149.

1225. Ross, S., and J. Berg. 1956. Stability of food dominance relationships in a flock of goats. *J. Mammal.* 37:129–131.

1226. Ross, S., and J. P. Scott. 1949. Relationship between dominance and control of movement in goats. *J. Comp. Physiol. Psychol.* 42:75–80.

1227. Ross, S., J. P. Scott, M. Cherner, and V. H. Denenberg. 1960. Effects of restraint and isolation on yelping in puppies. *Anim. Behav.* 8:1–5.

1228. Rossdale, P. D. 1967. Clinical studies on the newborn thoroughbred foal I: Perinatal behaviour. *Br. Vet. J.* 123:470–481.

1229. Rouda, R. R., D. M. Anderson, J. D. Wallace, and L. W. Murray. 1994. Free-ranging cattle water consumption in southcentral New Mexico. *Appl. Anim. Behav. Sci.* 39:29–38.

1230. Rowell, T. E. 1991. Till death us do part: Long-lasting bonds between ewes and their daughters. *Anim. Behav.* 42:681–682.

1231. Roy, J. H. B., K. W. G. Shillam, and J. Palmer. 1955. The outdoor rearing of calves on grass with special reference to growth rate and grazing behaviour. *J. Dairy Res.* 22:252–269.

1232. Rozin, P. 1967. Thiamine specific hunger. In *Handbook of Physiology. Section 6, Alimentary canal. Volume I, Control of Food and Water Intake*. C. F. Code and W. Heidel, eds. Washington, DC: American Physiological Society, pp. 411–431.

1233. Rozkowska, E., and E. Fonberg. 1973. Salivary reactions after ventromedial hypothalamic lesions in dogs. *Acta Neurobiol. Exp.* 33:553–562.

1234. Rubenstein, D. I., and M. A. Hack. 1992. Horse signals: The sounds and scents of fury. *Evol. Ecol.* 6:254–260.

1234a. Rubenstein, D. I. 1981. Behavioural ecology of island feral horses. *Eq. vet. J.* 13:27–34.

1235. Rubin, L., C. Oppegard, and H. F. Hintz. 1980. The effect of varying the temporal distribution of conditioning trials on equine learning behavior. *J. Anim. Sci.* 50:1184–1187.

1236. Ruckebusch, Y. 1972. The relevance of drowsiness in the circadian cycle of farm animals. *Anim. Behav.* 20:637–643.

1237. Ruckebusch, Y. 1974. Sleep deprivation in cattle. *Brain Res.* 78:4495–499.

1238. Ruckebusch, Y. 1975. The hypnogram as an index of adaptation of farm animals to changes in their environment. *Appl. Anim. Ethol.* 2:3–18.

1239. Ruckebusch, Y., R. W. Dougherty, and H. M. Cook. 1974. Jaw movements and rumen motility as criteria for measurement of deep sleep in cattle. *Am. J. Vet. Res.* 35:1309–1312.

1240. Ruckebusch, Y., and M. Gaujoux. 1976. Sleep patterns of the laboratory cat. *Electroenceph. clin. Neurophysiol.* 41:483–490.

1241. Ruckebusch, Y., M. Gaujoux, and B. Eghbali. 1977. Sleep cycles and kinesis in the foetal lamb. *Electroenceph. clin. Neurophysiol.* 42:226–237.

1242. Rudge, M. R. 1970. Mother and kid behaviour in feral goats (Capra hircus L.). *Z. Tierpsychol.* 27:687–692.

1243. Ruiz-Miranda, C. R. 1993. Use of pelage pigmentation in the recognition of mothers in a group by 2- to 4-month-old domestic goat kids. *Appl. Anim. Behav. Sci.* 36:317–326.

1244. Rushen, J. 1984. Stereotyped behaviour, adjunctive drinking and the feeding periods of tethered sows. *Anim. Behav.* 32:1059–1067.

1245. Rushen, J., and A. M. de Passillé. 1995. The motivation of non-nutritive sucking in calves, Bos taurus. *Anim. Behav.* 49:1503–1510.

1246. Rushen, J., A. M. de Passillé, and W. Schouten. 1990. Stereotypic behavior, endogenous opioids, and postfeeding hypoalgesia in pigs. *Physiol. Behav.* 48:91–96.

1247. Rushen, J., G. Foxcroft, and A. M. de Passillé. 1993. Nursing-induced changes in pain sensitivity, prolactin, and somatotropin in the pig. *Physiol. Behav.* 53:265–270.

1248. Rushen, J., J. Ladewig, and A. M. B. de Passillé. 1995. A novel environment inhibits milk ejection in the pig but not through HPA activity. *Appl. Anim. Behav. Sci.* 45:53–61.

1249. Russek, M., and P. J. Morgane. 1963. Anorexic effect of intraperitoneal glucose in the hypothalamic hyperphagic cat. *Nature* 199:1004–1005.

1250. Rutberg, A. T. 1990. Inter-group transfer in Assateague pony mares. *Anim. Behav.* 40:945–952.

1251. Rutberg, A. T., and S. A. Greenberg. 1990. Dominance, aggression frequencies and modes of aggressive competition in feral pony mares. *Anim. Behav.* 40:322–331.

1252. Rutberg, A. T., and R. R. Keiper. 1993. Proximate causes of natal dispersal in feral ponies: Some sex differences. *Anim. Behav.* 46:969–975.

1253. Ryder, M. L. 1976. Seasonal changes in the coat of the cat. *Res. Vet. Sci.* 21:280–283.

1254. Sacks, J. J., R. W. Sattin, and S. E. Bonzo. 1989. Dog bite-related fatalities from 1979 to 1988. *J. Am. Med. Assoc.* 262:1489–1492.

1254a. Salter, R. E. and R. J. Hudson. 1979. Feeding ecology of feral horses in western Alberta. *J. Range Manag.* 32:221–225.

1255. Salzinger, K., and M. B. Waller. 1962. The operant control of vocalization in the dog. *J. Exp. Anal. Behav.* 5:383–389.

1256. Sambraus, H. H., and D. Sambraus. 1975. Pragung von Nutztieren auf Menschen. *Z. Tierpsychol.* 38:1–17.

1257. Sandler, B. E., G. A. Van Gelder, W. B. Buck, and G. G. Karas. 1968. Effect of dieldrin exposure on detour behavior in sheep. *Psychol. Rep.* 23:451–455.

1258. Sandler, B. E., G. A. Van Gelder, D. D. Elsberg, G. G. Karas, and W. B. Buck. 1969. Dieldrin exposure and vigilance behavior in sheep. *Psychonom. Sci.* 15:261–262.

1259. Sandler, B. E., G. A. Van Gelder, G. G. Karas, and W. B. Buck. 1971. An operant feeding device for sheep. *J. Exp. Anal. Behav.* 15:95–96.

1260. Sappington, B. F., and L. Goldman. 1994. Discrimination learning and concept formation in the Arabian horse. *J. Anim. Sci.* 72:3080–3087.

1261. Sato, S. 1982. Leadership during actual grazing in a small herd of cattle. *Appl. Anim. Ethol.* 8:53–65.

1262. Sato, S., S. Sako, and A. Maeda. 1991. Social licking patterns in cattle (Bos taurus): Influence of environmental and social factors. *Appl. Anim. Behav. Sci.* 32:3–12.

1263. Sato, S., K. Tarumizu, and K. Hatae. 1993. The influence of social factors on allogrooming in cows. *Appl. Anim. Behav. Sci.* 38:235–244.

1264. Scarlett, J. M., S. Donoghue, J. Saidla, and J. Wills. 1994. Overweight cats: Prevalence and risk factors. *Int. J. Obesity* 18:S22–S28.

1265. Schafer, M. 1975. *The Language of the Horse.* New York, NY: Arco Publishing Co.

1266. Schake, L. M., and J. K. Riggs. 1969. Activities of lactating beef cows in confinement. *J. Anim. Sci.* 28:568–572.

1267. Scheepens, C. J. M., M. J. C. Hessing, E. Laarakker, W. G. P. Schouten, and M. J. M. Tielen. 1991. Influences of intermittent daily draught on the behaviour of weaned pigs. *Appl. Anim. Behav. Sci.* 31:69–82.

1268. Schein, M. W., and M. H. Fohrman. 1955. Social dominance relationships in a herd of dairy cattle. *Br. J. Anim. Behav.* 3:45–55.

1269. Schloeth, R. 1961. Das Sozialleben des Camargue-Rindes. Qualitative und quantitative Untersuchungen uber die sozialen Beziehungen -insbesondere die soziale Rangordnung—des halbwilden franzosischen SP—Kampfrindes. *Z. Tierpsychol.* 18:574–627.

1270. Schmidt, M. J., and H. Markowitz. 1977. Behavioral engineering as an aid in the maintenance of healthy zoo animals. *J. Am. Vet. Med. Assoc.* 171:966–969.

1270a. Schmisseur, W. E., J. L. Albright, W. M. Dillon, E. W. Kehrberg, and W. H. M. Morris. 1966. Animal behavior responses to loose and free stall housing. *J. Dairy Sci.* 49:102–104.

1271. Schneirla, T. C., J. S. Rosenblatt and E. Tobach. 1963. Maternal behavior in the cat. In *Maternal Behavior in Mammals.* H. L. Rheingold, ed. New York, NY: John Wiley & Sons, pp. 122–168.

1272. Schoen, A. M. S., E. M. Banks, and S. E. Curtis. 1976. Behavior of young Shetland and Welsh ponies (Equus caballus). *Biol. Behav.* 1:199–216.

1273. Schoen, A. M. S., S. W. Curtis, E. M. Banks, and H. W. Norton. 1974. Behavior and performance of swine subjected to preweaning handling [abstract]. *J. Anim. Sci.* 39:136–137.

1274. Schryver, H. F., M. T. Parker, P. D. Daniluk, K. I. Pagan, J. Williams, L. V. Soderholm, and H. F. Hintz. 1987. Salt consumption and the effect of salt on mineral metabolism in horses. *Cornell Vet.* 77:122–131.

1275. Schryver, H. F., S. VanWie, P. Daniluk, and H. F. Hintz. 1978. The voluntary intake of calcium by horses and ponies fed a calcium deficient diet. *J. Eq. Med. Surg.* 2:337–340.

1276. Schwartz, S. 1994. *Canine And Feline Behavior Problems. Instructions For Veterinary Clients.* American Veterinary Publications, Inc. 127 p. pp.

1277. Scott, D. W., R. W. Kirk, and J. Bentinck-Smith. 1979. Some effects of short-term methylprednisolone therapy in normal cats. *Cornell Vet.* 69:104–115.

1278. Scott, J. P. 1945. Social behavior, organization and leadership in a small flock of domestic sheep. *Comp. Psychol. Monogr.* 18:1–29.

1279. Scott, J. P. 1946. Dominance reaction in a small flock of goats. *Anat. Rec.* 94:380–381.

1280. Scott, J. P. 1948. Dominance and the frustration-aggression hypothesis. *Physiol. Zool.* 21:31–39.

1281. Scott, J. P. 1958. *Animal Behavior.* Chicago, IL: University of Chicago Press.

1282. Scott, J. P. 1958. *Aggression.* Chicago, IL: University of Chicago Press.

1283. Scott, J. P. 1962. Critical periods in behavioral development. *Science* 138:949–958.

1284. Scott, J. P., and J. L. Fuller. 1974. *Dog Behavior. The Genetic Basis.* Chicago, IL: University of Chicago Press.

1285. Scott, J. P., and M-V. Marston. 1950. Critical periods affecting the development of normal and maladjustive social behavior in puppies. *J. Genet. Psychol.* 77:25–60.

1286. Scott, M. D., and K. Causey. 1973. Ecology of feral dogs in Alabama. *J. Wildl. Manag.* 37:253–265.

1286a. Scott, P. P. 1970. Cats. In: *Reproduction and Breeding Techniques for Laboratory Animals.* Hafez, E.S.E., ed. Philadelphia: Lea & Febiger, pp. 192–208.

1287. Seabrook, M. K. 1972. A study to determine the influence of the herdsman's personality on milk yield. *J. Agric. Labour Sci.* 1:45–59.

1288. Seath, D. M., and G. D. Miller. 1946. Effect of warm weather on grazing performance of milking cows. *J. Dairy Sci.* 29:199–206.

1289. Sechzer, J. A., and J. L. Brown. 1964. Color discrimination in the cat. *Science* 144:427–429.

1290. Segerstad, C. H. A. F., and G. Hellekant. 1989. The sweet taste in the calf. I. Chorda tympani proper nerve responses to taste stimulation of the tongue. *Physiol. Behav.* 45:633–638.

1291. Seidel, W. F., T. Roth, T. Roehrs, F. Zorick, and W. Dement. 1984. Treatment of a 12-hour shift of sleep schedule with benzodiazepines. *Science* 224:1262–1264.

1292. Seitz, P. F. D. 1959. Infantile experiences and adult behavior in animal subjects. II. Age of separation from the mother and adult behavior in the cat. *Psychosom. Med.* 21:353–378.

1293. Selman, I. E., A. D. McEwan, and E. W. Fisher. 1970. Studies on natural suckling in cattle during the first eight hours post partum. I. Behavioural studies (dams). *Anim. Behav.* 18:276–283.

1294. Selman, I. E., A. D. McEwan, and E. W. Fisher. 1970. Studies on natural suckling in cattle during the first eight hours post partum. II. Behavioural studies (calves). *Anim. Behav.* 18:284–289.

1295. Senn, C. L., and J. D. Lewin. 1975. Barking dogs as an environmental problem. *J. Am. Vet. Med. Assoc.* 166:1065–1068.

1296. Seoane, J. R., and C. A. Baile. 1973. Feeding elicited by injections of Ca^{++} and Mg^{++} into the third ventricle of sheep. *Experientia* 29:61–62.

1297. Seoane, J. R., and C. A. Baile. 1973. Feeding behavior in sheep as related to the hypnotic activities of barbiturates injected into the third ventricle. *Pharm. Biochem. Behav.* 1:47–53.

1298. Seoane, J. R., C. A. Baile, and F. H. Martin. 1972. Humoral factors modifying feeding behavior of sheep. *Physiol. Behav.* 8:993–995.

1299. Serpell, J., and J. A. Jagoe. 1995. Early experience and the development of behavior. In *The Domestic Dog: Its Evolution, Behaviour And Interactions With People.* J. Serpell, ed. Cambridge, U.K.: Cambridge University Press, pp. 79–102

1300. Setchell, B. P. 1978. *The Mammalian Testis.* Ithaca, NY: Cornell University Press.

1301. Settle, R. H., B. A. Sommerville, J. McCormick, and D. M. Broom. 1994. Human scent matching using specially trained dogs. *Anim. Behav.* 48:1443–1448.

1302. Shackleton, D. M., and C. C. Shank. 1984. A review of the social behavior of feral and wild sheep and goats. *J. Anim. Sci.* 58:500–509.

1303. Shank, C. C. 1972. Some aspects of social behaviour in a population of feral goats (Capra hircus L.). *Z. Tierpsychol.* 30:488–528.

1304. Share, I., E. Martyniuk, and M. I. Grossman. 1952. Effect of prolonged intragastric feeding on oral food intake in dogs. *Am. J. Physiol.* 169:229–235.

1305. Shaw, E., and K. A. Houpt. 1985. Pre- and post–partum behaviour in mules impregnated by embryo transfer. *Eq. vet. J.* 17 (Suppl 3):73.

1306. Shaw, E. B., K. A. Houpt, and D. F. Holmes. 1988. Body temperature and behavior of mares during the last two weeks of pregnancy. *Eq. vet. J.* 20:199–202.

1307. Shaw, R. A. 1978. A time–controlled feeding system for cattle. *Anim. Prod.* 27:277–284.

1308. Sheppard, A. J., R. E. Blaser, and C. M. Kincaid. 1957. The grazing habits of beef cattle on pasture. *J. Anim. Sci.* 16:681–687.

1309. Sherry, C. J., T. J. Walters, G. G. Rodney,Jr., and P. J. Henry. 1994. Behavioral chaining in the goat (Capra hircus). *Appl. Anim. Behav. Sci.* 40:241–251.

1310. Sherry, C. J., J. M. Ziriax, T. J. Walters, R. L. Hamby, and G. G Rodney,Jr. 1994. Operant conditioning of the unrestrained goat (Capra hircus). *Sm. Ruminant Res.* 13:9–13.

1311. Shillito Walser, E. E. 1986. Recognition of the sow's voice by neonatal piglets. *Behaviour* 99:177–187.

1312. Shillito Walser, E., and P. Hague. 1980. Variations in the structure of bleats from sheep of four different breeds. *Behaviour* 75:22–35.

1313. Shillito Walser, E., and P. Hague. 1981. Field observations on a flock of ewes and lambs made of Clun Forest, Dalesbred and Jacob sheep. *Appl. Anim. Ethol.* 7:175–178.

1314. Shillito Walser, E., E. Walters, and P. Hague. 1981. A statistical analysis of the structure of bleats from shep of four different breeds. *Behaviour* 77:67–76.

1315. Shillito Walser, E., S. Willadsen, and P. Hague. 1982. Maternal vocal recognition in lambs born to Jacob and Dalesbred ewes after embryo transplantation between breeds. *Appl. Anim. Ethol.* 8:479–486.

1316. Shillito, E., and G. Alexander. 1975. Mutual recognition amongst ewes and lambs of four breeds of sheep (Ovis aries). *Appl. Anim. Ethol.* 1:151–165.

1317. Shillito, E. E. 1975. A comparison of the role of vision and hearing in lambs finding their own dams. *Appl. Anim. Ethol.* 1:369–377.

1318. Shillito, E. E., and V. J. Hoyland. 1971. Observations on parturition and maternal care in Soay sheep. *J. Zool.* 165:509–512.

1319. Shipka, M. P., and S. P. Ford. 1991. Relationship of circulating estrogen and progesterone concentrations during late pregnancy and the onset phase of maternal behavior in the ewe. *Appl. Anim. Behav. Sci.* 31:91–99.

1320. Shreffler, C., and W. D. Hohenboken. 1974. Dominance and mating behavior in ram lambs. *J. Anim. Sci.* 39:725–731.

1321. Shuleikina, K. V. 1976. Sensory mechanisms of learning in the behaviour of newborn kittens. *Activitas Nervosa Superior* 18:48–50.

1322. Signoret, J.-P. 1967. Duree du cycle oestrien et de l'oestrus chez la truie. Action du benzoate d'oestradiol chez la femelle ovariectomisee. *Ann. Biol. Anim. Biochem. Biophys.* 7:407–421.

1323. Signoret, J.-P. 1970. Sexual behaviour patterns in female domestic pigs (Sus scrofa L.) reared in isolation from males. *Anim. Behav.* 18:165–168.

1324. Signoret, J.-P. 1975. Influence of the sexual receptivity of a teaser ewe on the mating preference in the ram. *Appl. Anim. Ethol.* 1:229–232.

1325. Signoret, J.-P., B. A. Baldwin, D. Fraser and E. S. E. Hafez. 1975. The behaviour of swine. In *The Behaviour of domestic Animals.* 3rd ed. E. S. E. Hafez, ed. Baltimore, MD: Williams & Wilkins, pp. 295–329.

1326. Signoret, J.-P., and P. Mauleon. 1962. Action de l'ablation des bulbes olfactifs sur les mecanismes de la reproduction chez la truie. *Ann. Biol. Anim. Biochem. Biophys.* 2:167–174.

1327. Simpson, C. W., C. A. Baile, and L. F. Krabill. 1975. Neurochemical coding for feeding in sheep and steers. *J. Comp. Physiol. Psychol.* 88:176–182.

1328. Sivak, J., and D. B. Allen. 1975. An evaluation of the "ramp" retina of the horse eye. *Vision Res.* 15:1353–1356.

1329. Skinner, B. F. 1938. *The Behavior of Organisms.* New York, NY: D. Appleton–Century Co.

1330. Sly, J., and F. R. Bell. 1979. Experimental analysis of the seeking behaviour observed in ruminants when they are sodium deficient. *Physiol. Behav.* 22:499–505.

1331. Smith, B. L., J. H. Jones, G. P. Carlson, and J. R. Pascoe. 1994. Body position and direction preferences in horses during road transport. *Eq. Vet. J.* 26:374–377.

1332. Smith, F. V. 1965. Instinct and learning in the attachment of lamb and ewe. *Anim. Behav.* 13:84–86.

1333. Smith, F. V., C. Van-Toller, and T. Boyes. 1966. The "critical period" in the attachment of lambs and ewes. *Anim. Behav.* 14:120–125.

1333a. Sneva, F. A. 1970. Behavior of yearling cattle on eastern Oregon range. *J. Range Manage.* 23:155–158.

1334. Snowder, G. D., and A. D. Knight. 1995. Breed effects of foster lamb and foster dam on lamb viability and growth. *J. Anim. Sci.* 73:1559–1566.

1335. Sobocinska, J. 1978. Gastric distention and thirst: Relevance to the osmotic thirst threshold and metering of water intake. *Physiol. Behav.* 20:497–501.

1336. Sobocinska, J., and S. Kozlowski. 1987. Osmotic thirst suppression in dogs exposed to low ambient temperature. *Physiol. Behav.* 40:171–175.

1337. Soffie, M., G. Thines, and G. De Marneffe. 1976. Relation between milking order and dominance value in a group of dairy cows. *Appl. Anim. Ethol.* 2:271–276.

1338. Solomon, R. L., and L. C. Wynne. 1953. Traumatic avoidance learning: Acquisition in normal dogs. *Psychol. Monogr.* 67(4):1–19.

1339. Soltysik, S., and B. A. Baldwin. 1972. The performance of goats in triple choice delayed response tasks. *Acta Neurobiol. Exp.* 32:73–86.

1340. Someville, S. H., and B. G. Lowman. 1979. Observations on the nursing behaviour of beef cows suckling Charolais cross calves. *Appl. Anim. Ethol.* 5:369–373.

1341. Spencer, G. S. G. 1992. Immunization against cholecystokinin decreases appetite in lambs. *J. Anim. Sci.* 70:3820–3824.

1342. Spinka, M., and B. Algers. 1995. Functional view on udder massage after milk let–down in pigs. *Appl. Anim. Behav. Sci.* 43:197–212.

1343. Sprague, R. H., and J. J. Anisko. 1973. Elimination patterns in the laboratory beagle. *Behaviour* 47:257–267.

1344. Sprott, R. L. 1967. Barometric pressure fluctuations: Effects on the activity of laboratory mice. *Science* 157:1206–1207.

1344a. Squires, V. R. 1974. Grazing distribution and activity patterns of Merino sheep on a saltbush community in south-east Australia. *Appl. Anim. Ethol.* 1:17–30.

1345. Squires, V. R., and G. T. Daws. 1975. Leadership and dominance relationships in Merino and border Leicester sheep. *Appl. Anim. Ethol.* 1:263–274.

1346. Stahlbaum, C. C., and K. A. Houpt. 1989. The role of the flehmen response in the behavioral repertoire of the stallion. *Physiol. Behav.* 45:1207–1214.

1347. Stanford, T. L. 1981. Behavior of dogs entering a veterinary clinic. *Appl. Anim. Ethol.* 7:271–279.

1348. Stangel, G., and P. Jensen. 1991. Behaviour of semi–naturally kept sows and piglets (except suckling) during 10 days postpartum. *Appl. Anim. Behav. Sci.* 31:211–227.

1349. Stanley, W. C., W. E. Bacon, and C. Fehr. 1970. Discriminated instrumental learning in neonatal dogs. *J. Comp. Physiol. Psychol.* 70:335–343.

1350. Stanley, W. C., J. E. Barrett, and W. E. Bacon. 1974. Conditioning and extinction of avoidance and escape behavior in neonatal dogs. *J. Comp. Physiol. Psychol.* 87:163–172.

1351. Stanley, W. C., and O. Elliot. 1962. Differential human handling as reinforcing events and as treatments influencing later social behavior in basenji puppies. *Psychol. Rep.* 10:775–788.

1352. Stebbins, M. C. 1974. *Social Organization in Free-ranging Appaloosa Horses.* Pocatello, ID: M.S. Dissertation, Idaho State University.

1353. Stephens, D. B. 1974. Studies on the effect of social environment on the behaviour and growth rates of artificiallyreared British Friesian male calves. *Anim. Prod.* 18:23–24.

1354. Stephens, D. B. 1975. Effects of gastric loading on the sucking response and voluntary milk intake in neonatal piglets. *J. Comp. Physiol. Psychol.* 88:796–805.

1355. Stephens, D. B. 1980. The effects of 2–deoxy–D–glucose given via the jugular or hepatic portal vein on food intake and plasma glucose levels in pigs. *Physiol. Behav.* 25:691–697.

1356. Stephens, D. B., and B. A. Baldwin. 1971. Observations on the behavior of groups of artificial reared lambs. *Res. Vet. Sci.* 12:219–224.

1357. Stephens, D. B., D. L. Ingram, and D. F. Sharman. 1983. An investigation into some cerebral mechanisms involved in schedule–induced drinking in the pig. *Quart. J. Exptl. Physiol.* 68:653–660.

1358. Stephens, D. B., and J. L. Linzell. 1974. The development of sucking behaviour in the newborn goat. *Anim. Behav.* 22:628–633.

1359. Sterman, M. B., T. Knauss, D. Lehmann, and C. D. Clemente. 1965. Circadian sleep and waking patterns in the laboratory cat. *Electroenceph. clin. Neurophysiol.* 19:509–517.

1360. Stewart, J. C., and J. P. Scott. 1947. Lack of correlation between leadership and dominance relationships in a herd of goats. *J. Comp. Physiol. Psychol.* 40:255–264.

1361. Stone, C. C., M. S. Brown, and G. H. Waring. 1974. An ethological means to improve swine production [abstract]. *J. Anim. Sci.* 39:137.

1362. Stookey, J. M., and H. W. Gonyou. 1994. The effects of regrouping on behavioral and production parameters in finishing swine. *J. Anim. Sci.* 72:2804–2811.

1363. Strasia, C. A., M. Thorn, R. W. Rice, and D. R. Smith. 1970. Grazing habits, diet and performance of sheep on alpine ranges. *J. Range Manag.* 23:201–208.

1364. Stricklin, W. R., and H. Gonyou. 1981. Dominance and eating behavior of beef cattle fed from a single stall. *Appl. Anim. Behav. Sci.* 7:135–140.

1365. Stricklin, W. R., C. C. Kautz-Scanavy, and D. L. Greger. 1985. Determination of dominance–subordinance relationships among beef heifers in a dominance tube. *Appl. Anim. Behav. Sci.* 14:111–116.

1366. Stroup, W. W., M. K. Nielsen, and J. A. Gosey. 1987. Cyclic variation in cattle feed intake data: Characterization and implications for experimental design. *J. Anim. Sci.* 64:1638–1647.

1367. Sturgeon, R. D., P. D. Brophy, and R. A. Levitt. 1973. Drinking-elicited by intracranial microinjection of angiotensin in the cat. *Pharm. Biochem. Behav.* 1:353–355.

1368. Sufit, E., K. A. Houpt, and M. Sweeting. 1985. Physiological stimuli of thirst and drinking patterns in ponies. *Eq. vet. J.* 17:12–16.

1369. Sutherland, G. F. 1939. Salivary conditioned reflexes in swine [abstract]. *Am. J. Physiol.* 126:P640–641.

1370. Sweeting, M. P., C. E. Houpt, and K. A. Houpt. 1985. Social facilitation of feeding and time budgets in stabled ponies. *J. Anim. Sci.* 160:369–374.

1371. Sweetwood, H. L., D. F. Kripke, I. Grant, J. Yager, and M. S. Gerst. 1976. Sleep disorder and psychobiological symptomatology in male psychiatric outpatients and male nonpatients. *Psychosom. Med.* 38:373–378.

1372. Swenson, R. M., and W. Randall. 1977. Grooming behavior in cats with pontile lesions and cats with tectal lesions. *J. Comp. Physiol. Psychol.* 91:313–326.

1373. Syme, G. J., L. A. Syme, and T. P. Jefferson. 1974. A note on variations in the level of aggression within a herd of goats. *Anim. Prod.* 18:309–312.

1374. Syme, L. A., G. J. Syme, T. G. Waite, and A. J. Pearson. 1975. Spatial distribution and social status in a small herd of dairy cows. *Anim. Behav.* 23:609–614.

1375. Symoens, J., and M. van den Brande. 1969. Prevention and cure of aggressiveness in pigs using the sedative azaperone. *Vet. Rec.* 85:64–67.

1376. Symons, L. E. A., and D. R. Hennessy. 1981. Cholecystokinin and anorexia in sheep infected by the intestinal nematode Trichostrongylus colubriformis. *Int. J. Parasitol.* 11:55–58.

1377. Tan, S. S. L., and D. M. Shackleton. 1990. Effects of mixing unfamiliar individuals and of Azaperone on the social behaviour of finishing pigs. *Appl. Anim. Behav. Sci.* 26:157–168.

1378. Tanida, H., A. Miura, T. Tanaka, and T. Yoshimoto. 1995. Behavioral response to humans in individually handled weanling pigs. *Appl. Anim. Behav. Sci.* 42:249–259.

1379. Tanida, H., N. Miyazaki, T. Tanaka, and T. Yoshimoto. 1991. Selection of mating partners in boars and sows under multi-sire mating. *Appl. Anim. Behav. Sci.* 32:13–21.

1380. Tanzer, H., and N. Lyons. 1977. *Your Pet Isn't Sick. He Just Wants You to Think So.* New York, NY: Thomas Congdon Books.

1381. Tennessen, T., M. A. Price, and R. T. Berg. 1985. The social interactions of young bulls and steers after re-grouping. *Appl. Anim. Behav. Sci.* 14:37–47.

1382. Terlouw, E. M. C., A. B. Lawrence, and A. W. Illius. 1991. Influences of feeding level and physical restriction on development of stereotypies in sows. *Anim. Behav. Sci.* 42:981–991.

1383. Terlouw, E. M. C., A. Wiersma, A. B. Lawrence, and H. A. MacLeod. 1993. Ingestion of food facilitates the performance of stereotypies in sows. *Anim. Behav.* 46:939–950.

1384. Ternouth, J. H., and A. W. Beattie. 1970. A note on the voluntary food consumption and the sodium–potassium ratio of sheep after shearing. *Anim. Prod.* 12:343–346.

1385. Thiery, J. C., and J. P. Signoret. 1978. Effect of changing the teaser ewe on the sexual activity of the ram. *Appl. Anim. Ethol.* 4:87–90.

1386. Thinus-Blanc, C., B. Poucet, and N. Chapuis. 1982. Object permanence in cats: Analysis in locomotor space. *Behav. Proc.* 7:81–86.

1387. Thompson, D. S., and P. G. Schinckel. 1952. Incidence of oestrus in ewes. *Emp. J. Exp. Agric.* 20:77–79.

1388. Thompson, L. H., and J. S. Savage. 1978. Age at puberty and ovulation rate in gilts in confinement as influenced by exposure to a boar. *J. Anim. Sci.* 47:1141–1144.

1389. Thompson, W. R., and W. Heron. 1954. The effects of early restriction on activity in dogs. *J. Comp. Physiol. Psychol.* 47:77–82.

1390. Thompson, W. R., and W. Heron. 1954. The effects of restricting early experience on the problem-solving capacity of dogs. *Can. J. Psychol.* 8:17–31.

1391. Thompson, W. R., R. Melzack, and T. H. Scott. 1956. "Whirling behavior" in dogs as related to early experience. *Science* 123:939.

1392. Thorhallsdottir, A. G., F. D. Provenza, and D. F. Balph. 1990. The role of the mother in the intake of harmful foods by lambs. *Appl. Anim. Behav. Sci.* 25:35–44.

1393. Thorndike, E. L. 1911. *Animal Intelligence. Experimental Studies.* New York, NY: The Macmillan Co.

1394. Thornton, L. A. 1995. Animal behavior case of the month. *J. Am. Vet. Med. Assoc.* 206:1868–1870.

1395. Thorpe, W. H. 1963. *Learning and Instinct in Animals.* Cambridge, MA: Harvard University Press.

1396. Thrasher, T. N., C. J. Brown, L. C. Keil, and D. J. Ramsay. 1980. Thirst and vasopressin release in the dog: An osmoreceptor or sodium receptor mechanism? *Am. J. Physiol.* 238:R333–R339.

1397. Tilbrook, A. J. 1987. Physical and behavioural factors affecting the sexual "attractiveness" of the ewe. *Appl. Anim. Behav. Sci.* 17:109–115.

1398. Tilbrook, A. J., P. H. Hemsworth, J. S. Topp, and A. W. N. Cameron. 1990. Parallel changes in the proceptive and receptive behaviour of the ewe. *Appl. Anim. Behav. Sci.* 27:73–92.

1399. Tischner, M. 1982. Patterns of stallion sexual behaviour in the absence of mares. *J. Reprod. Fert. Suppl.* 32:65–70.

1400. Tischner, M., K. Kosiniak, and W. Bielanski. 1974. Analysis of the pattern of ejaculation in stallions. *J. Reprod. Fert.* 41:329–335.

1401. Titterington, R. W., and D. Fraser. 1975. The lying behaviour of sows and piglets during early lactation in relation to the position of the creep heater. *Appl. Anim. Ethol.* 2:47–53.

1402. Tobach, E., L. R. Aronson, and E. Shaw. 1971. *The Biopsychology of Development.* New York, NY: Academic Press.

1403. Todd, N. B. 1963. *The Catnip Response.* Cambridge, MA: Ph.D. Dissertation, Harvard University.

1404. Tomkins, T., and M. J. Bryant. 1974. Oestrous behaviour of the ewe and the influence of treatment with progestagen. *J. Reprod. Fert.* 41:121–132.

1405. Tomlinson, K. A., E. O. Price, and D. T. Torell. 1982. Responses of tranquilized post-partum ewes to alien lambs. *Appl. Anim. Ethol.* 8:109–117.

1406. Topel, D. G., G. M. Weiss, D. G. Siers, and J. H. Magilton. 1973. Comparison of blood source and diurnal variation on blood hydrocortisone, growth hormone, lactate, glucose and electrolytes in swine. *J. Anim. Sci.* 36:531–534.

1407. Tortora, D. F. 1980. Animal behavior therapy: The behavioral diagnosis and treatment of dominance-motivated aggression in canines: Part I. *Canine Pract.* 7:10–19.

1408. Tortora, D. F. 1980. Animal behavior therapy: The behavioral diagnosis and treatment of dominance-motivated aggression in canines: Part II. *Canine Pract.* 8:13–28.

1409. Toutain, P.-L., C. Toutain, A. J. F. Webster, and J. D. McDonald. 1977. Sleep and activity, age and fatness, and the energy expenditure of confined sheep. *Br. J. Nutr.* 38:445–454.

1410. Towbin, E. J. 1949. Gastric distention as a factor in the satiation of thirst in esophagostomized dogs. *Am. J. Physiol.* 159:533–541.

1411. Tribe, D. E. 1949. The importance of the sense of smell to the grazing sheep. *J. Agric. Sci. (Camb.)* 39:309–312.

1412. Trout, W. E., J. C. Pekas, and B. D. Schanbacher. 1989. Immune, growth and carcass responses of ram lambs to active immunization against desulfated cholecystokinin (CCK-8). *J. Anim. Sci.* 67:2709-2714.

1413. Trumler, E. 1959. Das "Rossigkeitgesicht" und ahnliches Ausdrucksverhalten bei Einhufern. *Z. Tierpsychol.* 16:478-488.

1414. Turek, F. W., and S. Losee-Olson. 1986. A benzodiazepine used in the treatment of insomnia phase-shifts the mammalian circadian clock. *Nature* 321:167-168.

1415. Turner, A. S., N. White, II, and J. Ismay. 1984. Modified Forssell's operation for crib biting in the horse. *J. Am. Vet. Med. Assoc.* 184:309-312.

1416. Turner, D. C., and P. Bateson. 1988. *The Domestic Cat. The Biology Of Its Behaviour*. New York, NY: Cambridge University Press.

1417. Turner, D. C., J. Feaver, M. Mendl, and P. Bateson. 1986. Variation in domestic cat behaviour towards humans: A paternal effect. *Anim. Behav.* 34:1890-1892.

1417a. Turner, R. R. 1961. Silage self-feeding. *Vet. Rec.* 73:1432-1436.

1418. Tyler, S. J. 1972. The behaviour and social organization of the New Forest ponies. *Anim. Behav. Monogr.* 5:85-196.

1419. Ursin, R. 1968. The two stages of slow wave sleep in the cat and their relation to REM sleep. *Brain Res.* 11:347-356.

1420. Ursin, R. 1970. Sleep stage relations within the sleep cycles of the cat. *Brain Res.* 20:91-97.

1421. Ursin, R., H. Cohen, S. Henriksen, G. Mitchell, and W. Dement. 1976. Effects on sleep of restricted sleep. A cat case study. *Electroenceph. clin. Neurophysiol.* 41:96-101.

1422. Vailes, L. D., and J. H. Britt. 1990. Influence of footing surface on mounting and other sexual behaviors of estrual holstein cows. *J. Anim. Sci.* 68:2333-2339.

1423. van der Borg, J. A. M., W. J. Netto, and D. J. U. Planta. 1991. Behavioural testing of dogs in animal shelters to predict problem behaviour. *Appl. Anim. Behav. Sci.* 32:237-251.

1424. van Miert, A. S. J. P. A. M., F. Kaya, and C. T. M. van Duin. 1992. Changes in food intake and forestomach motility of dwarf goats by recombinant bovine cytokine (IL-1ß, IL-2) and IFN-ß. *Physiol. Behav.* 52:859-864.

1425. Van Niekerk, A. I., J. F. D. Greenhalgh, and G. W. Reid. 1973. Importance of palatability in determining the feed intake of sheep offered chopped and pelleted hay. *Br. J. Nutr.* 30:95-105.

1426. Van Putten, G. 1969. An investigation into tail-biting among fattening pigs. *Br. Vet. J.* 125:511-517.

1427. Van Putten, G., and J. Dammers. 1976. A comparative study of the well-being of piglets reared conventionally and in cages. *Appl. Anim. Ethol.* 2:339-356.

1428. Vandenheede, M., and M. F. Bouissou. 1993. Sex differences in fear reactions in sheep. *Appl. Anim. Behav. Sci.* 37:39-55.

1429. Vecchiotti, G. G., and R. Galanti. 1987. Evidence of heredity of cribbing, weaving and stall walking. *Livestock Prod. Sci.* 14:91-95.

1430. Veeckman, J., and F. O. Ödberg. 1978. Preliminary studies on the behavioural detection of oestrus in Belgian "warm-blood" mares with acoustic and tactile stimuli. *Appl. Anim. Ethol.* 4:109-118.

1431. Veissier, I. 1993. Observational learning in cattle. *Appl. Anim. Behav. Sci.* 33:235-243.

1432. Veissier, I., D. Lamy, and P. Le Neindre. 1990. Social behaviour in domestic beef cattle when yearling calves are left with the cows for the next calving. *Appl. Anim. Behav. Sci.* 27:193-200.

1433. Veissier, I., and P. Le Neindre. 1989. Weaning in calves: Its effects on social organization. *Appl. Anim. Behav. Sci.* 24:43-54.

1434. Verberne, G., and J. De Boer. 1976. Chemocommunication among domestic cats, mediated by the olfactory and vomeronasal senses. *Z. Tierpsychol.* 42:86–109.

1435. Villablanca, J. R., and C. E. Olmstead. 1979. Neurological development of kittens. *Dev. Psychobiol.* 12:101–127.

1436. Vince, M. A. 1984. Teat seeking or presuckling behaviour in newly born lambs: Possible effects of maternal skin temperatures. *Anim. Behav.* 32:249–254.

1437. Vince, M. A., and M. W. Stanier. 1991. The effect of food intake on young Soay and Clun Forest lambs' response to touch on the face. *Appl. Anim. Behav. Sci.* 30:37–96.

1438. Vince, M. A., and T. M. Ward. 1984. The responses of newly born Clun Forest lambs to odour sources in the ewe. *Behaviour* 89:117–121.

1439. Vince, M. A., T. M. Ward, and M. Reader. 1984. Tactile stimulation and teat seeking behaviour in newly born lambs. *Anim. Behav.* 32:1179–1184.

1440. Vitale, A. F., M. Tenucci, M. Papini, and S. Lovari. 1986. Social behaviour of the calves of semi-wild Maremma cattle, Bos primigenius taurus. *Appl. Anim. Behav. Sci.* 16:217–231.

1441. Vogel, H. H.,Jr., J. P. Scott, and M.-V. Marston. 1950. Social facilitation and allelomimetic behavior in dogs. I. Social facilitation in a non-competitive situation. *Behaviour* 2:121–134.

1442. Voith, V. L., and P. L. Borchelt. 1985. Fears and phobias in companion animals. *Comp. Contin. Ed.* 7:209–218.

1443. Voith, V. L., and P. L. Borchelt. 1985. Elimination behavior and related problems in dogs. *Comp. Contin. Ed.* 7:537–546.

1444. Voith, V. L., J. C. Wright, and P. J. Danneman. 1992. Is there a relationship between canine behavior problems and spoiling activities, anthropomorphism, and obedience training? *Appl. Anim. Behav. Sci.* 34:263–272.

1445. Wagnon, K. A., R. G. Loy, W. C. Rollins, and F. D. Carroll. 1966. Social dominance in a herd of Angus, Hereford, and Shorthorn cows. *Anim. Behav.* 14:474–477.

1446. Walker, D. E. 1962. Suckling and grazing behaviour of beef heifers and calves. *N. Z. J. Agric. Res.* 5:331–338.

1447. Wallace, L. R. 1949. Observations of lambing behaviour in ewes. *Proc. N. Z. Soc. Anim. Prod.* 9:85–96.

1448. Waller, G. R., G. H. Price, and E. D. Mitchell. 1969. Feline attractant, cis,trans–nepetalactone: Metabolism in the domestic cat. *Science* 164:1281–1282.

1449. Waltl, B., M. C. Appleby, and J. Sölkner. 1995. Effects of relatedness on the suckling behaviour of calves in a herd of beef cattle rearing twins. *Appl. Anim. Behav. Sci.* 45:1–9.

1450. Wangsness, P. J., L. E. Chase, A. D. Peterson, T. G. Hartsock, D. J. Kellmel, and B. R. Baumgardt. 1976. System for monitoring feeding behavior of sheep. *J. Anim. Sci.* 42:1544–1549.

1451. Waran, N. K., and D. Cuddeford. 1995. Effects of loading and transport on the heart rate and behaviour of horses. *Appl. Anim. Behav. Sci.* 43:71–81.

1451a. Wardrop, J. C. 1953. Studies in the behaviour of dairy cows at pasture. *Br. J. Anim. Behav.* 1:23–31.

1452. Waring, G. H. 1982. Onset of behavior patterns in the newborn foal. *Eq. Pract.* 4:28–34.

1453. Waring, G. H. 1983. *Horse Behavior: The Behavioral Traits and Adaptations of Domestic and Wild Horses, Including Ponies.* Park Ridge, NJ: Noyes Publications.

1454. Waring, G. H., S. Wierzbowski and E. S. E. Hafez. 1975. The behaviour of horses. In *The Behaviour of Domestic Animals.* 3rd ed. E. S. E. Hafez, ed. Baltimore, MD: Williams & Wilkins, pp. 330–369.

1455. Warren, J. M., and A. Baron. 1956. The formation of learning sets by cats. *J. Comp. Physiol. Psychol.* 49:227–231.

1456. Warren, J. T., and I. Mysterud. 1993. Extensive ranging by sheep released onto an unfamiliar range. *Appl. Anim. Behav. Sci.* 38:67–73.

1457. Weary, D. M., and D. Fraser. 1995. Calling by domestic piglets: Reliable signals of need? *Anim. Behav.* 50:1047–1055.

1257a. Webb, F. M., V. N. Colenbrander, T. H. Blosser, and D. E. Waldern. 1963. Eating habits of dairy cows under drylot conditions. *J. Dairy Sci.* 46:1433–1435.

1458. Webster, A. S. F., J. S. Smith, and J. M. Brockway. 1972. Effects of isolation confinement and competition for feed on energy exchanges of growing lambs. *Anim. Prod.* 15:189–201.

1459. Weiguo, L., and C. J. C. Phillips. 1991. The effects of supplementary light on the behaviour and performance of calves. *Appl. Anim. Behav. Sci.* 30:27–34.

1460. Weir, W. C., and D. T. Torell. 1959. Selective grazing by sheep as shown by a comparison of the chemical composition of range and pasture forage obtained by hand clipping and that collected by esophageal–fistulated sheep. *J. Anim. Sci.* 18:641–649.

1461. Welch, A. R., and M. R. Baxter. 1986. Responses of newborn piglets to thermal and tactile properties of their environment. *Appl. Anim. Behav. Sci.* 15:203–215.

1462. Welch, R. A. S., and R. Kilgour. 1970. Mis–mothering among Romneys. *N. Z. J. Agric.* 121(4):26–27.

1463. Weldon, W. C., A. J. Lewis, G. F. Louis, J. L. Kovar, M. A. Giesemann, and P. S. Miller. 1994. Postpartum hypophagia in primiparous sows: I. Effects of gestation feeding level on feed intake, feeding behavior, and plasma metabolite concentrations during lactation. *J. Anim. Sci.* 72:387–394.

1464. Weller, R. F., and R. H. Phipps. 1985. The effect of silage preference on the performance of dairy cows. *Anim. Prod.* 42:435.

1465. Wells, S. M., and B. von Goldschmidt–Rothschild. 1979. Social behaviour and relationships in a herd of Camargue horses. *Z. Tierpsychol.* 49:363–380.

1466. Wesley, F., and F. D. Klopfer. 1962. Visual discrimination learning in swine. *Z. Tierpsychol.* 19:93–104.

1467. West, M. 1974. Social play in the domestic cat. *Am. Zool.* 14:427–436.

1468. West, M. J. 1977. Exploration and play with objects in domestic kittens. *Dev. Psychobiol.* 10:53–57.

1469. Whalen, R. E. 1963. The initiation of mating in naive female cats. *Anim. Behav.* 11:461–463.

1470. Whatson, T. S. 1985. Development of eliminative behaviour in piglets. *Appl. Anim. Behav. Sci.* 14:365–377.

1471. Whipp, S. C., R. L. Wood, and N. C. Lyon. 1970. Diurnal variation in concentrations of hydrocortisone in plasma of swine. *Am. J. Vet. Res.* 31:2105–2107.

1472. Whittlestone, W. G., and L. R. Cate. 1973. An animal activated feeding device for cattle. *J. Dairy Sci.* 56:1352–1353.

1473. Whittlestone, W. G., R. Kilgour, H. de Langen, and G. Duirs. 1970. Behavioral stress and the cell count of bovine milk. *J. Milk Food Technol.* 33:217–220.

1474. Whittlestone, W. G., M. Mullord, R. Kilgour, and L. R. Cate. 1975. Electric shocks during machine milking. *N. Z. Vet. J.* 23:105–108.

1475. Widdowson, E. M. 1971. Food intake and growth in the newly born. *Proc. Nutr. Soc.* 30:127–135.

1476. Widdowson, E. M., and R. A. McCance. 1963. The effect of finite periods of undernutrition at different ages on the composition and subsequent development of the rat. *Proc. Roy. Soc. B:* 158:329–342.

1477. Widowski, T. M., and S. W. Curtis. 1990. The influence of straw, cloth tassel, or both on the prepartum behavior of sows. *Appl. Anim. Behav. Sci.* 27:53–71.

1478. Wieckert, D. A. 1971. Social behavior in farm animals. *J. Anim. Sci.* 32:1274–1277.

1479. Wieckert, D. A., and G. R. Barr. 1966. Studies of learning ability in young pigs [abstract]. *J. Anim. Sci.* 25:1280.

1480. Wieckert, D. A., L. P. Johnson, K. P. Offord, and G. R. Barr. 1966. Measuring learning ability in dairy cows [abstract]. *J. Dairy Sci.* 49:729.

1481. Wiepkema, P. R., K. K. Van Hellemond, P. Roessingh, and H. Romberg. 1987. Behaviour and abomasal damage in individual veal calves. *Appl. Anim. Behav. Sci.* 18:257–268.

1482. Wierenga, H. K. 1990. Social dominance in dairy cattle and the influences of housing and management. *Appl. Anim. Behav. Sci.* 27:201–229.

1483. Wierenga, H. K., and H. Hopster. 1990. The significance of cublicles for the behaviour of dairy cows. *Appl. Anim. Behav. Sci.* 26:308–337.

1484. Wierzbowski, S. 1959. Odruchy plciowe ogierow. *Roczn. Nauk Rolnicz.* 73-B-4:753–788.

1485. Wierzbowski, S. 1978. The sexual behaviour of experimentally underfed bulls. *Appl. Anim. Ethol.* 4:55–60.

1486. Wikmar, G., and J. M. Warren. 1972. Delayed response learning by caged-reared normal and prefrontal cats. *Psychonom. Sci.* 26:243–245.

1487. Wilcox, S., K. Dusza, and K. A. Houpt. 1991. The relationship between recumbent rest and masturbation in stallions. *Eq. Vet. Sci.* 11(1):23–26.

1487a. Willard, J. G., J. C. Willard, S. A. Wolfram, and J. P. Baker. 1977. Effect of diet on cecal pH and feeding behavior of horses. *J. Anim. Sci.* 45:87–93.

1488. Willham, R. L., D. F. Cox, and G. G. Karas. 1963. Genetic variation in a measure of avoidance learning in swine. *J. Comp. Physiol. Psychol.* 56:294–297.

1489. Willham, R. L., G. G. Karas, and D. C. Henderson. 1964. Partial acquisition and extinction of an avoidance response in two breeds of swine. *J. Comp. Physiol. Psychol.* 57:117–122.

1490. Williamson, N. B., R. S. Morris, D. C. Blood, and C. M. Cannon. 1972. A study of oestrous behaviour and oestrus detection methods in a large commercial dairy herd. I. The relative efficiency of methods of oestrus detection. *Vet. Rec.* 91:50–58.

1491. Wilson, E. O. 1975. *Sociobiology. The New Synthesis.* Cambridge, MA: The Belknap Press of Harvard University Press.

1492. Wilson, M., J. M. Warren, and L. Abbott. 1965. Infantile stimulation, activity, and learning by cats. *Child Dev.* 36:843–853.

1493. Wilsson, E. 1984. The social interaction between mother and offspring during weaning in German Shepherd dogs: Individual differences between mothers and their effects on offspring. *Appl. Anim. Behav. Sci.* 13:101–112.

1494. Winchester, C. F. 1943. The energy cost of standing in horses. *Science* 97:24.

1495. Winfield, C. G., P. H. Hemsworth, M. R. Taverner, and P. D. Mullaney. 1974. Observations on the sucking behviour of piglets in litters of varying size. *Proc. Aust. Soc. Anim. Prod.* 10:307–310.

1496. Winfield, C. G., and R. Kilgour. 1976. A study of following behaviour in young lambs. *Appl. Anim. Ethol.* 2:235–243.

1497. Winfield, C. G., and A. W. Makin. 1978. A note on the effect of continuous contact with ewes showing regular oestrus and of post–weaning growth rate on the sexual activity of Corriedale rams. *Anim. Prod.* 27:361–364.

1498. Winfield, C. G., and P. D. Mullaney. 1973. A note on the social behaviour of a flock of Merino and Wiltshire horn sheep. *Anim. Prod.* 17:93–95.

1499. Winfield, C. G., G. J. Syme, and A. J. Pearson. 1981. Effect of familiarity with each and breed on the spatial behaviour of sheep in an open field. *Appl. Anim. Ethol.* 7:67–75.

1500. Winkler, W. G. 1977. Human deaths induced by dog bites, United States, 1974–1975. *Publ. Hlth. Rep.* 92:425–429.

1501. Winskill, L. C., N. K. Waran, and R. J. Young. 1996. The effect of a foraging device (a modified 'Edinburgh Foodball') on the behaviour of the stabled horse. *Appl. Anim. Behav. Sci.* 48:25–35.

1502. Winslow, C. N. 1944. The social behavior of cats. II. Competitive, aggressive, and food–sharing behavior when both competitors have access to the goal. *J. Comp. Psychol.* 37:315–326.

1503. Winter, A., and J. E. Hillerton. 1995. Behaviour associated with feeding and miling of early lactation cows housed in an experimental automatic milking system. *Appl. Anim. Behav. Sci.* 46:1–15.

1504. Wise, R. A., and V. Dawson. 1974. Diazepam–induced eating and lever pressing for food in sated rats. *J. Comp. Physiol. Psychol.* 86:930–941.

1505. Wolf, A. V. 1950. Osmometric analysis of thirst in man and dog. *Am. J. Physiol.* 161:75–86.

1506. Wolski, T. R. 1982. Social behavior of the cat. *Vet. Clin. North Am. Small Anim. Pract.* 12:693–706.

1507. Wolski, T. R., K. A. Houpt, and R. Aronson. 1980. The role of the senses in mare–foal recognition. *Appl. Anim. Ethol.* 6:121–138.

1508. Wood, M. T. 1977. Social grooming patterns in two herds of monozygotic twin dairy cows. *Anim. Behav.* 25:635–42.

1509. Wood, P. D. P., G. F. Smith, and M. F. Lisle. 1967. A survey of intersucking dairy cows in herds in England and Wales. *Vet. Rec.* 81:396–397.

1510. Wood, R. J., B. J. Rolls, and D. J. Ramsay. 1977. Drinking following intracarotid infusions of hypertonic solutions in dogs. *Am. J. Physiol.* 232:R88–R92.

1511. Wood, T., S. Stanley, and T. Tobin. 1989. Operant conditioning and its applications in equine pharmacology. *J. Eq. Vet. Sci.* 9:124–130.

1512. Wood–Gush, D. G. M. 1983. *Elements of Ethology*. London, UK: Chapman and Hall.

1513. Wood–Gush, D. G. M., and D. Csermely. 1981. A note on the diurnal activity of early weaned piglets in flat–deck cages at 3 and 6 weeks of age. *Anim. Prod.* 33:107–110.

1514. Wood–Gush, D. G. M., and K. Vestergaard. 1991. The seeking of novelty and its relation to play. *Anim. Behav.* 42:599–606.

1515. Woodbury, C. B. 1943. The learning of stimulus patterns by dogs. *J. Comp. Psychol.* 35:29–40.

1516. Woods, G. L., and K. A. Houpt. 1986. An abnormal facial gesture in an estrous mare. *Appl. Anim. Behav. Sci.* 16:199–202.

1517. Wright, J. C. 1980. Early development of exploratory and dominance in three litters of German shepherds. In *Early Experiences and Early Behavior*. New York, NY: Academic Press, pp. 181–206.

1518. Wright, J. C., and M. S. Nesselrote. 1987. Classification of behavior problems in dogs: Distributions of age, breed, sex and reproductive status. *Appl. Anim. Behav. Sci.* 19:169–178.

1519. Wyrwicka, W. 1978. Effects of electrical stimulation within the hypothalamus on gastric acid secretion and food intake in cats. *Exp. Neurol.* 60:286–303.

1520. Yamane, A., J. Emoto, and N. Ota. 1996. Factors affecting feeding order and social tolerance to kittens in the group-living feral cat (*Felis catus*). *Appl. Anim. Behav. Sci.* (In Press)

1521. Yang, T. S., B. Howard, and W. V. Macfarlane. 1981. Effects of food on drinking behaviour of growing pigs. *Appl. Anim. Ethol.* 7:259–270.

1522. Yang, T. S., M. A. Price, and F. X. Aherne. 1984. The effect of level of feeding on water turnover in growing pigs. *Appl. Anim. Behav. Sci.* 12:103–109.

1523. Yarney, T. A., G. W. Rahnefeld, R. J. Parker, and W. M. Palmer. 1982. Hourly distribution of time of parturition in beef cows. *Can. J. Anim. Sci.* 62:597–605.

1524. Yerkes, R. M. 1916. The mental life of monkeys and apes; a study of ideational behavior. *Behav. Monogr.* 3:145.

1525. Yerkes, R. M., and C. A. Coburn. 1915. A study of the behavior of the pig Sus scrofa by the multiple choice method. *J. Anim. Behav.* 5:185–225.

1526. Young, R. J., J. Carruthers, and A. B. Lawrence. 1994. The effect of a foraging device (The 'Edinburgh Foodball') on the behaviour of pigs. *Appl. Anim. Behav. Sci.* 39:237–247.

1527. Zablocka, T. 1975. Go-no go differentiation to visual stimuli in cats with different early visual experiences. *Acta Neurobiol. Exp.* 35:399–402.

1528. Zablocka, T., J. Konorski, and B. Zernicki. 1975. Visual discrimination learning in cats with different early visual experiences. *Acta Neurobiol. Exp.* 35:389–398.

1529. Zahorik, D. M., and K. A. Houpt. 1977. The concept of nutritional wisdom: Applicability of laboratory learning models to large herbivores. In *Learning Mechanisms in Food Selection.* L. M. Barker, M. R. Best, and M. Domjan, eds. Waco, TX: Baylor University Press, pp. 45–67.

1530. Zahorik, D. M., and K. A. Houpt. 1981. Species differences in feeding strategies, food hazards, and the ability to learn food aversions. In *Foraging Behavior.* A. C. Kamil and T. D. Sargent, eds. New York, NY: Garland STPM Press, pp. 289–310.

1531. Zapelin, H. 1989. Mammalian sleep. In *Principles and Practice of Sleep Medicine.* M. H. Kryger, T. Roth, and W. C. Dement, eds. Philadelphia, PA: W.B. Saunders Company, pp. 30–49.

1531a. Zemo, T. and J. O. Klemmedson. 1970. Behavior of fistulated steers on a desert grassland. *J. Range Manag.* 23:158–163.

1532. Zenchak, J. J., and G. C. Anderson. 1973. Discrimination learning in sheep [abstract]. *J. Anim. Sci.* 37:227).

1533. Zenchak, J. J., and G. C. Anderson. 1980. Sexual performance levels of rams (Ovis aries) as affected by social experiences during rearing. *J. Anim. Sci.* 50:167–174.

1534. Zimmerman, M. B., and E. M. Stricker. 1978. Water and NaCl intake after furosemide treatment in sheep (Ovis aries). *J. Comp. Physiol. Psychol.* 92:501–510.

INDEX

Acceptance of young. *See also* Mutual
    recognition
  cats, 204
  goats, 188, 189
  horses, 198–200
  sheep, 183–86
Acepromazine, 199, 281
Acorn poisoning, 313
Acral lick granuloma, 348
Activity patterns, 91–105
  biological rhythms, 83–91
  cats, 93–94
  cattle, 98–103, 121, 247
  clinical problems, 108–9
  dogs, 91–93
  goats, 105
  horses, 95–98
  pigs, 94–95
  reproductive cycle, 121
  sheep, 103–5
Acuity, visual, 4
Adipsia, 307
Adrenal gland, 87
Aerophagia, 316
Age
  circadian rhythms and, 88
  dominance and
    cattle, 40
    goats, 46
    horses, 47
    sheep, 45
Aggression, 33–81
  biological basis of, 36–38
  categories, 34–36
  cats, 231
    cat-directed, 80–81
    people-directed, 79–80
    posture, 22
    predatory, 36, 76–77, 79
    social, 74
    territorial, 74–76, 80
    vocalization, 21
  cattle, 248
    clinical cases, 43–44

  maternal, 43
  pain-induced, 35
  social, 38–39
crowding and, 37, 58, 59
defensive, 14, 23, 32
dogs, 61–73, 222
  castration effect on, 162–63
  toward children, 69–70
  fear-induced, 62–63, 70–72
  heritability, 36–37
  owner-directed, 64–69
  posture, 14
  predatory, 63
  pseudopregnancy, 209
  toward strangers, 70–72
  territorial, 62, 66, 70–72
  treatment, 67–69, 71–73
  urine marking, 20
  toward veterinarians, 72–73
fear-induced, 36, 62–63, 70–72, 80
food-related, 40, 57–58
goats, 46
hormonal control, 37–38
horses, 47–55
  maternal, 196, 198–200
  posture, 9, 11
hunger and, 37, 40
interspecies, 45, 48, 63–64
maternal, 36, 43, 68, 173, 196,
    198–200
pain-induced, 34–35
pigs, 56–60, 179
predatory, 36, 61, 63, 68, 70, 76–77, 79
reproductive cycle and, 121
serotonin levels, 87
sexual, 45, 58
sheep, 45–46
social, 34, 38–39, 56–58, 74. *See also*
    Dominance
temperament tests, 222, 231, 242
territorial, 34, 62, 66, 70–72, 74–76,
    80
Aldosterone, 331, 334
Alleomimetic behavior in sheep, 180

Altrenogest, 145
Amino acid imbalances, 295
Amitriptyline, 339
  aggression treatment, 69
  canine destructiveness, 348
  spraying, 27
  thunderphobia, 350
Amniotic fluid
  cats, 200
  cattle, 189–90, 193
  horses, 194
  sheep, 180, 184
Amphetamines, 108, 297, 304
Anal gland secretions
  cats, 25
  dogs, 18
Androstenone, 31
Androgens, 37–38, 113. *See also* Testos-
    terone
Anestrus in horses, 144
Angiotensin, 331, 334
Anorexia
  cats, 310
  dogs, 304–6
Anticonvulsants, 339, 340
Anxiety, separation. *See* Separation anx-
    iety
Apomorphine, 63, 306
Artificial insemination, 122, 124,
    129–30
Audition
  cats, 5, 20–21, 203, 226, 227
  dogs, 5, 13–14, 207, 215, 275
  horses, 7–8, 197, 200, 231
  maternal behavior
    cats, 203
    dogs, 207
    goats, 188
    horses, 197, 200
    pigs, 174, 176–78
    sheep, 183, 185
  overview, 5
  pigs, 29–31, 174, 176–78
  ruminants, 5, 183, 185, 188
Aural secretions, 18
Autoshaping, 258
Avoidance learning, 267, 268
Avoidance-response method, 263

Balanoposthitis, 129
Balking behavior, 51
Barbiturates, 329, 349

Barking
  dogs, 13, 19
  pigs, 29
Barometric pressure and circadian
    rhythms, 85
Barrier frustration, 342–43, 347–48, 355
Begging behavior, 16
Behavioral problems. *See* Clinical be-
    havior problems; *specific species*
Behavior modification of destructive-
    ness, 345–46
Benzodiazepines
  as appetite stimulant, 297–98, 329
  circadian rhythms and, 85
  disinhibition potential, 80
  phobia treatment, 350
Binocular vision. *See* Field of vision
Biological rhythms. *See* Rhythms, bio-
    logical
Blood volume and thirst, 331, 332
Body temperature
  biological rhythm, 84, 88, 94
  parturition
    cattle, 189
    dogs, 205–6
    horses, 193
Body weight maintenance
  cats, 309
  cattle, 324
  dogs, 303
  horses, 314
  pigs, 296–97
  sheep, 326
Body weight to brain weight ratio,
    260–62
Bombesin, 308
Brain development. *See* Neurological
    development
Breeding. *See* Sexual behavior
Bucking, 362
Buller steers, 130–31
Bunting behavior, 25, 38–39
Buspirone, 27
Buttercup poisoning, 313

Caffeine, 85
Cannibalism
  cats, 205
  pigs, 179
Car chasing, 350
Castration, 68
  cats, 26, 27, 74–76, 167–68, 309

dogs, 162–63
  aggression and, 162–63
  roaming, 351
  scent marking, 17–18
  weight gain, 302
goats, 140
hormonal effects, 112–13
horses, 152
pigs, 239
psychological, 130
rams, 138
sexual behavior, 112–13, 114
Catnip, 164
Cats
  activity patterns, 93–94
  aggression, 231
    toward other cats, 80–81
    pain-induced, 35
    toward people, 79–80
    predatory, 36, 76–77, 79
    social, 74
    temperament tests, 231
    territorial, 74–76, 80
  communication, 20–29
    audition, 20–21
    behavior problems, 25–29
    olfaction, 24–25
    vision, 21–24
  development of behavior, 222–31
    chart, 224–25
    clinical problems, 231
    human contact and, 230
    neurological development, 226–27
    play, 227–30
    sensitive periods, 222–26
    sleep, 227
  free-ranging, 73–74, 93, 163
  grooming, 77–78, 204
  learning, 283–85
    operant conditioning example,
      252–53
  maternal behavior, 200–205
    clinical problems, 204–5
    nesting, 203–4
    nursing, 201–3
    parturition, 200–201
  problems, clinical
    clawing, 28–29
    destructiveness, 353
    house soiling, 25–28
    ingestive, 310–12
    maternal, 204–5
    sexual, 165–68

  withdrawal, 353–54
  sexual behavior, 163–68
    female, 164–66
    male, 166–68
    play, 229–30
  sleep, 94, 227
Cattle
  activity patterns, 98–103, 247
  aggression, 248
    clinical cases, 43–44
    maternal, 43
    pain-induced, 35
    social, 38–39
  communication, 31–32
  development of behavior, 246–49
  free-ranging, 38
  grooming, 43
  learning, 275–79
    classical conditioning example, 252
  maternal behavior, 189–93
    bonding, 190
    clinical problems, 192–93
    nursing, 190–91, 192
    parturition, 189–90
    weaning, 191–92
  problems, clinical behavior, 363–64
    aggression, 43–44
    ingestive, 325
    maternal, 192–93
    sexual, 123, 128–31
  sexual behavior, 120–31
    female, 120–23
    male, 124–31
  sleep, 101–3, 104, 107
  social structure, 38–42
  vision, 4–5, 277
  vocalization, 32
CCK. See Cholecystokinin-pancreozymin (CCK)
Cervical stimulation
  goats, 188
  maternal behavior, 170
  sheep, 187
Chaining, 253–54
Cheek rubbing in cats, 25
Chewing. See Destructive behavior
Children, dog aggression toward, 69–70
Cholecystokinin-pancreozymin (CCK),
  287, 293
  cats, 308
  cattle, 324, 325
  dogs, 302
Circadian rhythms, 84–88

Circatrigentian rhythms, 88
Citronella, 19
Classical conditioning, 251–52
  cats, 284–85
  comparative intelligence measure-
    ment, 262
  dogs, 269
  goats, 279
  pigs, 266
  sheep, 279
Clawing, 28–29
Clenbuterol, 289, 324
Clinical behavior problems, 335–64. *See
  also specific species*
  activity and sleep, 105–9
  aggression
    cats, 79–81, 231
    cattle, 43
    dogs, 64–73
    horses, 52–55
    pigs, 59–60
  barking, excessive, 19
  car chasing, 350
  clawing and scratching, 28–29
  destructiveness, 338–48, 353
  digging, 351–52
  elimination
    cats, 25–28
    dogs, 19–20
  history and, 336–38
  ingestive behavior
    anorexia, 304–6, 310
    cats, 310–12
    cattle, 325
    dogs, 304–7
    obesity, 304, 310, 325, 329
    pica, 306, 311–12
    pigs, 299
    plant eating, 306, 310–11
  jumping up, 352
  maternal behavior
    cats, 204–5
    cattle, 192–93
    dogs, 208–9
    goats, 189
    horses, 198–200
    infanticide, 179, 204–5
    mismothering, 187, 205
    pigs, 179
    sheep, 186–88
  narcolepsy, 108–9
  nocturnal wakefulness, 109
  phobias, 349–50, 362

  roaming, 350–51
  self-mutilation, 348
  sexual behavior
    cats, 165–66, 167–68
    cattle, 123, 128–31
    dogs, 157–58, 160–63
    goats, 140
    horses, 143–46, 147–52
    pigs, 154, 156
    sheep, 139
  tail chasing, 349
  trailering problems, 356–61
  withdrawal, 353–54
  wool chewing, 364
Clomipramine, 312, 348
Clorazepate, 350
CNS and ingestive behavior, 297–98
  cats, 309
  cattle, 324
  dogs, 303–4
  pigs, 297–98
  sheep, 329–30
Colic, 288, 314, 318
Collars
  choke chain, 73
  electronic, 19, 55, 260
  Elizabethan, 348
Color vision, 4–5
  cats, 284
  lamb recognition, 185
  pigs, 268, 291
Colostrum feeder, 183
Communication, 3–32
  cats, 20–29
    audition, 20–21
    behavior problems, 25–29
    olfaction, 24–25
    vision, 21–24
  dogs, 13–20
    audition, 13–14
    behavior problems, 19–20
    olfaction, 16–18
    vision, 14–16
  horses, 7–13
    audition, 7–8
    olfaction, 12–13
    vision, 9–11
  overview, 3–7
    audition, 5
    olfaction, 7
    vision, 4–6
  pigs, 29–31
    audition, 29–30

olfaction, 31
  vision, 30–31
  ruminants, 31–32
Concaveation, 171, 186–87
Conceptual learning, 254–55, 285
Conditioned anxiety, 267
Conditioned avoidance in cattle,
    276–77
Coolidge effect, 130, 135–36
Coprophagia
  dogs, 306
  horses, 13, 234
Copulatory lock in dogs, 157, 159–60
Corticosteroids
  annual rhythm in cats, 89
  circadian rhythms, 87–88, 94
  food intake and, 324
Courtship behavior
  boars, 154–55
  cats, 165, 166–67
  cattle, 124
  dogs, 158–59, 159
  ewes, 132–33
  horses, 141
  rams, 134–36
  stallions, 146–47
Coyotes
  play, 221
  predation on sheep, 180
Cribbing, 282, 316–18
Critical periods. See Sensitive periods
Crossed extensor reflex, 214
Cross-fostering
  cats, 204
  cattle, 192–93
  sheep, 186–87
Crowding and aggression, 37, 58, 59
Cyproheptadine, 312, 329
Cystitis, 26
Cysts, ovarian, 144
Cytokines, 289

Debarking, 19
Declawing, 29
Defecation. See Elimination
Defensive aggression
  dog, 14
  sheep, 32
Delayed response method, 262–63
Denning, 346–47
Depo-Provera. See Progesterone
Desensitization

dogs, 71, 365–69
  habituation, 251
  horses, 357, 361–62
  phobias, 349–50
  separation anxiety, 365–69
Destructive behavior
  in cats, 353
  clinical cases, 339–40
  etiology, 340–42
  prevention, 343
  treatment, 343–48
  types, 342
Development of behavior, 211–49
  cats, 222–31
    chart, 224–25
    clinical problems, 231
    human contact and, 230
    neurological development, 226–27
    play, 227–30
    sensitive periods, 222–26
    sleep, 227
  cattle, 246–49
    activity patterns, 247
    human contact and, 248
    play, 246–47
  dogs, 211–22
    chart, 212–13
    human contact, 216–17
    neurological development, 218–19
    play, 215, 220–22
    sensitive periods, 211–18
    sleep, 219–20
  goats, 245–46
  horses, 231–39
    chart, 233
    play, 235–36, 238
    sleep, 235
  pigs, 239–42
    human contact and, 242
    play, 240–41
    sleep, 240
  sheep, 242–45
Dexamethasone, 145
Dextroamphetamine, 108
Diazepam
  as appetite stimulant, 268, 297–98,
    314
  lamb acceptance, 187
  phobia treatment, 350
  spraying, 27
  stallion impotence treatment, 150
Diet
  aggression and, 72

Diet (*continued*)
  sleep and, 95, 105
Digging behavior in dogs, 351–52
Dihydrotestosterone, 119
Dirofilaria immitis, 88
Discrimination learning, 254
Displacement behavior, 11
Dog doors, 347–48
Dogs
  activity patterns, 91–93
  aggression, 61–73
    toward children, 64–69
    fear-induced, 62–63, 70
    owner-directed, 64–69
    predatory, 63, 68, 70
    toward strangers, 70–72
    temperament tests, 222
    territorial, 62, 66, 70–72
    treatment, 67–69, 71–73
    toward veterinarians, 72–73
  communication, 13–20
    audition, 13–14
    behavior problems, 19–20
    olfaction, 16–18
    vision, 14–16
  development of behavior, 211–22
    chart, 212–13
    human contact, 216–17
    neurological development, 218–19
    play, 215, 220–22
    sensitive periods, 211–18
    sleep, 219–20
  dreams, 90, 91
  free-ranging, 61, 91–93, 157
  genetic influence on behavior, 36–37
  hyperactivity, 108
  learning, 268–75
    breed differences, 271–75
    classical conditioning examples,
      251–52
    cognitive dysfunction, geriatric,
      271
    housebreaking, 268–70
    learned helplessness, 260
  maternal behavior, 205–9
    clinical problems, 208–9
    nursing, 206–7
    parturition, 205–6
    weaning, 207–8
  narcolepsy, 108–9
  nocturnal wakefulness, 109
  problems, clinical, 338–52
    aggression, 64–73
    barking, 19

    car chasing, 350
    destructiveness, 338–48
    digging, 351–52
    hyperactivity, 108
    ingestive, 304–7
    jumping up, 352
    maternal, 208–9
    phobias, 349–50
    psychomotor epilepsy, 353
    psychosomatic, 353
    roaming, 350–51
    self-mutilation, 348
    separation anxiety, 365–69
    sexual, 157–58, 160–63
    submissive urination, 19–20
    tail chasing, 349
    urine marking, 20
    vomiting, 353
  sexual behavior, 156–63
    female, 156–58
    male, 158–63
    play, 221
  sleep, 91, 214–19
  socialization, 66
Dominance, 34
  cats, 76, 79, 163, 203
  cattle, 38–42, 130
  dogs, 61–69, 158
  genetic factors, 36, 42, 56, 65–66
  goats, 46
  horses, 46–49, 195, 235–36
  pigs, 56–58
  sheep, 45, 137–38
Dreams, 90, 91
Driving posture, 11
Dystocia in cattle, 190

Electrocorticogram (ECoG), 106–7
Electroencephalogram (EEG), 89–90,
  217
Electronic collar, 19, 55, 260
Elfazepam, 297–98
Elimination
  cats, inappropriate, 25–28
  dogs, 215, 268–70
    postures, 17–18
  pigs, 241
Elizabethan collar, 348
Enurination, 140
Epilepsy, psychomotor, 353
Estradiol, 118–19
Estrogen
  food intake and

cattle, 322
  pigs, 293
  sheep, 328
  hypothalamus and, 117
  maternal behavior, 170
  sexual behavior and, 113–14, 133
Estrous cycle
  cats, 164
  cattle, 120–22
  dogs, 156
  horses, 143–45
  sheep, 131
  sows, 153
Estrus detection in cattle, 122–23
Exploratory behavior
  calves, 247
  dogs, 222
  pigs, 241

Facial expression. *See also* Flehmen response
  cats, 23–24, 167, 226
  dogs, 221
  gape, 24, 31, 167, 226
  horses, 9–10, 50, 237–38
  snapping, 237–38
Farrowing crates, 173–74
Fat cow syndrome, 325
Fatty acids, 288
  cattle, 323
  goats, 330
  sheep, 326
Fear-induced aggression, 36, 62–63, 70–72, 80
Feeding. *See* Activity patterns; Grazing; Ingestive behavior
Feline ischemic syndrome, 79
Fermentation, rumenal, 322–23
Field of vision, 5, 6
Flehmen response
  cats, 167
  cattle, 124–25, 247
  horses, 10, 12, 146–47, 194, 237
  ruminants, 32
  sheep, 134–35
  testosterone and, 119
Flicker fusion, 4
Flocking in sheep, 44–45, 131
Fluorescein, 26
Fluoxetine (Prozac), 69, 298
Fly-catching behavior, 535
Foal heat, 143
Follicle-stimulating hormone (FSH), 117

Food intake. *See* Ingestive behavior
Fostering
  cats, 204
  cattle, 192–93
  sheep, 186–87
Founder, 288
Freemartins, 113, 119, 123
Furosemide, 331, 334

Gape expression
  cats, 24, 167, 226
  pigs, 31
Gastrocolic reflex, 269
Gastrointestinal factors in ingestive behavior
  cats, 308
  cattle, 322–23
  dogs, 301–2
  horses, 313–14
  pigs, 292–93, 294
  sheep, 326–27
Glucagon, 302
Glucose utilization
  cats, 309
  dogs, 303
  pigs, 295
  sheep, 328–29
Goats
  activity patterns, 105
  development of behavior, 245–46
  free-ranging, 46, 139–40
  ingestive behavior, 330
  learning, 279
  maternal behavior, 188–89
  olfaction, 32
  sexual behavior, 139–40
  temperament tests, 246
Gonadotropin-releasing hormone (GRH), 150–51
Granulosa cell tumor, 145–46
Grass eating
  cats, 310–11
  dogs, 306
Grazing. *See also* Activity patterns
  cattle, 100–101, 320–22
  horses, 97, 234, 313
  sheep, 103, 327–28
Greeting. *See* Communication
GRH (gonadotropin-releasing hormone), 150–51
Grooming
  cats, 77–78, 204
  cattle, 43

Grooming (*continued*)
  horses, 49, 52, 235
  pigs, 60
Growth curve, 287–88
Growth hormone
  circadian rhythms and, 88
  ultradian rhythms, 84
Guard dogs, 63–64, 66, 70–71

Habituation, 251
Hair coat and biological rhythms, 89
Handedness in horses, 362
Head shyness in horses, 361
Heart rate
  circadian rhythms and, 88
  high-frequency rhythms, 83–84
Heartworm, 88
Heat. *See* Estrous cycle; Sexual behavior
Hebb-Williams maze, 263–64
Herding posture in horses, 11
Hierarchy. *See* Social structure
High-frequency rhythms, 83–84
History, behavioral, 336–38
Homosexual behavior
  cattle, 121
  goats, 140
  rams, 139
Hormones. *See also specific hormones*
  aggression and, 43, 63
  gastrointestinal, 293, 302, 324
  maternal behavior, 169–70
  perinatal, 112–13
  sexual behavior and, 112–14, 117–20
Horses
  activity patterns, 95–98
  aggression, 47–55
    toward horses, 54–55
    pain-induced, 35
    toward people, 52–54
  communication, 7–13
    audition, 7–8
    olfaction, 12–13
    vision, 9–11
  development of behavior, 231–39
    chart, 233
    play, 235–36, 238
    sleep, 235
  dominance, 235–36
  foal rejection, 198–200
  free-ranging, 46–48, 97, 141
  grooming, 49, 52, 235
  learning, 280–83

    imprinting, 255–56
    influences on, 282–83
    maze learning, 281–82
    observational, 282
    operant conditioning, 280–82
    visual discrimination, 281
  maternal behavior, 193–200, 232
    clinical problems, 198–200
    mutual recognition, 194, 197
    nursing, 194–96
    parturition, 193–94
    weaning, 197
  narcolepsy, 108
  problems, clinical, 354–63
    aggression, 52–55
    ingestive, 314–19
    locomotion problems, 354–56
    maternal, 198–200
    phobias, 362
    saddle problems, 361–62
    sexual, 143–52
    trailering problems, 356–61
  sexual behavior, 141–52
    female, 141–46
    male, 146–52
  sleep, 95, 104, 107, 235
Housebreaking dogs, 268–70
House soiling in cats, 25–28
  castration effect on, 168
  causes, 26–27
  diagnosis, 25–26
  treatment, 26–27
Howling, 14
Hunger. *See* Ingestive behavior
Hypernatremia, 307
Hypothalamus
  electrical stimulation of, 284
  ingestive behavior and, 298, 309, 330
  melatonin and, 85
  oxytocin and, 170
  sexual behavior and, 117–20
  sodium status and, 334
  temperature regulation, 292
  water intake, 331
Hypothermia in pigs, 239

Identification of young. *See* Mutual
  recognition
Imipramine, 150
Imitation learning, 256, 285
Immobility response in sows, 153
Impotence

bulls, 128–30
dogs, 160
Imprinting, 255–56
Inappropriate elimination in cats, 25–28
Infanticide
  cats, 204–5
  pigs, 179
Infradian rhythms, 88
Ingestive behavior, 287–334
  biological rhythms, 84, 88
  cats, 307–12
    body weight maintenance, 309
    clinical problems, 310–12
    CNS and, 309
    gastrointestinal factors, 308
    glucose utilization, 309
    meal patterns, 307
    palatability, 307–8
    social facilitation, 307
    temperature, environmental, 380
  cattle, 319–25
    body weight maintenance, 324
    clinical problems, 325
    estrogen levels, 322
    grazing, 320–22
    meal patterns, 319
    palatability, 320
    social facilitation, 320
    temperature, environmental, 322
  dogs, 299–307
    body weight maintenance, 303
    clinical problems, 304–7
    CNS and, 303–4
    estrogen levels, 302
    gastrointestinal factors, 301–2
    glucose utilization, 303
    meal patterns, 299
    palatability, 299–301
    regurgitation feeding, 208
    social facilitation, 299
    temperature, environmental, 301
    vomiting, 353
  goats, 330
  horses, 312–19
    body weight maintenance, 314
    clinical problems, 314–19
    CNS and, 314
    gastrointestinal factors, 313–14
    grazing, 313
    meal patterns, 312
    palatability, 312
    social facilitation, 312

  pigs, 289–99
    amino acid imbalances, 295
    body weight maintenance, 296–97
    clinical problems, 299
    CNS and, 297–99
    estrogen levels, 293
    gastrointestinal factors, 292–93
    glucose utilization, 295–96
    meal patterns, 289–90
    palatability, 291
    parturition and, 295
    social facilitation, 290–91
    temperature, environmental, 291–92
  sheep, 325–30
    body weight maintenance, 326
    CNS and, 329–30
    estrogen levels, 328
    gastrointestinal factors, 326–27
    glucose utilization, 328–29
    grazing, 327–28
    meal patterns, 325
    palatability, 325
    social facilitation, 325
    temperature, environmental, 328
  water intake, 330–33
Inguinal wax, sheep, 182
Inhibin, 118
Injections and pain, 35
Intelligence. See Learning, comparative intelligence
Interviews, behavior, 335
Invisible fences, 259

Jumping up behavior, 352

Ketosis, 328

Lactation tetany, 216
Learned helplessness, 260, 268
Learning, 251–85
  cats, 283–85
  cattle, 275–79
    operant conditioning, 276–77
    taste aversion, 277
  comparative intelligence measures, 260–65
    avoidance response, 263
    brain to body weight ratio, 260–62
    classical conditioning, 262

Learning (*continued*)
    delayed response, 262–63
    learning rates, 262
    maze learning, 263–65
    multiple-choice, 263
    object permanence, 264
  dogs, 268–75
    breed differences, 271–75
    cognitive dysfunction, geriatric,
       271
    housebreaking, 268–70
  goats, 279
  horses, 280–83
    influences on, 282–83
    maze learning, 281–82
    observational learning, 282
    operant conditioning, 280–81
    visual discrimination, 281
  pigs, 265–68
    intraspecies differences, 266
    operant conditioning, 266–67
    visual discrimination, 268
  rates, 262
  sheep, 279–80
  state dependent, 359
  tasks, 257–60
    reinforcement schedules, 258–59
    rewards, 259–60
    shaping, 257–58
  types, 251
    chaining, 253–54, 257
    classical conditioning, 251–52, 253
    conceptual, 254–55
    discrimination, 254
    habituation, 251
    imitation, 256
    imprinting, 255
    operant conditioning, 252–53
    taste aversion, conditioned, 256–57
Leptin, 287, 289, 296
Leptospirosis, 20
LH (luteinizing hormone), 117, 123, 124
Libido. *See also* Sexual behavior
  bulls, 128–30
  experience and, 148–49
  pain and, 149
  rams, 137
  tests, 128, 137
Lick granuloma, 348
Licking behavior
  cats, 200–201
  cattle, 190
  dogs, 206–8

  horses, 194
  sheep, 180–82, 184
Light
  circadian rhythms and, 84–85
  reproductive cycle and, 131, 141, 143
  serotonin and, 87
Lignophagia, 318
Lissencephaly, 270, 271
Lithium, 63
  circadian rhythms and, 85
  taste aversion, 277
Litter, 27–28
Loading problems in horses, 356–59
Luteinizing hormone (LH), 117, 123,
    124

Magnus reflex, 214–15
Marking behavior. *See* Scent marking;
    Territorial behavior
Mastitis
  cattle, 193
  horses, 200
  pigs, 179
Masturbation
  cattle, 128–29
  dogs, 162
  horses, 147, 149
Mate preferences, 115–16
  cats, 167
  dogs, 158
  horses, 142
  pigs, 153, 154
  sheep, 135
  stallions, 151
Maternal behavior, 169–209
  aggression, 36, 43, 68, 173, 196,
    198–200
  cats, 200–205
    clinical problems, 204–5
    nesting, 203–4
    nursing, 201–3
    parturition, 200–201
    spaying effect, 166
  cattle, 43, 189–93
    bonding, 190
    clinical problems, 192–93
    nursing, 190–91, 192
    parturition, 189–90
    weaning, 191–92
  dogs, 205–9
    clinical problems, 208–9
    nursing, 206–7

parturition, 205–6
weaning, 207–8
general principles, 169–72
goats, 188–89
horses, 193–200, 232
clinical problems, 198–200
mutual recognition, 194, 197
nursing, 194–96
parturition, 193–94
weaning, 197
learning and, 171
pigs, 173–79
clinical problems, 179
defensive reaction, 178
mutual recognition, 177–78
nest building, 173, 174
nursing, 173, 175–77
parturition, 174–75
weaning, 178–79
sheep, 180–88
acceptance, 183–86
clinical problems, 186–88
licking, 180–82, 184
mutual recognition, 181, 184–86
nursing, 182–83
parturition, 180
Mating expression in horses, 10
Maze learning, 263–65
cattle, 278
dogs, 274
horses, 281–82
pigs, 266
sheep, 279
Meal patterns
cats, 307
cattle, 319
dogs, 299
goats, 330
horses, 312
pigs, 289–90
sheep, 325
Medroxyprogesterone, 27, 55
Megestrol acetate
aggression reduction, 69
spraying, 27
Melatonin, 85–87
Meningiomas, 79
Methylphenidate (Ritalin), 108, 281
Metritis, 179
Mibolerone, 209
Mismothering
cats, 205
horses, 198

sheep, 187
Mounting behavior. *See also* Sexual behavior
cats, 168
cattle
buller steers, 130–31
calves, 247
cows, 120–21
dogs, 162–63
Moving pups, 209
Mutual recognition
cats, 204
cattle, 190
dogs, 209
horses, 194, 197
pigs, 177–78
sheep, 181, 184–86
Muzzles
dogs, 68, 69, 72–73
horses, 316

Naloxone, 152, 302, 309, 316, 349
Naltrexone, 348
Narcolepsy, 108–9
Neighs, 7, 197
Nepetalactone, 164
Nesting
cats, 203–4
dogs, 205–7
pigs, 173, 174
Neurological development
cats, 226–27
dogs, 218–19
pigs, 239
Neurotransmitters. *See* CNS and ingestive behavior
Neutering. *See* Castration; Ovariohysterectomy
Nickers, 7–8, 197
Nipping, 220
Nocturnal wakefulness, 109
Nursing
cats, 201–3
cattle, 190–91, 192
dogs, 206–7
goats, 188
horses, 194–96
pigs, 173, 175–77
sheep, 182–83
Nymphomania
cattle, 123
mares, 145–46

Obedience training, 64, 66, 67, 352, 366
Obesity, 288
  cats, 309, 310
  chemical control, 297
  dogs, 304
  pigs, 297
  sheep, 329
Object permanence, 264
Obsessive compulsive disease, 348
Olfaction, 32
  catnip, 164
  cats, 24–25, 27, 166, 200–203
  dogs, 7, 16–18, 207, 300–301
  goats, 140, 188–89
  horses, 12–13, 147, 194, 197–99
  maternal behavior, 171
    cats, 200–203
    dogs, 207
    goats, 188–89
    horses, 194.197–99
    pigs, 175, 177–78
    sheep, 183–85, 187
  overview, 7
  pigs, 31, 153–55, 175, 177–78
  sexual behavior and, 117, 120, 126
    cats, 166
    cattle, 126
    goats, 140
    horses, 147
    pigs, 153–55
    sheep, 132, 134–35
  sheep, 132, 134–35, 183–85, 187
Olfactory bulbectomy, 27, 117, 153, 184,
    201
Onychectomy, 29
Operant conditioning, 252–53
  cats, 283–84
  cattle, 276
  dogs, 269, 339, 347
  goats, 279
  horses, 280–81, 356
  pigs, 266–67
  sheep, 279
Opiates
  cribbing, 316
  ingestive behavior and, 298, 302, 309,
    329–30
Osmotic pressure
  and satiety, 293
  and water intake, 331, 332
Ovaban, 27, 55
Ovariohysterectomy, 68
  cats, 165–66, 309

dogs, 302
  pseudopregnancy, 209
Oxytocin, 170, 199
  classical conditioning of release, 252
  pigs, 175–76

Pain
  aggression and, 34–35
  horses
    expression, 10
    foal rejection, 199
    libido decrease, 149
    vocalization, 8
  injections and, 35
  teeth grinding in pigs, 30
Palatability, food
  cats, 307–8
  cattle, 320
  dogs, 299–301
  horses, 312
  pigs, 291
  sheep, 326
Parasitism
  biological rhythms, 88
  cattle, 324–25
Parturition
  call, 180
  cats, 200–201
  cattle, 189–09
  dogs, 205–6
  goats, 188
  horses, 193–94
  pigs, 174–75, 295
  sheep, 180
Pavlov, Ivan Petrovich, 251–52
Pawing behavior in horses, 355
Phenobarbital, 69, 350
Phenylpropanolamine, 297
Pheromones
  anal sac secretions, 18
  cats, 165, 166
  dogs, 18, 158
  mate attractiveness, 116
  pigs, 31, 58, 59, 153–54
  ram effect, 132
  sexual behavior, 117–18
  urine, 18
Pherphenazine, 187
Phobias
  dogs, 349–50
  horses, 362
Phosphorus-specific hunger, 333

Pica
  cats, 311–12
  dogs, 306
Pigs
  activity patterns, 94–95
  aggression, 56–58, 242
  communication, 29–31
    audition, 29–30
    olfaction, 31
    vision, 30–31
  development of behavior, 239–42
    human contact and, 242
    play, 240–41
    sleep, 240
  free-ranging, 55, 173
  grooming, 60
  ingestive behavior, 289–99
  learning, 265–75
  maternal behavior, 173–79
    clinical problems, 179
    defensive reaction, 178
    mutual recognition, 177–78
    nest building, 173, 174
    parturition, 174–75
    weaning, 178–79
  problems, clinical, 364
    aggression, 59–60
    ingestive, 299
    maternal, 179
    sexual, 154, 156
  sexual behavior, 152–56
    female, 152–54
    male, 154–56
  sleep, 94, 104, 107, 240
  SPF (specific pathogen–free), 178
Piloerection
  cats, 23, 226
  dogs, 14, 16
Pineal gland, 85–87
Pit bulls, 64
Pituitary gland, 84, 170
Placentaphagia, 190
  cats, 201
  dogs, 206
  pigs, 175
Placing reactions in dogs, 219
Plant eating
  cats, 310–11
  dogs, 306
Play
  cats, 227–30
  cattle, 246
  dogs, 16, 215, 220–22

  fighting, 220–21, 238, 241
  goats, 245
  horses, 235–36, 238
  pigs, 240–41
  predatory, 229
  sexual, 221, 230, 235
  sheep, 243
Polydipsia
  horses, 318–19
  pigs, 299
  psychogenic, 307
Polyphagia, psychogenic, 306–7
Posture. See also Submissive posture
  cats, 21–24, 76, 78
  cattle, 38
  dogs, 14–16, 17–18, 61
  elimination, 17–18
  horses, 9, 11, 50–51
    estrous mare, 142
    recumbency, 96
  pigs, 30, 58
Predatory behavior, 36
  cats, 76–77, 79, 229
  dogs, 61, 63, 68, 70
Prefrontal lobotomy, 68
Pregnancy toxemia, 328
Problem behavior. See Clinical behavior
    problems; specific species
Proceptive behavior, 117
Progesterone, 55, 199
  aggression in dogs, 68
  food intake and, 322
  hypothalamus and, 117
  maternal behavior, 170
  psychic heat treatment, 145
  sexual behavior and, 113–14
  synchronization of estrus, 132
Progestins, 27. See also Progesterone
  aggression reduction, 69
  food intake and, 308
  horses, 55
  psychic heat treatment, 145
Prolactin, 42, 176
Promazine, 314
Propanolol, 350
Prostaglandins, 147
Prozac, 69, 298
Pseudopregnancy, 63, 68, 208–9
Psychic heat in mares, 145
Psychogenic polydipsia, 307, 318–19
Psychogenic polyphagia, 306–7
Psychomotor epilepsy, 353
Psychosomatic problems in dogs, 353

Punishment
  cats, 310
  dogs, 270, 351, 367
  horses, 316
  negative reinforcement compared,
      259–60
  separation anxiety, 367
Purring, 21
Pyometra, 209

Ram effect, 117, 132
Rapid eye movement (REM) sleep. *See*
  Sleep
Recognition. *See* Mutual recognition
Recumbancy response of foals, 196
Redirected aggression, 80
Reflexes, development of
  dog, 214, 219
  horses, 231
Regurgitation feeding, 208
Reinforcement, 258–59, 280. *See also* Re-
  wards
REM (rapid eye movement) sleep. *See*
  Sleep
Renin, 331
Reproduction. *See* Maternal behavior;
  Sexual behavior
Retrieval behavior
  cats, 203
  dogs, 207
Rewards, 259–60
  cats, 284
  cattle, 276
  dogs, 269, 272–73, 343–44
  pigs, 266
Rhythms, biological. *See also* Activity
    patterns; Sleep
  annual, 89
  circadian, 84–88
  clinical problems, 105
  feeding, 88
  high-frequency, 83–84
  miscellaneous, 88
  parasitic, 88
  ultradian, 84
Ritalin (methylphenidate), 108, 281
Ritualized aggression, 34
Roaming behavior
  cats, 168
  dogs, 162–63, 350–51
Roar, in horses, 8
Rooting
  pigs, 299

  puppies, 214–15
Rubbing in cats, 25
Rumen fermentation, 322–23

Salt hunger, 333–34
Scent marking
  dogs, 17–18
  horses, 12–13
Scratching, 28–29
Sedatives, 73
Self-mutilation in stallions, 151–52
Sensitive periods
  cats, 222–26
  dogs, 211–18
Separation anxiety, 365–69
  destructive behavior and, 342
  dogs, 19, 64
  obedience training, 64
  puppies, 217
Separation vocalization, 7
Serotonin, 37, 55
  aggression reduction, 69, 72
  circadian rhythms and, 87
  diet and, 72, 295
  food intake and, 329
Serving-capacity tests
  bulls, 128
  rams, 137
Sexual behavior, 111–68
  aggression, 45, 58
  biological rhythms, 89
  cats, 163–68
    female, 164–66
    male, 166–68
    play, 229–30
  cattle, 120–31
    female, 120–23
    male, 124–31
  CNS control, 117–20
  dogs, 156–63, 159
    female, 156–58
    male, 158–63
    play, 221
  environmental factors, 116, 131, 150,
      161–62
  experience and, 114–15, 148–49, 167
  goats, 139–40
  horses, 141–52
    female, 141–46
    male, 146–52
  mate preferences, 115–16
  physiological factors, 111–14
  pigs, 152–56

female, 152–54
    male, 154–56
play
    cats, 229–30
    dogs, 221
sheep, 131–39, 244
    female, 131–33
    male, 133–39
Shaping, 257–58
Sheep
    activity patterns, 103–5
    audition, 5
    communication, 32
    development of behavior, 242–45
    free-ranging, 44, 131–33
    guard dogs and, 63–64
    learning, 279–80
    maternal behavior, 180–88
        acceptance, 183–86
        clinical problems, 186–88
        licking, 180–82, 184
        mutual recognition, 181, 184–86
        nursing, 182–83
        parturition, 180
    problems, clinical
        maternal, 186–88
        sexual, 139
    sexual behavior, 131–39, 244
        female, 131–33
        male, 133–39
    temperament tests, 245
    vocalization, 32
    weaning, 243–44
Shock collar, 19, 55, 260
Shying, 362
Silent heats
    cattle, 122–23
    goats, 140
    mares, 144–45
Sinus arrhythmia, 84
Skinner box, 253, 254
Sleep, 89–90, 106–7
    cats, 94, 227
    cattle, 101–3, 104, 107
    dogs, 91, 214, 219–20
    dreams, 90, 91
    horses, 95, 104, 107, 235
    narcolepsy, 108–9
    pigs, 94, 104, 107, 240
    sheep, 104, 105, 107
    types, 89–90
Slow wave sleep (SWS). See Sleep
Snaking posture in horses, 11
Snapping in horses, 237–38

Snorts, 8
Social aggression, 34
    cats, 74
    cattle, 38–39
    pigs, 56–60
Social facilitation of food intake
    cats, 307
    cattle, 320
    dogs, 299
    horses, 312
    pigs, 290–91
    sheep, 325
Socialization
    dogs, 160–61, 215–18, 336–38
    pigs, 155–56
    sexual behavior, 114–15
Social structure
    cats, 73–76
    cattle, 38–42
    dogs, 61–62
    goats, 46
    horses, 46–49
    pigs, 55–58
    sheep, 44–45
Somatotropin, 176, 329
Spaying. See Ovariohysterectomy
SPF (specific pathogen–free) pigs, 178
Split estrus in horses, 143–44
Spraying. See House soiling in cats
Squeals
    horses, 8, 47
    pigs, 29
Stallion rings, 149
Stall kicking, 355–56
Stall walking in horses, 354–55
Stamping, 32
State dependent learning, 359
Stereotypic behavior, 314–15, 354
Stilbestrol, 322
Stimulation, environmental, 342, 344
Storm-related phobias, 349–50
Submissive posture
    cats, 24
    cattle, 39
    dogs, 15–16, 61
    horse, 9
    sheep, 32
Submissive urination, 15, 19–20
Succinylcholine, 54
Suckling. See also Nursing
    nonnutritive, 192, 205, 248
    reflex in horses, 231
Sugar preferences
    cats, 307–8

Sugar preferences (*continued*)
   cattle, 320
   dogs, 300
   goats, 330
   horses, 312
   pigs, 291
   sheep, 326
SWS (slow wave sleep). *See* Sleep

Tail biting in pigs, 59, 94
Tail chasing in dogs, 349
Taste aversion, 63
   cattle, 277
   conditioned, 256–57
   coprophagia treatment, 306
   fluoxetine (Prozac), 298
   sheep, 326
Taste preferences. *See* Palatability, food
Teasers. *See* Sexual behavior
Teat order
   kittens, 202
   pigs, 56, 240
Teeth grinding in pigs, 30
Temperament tests
   cats, 230
   cattle, 248
   dogs, 222
   goats, 246
   pigs, 242
   sheep, 245
Temperature and food intake, environ-
      mental
   cats, 308
   cattle, 322
   dogs, 301
   pigs, 291–92
   sheep, 328
   water intake and, 333
Territorial behavior
   aggression, 34, 62
   barking, 13, 19
   cats, 74–76, 80
      spraying, 24–25
   dogs, 62, 66, 70–72
      barking, 13, 19
      scent marking, 17–18
Testosterone, 38
   circadian rhythms, 87
   dihydrotestosterone, 119
   elimination posture in dogs, 17
   horses and social situations, 146
   hypothalamus and, 118

   sexual behavior in dogs, 113
Thermoregulation, 30–31, 85
Thirst, 330–33
Thunderphobia, 349–50
Tie, copulatory, 157, 159–60
Time budgets, 92. *See also* Activity pat-
      terns
   cattle, 92
   horses, 92, 99
   sheep, 92
Toxoplasmosis, 79
Trailering problems in horses, 356–61
Treading behavior, 202
Tryptophan, 37, 55, 72, 295

Ultradian rhythms, 84
Urinary calculi, 26
Urine marking, 17–18, 20
Urolithiasis, 330

Vaginal secretions in dogs, 18
Vasopressin, 87, 330
Vision. *See also* Color vision
   acuity, 4
   cats, 4, 6, 21–24
      development, 223–26, 227
      discrimination, 284, 285
      expression, 23, 24
      posture, 21–24
   cattle, 4–5, 277
   color vision, 4–5, 185, 268, 284, 291
   dogs, 4–5, 14–16
      development, 215
      posture, 14–16
   field of vision, 5, 6
   goats, 4–5, 279
   horses, 5, 6, 9–11
      development of, 231
      discrimination, 281
      expression, 9–11
      posture, 9, 11
   maternal behavior, 171
      cattle, 193
      goats, 188
      horses, 197–98, 199
      pigs, 179
      sheep, 184–86, 187
   pigs, 30–31, 268
   sexual behavior and
      cattle, 121, 126
      horses, 147

pigs, 155–56
sheep, 4–5, 279
Vocal cordectomy, 19
Vocalizations. *See also* Audition
  goats, 245
  kitten, 227
  puppies, 216
  separation, 7
Vomeronasal organ, 12
  cats, 24
  dogs, 159
  horse, 147
  sheep, 135

Watchdogs. *See* Guard dogs
Water intake, 330–33
Weaning
  cattle, 191–92
  dogs, 207–8
  horses, 197

pigs, 178–79
Weaving in horses, 354–55
Whining in dogs, 13–14
Whinny, 7, 197
Windsucking, 316
Wolves
  play, 221
  vocalization, 13–14
Wood chewing, 318
Wool chewing
  cats, 311
  sheep, 364

Xylazine, 73, 359

Yawing, 237

Zeitgebers, 84